Second International Virtual Conference on Environment & Natural Resources (IVCENR-2022)

IOP Conference Series: Earth and Environmental Science
Volume 1029

Al Diwaniyah, Iraq
19 – 20 January 2022

ISBN: 978-1-7138-5468-5
ISSN: 1755-1307

Printed from e-media with permission by:

Curran Associates, Inc.
57 Morehouse Lane
Red Hook, NY 12571

CURRAN ASSOCIATES INC.
proceedings.com

Some format issues inherent in the e-media version may also appear in this print version.

Institute of Physics
Dirac House, Temple Back
Bristol BS1 6BE UK

Phone: 44 1 17 929 7481
Fax: 44 1 17 920 0979

techtracking@iop.org

Additional copies of this publication are available from:

Curran Associates, Inc.
57 Morehouse Lane
Red Hook, NY 12571 USA
Phone: 845-758-0400
Fax: 845-758-2633
Email: curran@proceedings.com
Web: www.proceedings.com

TABLE OF CONTENTS

IOP Conf. Series: Earth and Environmental Science 1029 (2022) 011001 doi:10.1088/1755-1315/1029/1/011001

2nd International Virtual Conference on Environmentand Natural Resources (IVCENR-2022)
College of Science, University of Al-Qadisiyah
19th-20th January 2022

Preface

Welcome to the 2nd International Virtual Conference on Environment and Natural Resources (IVCENR-2022) That held on June 19th-20th 2022 at the College of Science, University of Al-Qadisiyah, Iraq.

In this Year the Unit of Environment and Protection from Pollution at the, College of Science, University of Al-Qadisiyah, Iraq, is holding its 2nd International Virtual Conference on Environment and Natural Resources and as the novel coronavirus outbreak shutters businesses and disrupts everyday life for billions around the globe, the annual conference has moved into virtual. This conference is established to provide an ideal platform for researchers to share views and experiences in Environmental areas. Over a period of two days at the Unit of Environment and Protection from Pollution, College of Science, University of Al-Qadisiyah, the event will include several activities such as guest speakers, lectures, and research articles.

The aims of the Conference is to bring together leading academic scientists, researchers, and knowledge-seekers to exchange and share their experiences and research results on all aspects of environmental sciences. It also provides a premier interdisciplinary platform for researchers and university professors to present and discuss the most recent innovations, trends, and concerns as well as practical challenges encountered, and solutions adopted in the fields of Environmental Sciences enable the collaboration between scientists and technologists in academic and industrial fields as well as international networks.

Hosted by College of Science, University of Al-Qadisiyah , the event was intended to provide scientific forum and research discussion on environment, renewable energy, ecology, and how technology is effectively employed for sustainable development of environment and green chemistry .

IOP Conf. Series: Earth and Environmental Science 1029 (2022) 011001 doi:10.1088/1755-1315/1029/1/011001

2nd International Virtual Conference on Environmentand Natural Resources (IVCENR-2022)
College of Science, University of Al-Qadisiyah
19th-20th January 2022

IVCENR-2022 recognizes that the face-to-face meeting is ideal for sharing research, ideas, and convening the community. Given the current challenges posed by COVID-19, we feel that the virtual forum format provides the best opportunity for the IVCENR-2022 community to meet and share information.

The conference was participated by 93 participants . The committee has selected 55 papers out of 93 papers to be published to IOP Conference series : Earth and Environmental Science.

The meeting was changed to a virtual format because covid-19 and travel restrictions, and for the safety of our participants, IVCENR-2022 held online (Zoom online conference without any physical participation). The online zoom of IVCENR-2022 consists of three parts:

1) Publish all the manuscripts in the conference proceeding .

2) Online presentation of articles by a pre-recorded video or ppt presentation with audio.

3) The papers was sent to high reputable reviewers for evaluation.

By removing some of the barriers to traditional forum attendance, including travel costs and time, the virtual forum format will enable greater participation from the overseas community-traditional and nontraditional. Ultimately, we decided that it is important for the community to gather and share information now, even if it is virtually, since it is safe and can be international easily.

2nd International Virtual Conference on Environmentand Natural Resources (IVCENR-2022)
College of Science, University of Al-Qadisiyah
19th-20th January 2022

Conference Photographs

2nd International Virtual Conference on Environmentand Natural Resources (IVCENR-2022)
College of Science, University of Al-Qadisiyah
19th-20th January 2022

Keynote Speakers

2nd International Virtual Conference on Environment andNatural Resources (IVCENR-2022)
19th-20th January 2022
College of Science, University of Al-Qadisiyah, Iraq

Amer Almadidy
American University of Iraq
Associate Professor & Director of the College of Health Sciences at Jumeira University.
Jumeira University
Massachusetts Institute of Technology.

Prof. Dr. Nadhir Al-Ansari
Department of Civil, Environmental and Natural Resources Engineering ,
Luleå University of Technology,
Sweden

Prof. Dr. Bengt Hansson
Department of Biology ,
Faculty of Science
Lund University, Sweden

Dr. Ayat Ghafori
Nottingham University Hospitals UK

Prof. Jeff King, Ed.D.
Director of Transformative Learning Universityof Central Oklahoma USA

IOP Conf. Series: Earth and Environmental Science 1029 (2022) 011001 doi:10.1088/1755-1315/1029/1/011001

2nd International Virtual Conference on Environmentand Natural Resources (IVCENR-2022)

19th-20th January 2022

Opening Session **Wednesday, January 12, 2022**		
National anthem		**10.00 AM**
University President	Prof. Dr. Kadhim J. Al-Jibouri	10.08 AM
Conference Chairman/ Dean of College of science	Assist. Prof Dr. Salwan Ali Abed	10.15 AM
Keynote Speakers	Amer Almadidy American University of Iraq	10.20 AM
	Nadhir Al-Ansari Lulea University of Technology Sweden	10.40 AM
	Bengt Hansson Lund University Sweden	11.00 AM
	Ayat Ghafouri NHS UK	11:20 AM
	Jeff King University of Central Oklahoma USA	11:40 AM

Environment and Pollution Section	**Session: 1, Time: 12.00 PM**
Wednesday, January 19, 2022	**Online Meeting ID:**
Session Chair: Prof. Dr. Hazim Abed Wale	**https://us02web.zoom.us/j/82223904505?pwd=Wkk2N3NNS2QybXp1VU1DbTdyVzM1UT09**
Session Rapporteur: Dr. Mohammed Kadhum Khwain	

1. **Stage - Discharge Relationships of Rounded Edge Broad Crested Weir Utilizing Dimensional Analysis and Incomplete Self-Similarly Technique**
Mohammed S.Shamkhi, Rawaa jabbar toama
12:00PM

2. **Body composition differences between males and females in diabetic patients**
Wasan Lafta Abdullah , Suha Talal Abd, Khalid Hamdan Gathwan
12:15PM

3. **Storm Water Management Model for Rainwater Networks of Al-Jaffryah Quarter and Kut Trade Center**
Mohammed Sewan Shamkhi, Marwaa kareem Azeez, Zahraa Hussein Obeid
12:30PM

4. **Evaluates the relationship between etiological agent of diarrhea (Rotavirus) and age group,**

type of feeding, and clinical features of the patients.
Zainab Abdul-Jabbar Faisal, Syoof Khowman AL-Ramahi
12:45PM

5. Coupling WetSpass and MODFLOW for Groundwater Recharge Assessment for East Wasit-Iraq
Mohammed S. Shamkhi, Hassan Jameel Al-Badry
1:00PM

6. A Numerical Study for Solving Mathematical Model in Chemistry
Mohammed S. Mechee, Ahlam A. Mahdi, Ali N. Al-Mousawi
1:15PM

7. Depositional Environments of the Mishrif Formation in Faihaa Oil Field, South Iraq
Hiba A.Kareem Mohammed, Manal Sh. Al-kubaisi, Ghazi Hasan Alshar'a
1:30PM

8. INVESTIGATION OF HEAVY METALS POLLUTION IN EUPHRATES RIVER (IRAQ) BY USING HEAVY METAL POLLUTION INDEX MODEL
Laith , Sadiq, Mahmood
1:45PM

9. Determination of chemical potential for Finasteride diffusion through SDS micelle suggested as alternative living membrane systems.
Ahmed Jassim Obaid , Oraas Adnan Hatem, ,
2:00PM

10. The effect of Hematological parameters on pregnancy outcome among pregnant women with Corona Virus -19 infection: a prospective cross-section study
Maitham G. Yousif, Shaymaa R. Al-Hasani, Alaa M. Sadeq, Samar M. Alfadhel
2:15PM

11. Psycho-immunological status of patients recovered from SARS-Cov-2
Kadhum J. Al-Jibouri, Maithm G. Yousif , Alaa M. Sadeq
2:30PM

12. Synthesis, Characterization and Application of Photocatalyst γ-Fe2O3-ZnO /MWCNTs For degradation of Cibacron Brilliant Yellow Dye
Hassan A. Alshamsi, Batool S. Hussein, Ahmed M. Abdullahb
2:45PM

13. Pd–Au nanoparticles supported on reduced graphene oxide nanoplatelets toward aerobic selective oxidation of benzoyl alcohol
Hadeel Altaee, Hassan Abbas Alshamsi
3:00PM

Green Chemistry and Waste Management Section	**Session: 2, Time: 10.00 AM**
Wednesday, January 20, 2022	**Online Meeting ID:**
Session Chair: Prof. Dr. Haider Mashkoor Hussein	**https://us02web.zoom.us/j/82223904505?pwd=Wkk2N3NNS2QybXp1VU1DbTdyVzM1UT09**
Session Rapporteur: Assist. Prof. Thair Abid Dasheesh	

14. Relation between epidemiological factors and seroprevalence of anti SARS-Cov-2 antibodies in COVID-19 patients from Al- Dewnia and Al-Najaf governorates
Adnan M. Melhan, Maitham G. Yousif
10:00AM

15. Facile one-step synthesis of Au-ZnO/MWCNTs for photocatalytic treatment of Reactive Blue dye in aqueous solution
Hassan A. Alshamsi, Qusay A. Nema, Salam H. Alwan
10:15PM

16. Determination of Polycyclic Aromatic Hydrocarbons in Selected Soils from Al-Diwaniyah governorate, Iraq-Case Study
Faiq F. Karam, Tania S. Kadhim
10:45AM

17. Determination of Polycyclic Aromatic Hydrocarbons in Surface Water of Al-Diwaniyah

River-Case Study Tania S. Kadhim, Faiq F. Karam 11:00AM
18. A Review on the Size Reduction Technology Using Mechanochemistry Approach Mohammad Ratib Alrbaihat 11:15AM
19. Thermal Synthesis of Au-ZnO Nanoparticles for Application in Photocatalytic Degradation of Reactive Blue 4 Dye in Aqueous Solution Hassan A. Alshamsi , Qusay A. Nema 11:30AM
20. Toxicity Remediation of Petroleum refinery Wastewater by Electrooxidation Techniques using graphite electrode Ehab Dheyaa radhy, Sarmad Taleb Najim 12:00PM
21. Isolation and identification of fungi contaminated with some local and imported dried fruite and the possibility of controlling one of the toxin-producing fungi by using some biological and chemical treatments. Ghasaq. M. R. AL-Barkawie , Abdul_Ameer S. Saadon 12:15PM
22. Response of Albizia (Albizia lebbeek) Seedlings growth for Nano fertilizers and applying methods Afak Ibrahim Jumaa Al-Jibouri, Abdel-Monnem Sadalaha. Kahlel, Hisham Hassan Naki 12:30PM
23. The Perfor mance study of Adsor ption of lead complex of synthesized crystalline structure of 4 (4-Benzophenol Azo pyrogallol) reagent, onto Activated Carbon Zeina M. Kadam, Nawras A Alwan 12:45PM
24. Adsorption/photo Reaction for Methyl Orange days Removal by TiO2 -Rutile impregnating with Graphite oxide and Multi walled carbon nanotube Farqad talib ahmed, Rawnaq B. Jimaa, Kalid Tuama Abdullah, Zena M. Jassem 1:00PM
25. Inference with gamma and inverse gamma prior densities in binary regression Esmaeel Ali Alselmawi 1:15PM
26. Study on Some Penicillin Derivatives by QSAR model: A Computational Chemistry Application of Physicochemical Descriptors Oraas Adnan Hatem, Zainab Adnan Hatem, Ban Adnan Hatem 1:30PM
27. Synthesis And Characterization of New Mn(II), Co(II), Cd(II) and Hg(II) Complexes with ligand [N-(pyrimidin-2-ylcarbamothioyl)benzamide] and their Anti-bacterial Study Bushra Mukhlif Fayyadh, Noor Alhuda Basim Abd2, Basima Muhsin Sarhan 1:45PM
28. Study of some colon bacteria in drinking water station in Al _Qassim city Hajer Abd Al_ameer Hussein, Russul jabbar Aljubouri, Ghaith Salah Al Mammori, Dhiaa Hameed 2:00PM
29. Mechanical and Structural Properties of Functionally Graded Epoxy Nano-composite by Centrifugal Method at Different Time of Rotation Adwaa Mohammed abdulmajeed, Ahmed Fadhil Hamzah 2:15PM
30. lecture title: Synthesis Techniques of Different Nonomaterials Dheyaa Jaber Al-zaidy 2:30PM

31. Estimation of Poly Aromatic Hydrocarbons (PAHs) in Samples of Figs Fruit by Using Gas Chromatography (GC) Ahmed Jalil Al-Safi, Arshed A. Ali Shihad 2:45:PM
32. Synthesis and biological activity of acetylene compounds derivatives of benzoin Moayad.N.Mohammed, Zahraa Hamza Najem 3:00PM
33. Synthesis of Imidazolidine 2,4 – Dione Derivatives Moayd Neama Mohamned, Khalid Jamal Hatem 3:15PM
34. Synthesis some Schiff bases compound derivative of benzoin and biological activity Moayd Neama Mohamned, Mustafa Hazim Jawad 3:30PM
35. synthesis some new of polymers derivatives of carboxylic pyrazine moayad nahma mohammed, Teba moayad nahma 3:45PM
36. synthesis some new of polymers derivatives of carboxylic acids oxazole moayad nahma mohammed, Sarah Masoud Sarheed 4:00PM
37. Effect of Biofertilizers on Vegetative Growth and Flower Yield of African Marigold *Tagetes erecta* L. c.v Pusa Narangi Gainda Razzaq Owayez Idan , Ali jabbar Abdulsada, Sabah Abd Fleih 4:15PM
38. Rapid detection of endocrine disrupting chemicals in watwer samples using flourescence technique Bassam F Alfarhani, Hyder A. Hamza 4:30PM

IOP Conf. Series: Earth and Environmental Science 1029 (2022) 011001 doi:10.1088/1755-1315/1029/1/011001

2nd International Virtual Conference on Environmentand Natural Resources (IVCENR-2022)
College of Science, University of Al-Qadisiyah
19th-20th January 2022

Technical Team

Dr. Salwan Ali Abed

Dr. Bassam Al-Farhani

Dr. Hameed J. Al-Jebouri

Esraa Raheem Alzaidi

IOP Conf. Series: Earth and Environmental Science 1029 (2022) 011001 doi:10.1088/1755-1315/1029/1/011001

2nd International Virtual Conference on Environmentand Natural Resources (IVCENR-2022)
College of Science, University of Al-Qadisiyah
19th-20th January 2022

IVCENR-2022 Editorial Board

IOP Conf. Series: Earth and Environmental Science 1029 (2022) 011001 doi:10.1088/1755-1315/1029/1/011001

2nd International Virtual Conference on Environmentand Natural Resources (IVCENR-2022)
College of Science, University of Al-Qadisiyah
19th-20th March 2022

Head of the Conference Assist. Prof. Dr. Salwan Ali Abed	
• Preparatory Committee	
Prof. Dr. Khadija A. Hammoud	Chairman
Assistant Prof. Dr. Bassam Faron Abd	Member
Assistant Prof. Dr. Prof. Aqeel Abbas Karim	Member
Assistant Prof. Saba Abdulamir Kazem	Member
Dr. Zahida Miran Hussain	Member
Dr. M.D. Muhammad Kazem Khwain	Member
Dr. Lujain Ibrahim Hussein	Member
• Scientific Committees	
Prof. Dr. Hazem Abd Wali	Chairman
Assistant Prof. Dr. Hamid Al-Jibbouri	Member
Assistant Prof. Dr. Ali Abdelhadi	Member
Assistant Prof. Thaer Abd Daishish	Member
Dr. Khatam Abbas Marhoon	Member
Dr. Fawzi Shaneen Abdul Hussein	Member
Dr. M. Haider Kamel Jabbar	Member
Dr. Karim Ghafil Muhajj	Member
• Technical Committees	
Assist. Prof. Dr. Salwan Ali Abed	Chairman
Assistant Prof. Dr. Bassam Faron Abd	Member
Assistant Prof. Dr. Hamid Al-Jibbouri	Member
Esraa Raheem Alzaidi	Member

Conference Chairman :

Dr. Salwan Ali Abed

College of Science

University of Al-Qadisiyah

Diwaniyah, Iraq

IOP Conf. Series: Earth and Environmental Science 1029 (2022) 011002 doi:10.1088/1755-1315/1029/1/011002

Peer review declaration

All papers published in this volume have been reviewed through processes administered by the Editors. Reviews were conducted by expert referees to the professional and scientific standards expected of a proceedings journal published by IOP Publishing.

- Type of peer review: Double Anonymous
- Conference submission management system: Morressier
- Number of submissions received: 55
- Number of submissions sent for review: 55
- Number of submissions accepted: 38
- Acceptance Rate (Submissions Accepted / Submissions Received × 100): 69.09 %
- Average number of reviews per paper: 2.00
- Total number of reviewers involved: 62
- Contact person for queries:

  Name: Dr. Salwan Ali Abed

  Affiliation: Dean of the College of Science, University of Al-Qadisiyah, Iraq

  Email: Salwan.abed@qu.edu.iq

A comparative environmental study of aqueous extracts of ginger and grapes to protect hepatocytes in Albino rabbits and a comparison of extracts in preserving Awassi lamb meat from oxidation

Zinah Ibrahim khaleel[1] , Abeer T . Abdulqader[2], Aseel Muthana Yousif AL-Sammarie[3], and Mohammed Ahmed Mustafa[4]

[1,4] Department of Pathological Analysis, College of Applied Sciences, University of Samarra, Iraq
[2] AL-Dour Technical Institute, Northern Technical University, Iraq .
[3] Department of Medical Laboratory Techniues , College of AL-Hadi, University Iraq
[4] Department of Medical Laboratory Techniques, College of Technology, University of Imam Jafar Al-Sadiq Dujail, Iraq

Abstract. The present study was conducted on white rabbits of Swiss origin demonstrate the effect of hydrogen peroxide on showing tissue (damage to body parts) of organs, and treatment with water-based ginger extract and water-based grape extract.

Histological signs appeared in hydrogen peroxide of liver tissues the occurrence of tissue changes in hepatic cell represented by of some cases of infiltration in lymphocytes, necrosis of the cytoplasm of cells, and degeneration of the nucleus. The results of the histological examination showed the group treated with aqueous ginger extract and a group of the aqueous grape extract showed the natural shape closest to the histological sections of the control group.

The study included a statement of the effect of adding an aqueous extract of ginger and grape seed extract on thigh meat taken from a calf immediately after slaughter.

That the meat of the Awassi lambs had the highest percentage of influence on the chemical characteristic (fat) at the degree of freezing (-18 m) of the cotton thigh piece in a group that would extract grapes and that the meat of the Awassi lambs showed that there was no significant difference in the chemical characteristic (ash) at the degree of freezing (- 18m) for the thigh piece at ($p \leq 0.05$) compared with the control treatment and the treatments of grape extract and ginger.

The highest fat percentage was in the group of grape extract, while the ash percentage was not significant, in the thigh meat.

And that this study has demonstrated stability in the composition of meat, which was reflected in the improvement of meat's ability to bind water, and it showed efficiency as an antioxidant in hindering fat oxidation and protecting meat pigment from oxidation after 90 days of frozen storage.

Keywords: aqueous grape , ginger extract, liver, Awassi ,oxidative stress

1. Introduction

It is a synthetic derivative of whole grape seed. The extract contains procyanidins. The quality of a grape seed extract is measured by its content of procyanidins. Generally, grape seed extract contains 95% procyanidins, but their effectiveness varies between products. Eating foods or drinks that contain a high percentage of it leads to a feeling of wrinkled mouth and dryness, as it is a vasoconstrictor.

Grape seeds contain a wide range of antioxidants that have a strong effect on diseases, and they also contain amino acids such as oleic acid, linoleic acid and palmitic acid, and grape seeds contain other substances such as phenols and steroids and a proportion of vitamin C and E, [1].

The grape plant classification

Kingdom: plants

Division: Seedlings

Gender: Vitis

Type: Grapes

Scientific name: Grape

1.1. Previous studies on grape plants and oxidative stress

A study conducted by [2] to study the inhibitory effect of the proanthocyanin compound on microorganisms proved that the poly-proanthocyanin compound (POCs) has a broad spectrum against infectious agents, and one of the molecular mechanisms is the formation of complexes with proteins by non-specific forces such as hydrogen bonds. In addition to forming covalent bonds [3]have shown that phenolic compounds have a role in reducing the virulence of pathogenic bacteria from the host cell receptors (the intestinal membrane during the failure of the intestinal toxin binding to the bacterium Helicobacter pylori) by forming a complex with the seeds after treatment with the extract of the grape extract. . The multi-proanthocyanin compound had a role in inhibiting the growth of anaerobic bacteria that cause gingivitis, based on measurements, which have more antioxidant activity than vitamin C, E.

Myrstad *et al* [4]indicated that the flavonoids present in grape seeds activate the activity of the enzyme γ-glutamyl-cysteine synthetase, which leads to stimulation and formation of GSH, and thus its level in the blood increases.

1.2. Active compounds in the grape plant

Seeds of Grape have a medically helpful because it contains active ingredients with related to the body function of living things effects, and contain of body-protecting chemical phenolic compounds known as polyphenol, as the active in them is Proanthocyanidins and Procyanidin, Tannins, and acids the most important of which are Linolic acid and Vitamin E.

The phenolic compound proanthocyanin has important physiological effects, as it has an antioxidant effect stronger than vit E and vit C. This is due to the plant's ability to remove oxidized agent and prevent the occurrence of fat peroxidation in cells in general . thus protecting the cell From the damage that affects cells and is based on oxidizing factors for its emergence, and it also increases the effectiveness of antioxidants in the body it a factor It protects against the effects of ultraviolet rays, so it is anti- malignant [5].

1.3. Classification of the ginger plant

Kingdom: a plant

Section: covered with seeds

Rank: Ginger

Family: Ginger

Gender: ginger

Type: Z. officinale

The scientific name Zingiber officinale

2. Materials and Methods

2.1. Preparing the water-based extract of grape seed

1-The water-based extract of grape seeds was prepared according to the [6]by crushing some clean seeds using an Ultra - Turrax device .

2-weighing 50 grams of dry powder and placing it in a 1 liter glass container Produced/showed water was added to it and the volume was completed to 1 liter, after which 3 ml of complete and total ethyl alcohol was added to prevent fungal growth.

3-The samples were left for half an hour in a flat/left-and-right vibrator at a medium speed, and the samples were left to settle for one hourthen the filtrate was introduced to a (device that spins something at a high speed) at 3000 rpm for 15 minutes to separate the small plankton, after which the water was disappeared. From the extract at a temperature of 40 Â° C using a rotary evaporator and placed in a (something that creates and grows life) at a temperature of 37 Â° C, then the extract was kept in a light-blocking/difficult to understand small glass container at room temperature appropriate for the prepared samples in conditions free of moisture until it was used in the study.

2.2. Extraction operation of ginger

The aqueous extraction of ginger root powder was carried out according to the [7]method by taking 20 grams of the powder and adding to it 600 ml of boiling water and leaving it for 10 minutes, then the mixture was filtered and the filtrate was taken as it was concentrated using a rotary evaporator at a temperature It ranged between (65-68 ° C) and then it was dried in an electric oven in glass dishes at a temperature of 40 ° C. A base solution was prepared from the dried crude extract at a concentration of 100 mg/ml, where 1 g of the dried extract was weighed and dissolved in 10 ml of distilled water.

Groups of laboratory animals were designed by dividing them into groups, and each group contained 4 rabbits **:**

Untreated group is 4 Rabbits dosed orally with normal saline solution, 0.1 mg/kg

1- 1st Group, dosed orally with peroxide hydrogen 0.5 mg / kg for 10 days.

2- 2nd Group, dosed orally with an aqueous extract of grape seeds 10 % mg/kg for 30 days.

3- 3ed Group , dosed orally with an aqueous extract of grape seeds 20 % mg/kg for 30 days.

4- 4th Group , dosed orally with an aqueous extract of Ginger 10 % mg/kg for 30 days.

5- 5th Group , dosed orally with an aqueous extract of Ginger 20 % mg/kg for 30 days.

2.3. Liver sections

The animals were autopsied and then the required organs were fixed in formalin at a ratio of 10% , It was placed in alcohol 70% for the purpose of removing water, and then it was cut and dyed as stated in [8]

2.4. Preparing meat for study

Local veal was obtained from the local market. The thigh meat weighing 3 kg was taken immediately after the slaughtering process, and it was stored in clean, sterilized, and cooled polyethylene covers in an ice box until it arrived at the laboratory, as it was kept in the refrigerator for 24 hours at 4 pm Until the conduct of the transactions for the study transactions of experience

1-(T1) the comparison treatment devoid of any additives.

2-(T2) Treatment of adding grape extract at a concentration of 50 mg / 100 gm

3-(T3) treatment of adding ginger 10 mg / 100 gm .

3. Results and Discussion

Laboratory animals were dosed with hydrogen peroxide for showed the presence of tissue represented by Death of living tissue, getting worse and Blood destruction. Sinusoids and Hepatocytes distention were also found there were also cases of small size of pyknotic centers of cells or atoms.

In the laboratory animals of Group II, the aqueous grape seed was dosed showed the presence of sudden, dangerous bleeding and infection-fighting cell invasion with the shedding of the cells lining the hepatic vein and a small number of cases of karyolysis with some centers of cells that suffered from small size and shortened (changed from gas to liquid) by Pyknotic Centers (of cells or atoms), and there were cases of improvement in hepatocyte of returning to the usual and regular shape that differs from the shape of cells in group fed with oxidizing hydrogen peroxide. Whereas, The tissue sections of the group III samples were shown hepatocytes and their regularity to the normal shape, as they were close to the normal shape, despite the presence of bleeding and invasion infection-fighting cell.

In the group of laboratory animals that were dosed with water-based ginger extract showed the presence of dangerous bleeding and infection-fighting cell invasion with the desquamation of the cells lining the central vein (desquamation) and also found a several cases of karyolysis with the presence of some nuclei that suffered from smallness The size and condensation of Pyknotic Nuclei also found cases of improvement in hepatio cyte of returning to regular shape differ from the group treated with hydrogen peroxide (figure 3). While the fifth group showed the regular/ healthy shape of the hepatic vein and regular or health size of hepatocytes contain normal-looking centers of cells, while in other areas it was observed that some cases of cell degeneration and infiltration of small numbers of lymphocytes and hypertrophy of a section of the centers (of cells or atoms) Auxes of Centers (of cells or atoms), which may be an regular or healthy condition and this Indication of cell division and hypertrophy (figure 4) .

Discussion:

When examining the histological sections of the liver, it was noted hemorrhages, degeneration, and lymphocyte infiltration were observed, ending with necrosis and cytolysis, and these results were consistent with what [9] found in case the mice were exposed to oxidative stress. Mediated by substances toxic to hepatocytes hydrogen oxide, Because it works to remove and inhibit radicals that have an oxidative effect that is harmful to tissues, the oxygen root and the hydroxyl radical. These roots are seen as their ability to interact with phospholipids and proteins involved in the (creation/combination) of cellular and (within a cell) membranes. Intracellular membrane and thus inhibit many of the vital activities of the membranes, such as the transport of substances necessary for the viability of cells, as well as interfering with the work of enzymes and so stopping cellular breathing that leads to a decrease in cell activities [10].

The interaction of body-damaging chemicals with phospholipids results in Peroxidation of harmful fats, which help produces a film – hemolytic reactions with a reduction in the energy of the mitochondrial membrane and the breakdown of the lysosomes membranes and so the cell's access to the death (or other living tissue) stage [11] and may be the cause of the damage. In liver is Detoxification is one of the main functions of the liver in the human body rid of the largest possible amount of toxic substances by breaking down the unwanted substance [12].

The study explained the repair that happened in the hepatocyte damaged by hydrogen peroxide and the return of these changes to the regular and healthy state treating animals with an extract water-based of the grape plant was due to the action of the parts/pieces of the extract in stoping the harmful activity of the substance by having many activities, especially the body-protecting chemical and so speeding up the process. Cellular repair and stimulation of hepatocyte to (release fluid) chemical attraction factors to attract insulting/swelling cells, which have been watched/followed to break into the area clear to eat/destroy the damaged tissue, as well as stimulating tissue to divide to make up for the damage by oxidative factors.

The current study showed that aqueous ginger extract may have a role in protecting liver tissue from the harmful effects of hydrogen peroxide, and this is exemplified by the improvement that occurred in it, especially in hepatocytes, their radial regularity, and the lack of lymphocyte infiltration, due to the role of vitamin C in increasing the body's immunity and strengthening components Nucleus and cellular organelles, especially cellular membranes [12].

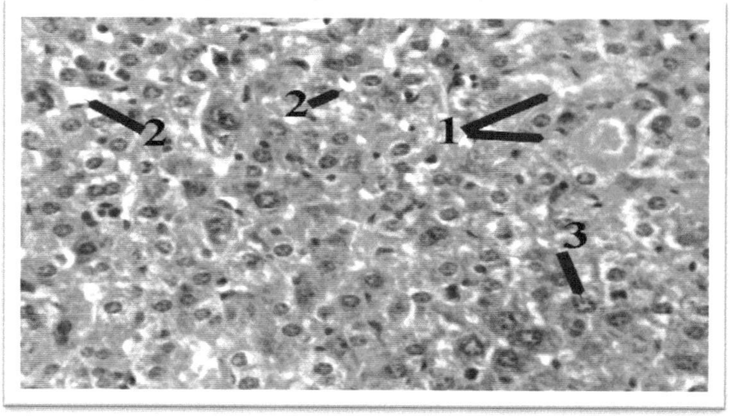

Figure 1. rabbit liver with hydrogen peroxide dose [2] central vein congestion CV, 2) hemorrhage in sinuses (S) H&E, 100X.

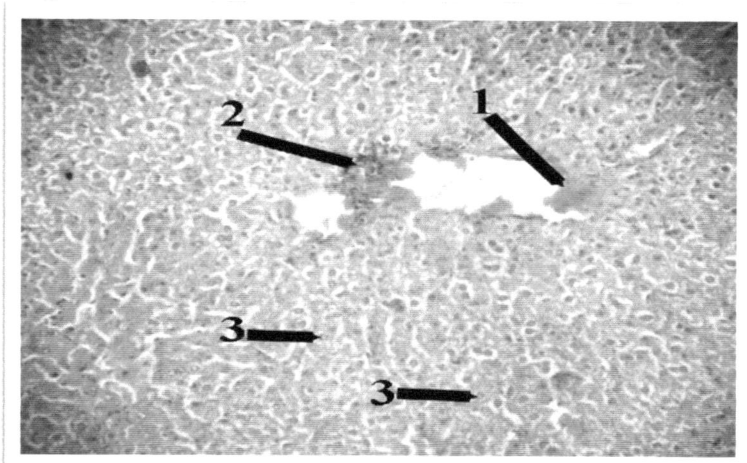

Figure 2. rabbit liver with a dose of group IV aqueous Grap extract [2] congestion CV, 2) infiltration of lymphocytes, [4] hepatic cell H&E, 100X

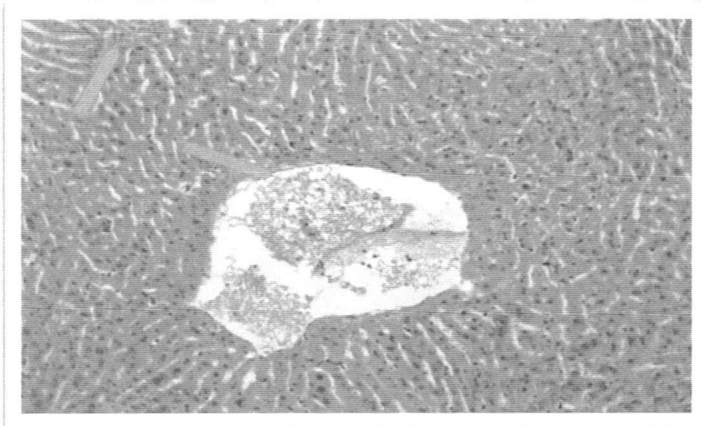

Figure 3. rabbit liver with a dose of group IV aqueous Grap extract [2] central vein CV, 2) infiltration of lymphocytes H&E, 100X

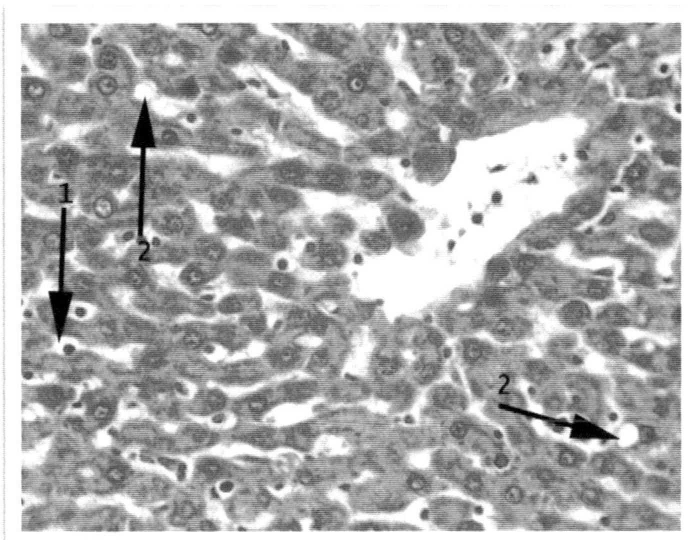

Figure 4. rabbit liver with a dose of group IV aqueous ginger extract [2] central vein CV, 2) infiltration of lymphocytes H&E, 100X

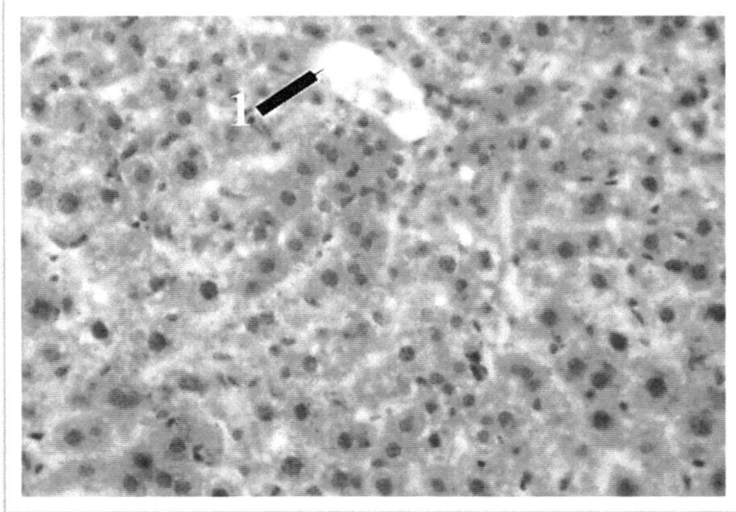

Figure 5. rabbit liver, group V, aqueous ginger extract at a dose of [2]central vein CV, 2) sinuses (S) H&00 X

The results of chemical tests for the thigh piece meat: moisture, protein, fat, ash.

It is evident from the figure (6)that the highest significant increase at ($p>0.05$) in Awassi lamb meat in the chemical characteristic (humidity) at the degree of freezing (-18 m) of the thigh meat piece of the ginger group (71.80) compared with the control group (68.50), and there was no significant difference ($p>0.05$) in the grape extract group (69.12) compared with the control treatment.

It is evident from the figure (6) that the highest significant increase at ($p>0.05$) in Awassi lamb meat in the chemical characteristic (protein) was at the degree of freezing (-18 m) of the thigh meat piece of the ginger group (19.40) compared with the control group (17.04) and there was no significant difference. At ($p>0.05$) in the grape extract group (17.28) compared with the control treatment.

Also there was lowest significant difference at ($p>0.05$) in Awassi lamb meat in the chemical characteristic (fat) at the degree of freezing (-18 m) of the thigh meat piece of the ginger sample (7.41) differ from Untreated totals (11.81) and there was no significant difference when ($p>0.05$) in the grape extract group (11.01) compared with the control treatment.

The Awassi lamb meat showed that there was no significant difference in the chemical characteristic (ash) at the degree of freezing (-18 m) of the thigh meat piece at ($p>0.05$) compared with the control treatment (1.1) and the treatments of ginger and grape extract (1.02) and (1.17). And (1.27), respectively.

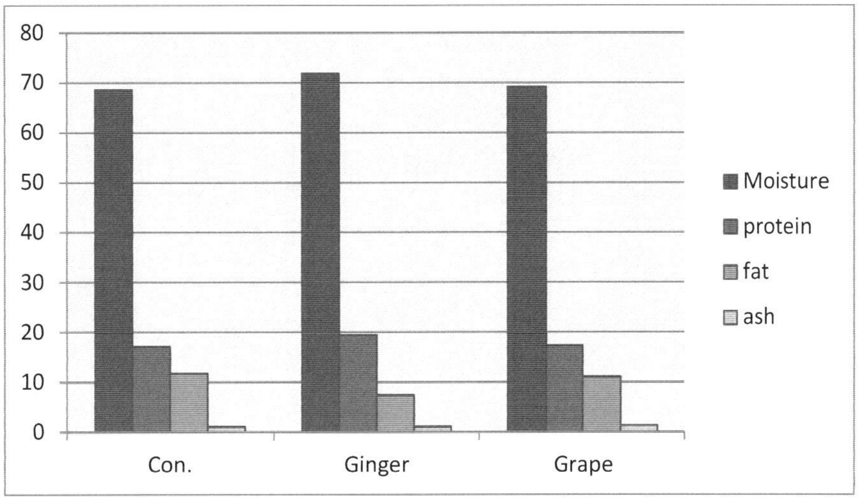

Figure 6. chemical characteristic characteristic for the thigh piece meat: moisture, protein, fat, ash

Results of physical examination of thigh piece meat: loss of cooking, loss of defoliation, water carrying capacity, pH.

It is evident from the figure(2) that the lowest significant difference at ($p>0.05$) in Awassi lamb meat in the chemical characteristic (loss of cooking) was at the degree of freezing (-18 m) of the thigh meat piece of the ginger group (20.31) compared with the control group (23.78) and there was no significant difference ($p>0.05$) in the grape extract group (23.05) compared with the control treatment.

The Awassi lamb meat showed that there was no significant difference in the physical characteristic (loss of defrosting) at the degree of freezing (-18 m) of the thigh meat piece at ($p>0.05$) compared with the control treatment (2.31) and that the slightest difference was at ($p>0.05$) It is shown in the treatments for the ginger and grape extract (1.91) and (1.93), respectively.

It is evident from the figure(7) that there is no significant difference ($p>0.05$) in the meat of Awassi lambs in the physical characteristic (water carrying capacity) at the degree of freezing (-18 m) of the thigh meat piece, the ginger group (42.60) compared with the control group (11.81), and the slightest difference was significant. At ($p\leq 0.05$) in the grape extract group (41.27) compared with the control treatment.

Awassi Lamb meat showed that there was no significant difference in the physical characteristic (pH) at the degree of freezing (-18 ° C) of the thigh meat at ($p>0.05$) in the group of ginger and grape extract treatments (5.56) and (5.49), respectively, compared with The control treatment (5.76).

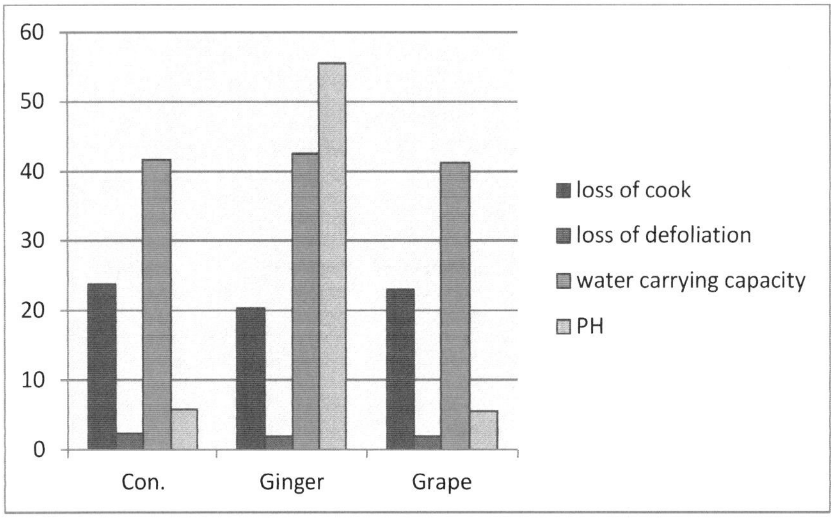

Figure 7. physical examination of thigh piece meat: loss of cooking, loss of defoliation, water carrying capacity, pH.

Through the results, we note that the additives do not have significant differences in the pH compared to the control group, and this may indicate the lack of decomposition of the meat components for the control treatment and the lack of dissolution of amino acids and peptides as a result of the absence of oxidation and decomposing enzymes , The meat of the Awassi lambs recorded the highest significant increase at (p) 0.05) in the meat of the Awassi lambs in the chemical characteristic (fat) at the degree of freezing (-18 m) of the thigh meat of the grape extract group (11.01) compared with the control group (11.81) .

This result is consistent with the findings of [13] that phenolic compounds contribute to suppressing free radicals within the fat membranes, as these compounds help protect the muscle tissue of meat from the effects of free radical damage and inhibit the products of fat oxidation reactions.

Also, the results show a high percentage of grape and ginger extract. In maintaining the stability of the meat cell membrane and inhibiting the formation of free radicals by preventing the transfer of hydrogen atom to the free root and this was reflected by the high susceptibility of proteins by binding water and reducing its loss compared to the control treatment. The results agreed with [14].

The ability of meat to carry water, as the safety and protection of these meat membranes and the reduction of their rupture contribute to preserving the cellular components of the meat, which leads to a lack of liquid exudation and an improvement in the ability of meat to hold water during storage,

It was previously indicated [15] that treating minced lamb meat with grape seed powder resulted in less loss of mature liquids when thawing, less loss during cooking, and a reflection of this in improving the ability of meat to hold water when storing these meats by freezing for three months [16] [17].

References

[1] Aalt Bast and Weseler ،Antje R.2017 ؛. Masquelier's grape seed extract: from basic flavonoid research to a well-characterized food supplement with health benefits. Nutrition Journal. 16 (1): 5. ISSN 1475-2891.

[2] Al-Jumaili, Essam Fadel, Al-Saffar, Muntaha Abdul-Karim, Kazem, Shaima Ismail, 2011, Study of the Inhibition Effectiveness of the Proanthocyanidine Compound Extract from Grape Seeds on Some Microorganisms, Iraqi Journal of Life Technologies ..

[3] Alberto, M. R.; Matias A.R.C.; Maria C. M. and De Nodra .(2008).Antimicrobial effect of polyphenols from apple skins on human bacterial pathogen. Electron J of Biotechnology .9 (3):issue 3-fultex-1 ..

[4] Myrstad, M.C.; Carlsen, H.; Nordstrom, O. (2002). Falvonoids increase the intracellular glutathione level by transactivation of gamma-glutamyl cysteine synthetase catalytical subunit promrter. Free Radical. Biol. Med., 32: 386-393..

[5] Katiyar, S. K. (2008). Grape seed proanthocyanidin and skin cancer prevention inhibition of oxidative stress and protection of immune system. Mol. Nutr. Food Res., 2, 52(s1), 71-76..

[6] Bayadar , N.G.; G.Ozkan and O.Sagdic 2004 .Total phenolic contents and antibacterial activities of grapes (Vitis vinifera L.) extracts. Food Control, 15(5): 335-339..

[7] Duh, P. and G. Yen. 1997. Antioxidative activity of three herbal water extracts. Food Chemistry, 60(4) : 639- 645..

[8] Ali et al (2020): Therapeutic efficacy of Punica granatum January 2020 Vol.23(1) ..

[9] Al-Sayediyah, Ahmed Muhammad Ali (2010). A pathological study of ginger's anti-hyperlipidemic effect in rats exposed to oxidative stress. Iraqi Journal of Veterinary Sciences. (108-103)..

[10]Krishna, V. (2004). Text book of pathology. 1st ed. Orient Longgman Private Limited.I ndia. Pp:538,564..

[11]Wendland, B.E., Aghdassi, E., Tam, C., Carrier, J., Steinhart, A.H., Wolman, S.L., Baron, D. and Allard, J.P.(2001). Lipid peroxidation and plasma antioxidant micronutrients in crohndisease. Am.J. Clin. Nutr.,74:259-264..

[12]Attyah, Ahmed M. and Ismail, Sajida H. 2012, Protective Effect of Ginger Extract Against Cisplatin-Induced Hepatotoxicity and Cardiotoxicity in Rats. Iraqi J Pharm Sci, Vol.21(1) ..

[13]Enginar, H.; Cemek, M.; Karaca, T.; Unak, P. (2007). Effect of grape seed extract on lipid peroxidation, antioxidant activity and peripheral blood lymphocytes in rats exposed to x-radiation. Phyt. Res. 21(11), 1029-1035..

[14]Coimbra, S. A., Santos-Silva, P., Rocha – Pereira, S. Rocha and Castro. E. (2007). Green tea consumption improves plasma Lipids profiles in adults Nutrition Research , Volume 26(11): 604-607.

[15]Saleh, Hatem Hassoun. (2007). Effect of Vitamins E and C and Grape Fruit Extract and Concentrate on some traits of ewe meat during frozen storage. PhD thesis, College of Agriculture, University of Baghdad..

[16]Mustafa, M.A & AL-Samarraie, M.Q . (2020) .SECONDARY MENOPAUSE and its RELATIONSHIP to HORMONAL LEVELS AMONG WOMEN at SALAH AL-DIN HOSPITAL . European Journal of Molecular & Clinical Medicine . Volume 7, Issue 09, PP 96-104..

[17]Mustafa, M.A., AL-Samarraie M.Q., Ahmed M. T. (2020). Molecular techniques of viral diagnosis, Science Archives, 1(3), 89-92 http://dx.doi.org/10.47587/SA.2020.1303 ..

[18]Bassam Faron Alfarhani et al 2019 J. Phys.: Conf. Ser. 1294 052045

[19]Bassam Alfarhani, Maha Al-Tameemi, Hector C. Goicoechea, Fernando Barbosa, Andres D. Campiglia, Direct analysis of benzo[a]pyrene metabolites with strong overlapping in both the spectral and lifetime domains, Microchemical Journal, 137 2018, 51-61,

[20]B. Alfarhani, M. Al-tameemi, A.V. Schenone, H.C. Goicoechea, F. Barbosa, A.D. Campiglia. Microchem. J., 2016 129 83–89. https://doi.org/10.1016/j.microc.2016.06.010.

[21]Zeina. M. Kadam (2021). Assessment of Electrochemical Activity of some Amino acids on the Modified Carbon Electrode Surface. Research Journal of Pharmacy and Technology, 14(10):5325-9. doi: 10.52711/0974-360X.2021.00928

[22]Z. M. Kadhim and A. H. Ali, (2018), Studying of Optical and Structural Properties of a Novel Ligand (5-ClCPAI) Thin Films by Spray Pyrolysis Method. Journal of Engineering and Applied Sciences, 13(14), 11032-11040.

[23]Z. M. Kadam, C. D. Gwenin, (2017), Polymer Membranes based on Ionophore-impregnated for Nutrients Detection by Electrochemical Methods, Der Pharma Chemica, 9(20):29-33.

[24]Alfarhani, B.F., Hammza, R.A. & Alzaidy, A.S. Potential effect of solvent and slit width on some properties of room temperature fluorescence of hydroxy polycyclic aromatic hydrocarbons. Chem. Pap. 75, 3915–3920 (2021). https://doi.org/10.1007/s11696-021-01602-1.

Antimicrobial and Environmental activity of biogenic CS-GO nanoparticles on Uropathogens

Fatima Hamza Alzubaidy[1] and Nawfal Hussein Aldujaili [2]

[1,2]Department of biology, faculty of science, Kufa University, Iraq

Abstract. Bioproduction options for nanoparticles are becoming a highly significant subject, with environmental and economic benefits over physical and chemical processes. The purpose of the current study is to detect the antimicrobial activity of chitosan-Graphene oxide nanoparticles that include The Minimum Inhibitory Concentration (MIC) is a method of determining antibacterial activity, Antibiofilm Activity by Tissue culture plate method, Antioxidant activity, and finally the hemolysis activity of mixing nanoparticles CS- GO that synthesis biologically. the result of Antibacterial activity showed the ID50 for inhibition is at a concentration (1 mg/ml), while the Antibiofilm activity result shows the high concentration of CS-GO nanoparticles (16 mg/ml) showed the higher activity in inhibition of biofilm formation while the less concentration (0.6mg/ml) appeared very low inhibition activity on biofilm formation. The antioxidant activity of nanoparticles showed the high scavenging ability at a concentration (600 mg/ml) as (54.06%) while the less concentration (100mg/ml) appeared scavenger ability as (29.42%) also the hemolysis activity on CS-GO nanoparticles on blood showed different ability at different concentration, the preferred concentration (1mg/ml) show no hemolysis on blood.

Keywords: Nanobiotechnology, Urinary tract infection, Chitosan-graphene oxide nanoparticles, MIC, Antibiofilm, Antioxidant, Hemolysis activity.

1. Introduction

Nanobiotechnology integrates biological concepts with physical and chemical processes to create nanoparticles with specialized functions at a cheaper cost than physical or chemical methods. Based on basic features such as size, distribution, and shape, nanoparticles have new or changed properties. The number of novels uses for NPs and nanomaterials is steadily rising [1]. Nanoparticles can be hollow or solid, and they can be formed of a variety of materials in a variety of layers, each with its function. A core functioning layer, a protective layer, and an outside layer that permits interaction with the biological environment are generally present [2]. The core functional layer usually has somebeneficial magnetic or optical activity, with fluorescence being the most frequent. The functional layer is protected from chemical damage caused by water, air, or cell componentsby the protective layer. as well as any harmful characteristics of the functional layer's chemicals. Nanoparticles can be identified using the outer layer [3].

In the production of NPS, Redox membrane proteins are used by bacteria and fungus for surface synthesis, and extracellular enzymes are used for extracellular synthesis [4]. Synthesizing NPS utilizing natural mechanisms such as carbohydrates, microbial enzymes, vitamins, polysaccharides, biodegradable polymers, microorganisms, and biological systemsis one approach to achieve this aim. The manufacture of NPs by bacteria is one method witha lot of promise [5]. The development of a controlled and scalable technique for thebiosynthesis of monodispersed and extremely stable NPS has been the focus of recent research. As a result, in green nanotechnology, Various bacterial species have been used to investigate alternative methods of NP production. [6]. Deacetylation of chitin, a protein foundpredominantly in crab exoskeletons, produces chitosan nanoparticles, a cationic biopolymer.
[7] Because of its biocompatibility, nontoxicity, mucoadhesion, biodegradability, and antibacterial activity, as well as its natural origin and inexpensive cost, this biopolymer hasbeen widely investigated for possible uses in several disciplines [8].

GO is a nanomaterial that has been utilized in a variety of applications for more than150 years [9]. It is the forerunner of graphene, a two-dimensional carbon allotrope that is one of the best in the world. Electrical conductivity [10], thermal properties [11], transparency [12] and mechanical strength [13]. Chitosan is a polysaccharide made up of glucosamine (2-amino- 2-deoxy-D-glucose) and N-acetyl glucosamine (2-acetamido-2-deoxy-D-glucose) units connected by (1-4) links. Chitosan has a cationic composition due to the presence of amino groups in its structure, which confers biological activity in low pH environments, particularly through association with Proteins, polysaccharides, and phospholipids are all negatively charged substances present in bacterial cells [14]. Research on chitosan modification is growing due to biomedical and pharmaceutical uses, intending to improve its already well- established antibacterial capabilities against *Staphylococcus aureus* and *Escherichia coli*. Because the biopolymer's bactericidal effectiveness has previously been proven, they are themost commonly seen microorganisms in scientific literature [15]. Increased free radical generation can overwhelm resistance enzymes like catalase, peroxidase, and superoxide dismutase, causing severe biological consequences (e.g., death) by oxidizing membrane lipids, cell proteins, enzymes, and DNA, affecting cellular respiration [16]. Lower-weight chitosan offers a number of advantages versus high-weight chitinoids when it comes to removing free radicals [17].Chitosan has been proven to prevent Gram-positive bacteria like *Staphylococcus aureus* and Gram-negative bacteria like *Escherichia coli* from forming biofilms.

Graphene has been used in medicine in recent years, notably for DNA sequencing [18],biosensor creation, and cell differentiation and proliferation [19]. Its functional derivative, GO, offers unique properties that make it more suited for biological uses. Its ability to disperse in a range of solvents distinguishes it, making it easier to handle [20]. GO is also usedto deliver anticancer drugs to biological cells [21], as well as aptamers for ATP probing in mouse epithelial cells and gene transfer [22]. The antibacterial activity of GO was investigatedby looking for an inhibitory zone around the GO disc, which showed bacterial toxicity against Staphylococcus aureus and Escherichia coli [23].

2. Material and procedure

2.1. Antioxidant Activity of CS-GO nanoparticles

Antioxidant activity of CS-GO Nanoparticles The DPPH technique was used to test the extract's ability to scavenge free radicals. The DPPH solution (0.006 percent w/v) was made in 95 percent methanol. Discoloration was detected at 517 after 30 minutes in the darkincubation, using freshly produced DPPH solution in test tubes, and mixed NPs (100, 200, 300, 400, 500, and 600 mg/ml) were added to each test tube until the total volume was 2 ml(UV visible spectrophotometer). Measurements were carried out in at least three different ways.

A DPPH solution with the same volume (without NPs) was used as a control, and 95 percentmethanol was utilized as a blank. The following equation was used to compute the percentage of DPPH free radical scavenging:

DPPH scavenging impact (%) = (Ao –A1) / Ao X 100

where Ao is the absorbance in the absence of mixed nanoparticles and A1 is the absorbancein the presence of mixed nanoparticles. The test chemicals' actual absorption reduction was compared to the positive controls [24].

2.2. Antibiofilm Activity of SC-GO NPs by Tissue Culture Plate Method

The gold standard test for identifying the formation of biofilms was TCPM [25]. In 10mL of trypticase soy broth with 1% glucose, a loopful of freshly cultivated isolates was injected. After that, the infected broth was maintained in the incubator for 24 hours at 37°C.Bacterial suspensions were further diluted in a new medium at a ratio of 1:100. A sterile polystyrene tissue culture plate with 96 flat bottom wells was used to pour the prepared bacterial suspension into individual wells. The tissue culture plate also contained control organisms. Sterility must be ensured, and non-specific binding must be discovered . the onlysterile broth was utilized. After 24 hours of incubation at 37°C, the plate was gently tappedto remove the contents of the wells, followed by washing with 200 l of phosphate buffer saline. To eliminate any free germs in the wells, The washing operation was carried out fourtimes in total. Sodium acetate (2%) was added to the wells and left for 30 minutes to repair the biofilms formed by bacteria adhering to them. Crystal violet was used to stain the

biofilms that had been repaired (0.1 percent). The wells were carefully cleaned withdeionized water after 30 minutes to eliminate any remaining discoloration.

The optical density (OD) of pigmented bacterial biofilms were assessed using a micro- ELISA reader after they had dried (at 570 nm wavelength). The test was repeated three times,with the average of the three OD findings. Optical densities revealed bacterial adherence tothe wells and biofilm formation. The OD values were computed, and biofilm development was classed as strong, moderate, or non/weak, as described in previous studies. (Table 1) [26] [27].

Table 1. Tissue culture plate technique for assessing biofilm development

Optical densities values	**Adherence**	**Biofilm formation**
< 0.120	Non	Non/weak
0.120- 0.240	Moderate	Moderate
> 0.240	Strong	Strong

2.3. Hemolysis Effect of CS-GO Nanocomposite

The percentage of hemolysis was used to assess the hemolytic toxicity of mixed NPs.On the blood of one healthy donor, hemolysis experiments were performed: 15 liters of nanoparticles at three concentrations (1, 5, and 10 mg/ml) were added to 285 liters of wholeblood. The positive control is triton X-100, whereas the negative control is blood only. On ashaking plate, the suspension is incubated for 24 hours at room temperature.

The suspension is centrifuged at 10000 for 5 minutes after the incubation period. X Mark (Biorad, USA) microplate scanning spectrophotometer analyzes the supernatant in a 96-wellplate at 550 nm.

H (percent) = (OD550nm sample – OD550nm Tyrode) x (OD550nm Triton X-100 1 percent

- OD550nm Tyrode)*100.

The controls (both positive and negative), respectively, produced 100% and 0% lysis [28].

2.4. Antibacterial activity of CS-GO nanocomposite

After a 24-hour incubation period at 37°C,the minimum inhibitory concentrations (MICs) were measured visually and spectrophotometrically. The minimum inhibitory concentration (MIC) for combined NPs was determined as the lowest NP concentration thathalted growth visually or reduced growth by 80% as compared to growth in the drug-free well. The 50 percent increase is noteworthy when compared to growth in the drug-free well.The IC 50 was determined as a 50% suppression of growth in comparison to growth in the drug-free well [29].

3. Result and discussion

3.1. Antibiofilm activity of CS-GO Nanocomposite on E.coli

The results of this study show that CS-GO nanoparticles can inhibit biofilm formationin *E. coli* bacterial isolates from UTI at various concentrations. For example, the high concentration of CS-GO (16mg/ml) has very high activity in inhibiting biofilm formation, whereas the inhibition titer decreases at lower concentrations (1mg/ml), implying that the effective concentration of CS-GO is higher. According to these findings, which are in line with prior study, the antibiofilm feature of chitosan is mostly owing to its polycationic nature, which is given by the functional amino groups (NH3 +) of N-acetylglucosamine units. The positive charge of CS interacts electrostatically with negatively charged biofilm components such as EPS, proteins, and DNA, preventing the production of bacterial biofilms[30] GO is an antibacterial carbon nanomaterial that may be employed alone or in conjunction with other substances [31].

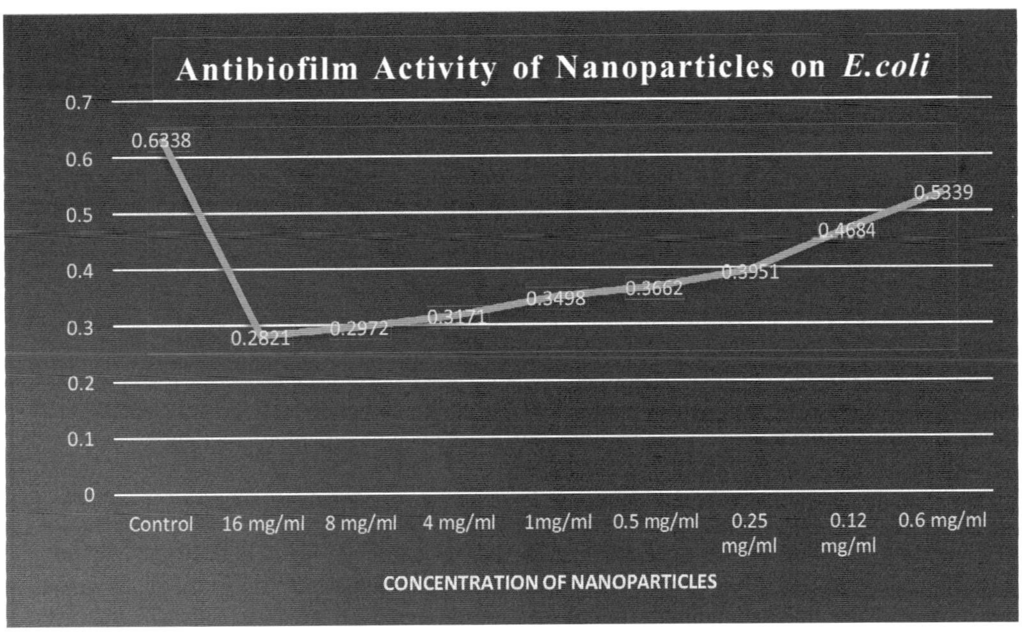

Figure 1. Antibiofilm activity of CS-GO Nanoparticles on biofilm formation in *E.coli* isolated from UTI

3.2. Antioxidant activity of CS-GO nanocomposite

According to the current study, Chitosan-Graphene oxide Nanoparticles can scavenge free radicals, and this ability increases as increase the concentration of CS-GO. As shown in Figure (2), the Scavenger ability is (29.42 percent) at 100 mg/ml, and it increases to (54.06 percent) at 600 mg/ml of CS-GO nanoparticles. The DPPH radical scavenging experiment result ascribed to [32] [33] revealed that graphene increases the antioxidative efficacy of nanoparticles at greater concentrations, but chitosan has very little influence on the activity at all doses. Electron transport [34] [35] and hydrogen donation from functional groups (-OH) linked to graphenic structure [36] are other essential factors. The reduction of DPPH, a stable free radical, is used in the DPPH assay technique. Antioxidants that scavenge DPPH radicalscan transfer hydrogen to free radicals, forming non-radical species [37].

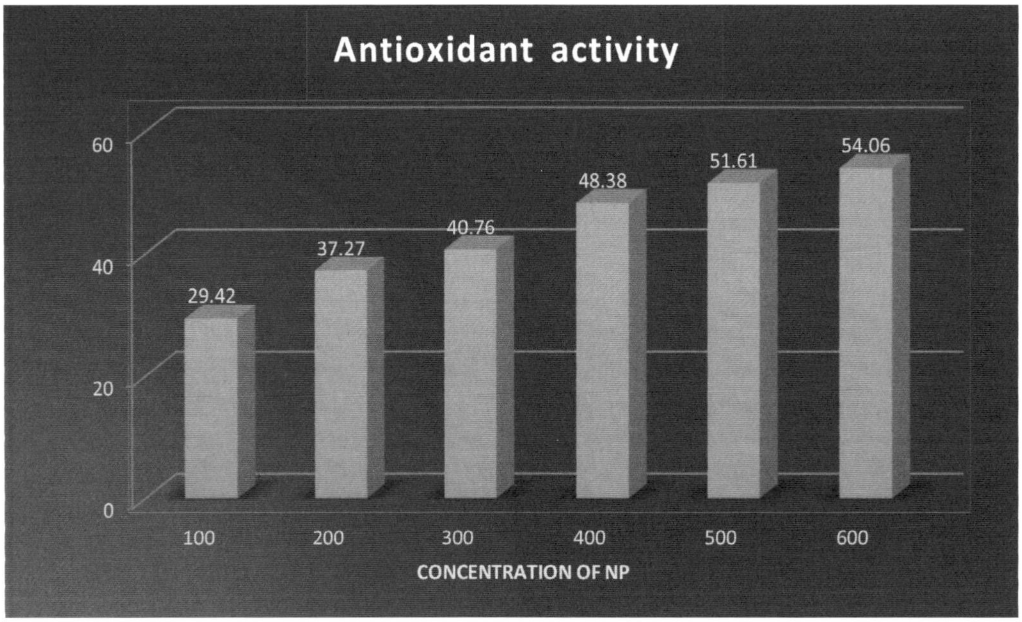

Figure 2. Antioxidant activity of CS-GO nanoparticles

3.3. Antibacterial activity of CS-GO nanocomposite by MIC

As demonstrated in Figure (3), the CS-GO nanoparticles had an ID50 of 1 mg/ml forinhibiting 80 percent of the growth in E. coli found in the urine after a urinary tract infection. According to the findings, the antibacterial activity of GO-CS nano-fibrillar was concentration-dependent. The GO-CS nano-hybrid also has antibacterial activity against *E. coli.* comparable to the broth test. These findings are consistent with earlier findings from other studies [38] [39].

When it comes to GO-CS nano-hybrids, the greatest antibacterial activity against *E. coli* was reported. It can be linked to the synergistic action that results from the combination of two of the most renowned antibacterial compounds, CS and GO. Graphene-basedmaterials' antimicrobial properties, as well as CS, has been reported to include both chemical and physical mechanisms. As a consequence, the crossing method suggests that the antibacterial presentation of GO and CS are combined. The combined interaction of GO functional groups and CS amine groups may boost nano-hybrid absorption and penetration,resulting in lipid molecule extraction damage [40] [41] [42]. The combined presence of GO and CS has a chemical consequence may be extrapolated to result in the formation Increased antibacterial activity of hybrids due to the production of more reactive oxygen species (ROS) [43].

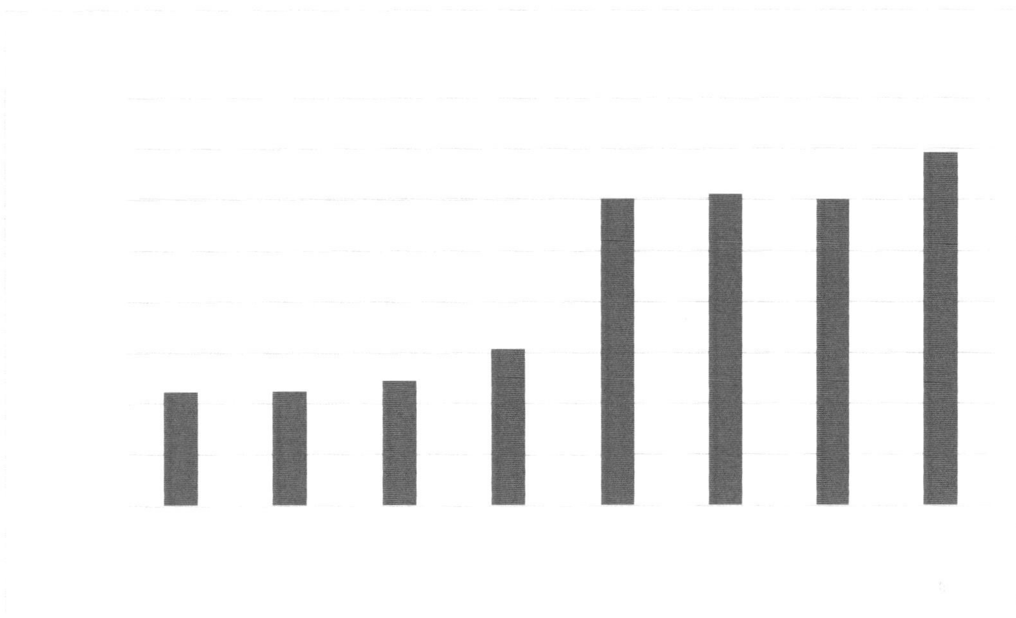

Figure 3. Minimum Inhibitory Concentration of CS-GO nanocomposite on *E. coli* isolated from UTI

3.4. Hemolysis activity of CS-GO nanocomposite

In this investigation, the hemolysis impact of nanoparticles on blood was not active atconcentrations of 1 mg/ml, but it was active at concentrations of 5,10 mg/ml, as shown in Figure (4). Yildirim et al. found that chitosan–graphene nanocomposites alter the membrane of blood cells, most likely by membrane deformation, resulting in membrane damage and hemoglobin release, regardless of surface modification. The proteins on the surface of erythrocytes can bind to the surface of chitosan–graphene films, reducing the lifetime of red blood cells dramatically [44]. According to the researchers, graphene oxide (GO) derivativesinteract with neutral, positively charged, and negatively charged lipid membranes (Liao et al.) The erythrocyte membrane can be disrupted by electrostatic interactions between GO and the positive groups of phosphatidylcholines located in the outer monolayer. Hemocompatibility may be improved via GO surface coatings. The erythrocyte membranehas previously been demonstrated to be damaged by GO and graphene sheets, resulting in adose-dependent hemolytic impact on RBCs [45].

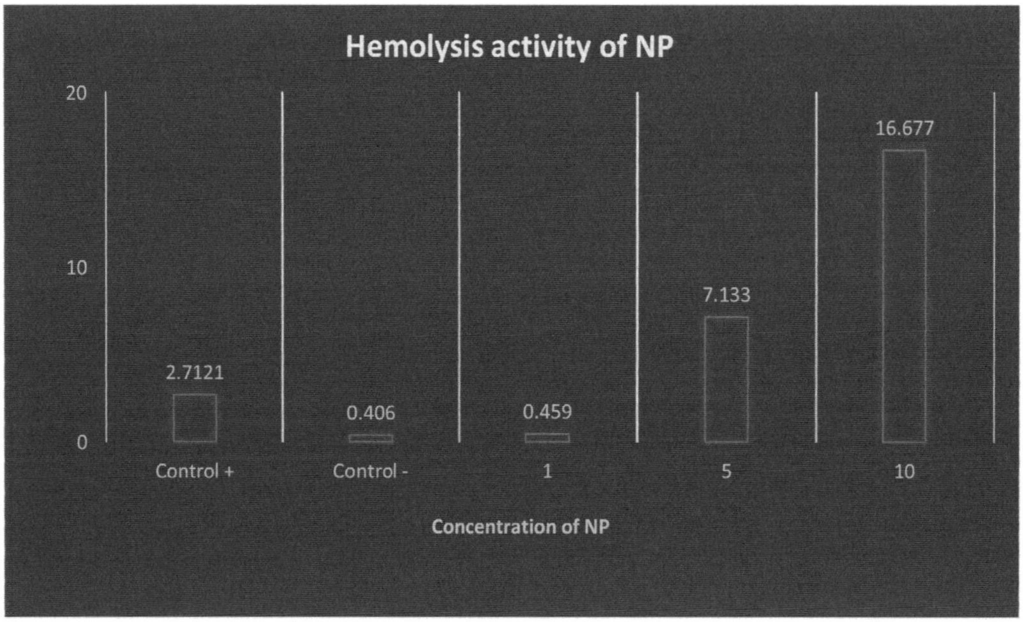

Figure 4. Hemolysis activity of CS-GO nanocomposite on healthy blood

References

[1] Ahmad, A., Mukherjee, P., Senapati, S., Mandal, D., Khan, M. I., Kumar, R., & Sastry, M. (2003). Extracellular biosynthesis of silver nanoparticles using the fungus Fusarium oxysporum. Colloids and surfaces B: Biointerfaces, 28(4), 313-318..

[2] Kashani, S. Y., Afzalian, A., Shirinichi, F., & Moraveji, M. K. (2021). Microfluidics for core–shell drug carrier particles–a review. RSC Advances, 11(1), 229-249..

[3] Clark, D., & Pazdernik, N. (2009). Molekulare Biotechnologie: Grundlagen und Anwendungen. Springer-Verlag..

[4] Dawadi, S., Katuwal, S., Gupta, A., Lamichhane, U., Thapa, R., Jaisi, S., ... & Parajuli, N. (2021). Current research on silver nanoparticles: Synthesis,characterization, and applications. Journal of Nanomaterials, 2021..

[5] Korbekandi, H., Iravani, S., & Abbasi, S. (2012). Optimization of biological synthesis of silver nanoparticles using Lactobacillus casei subsp. casei. Journal of Chemical Technology & Biotechnology, 87(7), 932-937..

[6] Kaviyarasu, K., Kanimozhi, K., Matinise, N., Magdalane, C. M., Mola, G. T., Kennedy, J., & Maaza, M. (2017). Antiproliferative effects on human lung cell lines A549 activity of cadmium selenide nanoparticles extracted from cytotoxic effects: investigation.

[7] Silveira, G. P., Nome, F., Gesser, J. C., Sá, M. M., & Terenzi, H. (2006). Estratégias utilizadas no combate a resistência bacteriana. Química Nova, 29, 844-855..

[8] Tang, R., Jiang, F., Wen, J., Deng, Y., & Sun, Y. (2016). Managing bacterial biofilms with chitosan-based polymeric nitric oxides: Inactivation of biofilm bacteria and synergistic effects with antibiotics. Journal of Bioactive and Compatible Polymers, 31(.

[9] Brodie, B. C. (1859). XIII. On the atomic weight of graphite. Philosophical transactions of the Royal Society of London, (149), 249-259..

[10] Wu, Z. S., Ren, W., Gao, L., Zhao, J., Chen, Z., Liu, B., ... & Cheng, H. M. (2009). Synthesis of graphene sheets with high electrical conductivity and good thermal stability by hydrogen arc discharge exfoliation. ACS nano, 3(2), 411-417..

[11] Ghosh, D. S., Calizo, I., Teweldebrhan, D., Pokatilov, E. P., Nika, D. L., Balandin, A. A., ... & Lau, C. N. (2008). Extremely high thermal conductivity of graphene: Prospects for thermal management applications in nanoelectronic circuits. Appli.

[12] Nair, R. R., Blake, P., Grigorenko, A. N., Novoselov, K. S., Booth, T. J., Stauber, T., ... & Geim, A. K. (2008). Fine structure constant defines visual transparency of graphene. Science, 320(5881), 1308-1308..

[13] Rafiee, M. A., Rafiee, J., Srivastava, I., Wang, Z., Song, H., Yu, Z. Z., & Koratkar, N. (2010). Fracture and fatigue in graphene nanocomposites. small, 6(2), 179-183..

[14] Costa, E. M., Silva, S., Vicente, S., Veiga, M., Tavaria, F., & Pintado, M. M. (2017). Chitosan as an effective inhibitor of multidrug resistant Acinetobacter baumannii. Carbohydrate polymers, 178, 347-351..

[15] Mahae, N., Chalat, C., & Muhamud, P. (2011). Antioxidant and antimicrobial properties of chitosan-sugar complex. International Food Research Journal, 18(4)..

[16] Schieber, M., & Chandel, N. S. (2014). ROS function in redox signaling and oxidative stress. Current biology, 24(10), R453-R462..

[17] Ngo, D. H., & Kim, S. K. (2014). Antioxidant effects of chitin, chitosan, and their derivatives. Advances in food and nutrition research, 73, 15-31..

[18] Min, S. K., Kim, W. Y., Cho, Y., & Kim, K. S. (2011). Fast DNA sequencing with a graphene-based nanochannel device. Nature nanotechnology, 6(3), 162-165..

[19] Wang, Y., Li, Z., Wang, J., Li, J., & Lin, Y. (2011). Graphene and graphene oxide: biofunctionalization and applications in biotechnology. Trends in biotechnology, 29(5), 205- 212..

[20] Paredes, J. I., Villar-Rodil, S., Martínez-Alonso, A., & Tascon, J. M. D. (2008). Graphene oxide dispersions in organic solvents. Langmuir, 24(19), 10560-10564..

[21] Chung, C., Kim, Y. K., Shin, D., Ryoo, S. R., Hong, B. H., & Min, D. H. (2013). Biomedical applications of graphene and graphene oxide. Accounts of chemical research, 46(10), 2211-2224..

[22] Wang, Y., Li, Z., Hu, D., Lin, C. T., Li, J., & Lin, Y. (2010). Aptamer/graphene oxide nanocomplex for in situ molecular probing in living cells. Journal of the American Chemical Society, 132(27), 9274-9276..

[23] Bao, Q., Zhang, D., & Qi, P. (2011). Synthesis and characterization of silver nanoparticle and graphene oxide nanosheet composites as a bactericidal agent for water disinfection. Journal of colloid and interface science, 360(2), 463-470..

[24] Goyal, A. K., Middha, S. K., & Sen, A. (2010). Evaluation of the DPPH radical scavenging activity, total phenols and antioxidant activities in Indian wild Bambusa vulgaris Vittata methanolic leaf extract. Journal of Natural Pharmaceuticals, 1(1)..

[25] Hassan, A., Usman, J., Kaleem, F., Omair, M., Khalid, A., & Iqbal, M. (2011). Evaluation of different detection methods of biofilm formation in the clinical isolates. Brazilian journal of infectious Diseases, 15(4), 305-311..

[26] Panda, P. S., Chaudhary, U., & Dube, S. K. (2016). Comparison of four different methods for detection of biofilm formation by uropathogens. Indian Journal of Pathology and Microbiology, 59(2), 177..

[27] Sultan, A. M., & Nabiel, Y. (2019). Tube method and Congo red agar versus tissue culture plate method for detection of biofilm production by uropathogens isolated from midstream urine: which one could be better. African Journal of Clinical and Experimenta.

[28] Laloy, J., Minet, V., Alpan, L., Mullier, F., Beken, S., Toussaint, O., ... & Dogné, J. M. (2014). Impact of silver nanoparticles on haemolysis, platelet function and coagulation. Nanobiomedicine, 1, 4..

[29] Kumar, R., Shukla, S. K., Pandey, A., Srivastava, S. K., & Dikshit, A. (2015). Copper oxide nanoparticles: an antidermatophytic agent for Trichophyton spp. Nanotechnology Reviews, 4(5), 401-409..

[30] Khan, F., Pham, D. T. N., Oloketuyi, S. F., Manivasagan, P., Oh, J., & Kim, Y. M. (2020). Chitosan and their derivatives: Antibiofilm drugs against pathogenic bacteria. Colloids and Surfaces B: Biointerfaces, 185, 110627..

[31] Cacaci, M., Martini, C., Guarino, C., Torelli, R., Bugli, F., & Sanguinetti, M. (2019). Graphene oxide coatings as tools to prevent microbial biofilm formation on medical device. Advances in Microbiology, Infectious Diseases and Public Health, 21-35..

[32] Galano, A. (2008). Carbon nanotubes as free-radical scavengers. The Journal of Physical Chemistry C, 112(24), 8922-8927..

[33] Burton, G. W., & Ingold, K. U. (1984). Beta-carotene: an unusual type of lipid antioxidant. Science, 224(4649), 569-573..

[34] Martínez, A., & Galano, A. (2010). Free radical scavenging activity of ultrashort single-walled carbon nanotubes with different structures through electron transfer reactions. The Journal of Physical Chemistry C, 114(18), 8184-8191..

[35] Morita, Y., Suzuki, S., Sato, K., & Takui, T. (2011). Synthetic organic spin chemistry for structurally well-defined open-shell graphene fragments. Nature chemistry, 3(3), 197- 204..

[36] Wright, J. S., Johnson, E. R., & DiLabio, G. A. (2001). Predicting the activity of phenolic antioxidants: theoretical method, analysis of substituent effects, and application to major families of antioxidants. Journal of the American Chemical Society, 123.

[37] Blois, M. S. (1958). Antioxidant determinations by the use of a stable free radical. Nature, 181(4617), 1199-1200..

[38] Ali A Alsudani et al 2019 J. Phys.: Conf. Ser. 1294 062099

[39] Ghaidaa Raheem Lateef Al-Awsi et al 2019 J. Phys.: Conf. Ser. 1294 062077.

[40] Ali A Alsudani and Ghaidaa Raheem Lateef Al-Awsi, 2020. Biocontrol of Rhizoctonia solani (Kühn) and Fusarium solani (Marti) causing damping-off disease in tomato with Azotobacter chroococcum and Pseudomonas fluorescens. Pakistan Journal of Biological Sciences, 23: 1456-1461.

[41] Ghaidaa Raheem Lateef Al-Awsi et al 2021 IOP Conf. Ser.: Earth Environ. Sci. 790 012013.

[42] Bassam F. ALfarhani et al 2021 IOP Conf. Ser.: Earth Environ. Sci. 790 012037

[43] Bassam F. ALfarhani et al 2021 IOP Conf. Ser.: Earth Environ. Sci. 790 012040

[44] Alfarhani, B.F., Hammza, R.A. & Alzaidy, A.S. Potential effect of solvent and slit width on some properties of room temperature fluorescence of hydroxy polycyclic aromatic hydrocarbons. Chem. Pap. 75, 3915–3920 (2021). https://doi.org/10.1007/s11696-021-01602-1

[45] Bassam Faron Alfarhani et al 2019 J. Phys.: Conf. Ser. 1294 052045

[46] Ahmed K Abass and Zeina M Kadhim (2019), Bioadsorption of Eriochrome Black T (EBT) onto Iraqi Rice Husk as Adsorbent, J. Phys.: Conf. Ser. 1294 052030

[47] Alaa Hussien Ali1, Khalid J Al-Adely1 and Zeina Mohammed Kadam (2019), Preparation and characterization of a novel ligand (5-MeTAQ)with some metal complexes of Cd(II),Zn(II),Hg(II).J. Phys.: Conf. Ser. 1294 052039

[48] Ristic, B. Z., Milenkovic, M. M., Dakic, I. R., Todorovic-Markovic, B. M., Milosavljevic, M. S., Budimir, M. D., ... & Trajkovic, V. S. (2014). Photodynamic antibacterial effect of graphene quantum dots. Biomaterials, 35(15), 4428-4435..

[49] Szunerits, S., & Boukherroub, R. (2016). Antibacterial activity of graphene-based materials.Journal of Materials Chemistry B, 4(43), 6892–6912..

[50] Je, J. Y., & Kim, S. K. (2006). Chitosan derivatives killed bacteria by disrupting the outer and inner membrane. Journal of agricultural and food chemistry, 54(18), 6629-6633..

[51] Ji, H., Sun, H., & Qu, X. (2016). Antibacterial applications of graphene-based nanomaterials: recent achievements and challenges. Advanced drug delivery reviews, 105, 176-189..

[52] Li, X. F., Feng, X. Q., Yang, S., Fu, G. Q., Wang, T. P., & Su, Z. X. (2010). Chitosan kills Escherichia coli through damage to be of cell membrane mechanism. Carbohydrate Polymers, 79(3), 493-499..

[53] Huang, X., Bao, X., Liu, Y., Wang, Z., & Hu, Q. (2017). Catechol-functional chitosan/silver nanoparticle composite as a highly effective antibacterial agent with species-specific mechanisms. Scientific reports, 7(1), 1-10..

[54] Yildirim, A., Ozgur, E., & Bayindir, M. (2013). Impact of mesoporous silica nanoparticle surface functionality on hemolytic activity, thrombogenicity and non-specific protein adsorption. Journal of Materials Chemistry B, 1(14), 1909-1920..

[55] Liao, K. H., Lin, Y. S., Macosko, C. W., & Haynes, C. L. (2011). Cytotoxicity of graphene oxide and graphene in human erythrocytes and skin fibroblasts. ACS applied materials & interfaces, 3(7), 2607-2615..

Facile one-step synthesis of Au-ZnO/MWCNTs for photocatalytic treatment of Reactive Blue dye in aqueous solution

Hassan A. Alshamsi1[1], Qusay A. Nema[2] and Salam H. Alwan[3]

[1]Department of Chemistry, College of Education, University of Al-Qadisiyah, Diwaniya, Iraq
[2]Al-Muthanna Education Directorate, Al-Muthanna Province, Samawa, Iraq
[3]College of Dentistry, University of Al-Qadisiyah, Diwaniya, Iraq

Abstract. In this work, the multiwall carbon nanotubes (MWCNTs) based gold-zinc oxide (Au-ZnO) nanocomposite has been synthesized for visible light driven photocatalytic decolorization of reactive blue-4 (RB4) dye. The Au-ZnO/MWCNTs nanophotocatalyst has been successfully synthesized via thermal treatment method at 600 0C. The physico-chemical properties of MWCNTs and Au-ZnO/MWCNTs have been investigated via powder X-ray diffraction (XRD) spectroscopy, UV-visible diffuse reflectance spectroscopy (UV-DRS), Fourier transform infrared spectroscopy (FTIR)scanning electron microscope (SEM), atomic force microscopy (AFM) and transmission electron microscopy (TEM) . The synergistic effect of MWCNTs and Au NPs doping could offers an interesting opportunity to design photocatalyst material with the maximum absorption shift to the visible region of the spectrum, where photocatalytic activity can be enhanced. The results exhibited the growth of Au and ZnO nanoparticles (NPs) on the MWCNTs surface with an average diameter of 20 – 30 nm. The photodecolorization of RB4 dye was investigated by changing pH (3-11), dosage of photocatalyst (0.2-1.4 g/L), and initial RB4 concentration (25–100 mg/L). Under best treatment conditions (pH 11 and catalyst dosage 1.0 g/L), the best photodecolorization efficiency was reached 94.91 %. Based to the obtained results, the photodecolorization efficiency adopting Au-ZnO/MWCNTs photocatalyst was stated to be a promising route for remediation the organic pollutants in aqueous solutions

Keywords: Thermal treatment, Au-ZnO, MWCNTs, Photocatalysis, RB4

1. Introduction

Now days, water pollution is one of the widespread problems for human health and and organisms that live in the aquatic environment. Water contaminations occur due to human activities including as urbanization, agricultural and industrialization activities which related to the rapid growing population and the continuous need for more energy, food and housing. The most serious source of water continuations comes from industrial activities which release large quantities of organic and inorganic pollutants, the most important of which are colorant materials, fertilizers, hydrocarbons, heavy metals, pesticides, insecticides, plasticizer, detergents, oils, greases, medicines and disinfectants [1] [2]. Industrial effluents such as liquid waste contaminated toxic pollutants like colored material residues are one of big environmental problems. Organic dyes have been spreadably used in different industries including textiles, paper processing, pharmaceuticals, cosmetics, leathers, foods, rubber, plastics, paints, inks and color photography [3] [4] [5]. In general and in most cases, artificial colorant materials are complex unsaturated ionizing and aromatic compounds with desirable properties like reactivity, color stability, brightness, solubility, light resistance and substantiveness [6].

Various conventional techniques including coagulation [7], coprecipitation [8], ozonation [9], adsorption [10], ion exchange [11], reverse osmosis [12] and advanced oxidation processes [13], sonocatalysis [14], ultrafiltration [15] and biodegradation [16] have been adopted for the remediation of hazardous organic pollutants from wastewater have attained great attention. In addition to the fact that most of these methods are limited, they are characterized by high cost and inflexible operation conditions. Photocatalysis is more attractive process because the photocatalyst can be recovered along with mineralization of organic pollutant into water and carbon dioxide [17]. Among various semiconductors, zinc oxide (ZnO) is the most widely used and most efficient in the field of photocatalytic treatment of water pollutants due to its, nontoxicity, cost effectiveness, high chemical stability, good photosensitivity and high activity [18]. Generally, semiconducting materials are described by a band structure where the electrons centered in the in the valence band (VB) can be photoinduced and move to the conduction band (e_{CB}^-) leaving empty states (positive holes, h_{VB}^+) in valance band. The photocatalytic activity of ZnO is affected very important by factors like optical band gap energy and charge carrier recombination (e^--h^+) [19] [20]. It well known that ZnO has a wide band gap energy (E_g=3.3 eV) which can only be photoactivated by light absorption in the UV region [21], hence this restricts its application in photocatalytic processes. Therefore, it has become an important issue to modifying ZnO structure and narrowing its band gap and hence extend the optical response to visible area for efficient photocatalytic applications under visible light or solar radiation. In this regard, doping with metals into ZnO crystal lattice has been used to modify the light absorption towards the visible region and to improve the photocatalytic efficiency [22]. The doping metals are normally created energy levels located between the VB and CB positions of ZnO. The photogenerated charge carrier pairs become more spatially separated via trapping at these surface sites, and the recombination rate is slowed, which subsequently improve the photocatalytic performance of ZnO. Additionally, to overcome disadvantages owing to recombination process, many attempts have made to loading ZnO NPs on solid support materials including bentonite [23], graphene [24], zeolite [25] and carbon nanotubes (CNTS) [26]. CNTs have some remarkable advantages, properties like superior chemical stability, strong adsorptive ability, and excellent electronic and conductive properties [27]. Additionally, the large electron storage capacity of CNTs improves the photocatalytic efficiency by accepting the photoinduced electrons from the CB of ZnO at the heterojunction interface of ZnO and CNT [28]. The synergistic effect of metals and CNTs on ZnO NPs can efficiently improve the interface charge separation and prevent the aggregation of ZnO NPs [29].

In the current study, Au-ZnO/MWCNTs was synthesized using thermal sintering of [Au (BIAHQ)Cl] Cl.H_2O and [Zn $(BIAHQ)_2$]. H_2O complexes. The synthesized nanostructure was obtained as a promise material for the photocatalytic decolorization of reactive blue 4 dye in aqueous solution. The physico-chemical characteristics involving morphological, structural, optical and topographical properties Au-ZnO/MWCNTs NC were investigated with various characterization techniques like XRD, FTIR, UV-Vis DRS, AFM, SEM and TEM. The effect of irradiation time, photocatalyst dosage, initial concentration of RB4 dye solution and initial pHs of solutions over photocatalytic decolorization efficiency of RB4 dye have been studied. Also, a suggested mechanism of the photocatalytic degradation/decolorization reactions of RB4 dye was reported.

2. Experimental

2.1. Materials

In this work, all obtained chemicalsand reagents were of high purity which used as received without any additional purification. Multi-wall carbon nanotubes (MWCNTs, 99 % purity) was supplied by Cheap Tubes Inc. (Brattleboro, VT, USA). Gold (III) chloride trihydrate ($HAuCl_4 \cdot 3H_2O$, purity of 99.9%), 2-aminobenzimidazole ($C_7H_7N_3$, 97% purity), reactive blue 4 dye ($C_{23}H_{14}Cl_2N_6O_8S_2$, 35 % dye content) and 8- hydroxy quinolone (C_9H_7NO, 99% purity) were supplied by Sigma-Aldrich. Zinc chloride ($ZnCl_2$, 99 %) was purchased from Merck. sodium hydroxide (NaOH, purity above 99%), sodium nitrite ($NaNO_2$, purity above 99%), and hydrochloric acid (HCl, 37%) were obtained from Scharlau Chemie.SA (Barcelona, Spain). Ethyl alcohol (C_2H_5OH, 99.5 %) was purchased from Baker. Ammonium acetate ($CH_3CO_2NH_4$, 99 %) was supplied by Fluka. Doubled distilled water (DDW) was obtained for preparation all aqueous solutions.

2.2. Instruments

XRD pattern of Au-ZnO/MWCNTs nanocomposite was collected in the region $2\theta = 10°$ to $80°$ using XRD-6000 diffractometer (Shimadzu-Japan) with Cuk radiation source (0.15405 nm). The FTIR spectrum was recorded using KBr disk method on the Shimadzu FTIR model 8400s (Japan) in the wavenumbers range of 4000- 400 cm^{-1}. JASCO V-570 spectrophotometer (Pfungstadt, Germany) was used to recording DRS of nanocomposites in wavelengths range 200-700 nm. The samples were investigated to determine the surface morphology by using Scanning Electron Microscope (KYKY-EM3200, China). The particle sizes and morphology of catalysts were evaluated using transmittance electron microscopy (TEM, Philips CM120 Cryo-electron microscope). Angstrom atomic force microscope (AFM, SPM-AA3000, Angstrom Advanced Inc, USA) was adopted to investigate the topographical characteristics of as synthesized nanomaterials. UV-1650PC, Shimadzu-Japan spectrophotometer was employed to record absorbance of aqueous dye solutions at λ_{max} of dye (590 nm). pH-meter 211-Instrument Hanna- Romania, was obtained to adjust solutions pH. Ultrasonic dispersion of ZnO NPs was carried out employing ultrasonic path (405 power sonic, Hwashin, Korea). A 125 W Osram lamp (Germany) was used as an excitation source in photodecolorization process.

2.3. Synthesis of the Azo Ligand, [Zn(BIAHQ)₂].H₂O and Au[BIAHQ)Cl]Cl.H₂O Complexes

The synthesis and characterization of 7-[2-(Benzimidazole) azo]-8-Hydroxy Quinolone (BIAHQ) ligand has been previously reported [30]. The $[Zn(BIAHQ)_2].H_2O$ complex was synthesized and characterized as previously reported [31]. In addition,the Au $[BIAHQ)Cl]Cl.H_2O$ complex was synthesized and characterized previously in mole ratio 1: 1 [M: L] [30].

2.4. Synthesis of Au-ZnO/MWCNTs photocatalyst

For synthesizing of Au-ZnO/MWCNTs photocatalyst [Au (BIAHQ) Cl] Cl.H_2O/[Zn $(BIAHQ)_2$]. H_2O mixture was prepared by mixing [Zn $(BIAHQ)_2$]. H_2O with [Au (BIAHQ)Cl] Cl.H_2O with a weight ratio of Au and Zn of 3:97. Afterward, MWCNTs (0.03 g) was added to mixture of complexes, and mixture was mixed well using handle mortar. Finally, the solid mixture was calcined at 600 ^{0}C for 2 hours under inert conditions [30] [31]. Meanwhile, a grayish-colored powder formation indicates the successful fabrication of Au-ZnO/MWCNTs nanocomposite.

2.5. Photocatalytic degradation of RB4 dye

The effectiveness of Au-ZnO/MWCNTs NC as photocatalyst under visible light irradiation was investigated using RB4 aqueous solution as an organic pollutant model. For this regard, a home-made photoreactor was employed as an enclosed chamber. The photoreactor was equipped with visible light source of 400 W Osram lamp and a temperature controller. The light source was positioned in the middle of the photoreactor about 10 cm above reaction vessel. Firstly, defined amount of the photocatalyst was suspended into 50 mL RB4 dye solution (50 ppm) and ultrasonically treated for 20 min to ensure homogeneity of the suspension. Afterwards, the mixture was subjected to continuous stirring in the darkness for 120 minutes at 25 °C to establish adsorption-desorption equilibrium of RB4 dye molecules onto surface of Au-ZnO/MWCNTs nanocomposite surface. Then, for photocatalytic performance, the synthesized photocatalyst was exposed to visible light irradiation for 120 min with continuous stirring at 25 °C. Zero-time degradation was considered the moment when the dye solution was illuminated with lamp light. Before irradiation, 4 mL of this solution was withdrawn centrifuged for 15 min at 6000 rpm to separate photocatalyst and then absorbance was recorded at 590 nm (λ_{max} of RB4). This was considered as an absorbance at zero time (A_o). Then, 4.0 mL samples have been withdrawn from the photoreactor vessel at regular time intervals (20 min) and centrifuged at 6000 rpm for 15 min to separate catalyst NPs. To determine the amount of RB4 remaining in the supernatant, the supernatant was extracted and the dye absorbance was determined at 590 nm. The photodecolorization efficiency [PDE (%)] of RB4 was calculated using the following expression:

$$PDE (\%) = (A_0\text{-}A_t/A_0) \times 100 \quad (1)$$

Where A_0 and A_t represent the absorbance of RB4 at zero and at time (t), respectively. In addition, blank experiments have been performed by illuminating the dye solutions without catalyst and there is no significance decrease of RB4 absorbance during the illumination has been noticed. The above-mentioned conditions were obtained in all runs unless otherwise mentioned. The effects of the photocatalyst dosage (0.2- 1.4 g/L), initial RB4 concentration (25-100 ppm) and initial pH (3-11) on the photodecolorization efficiency were considered. All photocatalytic decolorization investigations were conducted at a fixed stirring speed and constant operating temperature of 25°. With the assist of pH meter, the pH of suspensions was brought to exact value using 0.1N HCl or 0.1 N NaOH aqueous solutions.

3. Results and discussion

3.1. Characterization:

3.1.1. Powder XRD analysis

The phase analysis of Au-ZnO/MWCNTs and MWCNTs investigated by XRD technique as displayed Figure 1(a). The powder XRD pattern of Au-ZnO/MWCNTs exhibited the peaks assigned to ZnO at 2θ =31.75°, 34.45°, 38.33°, 48.03°, 56.52°, 63.93°, 65.98°, 68.11° and 69.25°, which indexed to (100), (002), (101), (102), (110), (103), (200), (112) and (201) lattice planes, respectively. As XRD data revealed, the ZnO NPs showed a sharp diffraction peaks are well indexed to the hexagonal ZnO wurtzite structure (JCPDS no. 36-1451) [32]. Au NPs showed four distinct peaks at 2θ = 37.96°, 44.30°, 64.39°, and 77.60°. All Au peaks assigned to standard Bragg reflections (111), (200), (220), and (311) of face center cubic (Fcc) lattice (JCPDS NO. 65-8601) [33]. The XRD pattern of Au-ZnO/MWCNTs nanocomposite show no diffraction peaks of MWCNTs can be noticed which may be due to the using of small ratio of MWCNT (1%). In contrast, the XRD pattern of pure MWCNTs show clear diffraction peaks at 25. 88° (002) and 44. 17° (101) which were related to the (002) and (101) planes of CNT, respectively [34]. The intensity of the diffraction peaks belongs to both ZnO and Au are strong indicating a high crystallinity of both materials. The crystallite size (D) of fabricated NPs was estimated adopting Debye-Scherrer equation (Eq. 2) obtaining the intense peak of NPs:

$$D= 0.9\lambda/\beta\cos\theta \qquad (2)$$

Where D = mean crystallite size, λ = X-ray wavelength (CuK_α=1.54056Å) and β=FWHM (full width at half maxima, in radians). The mean crystallite size of the MWCNTs and Au-ZnO/MWCNTs were found to be 5.29 and 27.75 nm which are derived from the FWHM of more intense peaks located at 25.81° (002), 38.54° (101), respectively.

3.1.2. FT-IR Anslysis

The FT-IR characterization of MWCNTs and Au-ZnO/MWCNTs at room temperature is displayed Figure 1 (b). As shown in FTIR spectrum of MWCNTs, there are bands at 1637 cm^{-1} 2920 cm^{-1} and 3455 cm^{-1} belong to the C=O, C-H and O-H stretching vibrations, respectively. The existence of O-H vibration is associated with the adsorption of water molecules onto MWCNTs surface. The band centered at 1615 cm^{-1} might be appointed to the C = C stretching vibration related to the skeletal vibration of aromatic rings. After loading ZnO with Au, there are important observations about intensity and location of characteristics bands especially that belongs to ZnO NPs. Figure 1b exhibited that the attained band located at 457 cm^{-1} is corresponds to zinc-oxygen bond (Zn-O) [35]. In addition, the centered band at 3430 cm^{-1} corresponds to the stretching vibration mode of O-H functional group which indicates the presence of moisture on Au-ZnO/MWCNTs surface. Also, the O-H bending vibration mode appears at 1625 cm^{-1} which might be mixed with aromatic C=C (sp^2) stretching vibrations.

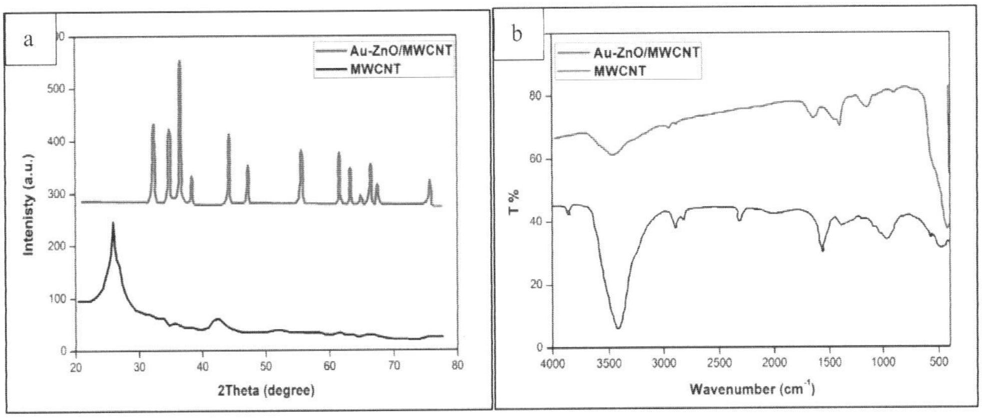

Figure 1. (a) Powder XRD analysis and (b) FT-IR characterization of MWCNTs and Au-ZnO/MWCNTs nanocomposite

3.1.3. UV–Vis DRS AQnalysis

To explore the optical characteristics, the UV–Vis DRS analysis and Tauc plot for energy band gap calculation of MWCNTs and Au-ZnO/MWCNTs are studied as represented in Figure 2 (a) and (b). It can be noticed that the absorption edge of pure MWCNTs is 1.8 eV. However, for Au-ZnO/MWCNTs the absorption edge was about 402 nm which exhibits an efficient red shift compared with typical ZnO NPs. In addition, in the region 400-700 nm, the Au plasmon absorbance at 576 nm is clearly noticed .Previously, the UV-Vis spectroscopy showed that the typical ZnO NPs have an absorption edge below 400 nm [36]. From the above results, it can be said that the shift in ZnO absorption edge can controlled by control the wt % of Au and MWCNTs. The red shift in ZnO absorption edge caused by doping of Au and MWCNTs indicates that the Au-ZnO/MWCNTs sample maybe exhibit a promise photocatalysis characteristics under the effect of visible light irradiation. In this regard, the MWCNTs can acts a photosensitizer, where the photogenerated electrons from MWCNTs can transfer to the conduction band of ZnO (CB) to enhance the absorption of light in in visible region [37]. As a result, the interfacial interaction between MWCNTs and ZnO NPs give rise in the narrowing of band gap energy.

In regard of Au NPs, Au NPs doped into ZnO structure lead to creation of the p-type electrical conductivity by the insertion of Au^+ ions into ZnO lattice sites, hence the Au sites maybe act as a traps for the photogenerated electrons because the Fermi energetic level of Au is lower than the CB of ZnO [38]. Thus, the electrons are injected from CB of ZnO to Fermi level of Au which improve the separation efficiency of charge carrier pairs [39]. This behavior suggests that the Au-ZnO/MWCNTs nanocomposite maybe presents promise photocatalytic performance when exposed to visible light.

a b

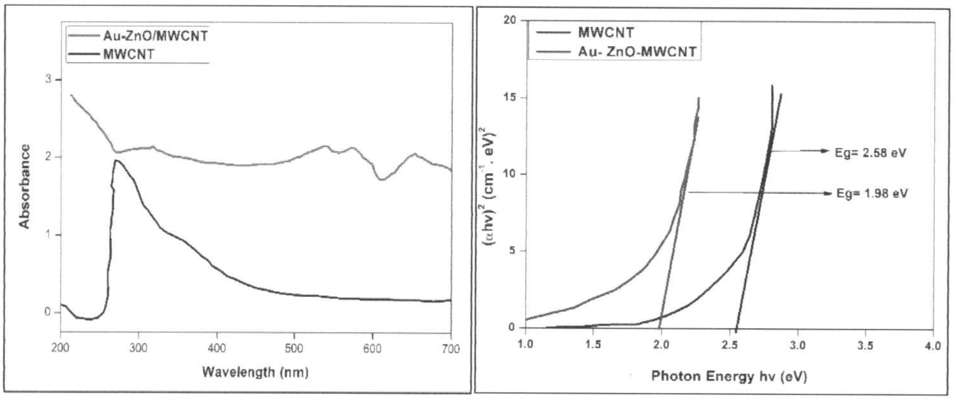

Figure 2. a. UV-Vis DRS analysis and b. Energy band gap of MWCNTs and Au-ZnO/MWCNTs nanocomposite

3.1.4. SEM Analysis

The morphologies and nanoparticle sizes of MWCNTs and Au-ZnO/MWCNTs samples were characterized by SEM analysis. Figures 3 (a) - (d) demonstrates the SEM microgram s of MWCNTs and Au-ZnO/MWCNTs. Figure 4(a) and (b) shows the smooth morphology nature of MWCNTs. Figure 4 (c) and (d) reveals that the MWCNTs act as substrates for the growth of Au and ZnO crystallites. The formation of spherical Au and ZnO NPs on the MWCNTs surface has happened simultaneously in the thermal treatment process. This confirms the fact that the existence of MWCNTs in the nanocomposite has improves the distribution of different sizer particles onto surface and reduces the particle aggregations [40]. According to SEM results, the average size of Au-ZnO/MWCNTs was estimated at 39.2-61.5 nm.

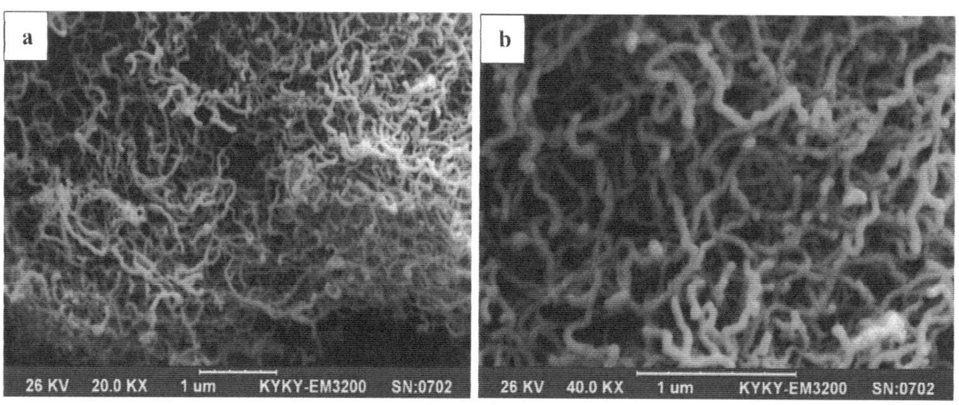

Figure 3. SEM micrograms of MWCNTs at (a) Low magnification and (b) high magnification. SEM micrograms of Au-ZnO/MWCNTs at (c) Low magnification and (d) high magnification.

3.1.5. TEM Images

Additional characteristic details such as shape, size and dispersion of MWCNTs and Au-ZnO/MWCNTs were investigated using transmission electron microscopy (TEM) technique. The TEM image of MWCNTs (Figure 4a) obviously desplyes that the strands of nanotubes tangled together in a spaghetti or interwoven -like structure with a diameter of the tubes is 13-18 nm. The TEM image of Au-ZnO/MWCNTs indicates that the growth of Au and ZnO NPs is almost uniformed distribution along the walls of the tubes and as noted, most particles have a size less than 60 nm. The TEM images showed that smaller Au and ZnO NPs combined into significant aggregates that were found on MWCNTs surface and this behavior is to be expected due to processing at high temperature. The dispersion of Au and ZnO nanoparticles on the MWCNTs surface confirm a strong interaction and binding between MWCNTs and Au and/or ZnO NPs. Therefore, it can be said that Au and ZnO were effectively loaded onto MWCNTs using the thermal treatment process, which is efficiently applicable route for the one step fabrication of Au-ZnO/MWCNTs nanocomposite.

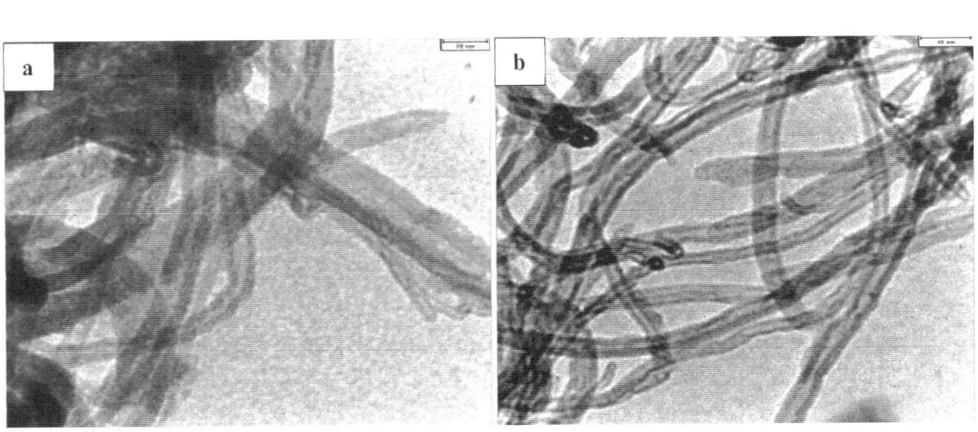

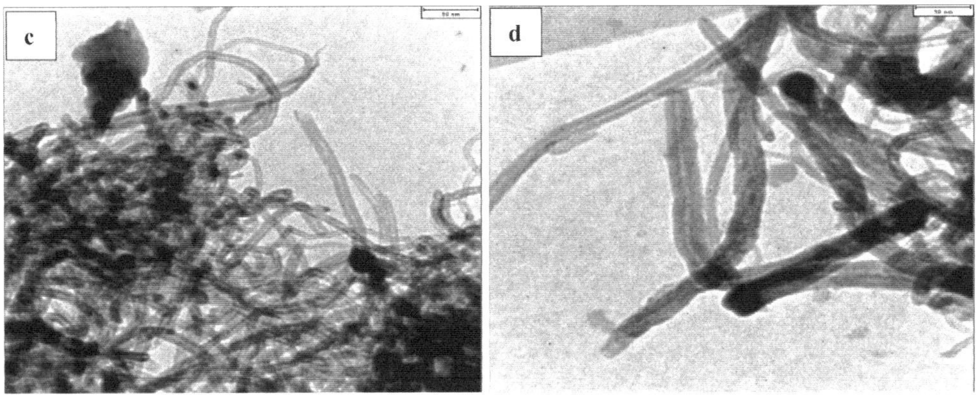

Figure 4. TEM images of MWCNTs (a and b) and Au-ZnO/MWCNTs nanocomposite (c and d)

3.1.6. AFM analysis

The external topography of MWCNTs and Au-ZnO/MWCNTS was tested by (AFM) analysis. AFM technique is an important tool to description of surface topography of catalysts. Figure 5 (a-b) depicts 3D images and granularity cumulation distribution charts of the surfaces of MWCNTs and Au-ZnO/MWCNTs. Table 1 illustrates the AFM statistical data of MWCNTs and Au-ZnO/MWCNTS. The AFM analysis displayed almost homogenous distribution of Au and ZnO NPS over MWCNTs. Also, the morphology of Au and ZnO surface confirms that the grains were evenly dispersed with existence of a significance agglomerations. Additionally, AFM analysis exhibits that most of the Au and ZnO NPs appear spherical and quazi-spherical like structures. Moreover, AFM results clearly demonstrate the slight increasing in roughness coefficient (R_a) and slight decreasing in thickness of Au-ZnO/MWCNTs NC in comparison with pristine MWCNTs sample. Also, as Table 1 shows, the AFM data exhibited that as the NPs thickness decreases as the average diameter decreases. Obviously, the enhanced roughness coefficient (R_a) of the surfaces is confirmed by the recorded negative values of surface skewness (R_{sk}) which mean the amount of surface valleys is more than peaks (hills) [41]. Additionally, kurtosis parameter (R_{ku}) is a statistical parameter obtained to characterize the flatness/sharpness of the surface height distribution. Due to MWCNTS has $R_{ku} < 3$; so it correspond to bumpy surface [42] and this suggests that Au-ZnO/MWCNTs has spiky surface duo to its $R_{ku\ is} >3$. R_{sk} and R_{ku} values can be used to describe the Au-ZnO/MWCNTs surface, where more negative value of R_{sk} and higher R_{ku} negative compared to bare MWCNTs suggests rougher surface.

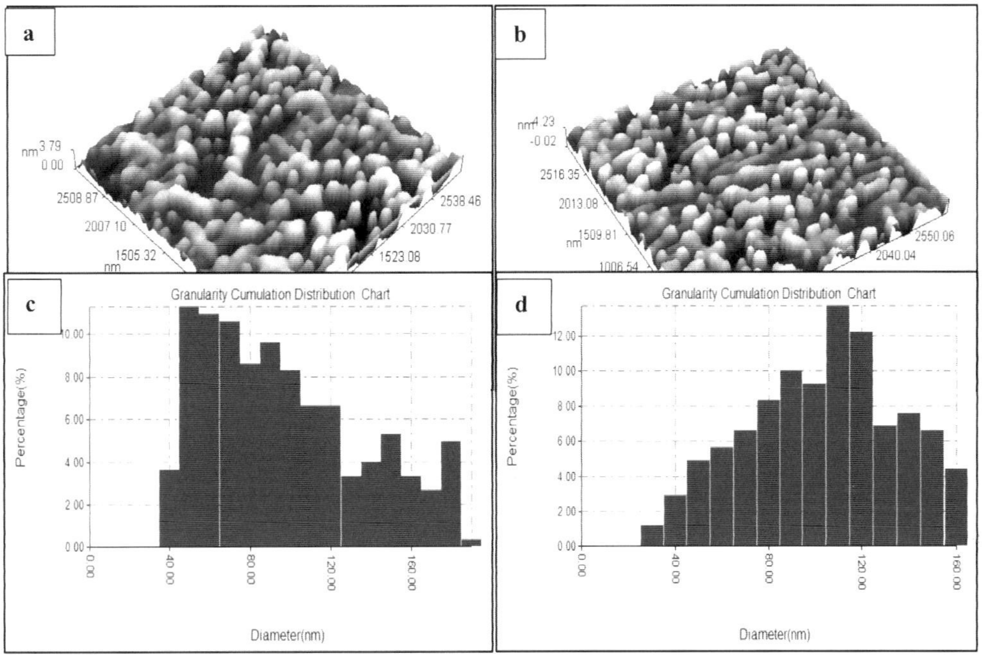

Figure 5. 3D AFM images of (a) MWCNTs and (b) Au-ZnO/MWCNTs and granularity cumulation distribution charts of the surfaces of (c) MWCNTs and (d) Au-ZnO-MWCNTs nanocomposite

Table 1. Surface parameters of MWCNTs and Au-ZnO/MWCNTs

Sample	R_a nm	Thickness nm	Avg. Diameter nm	R_{sk}	R_{ku}
MWCNTs	0.90	4.00	40-390	-0.197	1.69
Au-ZnO/MWCNTs	0.96	3.79	30-140	-0.217	3.06

3.2. Photodecolorization Efficiency
3.2.1. Influence of photocatalyst type
The photocatalytic performance of Au-ZnO/MWCNTs nanocomposite was investigated through color removal of RB4 dye under exposing to visible light. Preliminary photolysis experiments were conducted under irradiation with visible light only without the presence of catalyst (blank), where a negligible photodecolorization efficiency was found which indicates that the dye is stable under photolysis in the absence of photocatalyst. Also, dark adsorption experiments were conducted (absence of visible light) where the adsorption efficiency was 19.96 % and 29.26 % for the MWCNTs and Au-ZnO/MWCNTs after 120 min, respectively (Figure 6 a). After dark period, the dye solutions were exposed to direct visible light and the influence of photocatalysts on the decolorization of RB4 dye in aqueous solution was investigated. As illustrated in Figure, the obtained MWCNTs and Au-ZnO/MWCNTs photocatalysts were effectively degraded 28 % and 75 % of RB4, respectively, which is much more than the efficiency

obtained under adsorption process. The superiority of photocatalytic performance of Au-ZnO/MWCNTs is due to the synergistic roles of Au NPs and MWCNTs. Additionally, the high MWCNTs surface area may have contributed to the enhanced RB4 adsorption on the photocatalyst surface [43]. The above investigations were carried out under the conditions of 1.0 g/L of each photocatalyst, 50 ppm RB4, pH 7 and 25 ^{O}C.

3.2.2. Influence of catalyst dosage

Regarding the pollutants remediation from aqueous solutions, the photocatalyst dose dependence of photocatalysis performance against decolorization/degradation organic pollutants is very vital parameter. The influence of photocatalyst dosage was investigated on RB4 dye decolorization while other experimental parameter were kept constant (RB4 concentration 50 ppm, pH 7, illumination time 120 min). The photodecolorization efficiency of RB4 (50 ppm) using various Au-ZnO/MWCNTs photocatalyst doses was studied in the range 0.2 - 1.4 g/L. As displayed in Figure7 (b), it has been found that the dye photodecolorization efficiency rises with a rise in the Au-ZnO/MWCNTs dose up to 1.0 g/L, where the efficiency was increased from 16 % (0.2 g/L) to 75 % (1.0 g/L), respectively, which could be due the addition of more active sites to the reaction suspension [44]. Further increase in the photocatalyst dosage above 1.0 g/L up to 1.4 g/L leads to reduce the photodecolorization efficiency to 52%, which may be due to the phenomenon of light scattering and screening effects related to aggregation of particles [45], where the high turbidity of reaction medium decreased the penetration depth of visible light radiation.

3.2.3. Influence of initial RB4 dye concentration

The concentration of pollutants, including organic dyes, represents an essential parameter in the photocatalytic remediation processes in aqueous solutions. The photocatalytic decolorization efficiency was investigated in 25-100 ppm initial RB4 concentration in the presence of 1.0 g/L Au-ZnO/MWCNTs at pH of 7. The results obtained are demonstrated in Figure 6(c). After 120 min of exposure to visible light, the photodecolorization efficiency at 25, 50, 75 and 100 ppm were 88%, 75%, 62% and 40%, respectively. The results indicated that the photodecolorization efficiency of 25 ppm was faster among other concentrations. In general, the photodecolorization efficiency of RB4 dye is inversely proportional to its initial concentration. This behavior may be explained by the fact that at lower concentration, a maximum amount of photons will reach the surface of photocatalyst surface, ensuring a higher photodecolorization efficiency [46]. Alternatively, in the reactions with high RB4 concentration, a considerable amount of photons may absorbed by the RB4 molecules in suspension, hence reduced the penetration of light, thereby reducing the photocatalytic activity [47].

3.2.4. Influence of pH

In most cases, colored wastewater is characterized by a wide range of pH. Usually, the pH of dye solutions is one of the most critical factors affecting remediation processes. Due to the dye molecules tend to adsorb over the surface of catalysts, solution pH strongly controls the electrostatic surface charge of photocatalyst and thereby the photodecolorization of dye is greatly influenced by variation in pH. The influence of pH on the photocatalytic color removal from aqueous RB4 solution was investigated. For this purpose, the influence of initial pH was investigated in the initial pH range of 5-11, while the Au-ZnO/MWCNTs and initial dye concentrations were 1.0 g/L and 50 ppm, respectively. It is clear from the Figure 6(d) that the photodecolorization efficiency increases with increasing pH. Obviously, it was seen from the results that presented in Figure 6, the photodecolorization was most effective at pH 11 (94.91%). The influence of solution pH on the adsorption and hence photodecolorization efficiency could be interpret on the basis of zero point charge (pH_{zpc}) of ZnO and H^+ ions concentration [48]. This may be due to the fact that the RB4 is positively charged, whereas the surface ZnO surface is described as positively charged under pH of 9 and possess negative charge at pH values more than 9 as previously reported that ZnO possess pH_{zpc} 8.5~9 [49]. Generally, the photocatalytic reaction is faster in alkaline media which may be attributed to higher electrostatic attractive interactions between positive RB4 molecules and negatively charged photocatalyst surface.

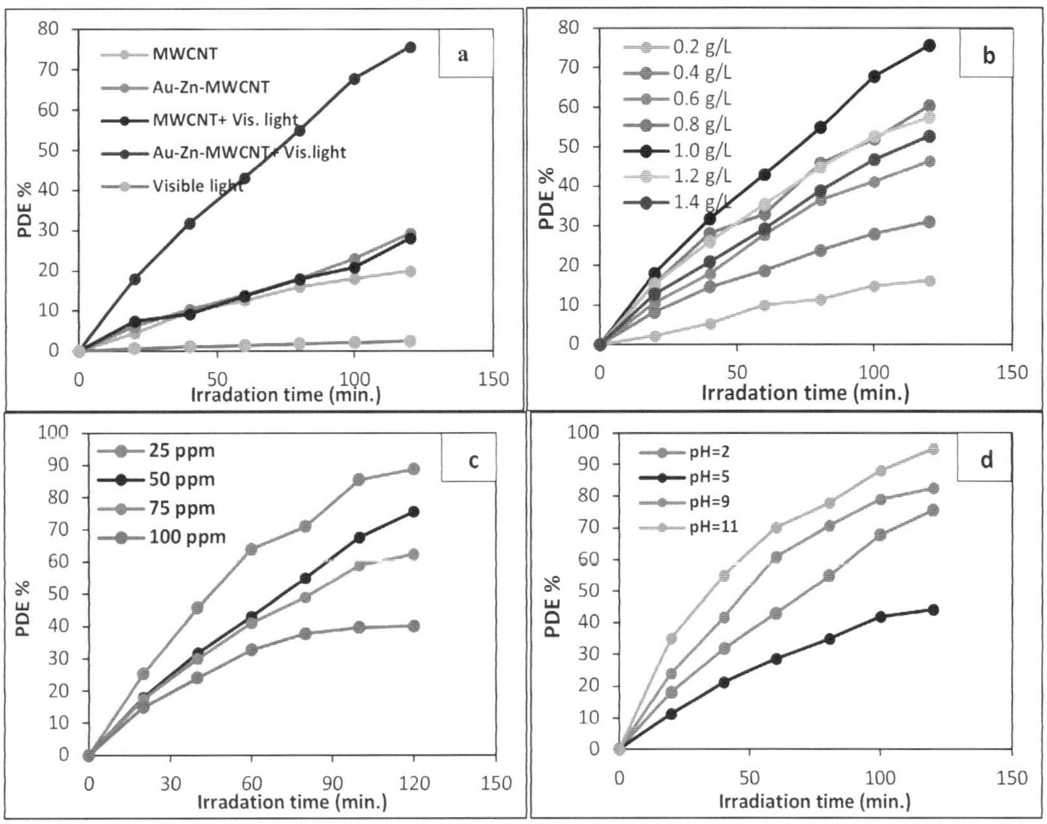

Figure 6. (a)Photolysis, photocatalytic degradation and adsorption curves of RB4. Photocatalytic activity of Au-ZnO/MWCNTs towards RB4 at different variables (b) Effect of Au-ZnO/MWCNTs dose, (c) influence of initial RB4 concentration and (d) influence of initial pH

4. Mechanism of phtodegradation of RB4 dye

The individual parts such as Au and MWCNT play serious functions in increasing the light induced degradation and shifting of photo response of ZnO NPs to visible region. The Au doped ZnO surface acts as effective catalyst under exposing to visible light. The characteristics of heterogeneous photocatalysis include activation of a photocatalyst by artificial or natural light. Initially, under visible light irradiation, ZnO undergoes charge separation process that leads to move of electrons from VB to CB leaving empty states (positive holes, h_{VB}^+) in VB.

$ZnO + h\nu \longrightarrow ZnO\ (e_{CB}^- + h_{VB}^+)$ **[3]**

The photogenerated electrons could react with oxygen molecules dissolved in the suspension and/or adsorbed over ZnO surface to produce high reactive superoxide species ($^{\cdot}O_2-$)[59]:

$e_{CB}^- + O_2 \longrightarrow {}^{\cdot}O_2^-$ **[4]**

The photogenerated holes react with OH^- oxidizing these species to $^{\cdot}OH$ [60]:

$h_{VB}^+ + OH^- \longrightarrow {}^{\cdot}OH$ **[5]**

The $^{\cdot}OH$ radicals are considered to be the most oxidizing agent employed for the destruction of organic targets like colored materials in aqueous solutions. The heterogeneous photocatalysis mechanism of the RB4 molecules under the contact with visible light irradiation and ZnO are summarized in equations 6-10. The molecules of RB4 are significantly adsorbed on the rough ZnO NPs surface and hence sensitization of ZnO photocatalyst. RB4 dye is very sensitive chromophore that intensely absorbs visible light, thus, RB4 molecules easily transfer to high energy excited status (RB4*) as Figure8 shows. The RB4* can easily injects electrons in the ZnO CB which then be transferred to different dopant energy levels. This process effectively contributed to improving the disappearance of RB4 via the self sensitized photolysis under visible light, as described below:

$RB4 + h\nu \longrightarrow RB4^*$ **[6]**

$RB4^* + ZnO \longrightarrow ZnO\ (e_{cb}^- + h_{VB}^+) + RB4^+$ **[7]**

The Au states generate a band level very close to ZnO CB edge causing the band gap of ZnO to narrow. Some photoelectrons transferred to ZnO CB may jump to Au at same time, which improve the separation efficiency of photogenerated charge carrier pairs and thus improve the photocatalytic performance of photocatalysts. Also, MWCNTs act as support for growth both ZnO and Au NPS where the photogenerated electrons in ZnO under visible light irradiation can transfer to the MWCNTs to improve the separation efficiency of charge carrier pairs (Figure 8). Accordingly both Au NPs and MWCNTs act as traps for the electrons and transfers it to oxygen molecules adsorbed on the catalyst surface:

$e-(\text{ or } e_{cb}^-) + O_2 \longrightarrow {}^{\cdot}O_2^-$ **[8]**

The active species ($^{\cdot}OH$, $^{\cdot}O_2^-$ and h^+) can oxidize the RB4 dye molecules to produce degradation and/or mineralization products [62]:

$^{\cdot}O_2^- + RB4/RB4^* \longrightarrow$ Degradation products $\longrightarrow CO_2 + H_2O$ **[9]**

$^{\cdot}OH + RB4/RB4^* \longrightarrow$ Degradation products $\longrightarrow CO_2 + H_2O$ **[10]**

$h_{VB}^+ + RB4/RB4^* \longrightarrow$ Degradation product $\longrightarrow CO_2 + H_2O$ **[11]**

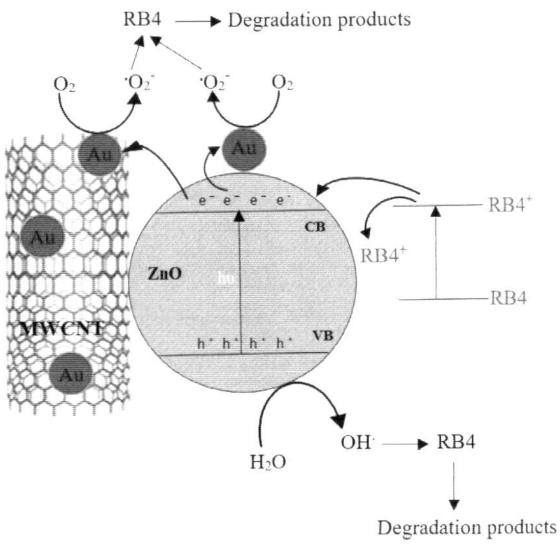

Figure 7. The suggested of RB4 photodegradation on Au-ZnO/MWCNTs photocatalyst.

5. Conclusions

In this research study, the photocatalyst Au-ZnO/MWCNTs was successfully synthesized by a facile and efficient thermal treatment process at high temperatures by using metal complexes as a starting material. The physico-chemical properties of as fabricated nanocomposite were studied using the techniques of XRD, UV-Vis DRS, FTIR, AFM, SEM and TEM. Based on the XRD results, the nanocomposite contains ZnO with hexagonal wurtzite structure and the size of ZnO was 20–50 nm. The FTIR analysis confirmed the existance of Zn-O bond in the as-synthesized photocatalyst. The band gap energy of Au-ZnO/MWCNTs was estimated to be about 1.97 eV, which it is necessary to induce the photocatalytic performance of the photocatalyst under by the effect of visible light from artificial source. The synthesized Au-ZnO/MWCNTs photocatalyst was adopted for color removal of reactive blue 4 dye under visible light irradiation and it exhibited a satisfactory photocatalytic property under visible irradiation. Specifically, the influences of three variables (photocatalyst dosage, initial RB4 concentration and initial pH of solution) on photodecolorization efficiency have been investigated. The best photodecolorization efficiency (94.91%) was achieved at optimum conditions of photocatalyst amount= 1.0 g/L, pH=11, illumination time= 120 min.

References

[1] M. Ilyas, W. Ahmad, H. Khan, S. Yousaf, M. Yasir and A. C. Khan, Environmental and health impacts of industrial wastewater effluents in Pakistan: a review, Rev Environ Health 34, 171–186 (2019)..
[2] M. R. Islam, M. G. Mostafa, Textile Dyeing Effluents and Environment Concerns - A Review, J. Environ. Sci. & Natural Resources, 11(1&2):131-144, 2018..
[3] R. Javaid, U. Y. Qazi, Catalytic Oxidation Process for the Degradation of Synthetic Dyes: An Overview, Int. J. Environ. Res. Public Health 16, 2066 (2019)..
[4] K. G. Pavithra , P. S. Kumar, V. Jaikumar and P. S. Rajan , Removal of Colorants from Wastewater: A

Review on Sources and Treatment Strategies, Journal of Industrial and Engineering Chemistry 78, 1-19 (2019)..
[5] Ratna, B. S. Padhi, Pollution due to synthetic dyes toxicity & carcinogenicity studies and remediation, Internatinal Journal of Environmental Sciences 3(3), 2012, 940-955..
[6] N. M. Julkapli, S. Bagheri, S. B. A. Hamid, Recent Advances in Heterogeneous Photocatalytic Decolorization of Synthetic Dyes, 2014, ,2014 6923077..
[7] M. R. Gadekar and M. M. Ahammed,Coagulation/flocculation process for dye removal using water treatment residuals: modelling through artificial neural networks, Desalination and Water Treatment 57, 26392-26400 (2016)..
[8] S.M. Ibrahim, A.A. Badawy and H. A. Essawy, Improvement of dyes removal from aqueous solution by Nanosized cobalt ferrite treated with humic acid during coprecipitation, J Nanostruct Chem 9, 281–298 (2019)..
[9] H. A. Habeeb and I. J. Sahib, Photo oxidation of (Azure A) dye by UV-enhanced ozonation: optimization and kinetics study, Al-Qadisiyah Journal of Pure Science 19,129-141 (2014)..
[10] N. Hussain, S. Alwan, H. Alshamsi, I. Sahib, Green Synthesis of S- and N-Codoped Carbon Nanospheres and Application as Adsorbent of Pb (II) from Aqueous solution, International Journal of Chemical Engineering 2020, 9068358 (2020)..
[11] M. Wawrzkiewicz, Z. Hubicki, Remazol Black B removal from aqueous solutions and wastewater using weakly basic anion exchange resins. Central Eur J Chem 2011; 9(5): 867–76.
[12] S. E. Ebrahim, T. J. Mohammed, H. O. Oleiwi, Removal of Acid Blue Dye from Industrial Wastewater by using Reverse Osmosis Technology, Association of Arab Universities Journal of Engineering Sciences 25,29-40 (2018)..
[13] H. A. Habeeb and A. S. Waheeb, Study of chemical oxygen demand (COD) and Removal of azure B dye from aqueous solution by using Advanced Oxidation Processes, AL-Qadisiyah Journal of Pure Science 18, 162-180 (2013)..
[14] 10. H. A. H. Alshamsi and A. T. Hamza, Degradation of Thiazine Dyes Azure B and C by Sonolysis, Sonophotolysis and Sonocatalysis, Asian Journal of Chemistry 27, 1935-1940 (2015)..
[15] M.Y. Ghadhban , H. S. Majdi , K. T. Rashid , Q. F. Alsalhy , D. S. Lakshmi , I. K. Salih and A. Figoli, Removal of Dye from a Leather Tanning Factory by Flat-Sheet Blend Ultrafiltration (UF) Membrane, Membranes 10, 47 (2020)..
[16] P. Saranraj, Bacterial biodegradation and decolourization of toxic textile azo dyes. Af- rican J Microbiol Res 2013; 7(30):3885–3890..
[17] H. A. Alshamsi, F. Beshkar, O. Amiri, M. Salavati-Niasari, Porous hollow Ag/Ag2S/Ag3PO4 nanocomposites as highly efficient heterojunction photocatalysts for the removal of antibiotics under simulated sunlight irradiation,Chemosphere,Volume 274,2021,12976.
[18] M. A. Al-Bedairy and H. A. H. Alshamsi, Environmentally Friendly Preparation of Zinc Oxide, Study Catalytic Performance of Photodegradation by Sunlight for Rhodamine B Dye, Eurasian Journal of Analytical Chemistry 13, em72 (2018)..
[19] M. Karami, M. Ghanbari, H. A. Alshamsi, S. Rashki, M. Salavati-Niasari, Facile fabrication of Tl4HgI6 nanostructures as novel antibacterial and antibiofilm agents and photocatalysts in the degradation of organic pollutants, Inorganic Chemistry Frontiers 8.
[20] H. A. H. Alshamsi, A. A. JAFFER, Microwave-assisted synthesis of ZnO nanoparticles and its photocatalytic activity in degradation of Rhodamine B dye, International Journal of Pharmaceutical Research,12(1), 2020, 201-210..
[21] Hairui Liu1, Haifa Zhai, Chunjie Hu, Jien Yang Zhiyong Liu, Hydrothermal synthesis of In2O3 nanoparticles hybrid twins hexagonal disk ZnO heterostructures for enhanced photocatalytic activities and stability, Nanoscale Research Letters (2017) 12:466..

[22] H. A. H. Alshamsi and B. S. Hussein, Hydrothermal preparation of silver doping zinc oxide nanoparticles: study the characterization and photocatalytic activity, Orient J Chem 34, 1898-1907 (2018)..
[23] S. P. Sasikala, T. A. Nibila, K. B. Babitha, A. P. Mohamed, A. Solaiappan. Competitive photo-degradation performance of ZnO modified bentonite clay in water containing both organic and inorganic contaminants. Sustain Environ Res 29, 1 (2019)..
[24] R. T. Ngaloy, A. M. Fontanilla, M. S. R. Soriano, C. S. Pascua, Y. Matsushita, I. J. A. Agulo, Highly Efficient Photocatalysis by Zinc Oxide-Reduced Graphene Oxide (ZnO-rGO) Composite Synthesized via One-Pot Room-Temperature Chemical Deposition Method,.
[25] R. Gayatri, T. E. Agustina, R. Moeksin, D. Bahrin, G. Gustini, Preparation and Characterization of ZnO-Zeolite Nanocomposite for Photocatalytic Degradation by Ultraviolet Light. Journal of Ecological Engineering. 2021; 22(2):178-186..
[26] N. Arsalani, S. Bazazi, M. Abuali, S. Jodeyri, A new method for preparing ZnO/CNT nanocomposites with enhanced photocatalytic degradation of malachite green under visible light, Journal of Photochemistry and Photobiology A: Chemistry, Volume 389,2020, 112.
[27] N. Anzar, R. Hasan, M. Tyagi, N. Yadav, J. Narang, Carbon nanotube - A review on Synthesis, Properties and plethora of applications in the field of biomedical science, Sensors International, Volume 1,2020,100003.
[28] M. Elias, M. N. Uddin, J. K. Saha,M. A. Hossain, D. R. Sarker,S. Akter S,I. A. Siddiquey, J. A. Uddin, A Highly Efficient and Stable Photocatalyst; N-Doped ZnO/CNT Composite Thin Film Synthesized via Simple Sol-Gel Drop Coating Method. Molecules. 2021; 26.
[29] M. A. Mohd Adnan, N. M. Julkapli, S. N. Muhd and S. B. A. Hamid, Review on ZnO hybrid photocatalyst: impact on photocatalytic activities of water pollutant degradation: Reviews in Inorganic Chemistry, vol. 36, no. 2, 2016, pp. 77-104..
[30] Q. A. Nema and H. A. Alshamsi, Synthesis, Characterization, Synthesis, Spectroscopic and Thermal Study of Azo Dye 7-[2-(BenzImidazolyl)Azo]- 8-Hydroxy Quinolineand its Gold(III) Complex , International Journal of Science and Research , 6, 235-240 (2017)..
[31] H. A. Alshamsi, Q. A. Nema, New approach for synthesis of ZnO nanoparticles via thermal decomposition of [Zn(BIAHQ)2].H2O complex: characterization and photocatalytic degradtion of reactive blue 4 dye..
[32] H. M. Kamari, N. M. Al-Hada, E. Saion, A. H. Shaari, Z. A. Talib, M. H. Flaifel, A. A. A. Ahmed, A.A.A. Calcined Solution-Based PVP Influence on ZnO Semiconductor Nanoparticle Properties. Crystals 2017, 7, 2..
[33] S. Krishnamurthy, A. Esterle, N. C. Sharma, S. V. Sahi. Yucca-derived synthesis of gold nanomaterial and their catalytic potential. Nanoscale Res Lett 9, 627 (2014)..
[34] H. K. Nguyen, J. Bae, J. Hur, S. J. Park, M. S. Park, I. T. Kim, Tailoring of Aqueous-Based Carbon Nanotube–Nanocellulose Films as Self-Standing Flexible Anodes for Lithium-Ion Storage. Nanomaterials 2019, 9, 655..
[35] H. A. Alshamsi, A. M. Chessab, Photo Catalytic Degradation of Reactive Blue 4 Dye using Hydrothermal Synthesized ZnO Nanoparticles, Journal of Global Pharmacy Technology. 2017; 12(9):257-264..
[36] S. H. Jaafar, M. H. M. Zaid, K. A. Matori, S. H. A. Aziz, H. M. Kamari, S. Honda, Y. Iwamoto, Influence of Calcination Temperature on Crystal Growth and Optical Characteristics of Eu3+ Doped ZnO/Zn2SiO4 Composites Fabricated via Simple Thermal Treatment M.
[37] A. W. Yibeltal, B. B. Beyene, S. Admassie, A. M. Taddesse, MWCNTs/Ag-ZnO Nanocomposite for Efficient Photocatalytic Degradation of Congo Red, Bull. Chem. Soc. Ethiop. 2020, 34(1), 55-66..
[38] V. Perumal, U. Hashim, S. C. B. Gopinath, H. R. Prasad, L. Wei-Wen, S. R. Balakrishnan, T.

Vijayakumar, R. Abdul Rahim. Characterization of Gold-Sputtered Zinc Oxide Nanorods—a Potential Hybrid Material. Nanoscale Res Lett 11, 31 (2016)..

[39] H. Yu, H. Ming, J. Gong, H. Li, H. Huang, K. Pan, Y. Liu, Z. Kang, J. Wei, D. Wang, Facile synthesis of Au/ZnO nanoparticles and their enhanced photocatalytic activity for hydroxylation of benzene, Bull. Mater. Sci., Vol. 36, No. 3, June 2013, pp. 367–372.

[40] M.H. Ahmadi Azqhandi, M. Shekari, B. Ghalami-Choobar, Synthesis of carbon nanotube-based nanocomposite and application for wastewater treatment by ultrasonicated adsorption process, Appl Organometal Chem. 2018;e4410..

[41] R. Alipour, M. T. Hosseinnejad, A. S. Salar Elahi and M. Ghoranneviss,New perspective on morphological features of the zinc oxide thin films as a gas sensor, J Alloys Compd 687, 72-78(2016)..

[42] S. H. Altaa, H. A. H. Alshamsi and L. S. Al-Hayder, Synthesis and characterization of rGO/ CO3O4 composite as nanoadsorbent for Rhodamine 6G-dye removal , Desalin Water Treat 114,320-331(2018)..

[43] M. Bagheri, N. R. Najafabadi, E. Borna, Removal of reactive blue 203 dye photocatalytic using ZnO nanoparticles stabilized on functionalized MWCNTs, Journal of King Saud University – Science 32 (2020) 799–804..

[44] S. M. EL-Dafrawy, M. Tarek, S. Samra, S. M.Hassan, Synthesis, photocatalytic and antidiabetic properties of ZnO/PVA nanoparticles. Sci Rep 11, 11404 (2021)..

[45] A. V. Karim, A. Shriwastav, Degradation of amoxicillin with sono, photo, and sonophotocatalytic oxidation under low-frequency ultrasound and visible light, Environmental Research,Volume 200,2021,111515..

[46] R. M. Mohamed, I. A. Mkhalid, E. S. Baeissa, M. A. Al-Rayyani, Photocatalytic Degradation of Methylene Blue by Fe/ZnO/SiO2 Nanoparticles under Visiblelight, Journal of Nanotechnology Volume 2012, Article ID 329082..

[47] K. A. Isai, V. S. Shrivastava, Photocatalytic degradation of methylene blue using ZnO and 2%Fe–ZnO semiconductor nanomaterials synthesized by sol–gel method: a comparative study. SN Appl. Sci. 1, 1247 (2019)..

[48] N. Daneshvar, S. Aber, M. S. Seyed Dorraji, A. R. Khataee, and M. H. Rasoulifard, Photocatalytic degradation of the insecticide diazinon in the presence of prepared nanocrystalline ZnO powders under irradiation of UV-C light, Separation and Purification.

[49] H. A. Alshamsi, M. A. Al Bedairy, S. H. Alwan, Visible light assisted photocatalytic degradation of Rhodamine B dye on CdSe-ZnO nanocomposite: Characterization and kinetic studies, IOP Conf. Series: Earth and Environmental Science 722 (2021) 012005..

Thermal Synthesis of Au-ZnO Nanoparticles for Application in Photocatalytic Degradation of Reactive blue 4 Dye in Aqueous Solution

Hassan A. Alshamsi[1] and Qusay A. Nema[2]

[1]Department of Chemistry, College of Education, University of Al-Qadisiyah, Diwaniya, IRAQ

[2]Al-Muthanna Education Directorate, Al-Muthanna Province, Samawa, Iraq

Abstract. In the present study, zinc oxide (ZnO), gold (Au) and gold doped zinc oxide (Au-ZnO) nanoparticles (NPs) were successfully synthesized through thermal decomposition of [Au(L)Cl]Cl.H_2O and [Zn(L)$_2$].H_2O complexes ,where L is 7-[2-(BenzImidazolyl) Azo]- 8-Hydroxy Quinoline (BIAHQ). Several characterization techniques such as XRD, FT-IR), UV–Vis DRS, SEM, TEM, AFM) were employed to investigate the phase structure, crystallite size, chemical structure, optical features, particles morphology and surface topography of as synthesized nanomaterials . The visible light photocatalytic activities of undoped ZnO and Au-ZnO NPs were compared for the decolorization of reactive blue 4 dye (RB4) aqueous solution. Factors influencing the photocatalytic degradation process including irradiation time, catalyst dosage, initial RB4 concentration and initial pH have been evaluated. The photocatalytic degradation results showed that the Au-ZnO NPs exhibited excellent photodecolorization efficiency towards cationic RB4 dye decolorization with 97.36 removal efficiency under visible light illumination for 160 min

Keyword: Au-ZnO; Thermal synthesis;[Zn (BIAHQ)$_2$].H_2O complex ; XRD; Visible light; Photocatalytic degradation; RB4 dye.

1. Introduction

Artificial aromatic dyes, which are widely used with large quantities in textile, food, paper, pharmaceuticals, cosmetics, leathers, rubber, varnish, plastics and paints industries as well as color photography and printing inks are sufficiently stable, thermally and chemically which causing a serious damages to the environment [1]. The colored discharges causing an irreversible disruption of the aquatic ecosystem, if they directly released into sustainable water bodies without conventional treatment processes [2-3]. Due to consume large amounts of water, the dyeing and coloring industry has caused a big pollution problems as it is one of the most widespread modern industries [4].

The wastewater from textile industries is classified as the most polluting of all the industrial sectors, taking into account the volume released as well as the discharges chemical content [5]. Various conventional methods including biological treatment [6], adsorption [7], sonocatalysis [8], electrocoagulation [9], precipitation [10], oxidation [11], flotation [12], flocculation [13], reverse osmosis [14] so forth have been employed for the remediation of wastewaters. However, in addition to that fact that most of these treatment methods often involve high capital and operational costs, it have been found to be limited efficacy and cannot completely degrade most of the water pollutants. In contrast, compared with these methods, photocatalytic technology can completely degrade the water pollutants under visible or solar radiation photocatalysis [15-17] In addition, photocatalysis offers several advantages such as mild reaction conditions, easy operation, high treatment efficiency, less energy consumption, high chemical and thermal stability and minimal secondary pollution [18-20]. Nowadays, studies on semiconductors nanocomposites based photocatalytic technology have aroused more attention due to its potential application prospects in the fields of air purification [21], wastewater treatment [22], hydrogen production [23], solar energy conversion [24] and cancer and microorganisms treatment [25, 26]. The photocatalytic degradation of colored pollutants using metal oxides is a promising way for water remediation because it exhibits remarkable activity for degrade many organic pollutants in wastewater into less toxic products without additional chemical or energy input. Furthermore, photocatalytic degradation under visible light illumination is a proper route for the removing of hazardous organic pollutants from aquatic environment [27].

Since the discovery of water splitting into hydrogen and oxygen using titanium dioxide (TiO_2) in 1972, where, Honda and Fujishima used a TiO_2 photoanode to split water by UV light [28], the photocatalysis technology has been extensively investigated. In recent years, ZnO nanostructures have become one of the most popular nanomaterials for a number of applications, especially, in photocatalytic degradation of pollutants, solar energy conversion, sunscreens, gas sensors, antibacterial agents, electrical devices, corrosion protection and optical coatings [29]. ZnO has become a widely used material for various applications due to its excellent characteristics like a wide range of radiation absorption, environmentally friendly, cost-effectiveness, high chemical stability, high electron mobility, high efficiency, high photochemical stability and great transparency in solar and visible range spectra [30, 31].

ZnO is abundant-in-nature material and classified as n-type metal oxide semiconductor having a wide direct band gap (3.3 eV) and excitation binding energy (60 meV) which makes it relevance for short wavelength electronic and optoelectronic applications [32]. The photo-excitation of ZnO upon UV irradiation, induces the formation of charge carrier pairs (electron–hole, e_{cb}^{-}-h_{vb}^{+},) in the conduction band (CB) and valance band (VB). The reaction of the positive holes and electrons on the surface of a semiconductor can generate reactive species such as hydroxyl radicals ($^{\bullet}OH$) superoxide radical anion ($^{\bullet}O_2-$). The $^{\bullet}OH$ radicals are effective oxidizing species that react with organic contaminations and degrade it into water and carbon dioxide. The fast recombination of photogenerated electron/hole pairs is important factor that impedes further applications of pure ZnO NPs. Additionally, due to its wide E_g, ZnO can't be employed as an effective photocatalyst under visible and solar irradiation [33].

In order to improvement the photocatalytic performance of ZnO, constructing heterostructures could enhance efficient separation of photogenerated charge carriers which could inhibit the recombination of electron–hole pairs in electron transfer process. Various processes have been developed to narrowing ZnO band gap or dividing it into many subgaps , hence, extend the light absorption to visible light (400–800 nm) for photocatalytic activity enhancement, which include semiconductor surface photosensitization [34], surface noble metal deposition [35], metal doping [36], nonmetal doping [37], hybridization with nanomaterials [38], heterogeneous compositing [39] and surface modification [40].

The doping of transition metals such as (Au, Ag, Cu, Pd, Fe, etc.) in the structure of ZnO has been considered as one of the most efficient processes used for narrowing the band gap. The doping metals are normally create energy levels located between the VB and CB positions of ZnO. The photogenerated charge carriers become more spatially separated via trapping at surface sites, and the rate of recombination is slowed, which subsequently enhance the overall photocatalytic performance of ZnO [41]. In this regard, one of the most efficient ways for improving photocatalytic performance of ZnO NPs is the doping with gold nanoparticles (Au NPs) [42]. Au is one of the important noble metals which can directly gain energy from light irradiation (not have a band gap) and does not undergo corrosion process [43]. Among various hybrids, Au-ZnO have been shown to be one of the most efficient photocatalysts due to its efficient absorption of visible light which make it appropriate for photocatalysis, photodegradation and optoelectronic devices applications [44]. There are many works that have reported the fabrication of Au-ZnO nanocomposite such as solution-phase [45], hydrothermal [46], one-step flame spray pyrolysis (FSP) [47], pulsed laser ablation [48] and green approach [49]. For the best of our knowledge, there is no previous work reported the synthesis of Au-ZnO via thermal decomposition of transition metal complexes.

In the current study, Au, ZnO and Au-ZnO NPs have been synthesized using thermal sintering of [Au(BIAHQ)Cl]Cl.H_2O and [Zn(BIAHQ)$_2$].H_2O complexes. ZnO and Au-ZnO were obtained as catalysts for the photocatalytic decolorization of reactive blue 4 dye in aqueous solution. The crystal structural, optical, morphological and topographical characteristics of Au, ZnO and Au-ZnO NPs were characterized with various characterization techniques like XRD, FTIR, UV-Vis DRS, AFM, SEM and TEM. The effect of irradiation time, catalyst dose, RB4 concentration and pH of the dye solution on the photocatalytic color removal efficiency of RB4 dye have been studied. Also, the photocatalytic degradation mechanism of dye was suggested.

2. Experimental

2.1. Chemicals

In this work, all solvents and chemicals obtained were of high purity analytical grade and used as received without additional purification. Gold (III) chloride trihydrate ($HAuCl_4 \cdot 3H_2O$, purity of 99.9%), zinc chloride ($ZnCl_2$, purity of 99 %) , 2-aminobenzimidazole ($C_7H_7N_3$, purity of 97%), reactive blue 4 dye ($C_{23}H_{14}Cl_2N_6O_8S_2$, dye content 35 %) and 8- hydroxy quinolone (C_9H_7NO, purity of 99%) were supplied by Merck. Sodium hydroxide (NaOH, purity of 99%), sodium nitrite ($NaNO_2$, purity of 99%) and hydrochloric acid (HCl, 37%) were supplied by B.D.H. Ethyl alcohol (C_2H_5OH, purity of 99.5 %) was purchased from Baker. Ammonium acetate ($CH_3CO_2NH_4$, purity of 99 %) was supplied by Fluka.

2.2. Instruments

The powder X-ray diffraction (XRD) diffractgrams of Au, undoped ZnO and Au doped ZnO NPs were recorded with XRD-6000 diffractometer (Shimadzu, Japan) with Cuk radiation source (0.15405 nm) and a scanning range of 25° to 80°. FTIR analysis was performed using instrument FTIR-8400, Shimadzu- Japan, adopting KBr disks method with scanning in the range between 4000 and 400 cm^{-1}. The powder diffuse reflectance spectra (UV-Vis DRS) of as synthesized nanoparticles were scanned in the wavelength range of 200 - 700 nm by a Shimadzu UV-1800 UV-Visible spectrophotometer to estimate their optical band gap. Scanning electron microscopy (KYKY-EM3200, China) was applied for investigation of morphology of synthesized NPs. The sizes of synthesized NPs were determined with the use of a transmittance electron microscopy (Philips CM120 Cryo-electron microscope). The surface topography of as synthesized NPs was investigated using atomic force microscopy (AA300, Angstrom Advanced Inc, USA). The pH of solutions was adjusted using pH-meter 211-Instrument Hanna, Romania. Ultrasonic dispersion of synthesized NPs was carried out employing ultrasonic path (405 power sonic, Hwashin, Korea). A 125 W Osram lamp (Germany) was obtained as an artificial irradiation source in photodecolorization experiments.

2.3. Synthesis of Ligand and Complexes

The 7-[2-(Benzimidazole) azo]-8-Hydroxy Quinolone (BIAHQ) ligand has been synthesized according the previously reported method [50]. The $[Zn(BIAHQ)_2].H_2O$ complex has also been synthesized as recently reported [51]. The $Au[BIAHQ)Cl]Cl.H_2O$ complex was synthesized by adding BIAHQ to solution of hydrogen tetrachloroaurate(III) hydrate metal with mole ratio 1: 1 [M: L] followed by continuous stirring. The metal precursor was pre-dissolved in 50 mL of ethanoic buffer solution at pH = 7.0).After completion the addition, the color is immediately changed to the earthy yellow. Afterwards, the reaction mixture was heated for 30 min at 60 ^{0}C, then the precipitation was repeatedly washed with DDW and ethanolic alcohol to eliminate the non-reacted organic materials. Finally, the product was dried at a maximum temperature of 90 ^{0}C for 4 hours.

IOP Conf. Series: Earth and Environmental Science 1029 (2022) 012004 doi:10.1088/1755-1315/1029/1/012004

2.4. Synthesis of ZnO, Au and Au-ZnO NPs

For synthesizing the ZnO NPs, [Zn (BIAHQ) $_2$] .H_2O complex was calcined at 700 ^{0}C for 2 hours under inert conditions [50, 52]. Meanwhile, a white purple color formation indicates the formation ZnO NPs. The same process was followed for the preparation of Au and Au doped ZnO samples using [Au(BIAHQ)Cl]Cl.H_2O and [Au(BIAHQ)Cl]Cl.H_2O/ [Zn(BIAHQ)$_2$].H_2O mixture, respectively. For synthesis of Au-ZnO nanocomposite, the [Au(BIAHQ)Cl]Cl.H_2O/ [Zn(BIAHQ)$_2$].H_2O mixture was prepared by mixing Zn (BIAHQ) $_2$] .H_2O with [Au(BIAHQ)Cl]Cl.H_2O with a Au/Zn weight ratio of 3/97.

2.5. Photodecolorization of RB4 dye

To study the photocatalytic activity of undoped and Au doped ZnO photocatalysts towards color removal of RB4 dye under visible light illumination, a batch scale photoreactor of 100 mL capacity was employed as an enclosed chamber. The photoreactor was equipped with a 125 W Osram lamp and temperature controller. The lamp was positioned in the middle of the photoreactor about 10 cm above reaction vessel. For this regard, 50 mL of an aqueous dye solution containing a known dose of catalyst was allowed to reach adsorption–desorption equilibrium state for dye molecules onto catalyst surface under constant stirring for 120 min at 25 ^{0}C in the darkness. Hence, prior to all photocatalysis experiments a 120 min adsorption was done in the dark.

Afterwards, zero time decolorization was considered the moment when the dye solution was illuminated with lamp light, whereas the mixture was kept suspended with magnetic stirring for 120 min at 25 °C .Then, 5.0 mL samples were taken at regular time intervals (20 min) and directly centrifuged at 6000 rpm for 15 min to separate catalyst particles. To determine the amount of RB4 remaining in the supernatant, the supernatant was extracted and the dye absorbance was recorded at 590 nm (λ_{max} of RB4). The photodecolorization efficiency [PDE (%)] of RB4 was calculated by from the following expression:

$$PDE\ (\%) = (A_0\text{-}A_t/A_0)\ x100 \qquad [1]$$

Where A_0 and A_t represent the initial absorbance and the absorbance after decolorizationat time (t), respectively. Dye absorbance determination was carried out in duplicate on two independently. Except where otherwise noted, the above procedure was obtained in all experiments. The effect of different operational factors such as irradiation time (0-120 min), photocatalyst dose (0.1– 1.3 g/L), initial pH (5-11) and initial RB4 dye concentration (25-100 ppm) on the photodecolorization efficiency were tested. The pH of the RB4 solutions was adjusted by addition either 0.1 M HCl or 0.1 M NaOH solution and controlled with assist of digital pH meter.

3. Results and discussion

3.1. X-ray powder diffraction

In order to examine the crystallinity of Au, ZnO and Au doped ZnO NPs, XRD patterns have been collected at 2θ from ~25°–80°. As shown in Figure 1, the formed ZnO NPs exhibited nine distinguished peaks at 2θ angles centered at 31.75° , 34.55° , 36.23° , 47.97° , 56.58°◦ , 63.24° , 66.73° , 68.08° and 69.07° , which assigned to (100), (002), (101), (102), (110), (103), (200) ,(112) and (201) lattice planes of hexagonal ZnO wurtzite structure (JCPDS no. 36–1451) [53], respectively . The XRD pattern of Au NPs shows characteristic diffraction peaks at 2θ = 38.27°, 44.36°, 64.61°, and 77.60°. All the four peaks are assigned to correspond to the crystalline face center cubic (fcc) lattice (JCPDS NO. 65-8601) standard which assigned to (111), (200), (220), and (311) standard Bragg reflections [54]. No further peaks can be found confirming the formation of single ZnO and Au phases. The XRD diffractogram of Au–ZnO nanocomposite exhibited two different sets of peaks associated with Au and ZnO NPs. The first is due to ZnO positions correspond to (100), (002), (101), (110), (103), (200), (112), (201), and (004) planes (JCPDS NO. 05-0664) [44]. The second set corresponding to the Au positions indicates the presence of the four characteristics peaks at 38.15°, 43.97°, 64.53° and 77.39° clearly assigned to (111), (200), (220) and (311) planes confirmed the presence of Au in the sample . The XRD pattern of Au-ZnO exhibited slight shift in related peaks suggesting that the Au NPs are localized onto surfaces of crystalline ZnO NPs. In general, the XRD results show that the Au-ZnO nanocomposite was efficiently formed.

The apparent nanocrystallite sizes (D) of synthesized nanomaterials were calculated using Debye–Scherrer equation (Eq. 2) obtaining the most intense diffraction:

$$D= 0.9\lambda/\beta\cos\theta \qquad [2]$$

Where λ is a wavelength of X-ray obtained from Cu kα radiation (λ=0.15406 Å), and β is a full line width at the half-maximum of the reflection peak (FWHM). The average particle size of undoped ZnO, Au and Au doped ZnO nanoparticles were found to be 20.53, 39.98 and 24.00 which are derived from the FWHM of more intense peaks located at 36.42° (101), 38.27° (111) and 38.01° (111) ,respectively.

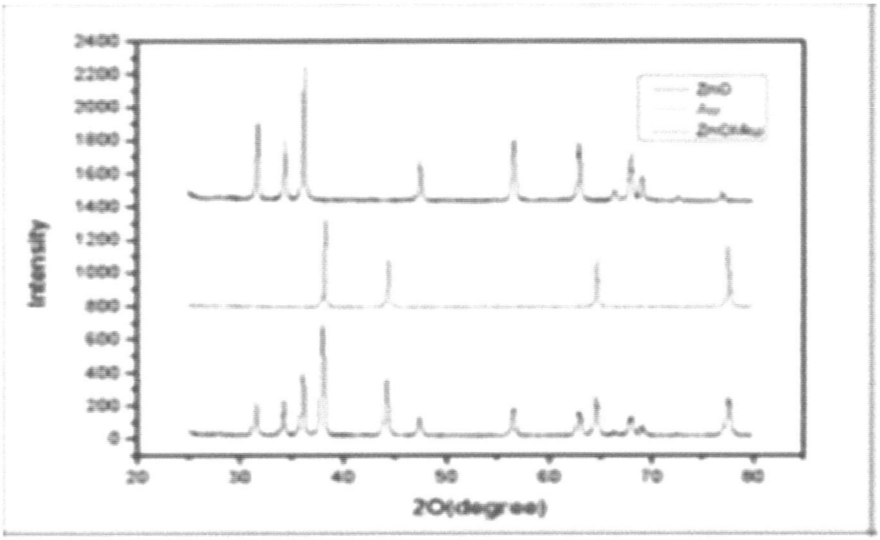

Figure 1: XRD diffractograms of , Au , pure ZnO and Au doped ZnO NPs

3.2. FTIR Spectroscopy

To detect the characteristic functional groups existing in the fabricated undoped and Au doped ZnO NPs, FT-IR analysis was conducted. The FT-IR spectra of synthesized nanomaterials in the wavenubers range from 400 to 4000 cm^{-1} are displayed in Figure 2(a). It can be seen that the ZnO has absorption bands at 3415 cm^{-1}, 1595 cm^{-1}, 1330 cm^{-1} and 1118.1 cm^{-1}. The broad absorption band at 3415 cm^{-1} is in agreement with the O–H stretching vibration which is due to adsorption of atmospheric moisture. The vibration modes centered at 1330 cm^{-1} and 1595 cm^{-1} are corresponding to bending vibration related to the hydroxyl groups.The sharp and intense band at 440 cm^{-1} is corresponding to the Zn-O stretching vibration mode which indicates the successful formation of ZnO NPs [55]. In the case of Au–ZnO Nps, the positions of bands are slightly shifted to lower wavelengths as a result of Au NPs formation onto ZnO surface. The broad band centered at 3410 cm^{-1} is corresponds to the stretching vibration mode of the moisture O–H. In addition, the vibration modes centered at 1330 cm^{-1} and 1591 cm^{-1} are corresponding to bending vibration of hydroxyl groups. The absorption band related to the vibrational characteristics of Zn-O has been shifted to 435 cm^{-1} due to presence of Au NPs which is related with the compressive stress acting on ZnO NPs [56].

3.3. UV-Vis diffuse reflectance spectroscopy

The optical properties of undoped and Au doped ZnO NPs were investigated uing UV–Vis diffuse reflectance spectroscopy. Figure 3 shows the DRS of the synthesized samples. It has been observed that the ZnO sample respond to UV light due to it,s large band gap energy (E_g). ZnO NPs exhibit a sharp and intense absorption edge at 380 nm belongs to the ground excitonic state of ZnO which being in good adjustment with previous literatures, where the absorption spectra of ZnO nanoparticles were observed between 330 nm and 380 nm [17, 34]. After Au reduction onto ZnO surface, the Au-ZnO exhibits a red shift as compare to unmodified ZnO. In context, after Au doping, a high absorption was observed in the visible region (400 -800 nm). The Au-ZnO spectrum displays a strong absorption edge at 405 nm. The visible light absorption of Au-ZnO is also reflected by the color change of the sample from white to beige, which further confirms the formation of Au-ZnO nanocomposite. A classical Tauc's plot was further employed to estimate the indirect band gap of pure and Au doped ZnO NPs obtaining the following expression [57]:

$$E_g\,(eV) = 1240/\lambda(nm) \qquad (3)$$

Where E_g is the indirect band-gap energy (eV) and λ is the absorption edge wavelength (nm). The ZnO NPs exhibits the indirect E_g of 3.26 eV which is ia good agreement with previous work where the E_g was reported to be 3.37 eV (368 nm) [58]. For Au-ZnO nanocomposite, it has been found that E_g was 3.06 eV. The insertion of Au^+ ions into ZnO lattice sites lead toe formation of the p-type conductivity, and hence leads to narrowing band gap. The results revealed that the doping gold is an effective way to reduce E_g value from 3.26 eV to 3.06 eV.

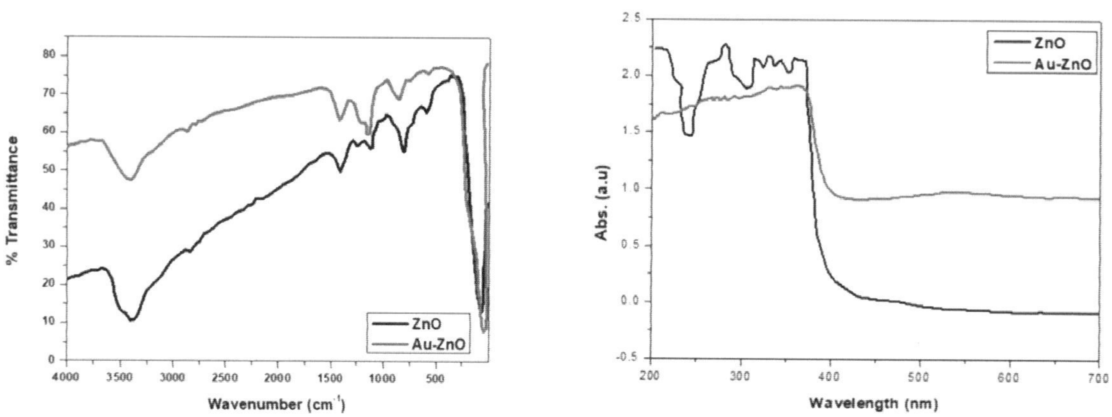

a b

Figure 2. (a) FT-IR spectra and (b) DRS curves of undoped and Au doped -ZnO.

3.4. SEM and TEM

SEM technique is used to visualize surface morphology and size of fabricated NPs. Figure 4 depicts the SEM images of as synthesized Au, undoped and Au doped ZnO samples. It could be seen that the all nanomaterials are in nanometer scale and clearly demonstrated the presence of spherical and quazi-spherical shaped with the average diameter of 43-85 nm (ZnO), 31-58 nm (Au) and 39-54 (Au-ZnO). Remarkably, all the synthesized NPs are slightly agglomerated which is expected with the thermal synthesis of nanoparticles at high temperatures. The above data suggest that the Au NPs grown in an early stage of the reaction and enhance the nucleation of ZnO NPs. In addition to SEM analysis, the structure and morphology of as synthesized samples were investigated TEM technique and the images are demonstrated in Figure 3(e, g and h). It can be seen that ZnO TEM microgram shows almost uniform distribution of nearly spherical nanoparticles with weak agglomeration. The Au TEM imge revealed tht the particles have spherical shapes with a particle size distribution in the range of 5 to 15 nm with an average particle size of 12 nm The TEM data of synthesized Au-ZnO reveals highly crystalline of composite with nearly uniform Spherical shapes with a mean size of average particle size of 14.0 nm.

Figure 3. SEM micrograms of Au, undoed and doped ZnO NPs.

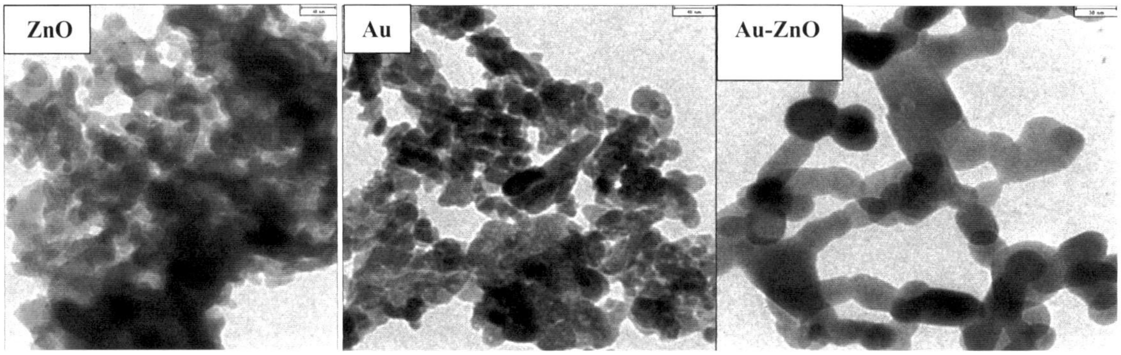

Figure 4. TEM micrograms of Au, undoped and Au doped ZnO NPs.

3.5. AFM analysis

Atomic force microscopy (AFM) is an influential surface analysis technique and often used for description the topography of synthesized NPs surface. Figure 4 provides a 3D images and granularity cumulation distribution charts of Au, undoped and Au doped-ZnO samples. The AFM data displayed almost homogenous distribution of as synthesized NPs. Also, the morphology of ZnO surface confirms that the grains were evenly dispersed with existence of a significance agglomerations. Additionally, AFM analysis exhibits that most of the NPs appear spherical and quazi-spherical like structures. Moreover, tThe AFM results clearly demonstrate the rough nanoparticles and a significance agglomerate particles exist. The AFM data reveal that surface roughness of Au doped ZnO composite is more than individual Au and ZnO samples. Also the data demonstrate that as the NPs thickness decreases as the grain size increases. . As Table 1 displays, the roughness coefficient (R_a) of the three samples is associate with negative values of surface skewness (R_{sk}) which confirms that the surface have valleys more than peaks (hills) [59]. Also, kurtosis parameter (R_{ku}) is a statistical analysis used to desribe the sharpness of the roughness profile and it qualifies the flatness of the height distribution. Due to the three synthesized samples have $R_{ku} > 3$; so all samples correspond to spiky surface [60].

TABLE 1. Topographical characteristics of ZnO, Au Au-ZnO

Sample	R_a nm	Thickness nm	Avg. Diameter nm	R_{sk}	R_{ku}
ZnO	2.56	20.84	40-390	-0.174	11.9
Au	0.61	3.10	26-140	-0.424	3.06
Au-ZnO	3.03	3.75	20-146	-0.448	3.66

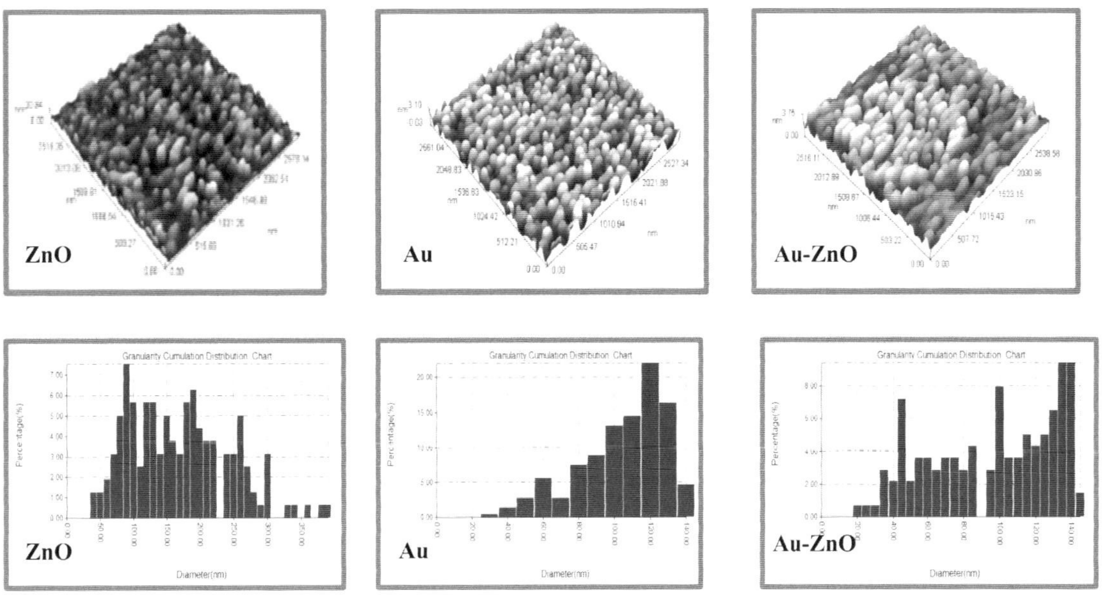

Figure 5. 3D images and granularity accumulation distribution charts of Au, undoped and Au doped ZnO samples.

3.6. Photocatalytic activity

The decolorization of RB4 dye using ZnO and Au-ZnO sample was conducted photocatalytically under direct visible light illumination. The photocatalytic decolorization tests were carried out at under1.0 g/L of each catalyst, 50 ppm RB4, pH 7 and 25 ^{0}C. A preliminary experiments were carried out to explore the influrnce of visible light irradiation on the efficiency of photocatalytic decolorization process in absence of photocatalyst. According to the results obtained, almost negligible photodecolorization efficiency was found indicating that the features of RB4 are steadier under visible light photolysis. In parallel, the adsorption in the dark was carried out, the mixed suspension of catalyst and dye was constantly agitated for 120 min to achieve the adsorption/desorption equilibration between catalyst and dye molecules. In contrast of photolysis result, after 120 min, the adsorption efficiency was 23.86 % and 35.26 % for the undoped and Au doped ZnO samples, respectively. After the dark period, the dye solutions were irradiated by visible light and the effect of photocatalysts on the ability to degrade RB4 in aqueous solution was investigated. Figure 6(a) displays the photodecolorization efficiency of as-synthesized nanomaterials. It is clear that bare ZnO is slightly effective for RB4 removal with photodecolorization efficiency being about 21.09 %. On the contrary, Au-ZnO nanocomposite seems to be more effective in comparison with bare ZnO. However, after 120 min of visible light illumination, 60.42 % photocatalysis of RB4 was achieved using Au-ZnO nanocomposite.

3.7. Effect of photocatalyst dose

The photocatalyst dose is one of the most crucial parameters that influences the efficiency of photocatalytic degradation of polluted water contaminations. For an efficient remediation of dye, the optimum photocatalyst dose must be find out to avoid use of excess material. At a fixed operating temperature of 25 ^{0}C and a pH of 7, the photodecolorization of BR4 was evaluated with different dosages of Au-ZnO ranging from 0.1 to 1.3 g/L and the results are shown in Figure 6(b). The increase of photocatalyst dosage from 0.1 to 0.9 g/L results in the increased photodecolorization efficiency sharply from 17.01 % to 72.49 % after 120 min irradiation, whereas the photodecolorization efficiency declines with further increase of photocatalyst amount. The optimum photodecolorization efficiency was achieved for Au-ZnO (72.49 %). The increase in the photodecolorization efficiency of Au-ZnO is attributed to the synergic role of Au NPs which improves more surface active sites on the photocatalyst surface for dye molecule adsorption and for formation of highly reactive oxygen species ($\bullet O_2^-$ and $\bullet OH$) ,thus enhancing the amount of adsorbed RB4 and improve the photocatalytic activity [20] . However, as the photocatalyst dose was further increase up to 1.3 g/L, the photocatalytic activity decrease , where the suspension is becomes more turbid and thus increase in thickness of particles in suspension prevent from reaching irradiation and efficiency decreased [61]. In addition, the reduction of color removal efficiency can be attributed to reducing of surface area and blocking most of the active surface sites via aggregation and agglomeration of catalyst particles in the reaction suspension and as a result, a decrease in the production of reactive species that drive the photocatalytic decolorization process[17, 18]..

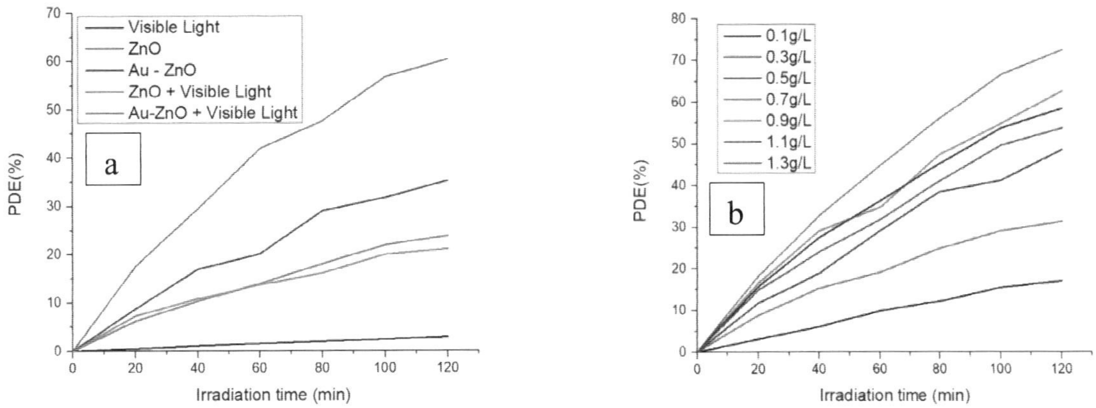

Figure 6. (a) Primlinary experiments and (b) effect of Au-ZnO dose on the efficiency of photocatalytic decolorization.

3.8. Effect of Initial Dye Concentration

In general and from an application point of view, the initial polllutant concentration plays a vital role in the photocatalytic degradation process. Fig. 7(a) expresses the photodecolorization efficiency in the presence of 25, 50, 75 and 100 ppm of aqueous dye solution. During the experiments, the Au-ZnO dosage (0.9 g/L), initial pH (7) and temperature (25 ^{0}C) were kept constant. As results reveal, the color removal is gradually declined as the initial RB4 concentration was increased. It is clear that the RB4 dye is sufficiently photodecoloring at initial low concentrations. This may be due to the available number of vacant active sites exposed to the light compared to the smaller number of dye molecules adsorbed on the catalyst surface [62]. Approximately 43.62 % dye degradation of 100 ppm RB4 was achieved with 120 min, it was 90.03 % with 25 ppm. The most likely interpretation for this behavior is that as the Rb4 concentration increases, the light penetration entering the suspension that reaches both dye molecules and ZnO NPs decreases due to the declining transparency of the suspension [63].

3.9. Effect of pH

It has been proven that the solution alkalinity or acidity is a major controlling parameter in the photocatalytic decolorization of various water pollutants using semiconductor composites as photocatalysts. Initial pH is an importance condition that affects the loading of dye onto catalyst surface due to its influence on the catalyst surface charge [64]. The photocatalyst experiments were conducted at 5, 7, 9 and 11 which. Figure 7(b) shows the photodecolorization efficiency of the Au-ZnO NPs towards RB4 dye at constant catalyst amount (0.9 g/L), primary RB4 dye concentration (50 ppm) and temperature of 25 ^{0}C under visible light illumination for 120 min. It has been noticed that the color removal efficiency arises with an arise in pH up to 11. The photodecolorization efficiencies obtained were 45.37, 72.49,

80.09 and 91.65 % at pH values of 5, 7, 9 and 11, respectively. Thus, an increase in pH will enhance the photocatalytic decolorization, whereas the optimal pH was noticed at about 11. This can be clarified based on electrostatic attractive/repulsive forces between the surface of the photocatalyst NPs and the RB4 dye molecules. The possible interpretation may be due to the positive charge (acidic nature) of RB4 dye combined with the amphoteric features of Au-ZnO surface. The efficacy of Au-ZnO NPs for RB4 photocatalytic decolorization was found to be higher in basic media which must be depend on the Au-ZnO surface charge properties (zero-point charge (zpc) of ZnO ~8.7) [65]. The effect of alkaline pH medium on the ionization state of the Au-TiO_2 surface becomes negatively charged, and RB4 adsorption is favored in alkaline solution because it has a cationic configuration. The Au-ZnO surface is positively charged at pHs less than pH_{zpc}, neutrally at pH_{zpc} and negatively at pHs beyond pH_{zpc} value. The negatively charged Au-ZnO improves its interaction with the RH4 molecules via an electrostatic attraction. As Figure 6-d shows, the time for almost complete decolorization would be longer than 120 min. Thus, after 160 min of visible light irradiation, ~ 97.36 % of RB4 photodecolorization efficiency was reached using 0.9 g/L Au-ZnO, pH of 9 and 50 ppm dye concentration. These results indicate that optimizing irradiation time for the complete decolorization process is required.

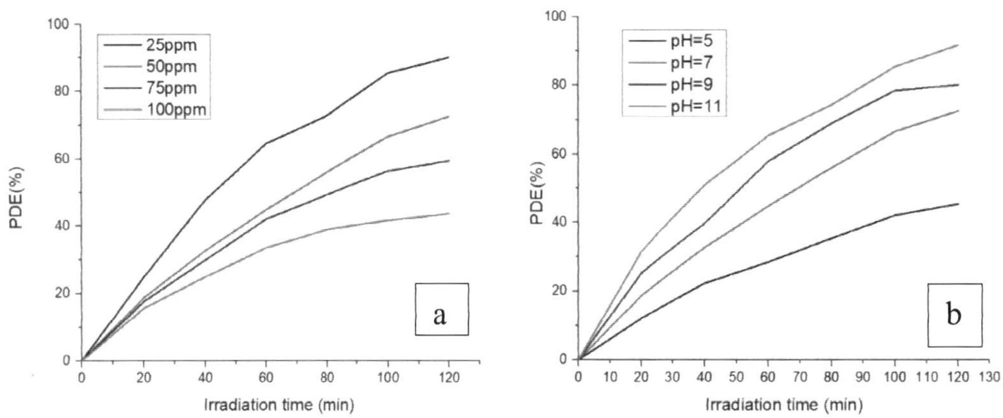

Figure 7. Photocatalytic decolorization efficiency under effect of (a) initial dye concentration and (b) pH.

3.10. A Suggested Photocatalytic degradation Mechanism of RB4 on Au-ZnO nanocomposite

The individual components such as Au and ZnO play a synergic roles function in increasing the light induced decolorization process and shifting the photo response of Au-ZnO to visible area. The Au doped ZnO nanostructure act as effective catalyst upon direct visible light illumination. The characteristics of designed heterogeneous photocatalysis systems include the energizing of a photocatalyst either by artificial or natural light sources. Initially, upon visible light illumination, Au-ZnO undergoes charge separation process that leads to transfer of exited electrons from valence band to conduction band (e_{CB}^-) leaving behind positively charge holes in the valence band (h_{VB}^+).

$ZnO + hv \longrightarrow ZnO\ (e_{CB}^{-} + h_{VB}^{+})$ [4]

The photogenerated electrons could react with oxygen molecules dissolved in the suspension and/or loaded onto ZnO surface to formation high reactive superoxide radical anion ($^{\cdot}O_2-$)[59]:

$e_{CB}^{-} + O_2 \longrightarrow {}^{\bullet}O_2^{-}$ [5]

The photogenerated holes react with OH^- oxidizing these species to $^{\bullet}OH$ [60]:

$h_{VB}^{+} + OH^{-} \longrightarrow {}^{\bullet}OH$ [6]

The $^{\bullet}OH$ radicals are considered to be the most oxidizing agent employed for the destruction of organic targets like colored materials in aqueous solutions. The heterogeneous photocatalysis mechanism of the RB4 molecules under the contact with visible light irradiation and ZnO are summarized in equations 7-11. The molecules of RB4 are significantly adsorbed on the rough ZnO NPs surface and hence sensitization of ZnO photocatalyst. RB4 dye is one of the chromophores that has a high to a wide range of wavelengths, especially the visible region, thus the RB4 molecules can easily be excited under the effect of visible light and go in high energy excitation state. Electrons can jump from the energy level of the excited dye molecules to the conduction band of the ZnO NPs and then be transferred to various dopant levels.

This process plays a very crucial role in the disappearance of RB4 via self sensitized photolysis under visible light exposure, as described below:

$RB4+hv \longrightarrow RB4^{*}$ [7]

$RB4^{*} + ZnO \longrightarrow ZnO\ (e_{cb}^{-} + h_{VB}^{+}) + RB4^{+}$ [8]

The Au states generate a band level very close to ZnO CB edge, hence the ZnO band gap will narrowed. The electrons transferred to the ZnO CB tend to move to Au NPs at the same time which improve the separation photogenerated of e-/h+ pairs and hence elevate the photocatalytic performancey. The Fermi level of gold NPs is located below the conduction band of ZnO where the electrons will jump to gold particles, hence the Au particles act as electron traps. The electrons are transfer to the molecular oxygen adsorbed on the surface of catalyst, where reactive superoxide species are formed. In the time, the photo induced positive hole can migrate to the surface and react with adsorbed water molecules and/or hydroxide ions to produce highly reactive hydroxyl radicals:

$e-(\text{ or } e_{cb}^{-})+O_2 \longrightarrow {}^{\bullet}O_2^{-}$

and

$h^+ + H_2O \longrightarrow + OH\bullet + H^+$

The active species ($^{\bullet}OH$, $^{\bullet}O_2^-$ and h^+) can oxidize the RB4 dye molecules to form degradation and/or mineralization products [62]:

$^{\bullet}O_2^- + RB4/RB4^* \longrightarrow$ Degradation products $\longrightarrow CO_2 + H_2O$ [9]

$^{\bullet}OH + RB4/RB4^* \longrightarrow$ Degradation products $\longrightarrow CO_2 + H_2O$ [10]

$h_{VB}^+ + RB4/RB4^* \longrightarrow$ Degradation product $\longrightarrow CO_2 + H_2O$ [11]

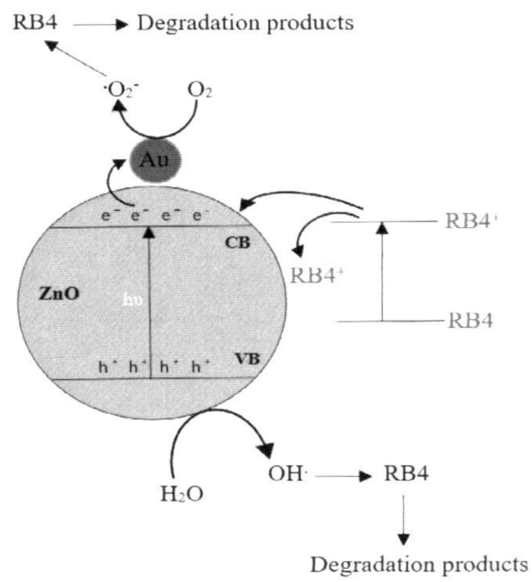

Figure 8. Photodegradation mechanism of RB4 dye

4. Conclusions

In the current work, Au, undoped and Au doped ZnO NPs have been successfully fabricated using the simple thermal decomposition at 700 ^{0}C of Zn(II) and Au(III) complexes with azo ligand 7-[2-(Benzimidazole) azo]-8-Hydroxy Quinolone . The as-synthesized nanomaterials were investigated by various techniques including XRD, FT-IR, UV-Vis DRS, SEM, TEM and AFM. All these techniques confirmed that the thermal decomposition process is an efficient route for synthesis NPs under controllable conditions. The synthesized samples were examine as photocatalysts for photocatalytic color removal of RB4 dye under visible light irradiation. The Au-ZnO sample exhibited the maximum photodecolorization efficiency (91.65 %) under optimum operating conditions (irradiation time 120 min, 0.9 g/L of catalyst and pH 11) with using of dye concentration of 50 ppm, while dye solution was almost completely degraded (~97.36 %) after irradiation for 2.5 hours.

References

[1] K. G. Pavithra , P. S. Kumar, V. Jaikumar, P. S. Rajan , Removal of Colorants from Wastewater: A Review on Sources and Treatment Strategies, Journal of Industrial and Engineering Chemistry 78, 1-19 (2019).

[2] R. Javaid, U. Y. Qazi, Catalytic Oxidation Process for the Degradation of Synthetic Dyes: An Overview, Int. J. Environ. Res. Public Health 16, 2066 (2019).

[3] M. Ilyas, W. Ahmad, H. Khan, S. Yousaf, M. Yasir and A. C. Khan, Environmental and health impacts of industrial wastewater effluents in Pakistan: a review, Rev Environ Health 34, 171–186 (2019).

[4] R. Kant, Textile dyeing industry an environmental hazard. *Natural Science*, 4, 22-26(2012).

[5] M. R. Islam, M. G. Mostafa, Textile Dyeing Effluents and Environment Concerns - A Review, J. Environ. Sci. & Natural Resources, 11(1&2):131-144, 2018.

[6] B. Liang, Q. Yao, H. Cheng, S. Gao, F. Kong, D. Cui, Y. Guo, N. Ren, A. Wang. Enhanced degradation of azo dye alizarin yellow R in a combined process of iron- carbon microelectrolysis and aerobic bio-contact oxidation. Environmental Science and Pollution Research, 2012, 19(5), 1385-1391.

[7] N. Hussain, S. Alwan , H. Alshamsi and I. Sahib, Green Synthesis of S- and N-Codoped Carbon Nanospheres and Application as Adsorbent of Pb (II) from Aqueous solution , International Journal of Chemical Engineering 2020, 9068358 (2020).

[8] H. A. H. Alshamsi and A. T. Hamza, Degradation of Thiazine Dyes Azure B and C by Sonolysis, Sonophotolysis and Sonocatalysis, Asian Journal of Chemistry 27, 1935-1940 (2015).

[9] K. Gautam, S. Kamsonlian, S. Kumar , Removal of Reactive Red 120 dye from wastewater using electrocoagulation:optimization using multivariate approach, economic analysis, and sludge characterization, Separation Science and Technology 55, 3412-3426 (2020).

[10] S. M. Ibrahim, A. A. Badawy and H. A. Essawy, Improvement of dyes removal from aqueous solution by Nanosized cobalt ferrite treated with humic acid during coprecipitation, J Nanostruct Chem 9, 281–298 (2019).

[11] H. A. Habeeb and A. S. Waheeb, Study of chemical oxygen demand (COD) and Removal of azure B dye from aqueous solution by using Advanced Oxidation Processes, AL-Qadisiyah Journal of Pure Science 18, 162-180 (2013).

[12] F. Wei, M. J. Shahid, G. S. H. Alnusairi, M. Afzal, A. Khan, M. A. El-Esawi, Z. Abbas, K. Wei, I. E. Zaheer, M. Rizwan, S. Ali, Implementation of Floating Treatment Wetlands for Textile Wastewater Management: A Review.

[13] M. R. Gadekar and M. M. Ahammed,Coagulation/flocculation process for dye removal using water treatment residuals: modelling through artificial neural networks, Desalination and Water Treatment 57, 26392-26400 (2016).

[14] S. E. Ebrahim, T. J. Mohammed, H. O. Oleiwi, Removal of Acid Blue Dye from Industrial Wastewater by using Reverse Osmosis Technology, Association of Arab Universities Journal of Engineering Sciences 25,29-40 (2018).

[15] H. A. Alshamsi and B. S. Hussein, Synthesis, Characterization and Photocatalysis of g-Fe_2O_3 Nanoparticles for Degradation of Cibacron Brilliant Yellow 3G-P, Asian Journal of Chemistry 30, 273-279 (2018).

[16] B. S. Hussein, A. S. Waheeb, A. M. Yasir and H. A. Alshamsi, Color Removal of Azor A Dye in Aqueous Solution by ZnO and Hydrogen Peroxide under Solar Irradiation, Journal of Physics: Conf. Ser. 1294, 052048 (2019).

[17] M. A. Al-Bedairy and H. A. H. Alshamsi, Environmentally Friendly Preparation of Zinc Oxide, Study Catalytic Performance of Photodegradation by Sunlight for Rhodamine B Dye, Eurasian Journal of Analytical Chemistry 13, em72 (2018).

[18] H. A. Alshamsi, F. Beshkar, O. Amiri, M. Salavati-Niasari, Porous hollow $Ag/Ag_2S/Ag_3PO_4$ nanocomposites as highly efficient heterojunction photocatalysts for the removal of antibiotics under simulated sunlight irradiation,Chemosphere,Volume 274,2021,129765.

[19] M. Karami, M. Ghanbari, H. A. Alshamsi, S. Rashki, M. Salavati-Niasari, Facile fabrication of Tl4HgI6 nanostructures as novel antibacterial and antibiofilm agents and photocatalysts in the degradation of organic pollutants, Inorganic Chemistry Frontiers 8 (10), 2442-2460.

[20] M. Dara, M. Hassanpour, H. A. Alshamsi, M. Baladi and M. Salavati-Niasari, Green sol–gel auto combustion synthesis and characterization of double perovskite Tb_2ZnMnO_6 nanoparticles and a brief study of photocatalytic activity, RSC Adv., 2021, 11, 8228-8238.

[21] F. He, W. Jeon, W. Choi, Photocatalytic air purification mimicking the self-cleaning process of the atmosphere. *Nat Commun* 12, 2528 (2021).

[22] S. R. Yousefi, H. A. Alshamsi, O. Amiri, M. Salavati-Niasari, Synthesis, characterization and application of Co/Co_3O_4 nanocomposites as an effective photocatalyst for discoloration of organic dye contaminants in wastewater and antibacterial properties, Journal of Molecular Liquids, Volume 337, 2021,116405.

[23] M. Karami, M. Ghanbari, H. A. Alshamsi, M. Ghiyasiyan-Aranı, M. Salavati-Niasari,The effect of CuI–PbI2 nanocomposite fabricated by the sonochemical route on electrochemical hydrogen storage characteristics, International Journal of Hydrogen Energy,Volume 46, Issue 36,2021,Pages 19074-19084.

[24] Q. A. Yousif, K. M. Mahdi, H. A. Alshamsi, TiO2/graphene and MWCNT/PEDOT:PSS nanocomposite-based dye-sensitized solar cell: Design, fabrication, characterization, and investigation, Optik, Volume 219, 2020,165294.

[25] M. Goudarzi, H. A. Alshamsi, M. Amiri, M. Salavati-Niasari, ZnCo2O4/ZnO nanocomposite: Facile one-step green solid-state thermal decomposition synthesis using Dactylopius Coccus as

capping agent, characterization and its 4T1 cells cytotoxicity investigation and anticancer activity, Arabian Journal of Chemistry, Volume 14, Issue 9, 2021,103316.

[26] S. Rashki, H. A. Alshamsi, O. Amiri, H. Safardoust-Hojaghan, M. Salavati-Niasari, A. Nazari-Alam, A. Khaledi, Eco-friendly green synthesis of ZnO/GQD nanocomposites using Protoparmeliopsis muralis extract for their antibacterial and antibiofilm activity, Journal of Molecular Liquids, Volume 335, 2021, 116195.

[27] G. Ren, H. Han, Y. Wang, S. Liu, J. Zhao, X. Meng, Z. Li, Recent Advances of Photocatalytic Application in Water Treatment: A Review. Nanomaterials 2021, 11, 1804.

A. Fujishima, K. Honda, Electrochemical photolysis of water at a semiconductor electrode. Nature 1972, 238, 37–38.

[28] Kołodziejczak-Radzimska, T. Jesionowski, (2014). Zinc Oxide-From Synthesis to Application: A Review. Materials (Basel, Switzerland), *7*(4), 2833–2881.

[29] M. A. Borysiewicz, ZnO as a Functional Material, a Review. Crystals 2019, *9*, 505.

[30] P. Basnet, T. I. Chanu, D. Samanta, S. Chatterjee, a review on bio-synthesized zinc oxide nanoparticles using plant extracts as reductants and stabilizing agents. Journal of Photochemistry and Photobiology B: Biology, Volume 183, 2018, Pages 201-221.

[31] F. Rahman, Zinc oxide light-emitting diodes: a review, Opt. Eng. 58(1), 010901 (2019).

[32] M. Adeel, M Saeed, I. Khan, M. Muneer N. Akram. Synthesis and Characterization of Co-ZnO and Evaluation of Its Photocatalytic Activity for Photodegradation of Methyl Orange. *ACS Omega*. 2021; 6(2):1426-1435.

[33] L. Saikia, D. Bhuyan, M. Saikia, B. Malakar, D. K. Dutta, P. Sengupta, Photocatalytic performance of ZnO nanomaterials for self sensitized degradation of malachite green dye under solar light, Applied Catalysis A: General, Volume 490,2015,Pages 42-49.

[34] P. S. S. Kumar, M. R. Raj, Sambandam Anandan, Meifang Zhou, Muthupandian Ashokkumar; Visible light assisted photocatalytic degradation of acid red 88 using Au–ZnO nanophotocatalysts. *Water Sci Technol* 1 September 2009; 60 (6): 1589–1596.

[35] Ahmad, E. Ahmed, M. Ahmad, The excellent photocatalytic performances of silver doped ZnO nanoparticles for hydrogen evolution. *SN Appl. Sci.* 1, 327 (2019).

[36] V. Kumari, A. Mittal, J. Jindal, S. Yadav, N. Kumar, S-, N- and C-doped ZnO as semiconductor photocatalysts: A review, Front. Mater. Sci. 2019, 13(1): 1–22.

[37] N. Mirikaram, A. Pérez-Molina, S. Morales-Torres, A. Salemi, F. J. Maldonado-Hódar, L. M. Pastrana-Martínez, Photocatalytic Perfomance of ZnO-Graphene Oxide Composites towards the Degradation of Vanillic Acid under Solar Radiation and Visible-LED. Nanomaterials 2021, 11, 1576.

[38] W. Zou,Y. F. Rao, A. Alyaman,W. ChuM. J. Chen, D. A. Patterson E. A. Emanuelsson,W. Gao, Heterogeneous lollipop-like V2O5/ZnO array: a promising composite nanostructure for visible light photocatalysis. Langmuir. 2010 Jul 20; 26(14):11615-20.

[39] R.Y. Hong, J.H. Li, L.L. Chen, D.Q. Liu, H.Z. Li, Y. Zheng, J. Ding Synthesis, surface modification and photocatalytic property of ZnO nanoparticles, Powder Technology,Volume 189, Issue 3,2009, Pages 426-432.

[40] M. Samadi, M. Zirak, A. Naseri, E. Khorashadizade, A. Z. Moshfegh, Recent progress on doped ZnO nanostructures for visible-light photocatalysis, Thin Solid Films, Volume 605, 2016, Pages 2-19.

[41] Peychev, P. Vasileva, Novel starch-mediated synthesis of Au/ZnO nanocrystals and their photocatalytic properties, Heliyon, Volume 7, Issue 6,2021,e07402.

[42] Y. Lin, Y. Cao, Q. Yao, O. J. H. Chai, J. Xie, Engineering Noble Metal Nanomaterials for Pollutant Decomposition, Ind. Eng. Chem. Res. 2020, 59, 47.

[43] Ma, X. Wang, S. Zhan, X. Li, X. Liu, Y. Chai, R. Xing, H. Liu, Photocatalytic Activity of Monosized AuZnO Composite Nanoparticles. *Appl. Sci.* 2019, *9*, 111.

[44] L. Sun, G. Wei, Y. Song, Z. Liu, L. Wang, Z. Li, Solution-phase synthesis of Au@ZnO core-shell composites, Materials Letters, Volume 60, Issues 9–10 ,2006,Pages 1291-1295.

[45] P. K. Chen, G. J. Lee, S. H. Davies, S. J. Masten, R. Amutha, J. J. Wu, Hydrothermal synthesis of coral-like Au/ZnO catalyst and photocatalytic degradation of Orange II dye, Materials Research Bulletin,Volume 48, Issue 6, 2013,Pages 2375-2382.

[46] P. Pawinrat, O. Mekasuwandumrong, J. Panpranot, Synthesis of Au–ZnO and Pt–ZnO nanocomposites by one-step flame spray pyrolysis and its application for photocatalytic degradation of dyes, Catalysis Communications, Volume 10, Issue 10, 2009,Pages 1380-1385.

A. M. Mostafa, E. A. Mwafy, Synthesis of ZnO and Au@ZnO core/shell nano-catalysts by pulsed laser ablation in different liquid media ,Journal of Materials Research and Technology, Volume 9, Issue 3, 2020, Pages 3241-3248.

[47] K. Bharti, S. A. Lone, A. Singh, S. Nathani, P. Roy , K. K. Sadhu, Green Synthesis of Luminescent Gold-Zinc Oxide Nanocomposites: Cell Imaging and Visible Light–Induced Dye Degradation, Frontiers in Chemistry,2021, 9, 105.

[48] Q. A. Nema and H. A. Alshamsi, Synthesis, Characterization, Synthesis, Spectroscopic and Thermal Study of Azo Dye 7-[2-(BenzImidazolyl)Azo]- 8-Hydroxy Quinolineand its Gold(III) Complex , International Journal of Science and Research , 6, 235-240 (2017).

[49] H. A. Alshamsi, Q. A. Nema, New approach for synthesis of ZnO nanoparticles via thermal decomposition of $[Zn(BIAHQ)_2].H_2O$ complex: characterization and photocatalytic degradtion of reactive blue 4 dye ,

[50] K. J. AL-Adilee, H. A. H. Alshamsi and M. N. Dawood, Synthesis , Spectral Characterization and photo Thermal Decomposition Studies of New Hetrocyclic Azo Dye Compound Derived From Imidazole with Some Transition Metal Complexes, Research Journal of Pharmaceutical, Biological and Chemical Sciences 7,2882-2905 (2016).

[51] H. M. Kamari, N. M. Al-Hada, E. Saion, A. H. Shaari, Z. A. Talib, M. H. Flaifel, A. A. A. Ahmed, A.A.A. Calcined Solution-Based PVP Influence on ZnO Semiconductor Nanoparticle Properties. *Crystals* 2017, *7*, 2.

[52] S. Krishnamurthy, A. Esterle, N. C. Sharma, S. V. Sahi. *Yucca*-derived synthesis of gold nanomaterial and their catalytic potential. *Nanoscale Res Lett* 9, 627 (2014).

[53] N. Belachew, M.H. Kahsay, A. Tadesse, K. Basavaiah, Green synthesis of reduced graphene oxide grafted Ag/ZnO for photocatalytic abatement of methylene blue and antibacterial activities, Journal of Environmental Chemical Engineering, 8 (2020) 104106.

[54] K. S. Babu, A. R. Reddy, C. Sujatha, K. V. Reddy, A. N. Mallika, Synthesis and optical characterization of porous ZnO , Journal of Advanced Ceramics 2013, 2(3): 260–265.

[55] H. A. H. Alshamsi and B. S. Hussein, Hydrothermal preparation of silver doping zinc oxide nanoparticles: study the characterization and photocatalytic activity, Orient J Chem 34, 1898-1907 (2018).

[56] Q. A. Yousif , K. M. Mahdi and H. A. Alshamsi, Enhanced photovoltaic performance of dye-sensitized solar cell based on ZnO nanoparticles and ZnO/graphene nanocomposites, Journal of the Chinease Chemical society, 2021.

[57] S. H. Altaa, H. A. H. Alshamsi and L. S. Al-Hayder, Synthesis and characterization of rGO/ CO3O4 composite as nanoadsorbent for Rhodamine 6G-dye removal , Desalin Water Treat 114,320-331(2018).

[58] R. Alipour, M. T. Hosseinnejad, A. S. Salar Elahi and M. Ghoranneviss,New perspective on morphological features of the zinc oxide thin films as a gas sensor, J Alloys Compd 687, 72-78(2016).

[59] M. Ghiyasiyan-Arani, M. Masjedi-Arani, Size controllable synthesis of cobalt vanadate nanostructures with enhanced photocatalytic activity for the degradation of organic dyes, Journal of Molecular Catalysis A: Chemical, 425 (2016) 31-42.

[60] H. A. Alshamsi1, M. A. Al Bedairy, S. H. Alwan, Visible light assisted photocatalytic degradation of Rhodamine B dye on CdSe-ZnO nanocomposite: Characterization and kinetic studies, IOP Conf. Series: Earth and Environmental Science 722 (2021) 012005.

[61] S. Alkaykh, A. Mbarek, E. E. Ali-Shattle, Photocatalytic degradation of methylene blue dye in aqueous solution by MnTiO3 nanoparticles under sunlight irradiation, Heliyon, 2020, 6, e03663.

[62] D. Thongam and H. Chaturvedi, Advances in nanomaterials for heterogeneous photocatalysis, Nano Express 2 (2021) 012005.

[63] H. B. Hadjltaief, S. B. Ameur, P. D. Costa, M. B. Zina and M. E. G.,Photocatalytic decolorization of cationic and anionic dyes over ZnO nanoparticle immobilized on natural Tunisian clay, Applied Clay Science 152, 14-157 (2018).

Environmental Study Comparison between normal and standard culture media for the growth of fungi

Hazim A.Walli[1] and Wisam jasim Abed Ali[2]

[1]University of Al-Qadisiyah
[2]University of Sumer

Abstract. The study aims to evaluate the use of rice and starch medium in preparing a culture medium for the development of some fungi isolated from the soil and compare it with standard medium PDA, where the fungi were identified microscopically. The media was prepared with the same components as the standard medium. The results showed that the rice and starch medium were significantly superior in the growth of fungi compared to the standard medium, as the rice and starch medium gave the average growth diameter of Aspergillus niger that reached 4.3-4.6 cm compared with the standard medium PDA which reached 4.1-3.8 cm. It was followed by Alternaria alternate at a rate of 3.6-3.4 cm compared to growth on standard medium 3.3-3.1cm, then Rhizopus stolonifera at a rate of 3.5-3.2 cm compared with growth on standard medium 3.2-3.1 cm, in growth on rice and starch culture medium, while it was less growth of Aspergillus flavus at a rate of 4.3-3.9cm compared with growth rates on the standard culture medium 4.4-4.2 cm.

Key words: fungi , culture media , PDA , and rice and starch medium.

1. Introduction

Culture media contain all the elements that most fungi need for growth and are not selective, so they areused for the general cultivation and maintenance of fungi kept in laboratory culture collections [1].

A defined medium that has enough ingredients for support growth called a minimum media. The number of that ingredients be added for a minimum media of varies enormously depending on microorganisms are grown. Minimum media are contain of minimum nutrients possible to growth of colony, generally without of presence that amino acids, often used with microbiologist which grow wild type of microorganism. Minimum media also be used for select or against of recombinants [2].

Selective media are used for growth only selected of microorganisms. example, if microorganisms are resistant to antibiotic, such as tetracycline, that antibiotic can be added to medium for prevent other cells, that do not possess that resistance to growing [3]. Differential media distinguish of one microorganism from another type which growing on same media. This media used biochemical characteristics for grow of microorganism in presence that specific nutrients which added to media [4].

doi:10.1088/1755-1315/1029/1/012005

2. Material and methods

Soil samples were collected in sterile plastic containers and then transferred to the laboratory to isolate some types of fungi from them, where 10 gm of soil were weighed and placed in beaker 250 ml, where 90 ml of distilled water was added to it and mixed well and then a group of decimal dilutions were preparedfor the purpose of cultivation and purification on standard PDA medium, rice and starch medium for comparison. The standard medium was prepared by adding 39 g of PDA per 1000 ml of distilled water and was sterilized by a sterilizer according to the manufacturer's instructions, at a temperature of 121 degrees and a pressure of 1.5 bar for 15 minutes [5].

Fungus diagnosis:

Fungi grown on PDA were identified phenotypically based on colony shape and color andmicroscopically based on conidia and their carriers.

Preparation of culture media:

The standard medium (PDA) was prepared according to the manufacturer's instructions, while the comparison medium was prepared by dissolving 39 gm of rice flour and 20 gm of starch flour, which is a substitute for agar used to harden the culture medium in addition to being a nutritious substance, then 20 gm of dextrose sugar was added in 1000 ml of distilled water, after that the culture media was sterilized with autoclave at a temperature of 121°C and a pressure of 1.5 bar for 15 minutes [4].

Statistical Analysis

The SPSS program was used to analyze the data for the results of the current study using the Analysis of variance test, then significant differences were extracted using the least significant difference between the means at the level of $p \leq 0.05$ [6].

3. Results and discussion

The results of agricultural diagnosis showed the presence of several types of fungi in the soil samples, according to the table below (1)

Table 1. Fungi isolated from soil

No.	Fungi isolation	Frequency
1	*Aspergillus flavus*	55
2	*Aspergillus niger*	48
3	*Alternaria alternate*	37
4	*Rhizopus stolonifer*	22

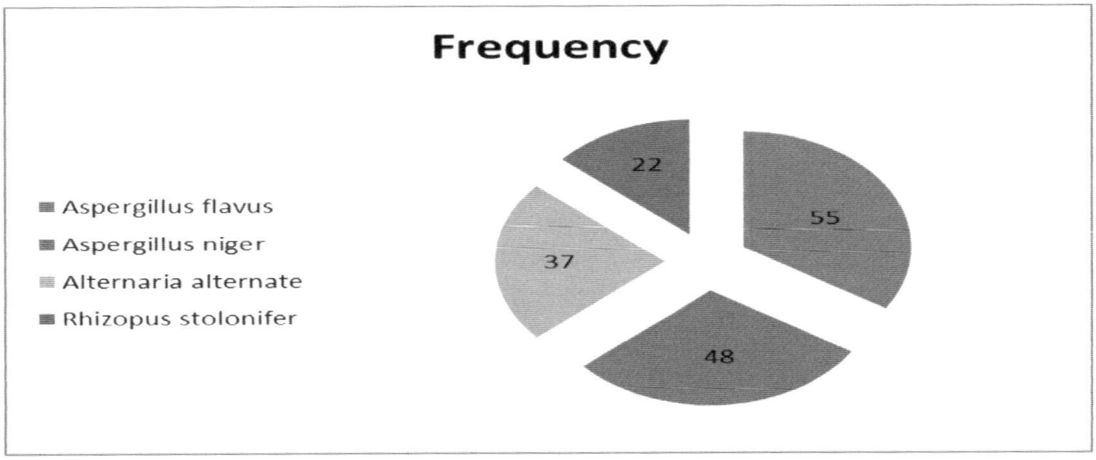

figure 1. Fungi isolated from soil.

Four types of fungal species that were diagnosed were selected for comparison in their growth on rice and starch culture media compared with the standard medium PDA, where the results showed significant differences in the growth of fungi between the two media, where the culture media for rice and starch outperformed in the growth of fungi according to the growth diameters shown in the table (2) Aspergillus niger, followed by Alternaria alternate, then Rhizopus stolonifera, while it was lower for Aspergillus flavus

Table 2. growth diameters on culture media

No.	Fungi isolation	growth diameter	
		On PDA media	**On rice and starch media**
1	***Aspergillus flavus***	**4.4 - 4.2**	**4.3-3.9**
2	***Aspergillus niger***	**4.1-3.8**	**4.6-4.3**
3	***Alternaria alternate***	**3.3-3.1**	**3.6-3.4**
4	***Rhizopus stolonifer***	**3.2-3.1**	**3.5-3.2**

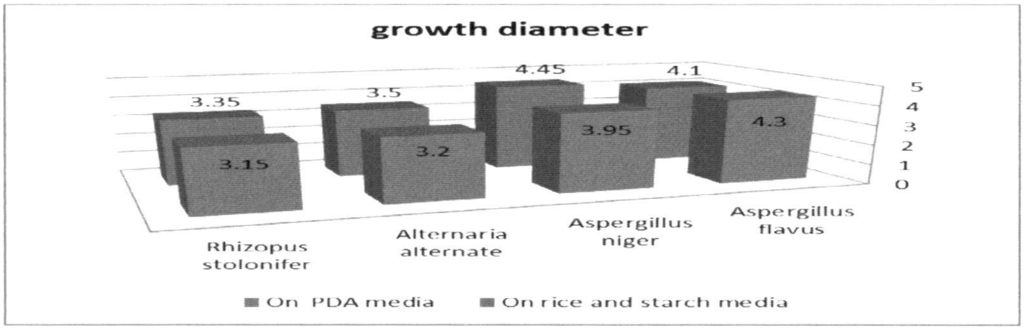

Figure 2. growth diameters on culture media

Through the results, the natural medium of rice and starch proved highly effective in the growth of fungi and at a low financial cost, through the use of starch instead of agar as a hardening agent for the culture medium, which helped reduce the financial cost of the culture medium in addition to its role in feeding the developing microorganisms [7] [8] [9].

Natural media are complex natural plant extracts with known composition, quantity and quality, often prepared from plant tissues such as fruits, vegetables, yeast extracts, barley, body fluids, and others [10] [11]. These media are characterized by the fact that they are similar in composition to the medium on which mushrooms grow in nature, in addition to containing vitamins and other elements that are not found in the rest of the media[8,10]. Examples include extracts of plant parts such as potatoes, carrots, beans, barley andrice. Rice, as indicated by the US Department of Agriculture, contains the following food ingredients: carbohydrates in a large proportion. Each 100 gm of rice contains 31% carbohydrates, 3.5% proteins, and 1.2% fiber. Whereas, starch consists of carbohydrates by a large percentage of up to 95%, in addition to theimportant elements [12] [13].

References

[1] Chen, Y. (2006). Integrating organic matter into plant nutrient management..
[2] Didiek, H.G. Siswanto and Y. Sugiarto (2002). Bioactivation of Poorly Soluble Phosphate Rocks with A phosphorous Solubilrizng Fungus. Soil Sci. Soc. of Am. J. 64: 927-932..
[3] Duncan, D. (1955). Multiple range and multiple F-test. Biometric 11: 1-42 .Gervais, P.; i. P. Molin; W. Grajek and M. Bensoussan (1988). Influence of the water activity of a ii.solid substrate on the growth rate and sporogenesis of filamentous fungi. Biot.
[4] Garrett, S.D. (1963). Soil Fungi and Soil Fertility. Pergamon Press. Oxford. England 165pp..
[5] Weil, R.R. and F. Magdoff (2004). Significance of soil organic matter to soil quality and health. pp. 1-43. Cited in: F.R. Magdoff and R.R. Weil (eds.(. Soil Organic. CRC Press, New York. USA..
[6] Gurbatl, J. (1997). Essentials of food microbiology. 1sted. Arnold: London, Sydney..
[7] Harrigan, W.F. and M.E. McCance (1976). Laboratory Methods in Food and Dairy Microbiology. Academic Press: London, New York..
[8] IFA Agriculture Conference, Optimizing Resource of Agriculture, for Sustainable Intensification of

Agriculture, Kunming, China. 27 Februray– 2 March..
[9] Okumura, Y.; Y. Ogawa and T. Nikai (2004). Elastase and elastase inhibitor form Aspergillus fumigatus, Aspergillus flavus and Aspergillus niger. J. Med. Microbiol. 53: 351-354..
[10] Hoitink, H.A.J. and M.J. Boehm (1999). Biocontrol within the context of soil microbial communities: A substrate – dependent phenomeon. Ann. Rev. Phytopathol. 37: 427- 446..
[11] Ravimannan, N., Arulanantham, R., Pathmanathan, S., & Niranjan, K. (2014). Alternative culture media for fungal growth using different formulation of protein sources. Annals of Biological Research, 5(1), 36-39..
[12] Meloan, C. and Y. Pomeranz (1973). Food analysis laboratory experiments. The AVI Publishing Inc: Westport, Connecticut. USA..
[13] Vonberg, R. and P. Gastmeier (2006). Nosocomial aspergillosis in outbreak settings. J. Hosp. Infect. 63: 246-254..
[14] Reeslev, M. and A. Kioller (1995). Comparison of biomass dry weights and radial growth rates of fungal colonies on media solidified with different gelling compounds. Appl. Environ. Microbiol. 61: 4236-4239..
[15] Watanab.J.(2002). Pictorial atlas of soil and seed fungi morphologies of cultured fungi and keytos species 2 ISSN:8493-1118..
[16] Bassam Alfarhani, Maha Al-Tameemi, Hector C. Goicoechea, Fernando Barbosa, Andres D. Campiglia, Direct analysis of benzo[a]pyrene metabolites with strong overlapping in both the spectral and lifetime domains, Microchemical Journal, 137 2018, 51-61,
[17] B. Alfarhani, M. Al-tameemi, A.V. Schenone, H.C. Goicoechea, F. Barbosa, A.D. Campiglia. Microchem. J., 2016 129 83–89. https://doi.org/10.1016/j.microc.2016.06.010.
[18] Zeina. M. Kadam (2021). Assessment of Electrochemical Activity of some Amino acids on the Modified Carbon Electrode Surface. Research Journal of Pharmacy and Technology, 14(10):5325-9. doi: 10.52711/0974-360X.2021.00928
[19] Z. M. Kadhim and A. H. Ali, (2018), Studying of Optical and Structural Properties of a Novel Ligand (5-ClCPAI) Thin Films by Spray Pyrolysis Method. Journal of Engineering and Applied Sciences, 13(14), 11032-11040.
[20] Z. M. Kadam, C. D. Gwenin, (2017), Polymer Membranes based on Ionophore-impregnated for Nutrients Detection by Electrochemical Methods, Der Pharma Chemica, 9(20):29-33.
[21] Yu, J.; T. Cheveland; W. Nierman and Bennet (2005). Aspergillus flavus genomics: gateway to human and animal health, food safety and crop resistance to diseases. Rev Iberoam Micol . 22: 194-202.

Isolation and Identification of *Aspergillus niger* from eyes

Hazim A.Walli[1] and Wisam jasim Abed Ali[2]

[1]University of Al-Qadisiyah
[2]University of Sumer

Abstract. This study was carried out to detect the contamination of the eyes with the black fungus and its presence in the eyes, where swabs were taken with the number of 50 swabs from the eyes, of twenty-five healthy people aged 15-30, and twenty-five elderly people aged 50-70 with diabetes. The results showed significant differences in the rate of isolation between healthy people and people suffering from chronic diseases, the results showed that percentage of isolation from elderly was 62.7 %, while the percentage of isolation from healthy people was 37.3% , also three other species were isolated (*A.flavus* , *A.ochracus* , *A. terreus*) in the rate of (16.9, 10.2 , 3.4)% respectively in elderly people , but other species were isolated (*A.flavus* , *A.ochracus*) in the rate of (11.9, 6.8)% respectively in healthy people.

Key words : *Aspergillus niger* , *Aspergillus flavus* , *Aspergillus terreus*

1. Introduction

Fungi are eukaryotic living organisms that spread in all environments, some of them are restored and some cause diseases to humans and animals [1]. People with weak immunity are exposed to opportunistic fungi that invade the body and cause infections in various areas, including the respiratory system, eyes, wounds, and others [2].

Fungi are microorganisms that are found everywhere in the air, water, and soil [3]. Most are restored, some are pathogenic, and some are opportunistic [4]. It is susceptible to air pollution with dust laden with dust-filled and fungal spores , which it carried and transported by wind from one place to another [5]. Fungi have become pathogens that affect the respiratory system due to their presence in the air [6]. Which causes allergies and irritation of the respiratory system , when inhaled fungal spores in high quantities during breathing [7]. As it revealed three mechanisms of the disease that can be caused by fungi, which are allergies, infection and food poisoning [8]. Also published research indicated that increased exposure to fungi in the air causes asthma [9]. All parts of fungus cause allergies when inhaled (spores or mycelium), And it has the ability to induce antibodies in the body [10].

2. Material and methods

Swabs were collected from 25 healthy people and 25 elderly people suffering from diabetes. The people were randomly selected from different areas in the city of Al-Diwaniyah.

The swab samples were transferred to petri dishes containing Potato Dextrose Agar medium in triplicate. The dishes were incubated at 25 ° C for 5 to7 days until the fungal colonies appeared to identify fungal isolates, a portion of the growing fungal colonies on the PDA medium were moved to new dishes. The identification was based on morphological and characteristics under microscopical of the fungus according to [11]

3. Prepare the culture medium

The medium of potato and dextrose was prepared by dissolving 39 g of ready-made powdered media from the Indian company Himedia, adjusting the pH to 6.5, then adding 50 ppm of streptomycin. It is placed in a conical flask and sterilized with a sterilizer under pressure of 1 atmosphere and at a temperature of 121 ° C and then distributed in Petri dishes [12].

4. Result and discussion

This study was carried out to detect the contamination of the eyes with the black fungus and its presence in the eyes, where swabs were taken with the number of 50 swabs from the eyes, of twenty-five healthy people aged 15-30, and twenty-five elderly people aged 50-70 with diabetes. The results showed significant differences in the rate of isolation between healthy people and people suffering from chronic diseases, the results showed that percentage of isolation from elderly was **62.7** %, while the percentage of isolation from healthy people was **37.3%** , also three other species were isolated (*A.flavus* , *A.ochracus* , *A.*

terreus) in the rate of (**16.9**, **10.2** , **3.4**)% respectively in elderly people , but other species were isolated (*A.flavus* , *A.ochracus*) in the rate of (**11.9**, **6.8**)% respectively in healthy people.

Table 1. The number of isolates and percentage of isolated

No.	Isolated fungi	Number of isolates				Total	percentage	L.S.D.
		healthy people	**percentage**	**Unhealthy people**	**percentage**			
1	***Aspergillus niger***	11	18.6	19	32.2	30	50.8	2.75
2	***A. flavus***	7	11.9	10	16.9	17	28.8	1.94
3	*A. ochracus*	4	6.8	6	10.2	10	16.9	0.96
4	*A. terreus*	0	0	2	3.4	2	3.4	0.52
5	**Total**	22	37.3	37	62.7	59	100	

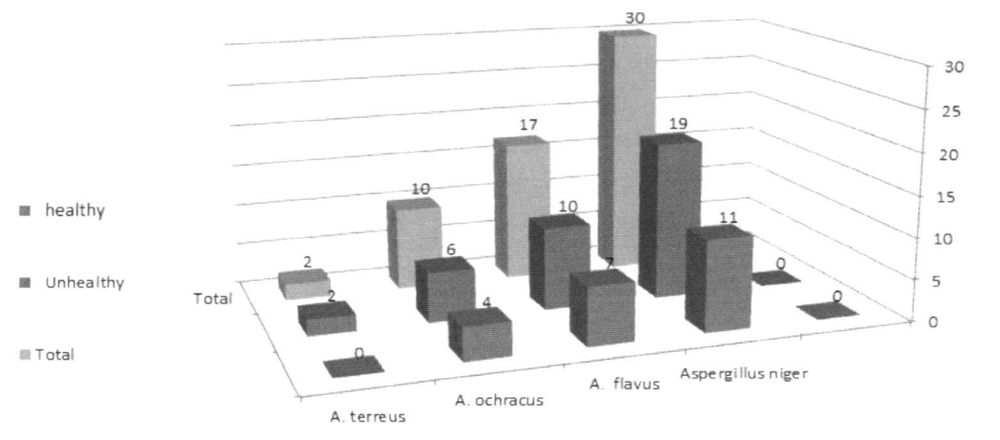

Figure 1. number of isolates

The results showed significant differences in the number of fungal isolates that were isolated from healthy people by rate **37.3 %** , from unhealthy people by rate **62.7** with chronic diabetes (table 1).

These results are consistent with his findings [13], which indicated that the fungus *Aspergillus* spp. was the most common fungus it spreads in all environments . The results also agree with his findings [14] that the fungi in the air are an essential factor for the occurrence of anaphylaxis and the production of antibodies in the body. While [15] indicated that fungi in dust cause respiratory diseases such as bronchopul-monaryaspergillosis. Air does not consider the environment for the growth of microorganisms, including fungi [16]. Rather, it is present in them and their numbers are affected by the speed of wind, humidity, temperature, distance and proximity to agricultural areas and foodstuffs that are consumed in buildings [17].

References

[1] Uzochukwu, O.V.; Nkpouto, N. (2013). Airborne fungi in the indoor and outdoor environments of a higher institution in Nigeria. I.J.A.B.R., 3(1), 2013, 9-12..

[2] Pavan, R.; Manjunath, K. (2014). Indoor study on airborne fungi in swine house of Bangalore,India. Int. J. Curr. Sci., 9, 77-82..

[3] Hedayati, M.; Mayahi, S.; Aghili, R.; Goharimoghadam, K. (2005). Airborne fungi in indoor and outdoor of asthmatic patients home, living in the city of Sari. Iranian J. Allerg. Asthma and Immunol. 4(4), 189-191..

[4] Charles, B.; Freddy, P.; Minati, D.; Jay, P. (2009). Alternaria and Cladosporium fungal allergen epitopes are denatured by sodium hypochlorite. World Allergy Organ J. 2(12), 296–302..

[5] Begum, E.; Shahidul, A.; Shah, A. (2009). Incedence of airborne fungi in Rajshahi metropolitan city in relation to seasonal fluctuation . J. Life . Earth . Sci. 34, 37-41..

[6] Black, P. N.; Udy, A. A.; Brodie, S. M. (2000). Sensitivity to fungal allergens in a risk factor for life threatening asthma. Allergy. 55(5), 501-504..

[7] Hogaboam, C. M.; Carpenter, K. J.; Schuh, G. M.; Buckland, K.F. (2002). Aspergillosis and asthma any link. Med. Mycol. 20 (3), 141-144..

[8] Panagopoulo, P. (2003). Environment surveillance of filamentous fungi in three tertiary care hospitals in Greece. J. Hospital Infection. 52(3),185-191..

[9] Al-Bader, S.M.; Ahmod, S.A.; Al-Hamdani, M. M. (2007). Taxonomic study on.

[10]Kim, K.Y. and Kim, C.N. (2007).Airborne microbiological characteristics in public buildings of Korea. Building and Environment.(42);2188-2196..

[11]Shelton, B. G.; Kirkland, K. H.; Flanders, W. D.; Morris, G. K. (2002). Profile of airborne fungi in building and outdoor environments in the United State. Appl. Envirn. Microbial. 68,1743- 53..

[12]Haas, D., J. Habib, H. Galler, W. Buzina, R. Schlacher; E. Marth and F. F. Reinthaler(2007). Assessment of indoor air in Austrian apartments with and without visible mold growth. Atmospheric Environment ,41: 5192-5201..

[13]Horner, W. E., A. G. Worthan and P. R. Morey(2004). Air- and dust borne mycoflora in houses free of water damage and fungal growth. Applied and Environmental Microbiology, 70 (11) : 6394 – 6400..

[14]Lee, Ho-chingand Lin, Chung-Ming (2004): Coping with dust storm events: Information, Impacts and policy making in Taiwan, TAO. 15(5): 10351060..

[15]M. A. Klich, Identification of common Aspergillus species (Utrecht, Netherlands:Centraalbureau voor Schimmelcultures, 2002)..

[16]N. Nolard, M. Detandt, H. Beguin, Ecology of Aspergillus species in the human environment. In Aspergillus and aspergillosis (Springer, Boston, MA, 1988), pp. 35-41..

[17]F. M. Bokhari, Mycobiology, 35 (2), 47-53 (2007)..

Environmental Removal of brilliant yellow Dye onto Porous Adsorbent Derived from Pomegranate peels Waste

Aseel M. Aljeboree[1], Holya A. Lafta[2], Mohammed Mustafa Abdulkadhm[3], Baraa A. Hamad[4], Noor B. Dheyaa[5], Aiman Mohammed Baqir Al-Dhalimy[6,7] , Ayad F. Alkaim[8] and Salwan Ali Abed [9]

[1,4,5,8] Department of Chemistry, College of Sciences for Girls, University of Babylon, Hilla, Iraq

[2]Al-Nisour University College/Baghdad/Iraq

[3]Al-Manara College For Medical Sciences/Misan/Iraq

[6] College of Nursing, Altoosi University College, Najaf, Iraq.

[7] The Islamic University, Najaf, Iraq.

[9] College of Science, University of Al-Qadisiyah, Iraq

Abstract. Recently, the adsorption process relied on the use of low-cost, available and eco- friendly materials to reduce costs. Pomegranate peels(PPs) have been used as environmental and inexpensive bio-sorbent material to remove dye (brilliant yellow) from aqueous solution. characterization of Pomegranate peels(PPs) via two techniques like FTIR,FESEM to analyze the morphology and chemical groups of Pomegranate peels(PPs) before and after sorption of brilliant yellow dye , Adsorption method is occurred under some optimum conditions like initial dye concentration, mass of adsorbent and contact time Where the Pomegranate peels(PPs) un- activated, the ability to be recycled and used again, by washing it (water, acid and base). Isotherm Freundlich and isotherm Langmuir as models were utilized to investigate the adsorption method. It was find that the model Langmuir better described the experimental result .

Keyword: Adsorption , Removal , Brilliant yellow Dye, Pomegranate peels, Isotherm model

1. Introduction

Adsorption is the utmost effective technique utilized to remove pollutant from aqueous solution compared to other techniques (coagulation, ion exchange, photo-catalysis, hydrogen peroxide ,flocculation , reverse osmosis, membrane filtration, irradiation, oxidation, advanced oxidation, precipitation and adsorption) due to their ease of operation, high efficiency, low-energy requiring technology, avail-ability of several adsorbents, effectiveness in regeneration and reuse of the adsorbent[1-5] . In this context, several researchers developed low-cost substance like agricultural wastes to removal the contamination from waste waters. Agricultural wastes have major components: lignin, cellulose, lipids, hemicelluloses, proteins, etc. [6, 7]. in addition are avail-able and abundant in large quantities and have high potential sorbents due to the variety of functional groups (—C–O, C=O, OH, –NH2) on their surfaces . Many agricultural wastes are used to remove dyes from the aqueous solution, such as date stones, palm fronds, palm fiber, jujube husks, argan skin, and olive pits.

Platanus leaves, walnut peel, mandarin peel, banana peel, fig leaf, jojoba residue, and citrus waste. [8-13]. Pomegranate is one of the most famous and important fruits because it has a very high nutritional value, wonderful taste, and many medicinal properties. Pomegranate peels are considered as solid waste and are disposed of as inexpensive and renewable toxic waste of biomass [14-16]. Pomegranate peels are used as a cheap and available absorbent material to treat pollutants in drinking water [17-20]. In this study, pomegranate peels were prepared without any chemical treatment as environmentally friendly surfaces to remove the dye from its aqueous solution, and also study adsorption isotherms .

2. Materials and method

2.1. Biosorbent preparation

Fresh pomegranate peels(PPs) are washed with hot water several times to get rid of soluble substances also to remove the red color , then cut into small pieces , washed with distilled water and dried at a temperature 70 oC for a period of 24 h , after that the dried pomegranate peels are ground and sieved to particle size 25nm. To obtain the powder used in the experiment as appear in Schema 1.

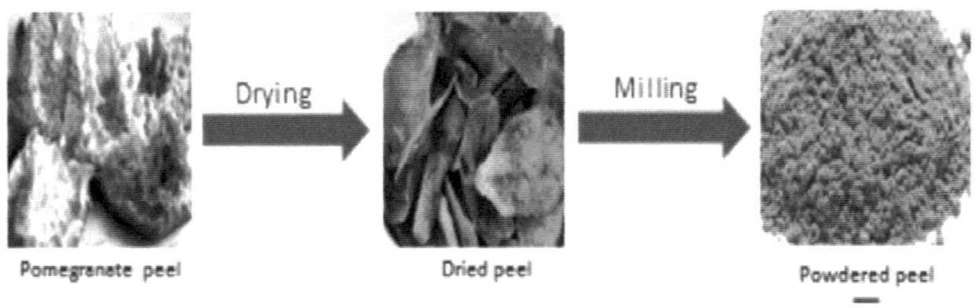

Schema 1. Preparation of pomegranate peels

2.2. Preparation of brilliant yellow solution

A standard solution 1000 mg/L was prepared by dissolving (1g) from brilliant yellow Dye in 1000ml distilled water, the calibration curve depends by preparing a series of concentrations of brilliant yellow Dye from (10-100 mg/L), through the results it was found that it obeys Lambert-Beer's law.

2.3. Effect different parameter

2.4. Effect of Initial pH solution

The effect of primary pH on the removal of **brilliant yellow** was analyzed at (2-10) . 100 ml of solution **brilliant yellow** were put in elementary flask at (25°C) and were agitated at 120 rpm by use 0.1 g weight of pomegranate peels(PPs) .

2.5. Effect of weight pomegranate peels

The effect of pomegranate peels (PPs)dose on the quantity of **brilliant yellow** adsorbed was find via contacting 100 ml of solution **brilliant yellow** with primary concentration 50 mg/L with several quantity of pomegranate peels (0.025-0.2) put in 100 ml elementary flask and at (25°C). The flasks were agitated at 120 rpm.

2.6. Effect of reactivate of pomegranate peels PPs (Desorption)

The desorption of **brilliant yellow** dye was performed via batch mode at 25°C utilizing three several conditions of desorption: NaOH 0.1N , CH3COOH 0.1 N and HCl 0.1 N . at 50 mg/L of **brilliant yellow dye** and 0.1 g of **pomegranate peels**(PPs).

3. Results and Discussions

3.1. FTIR

The IR spectra of adsorbent PPs among 3500 and 500 cm–1 is find to study the characteristics surface of adsorbent. FTIR spectrum of PPs particles taken before adsorption of Brilliant yellow appear the involvement of functional groups present on the of PPs surface[15, 21] . In the spectrum of PPs before adsorption there is a strong peak at 2450 cm−1 representing the –CH stretching as appear in Figure 1.

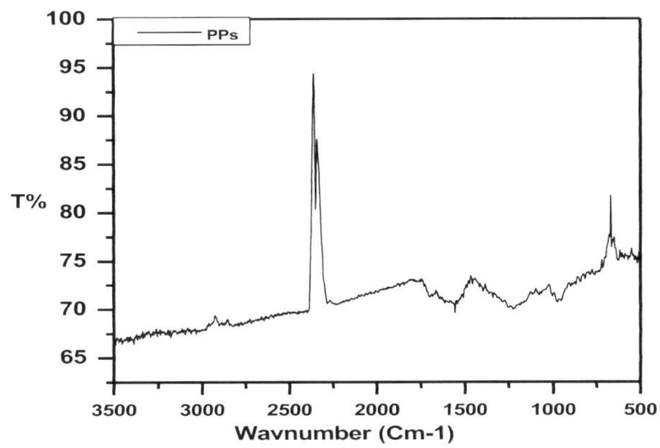

Figure 1. FT-IR image of PPs before adsorption

3.2. FESEM

study of technique (FESEM) to determine the surface physical properties , That note the surface before the Adsorption contains many cavities and many non-active sites, after adsorption method note full all active sites ,because loading **brilliant yellow dye** on the surface **pomegranate peels**(PPs) as shown in the figure 2

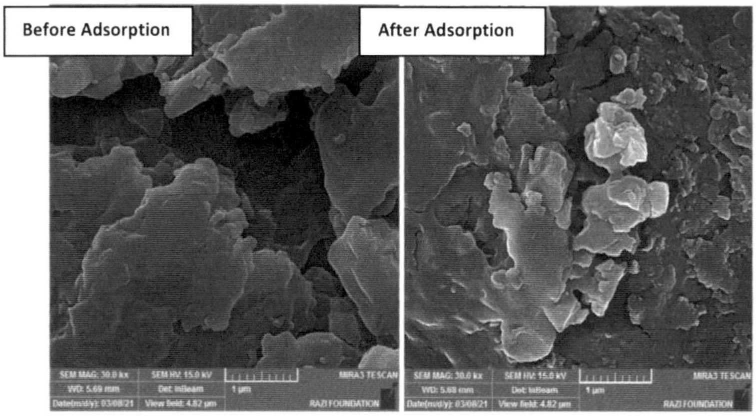

Figure 2. FESEM Image of pomegranate peels(PPs) before and after adsorption

3.3. Effect of weight pomegranate peels

The effect of weight pomegranate peels(PPs) on the removal percentage% of **brilliant yellow** as appear in Fig. 3. The data explain that an increase in the pomegranate peels(PPs) amount causes a increase in the remaining **brilliant yellow** concentration at contact time and consequently an increase in the bios-orption removal capacity . certainly, the bio-sorption percentage removal values raise from 54.23 – 94.87 % as the weight of pomegranate peels(PPs) was raised from 0.025 –0.2 g/100 mL. conversely, the quantity of sorbet dye per unit weight of pomegranate peels(PPs) decreased with the raise in sorbent dose [22]. The bio-sorption efficiency values increased from 29.33- 109.12 mg/g as the weight of pomegranate peels (PPs)was increased from 0.025 -0.2 g/100 mL. [23, 24]

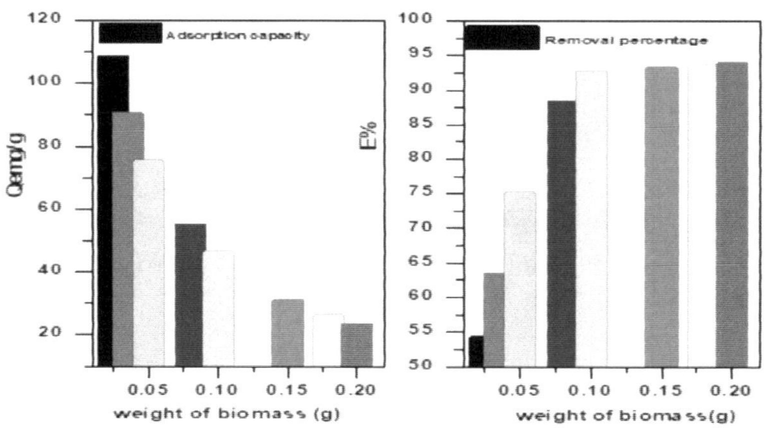

Figure 3. Effects of the mass pomegranate peels on the quantity of adsorbed brilliant yellow (Initial dye concentration = 50 mg. L^{-1}, contact time 60 minute and Temp . = 25°C)

3.4. Effect of solution pH

The pH value is known as one of the utmost significant parameter affecting adsorption efficiency of pomegranate peels. The effect of pH on the adsorption of dye **brilliant yellow** onto pomegranate peels was estimated via several solution pH (2 – 10) as appear in (Fig. 4). The experimental result explain that pH solution effect on adsorption of dye **brilliant yellow** on to pomegranate peels(PPs) , adsorption capacity decrease with increase in solution pH and best adsorption capacity at pH 2 (49.78 mg/g) and minimum adsorption capacity at pH 10 (16.45 mg/g) [25, 26].

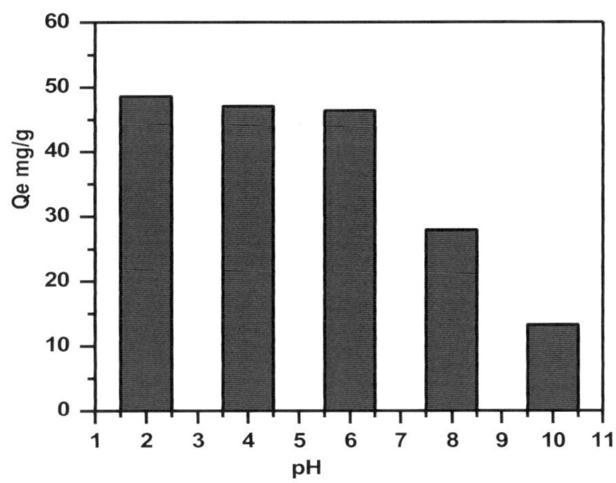

Figure 4. Effects of solution pH on the brilliant yellow adsorption onto pomegranate peels. (Exp. Condition: conc. Of dye = 50 $mg.L^{-1}$, Temp. = 25°C, and mass of surface 0.1 gm).

3.5. Desorption study

The desorption of **brilliant yellow** was performed via batch mode at 25°C utilizing three several conditions of desorption: NaOH 0.1N , CH3COOH 0.1 N and HCl 0.1 N . Batch desorption study was carry out via agitate 100ml of **brilliant yellow** solution of 50 mg/L and 0.1 g of **pomegranate peels (PPs)**. contact time about 60 min and the supernatant **brilliant yellow** solution was discarded. The quantity of dye-loaded **pomegranate peels** was washed firstly by water for removing non bio-sorbed dye [27]. The washed **pomegranate peels** was additional about NaOH 0.1N , CH3COOH 0.1 N and HCl 0.1 N at 25 oC. The desorbed **brilliant yellow** was separated and determine as before. The quantity of **brilliant yellow** taken up via the **pomegranate peels**(PPs) was estimation via Equation. (2). The better percentage desorption was find by CH3COOH (it percentage desorption 90.11% of **brilliant yellow** from **pomegranate peels**), whereas utilizing NaOH and HCl (0.1N) and H2O solution makes it probable to percentage desorption (55.55% 80.133% and 77.72%)of **brilliant yellow** onto **pomegranate peels** as appear in Figure 5.

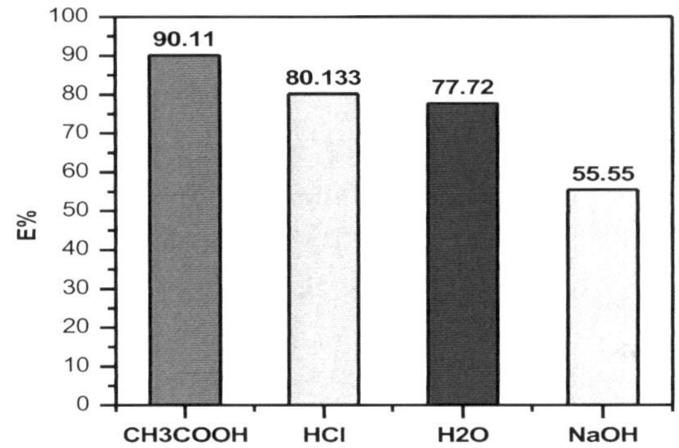

Figure 5. Comparison of desorption capacity of several Solution of the dye from surface of pomegranate peels

3.6. Adsorption model

Figure 6 appear the Isotherm Freundlich and Langmuir for the uptake of **brilliant yellow** on to **pomegranate peels** and Table 1 lists the different model factor as estimate from the slopes and intercepts of the plots. comparing the values R2 of the isotherms utilized, the adsorption result fitted better in to model Freundlich for **pomegranate peels**(PPs) with a value R2 =0.9855, and model Langmuir for **pomegranate peels** (PPs)with value R2 =0.9498 [28, 29].

The Langmuir model nonlinear equation given in equation :

$$Q_e = \frac{Q_m K_L C_e}{1 + K_L C_e} \qquad (1)$$

The isotherm Freundlich model was applic-able in hetero-geneous surface adsorption and The Freundlich model nonlinear equation given in equation :

$$Q_e = k_f C_e^{\frac{1}{n}} \qquad (2)$$

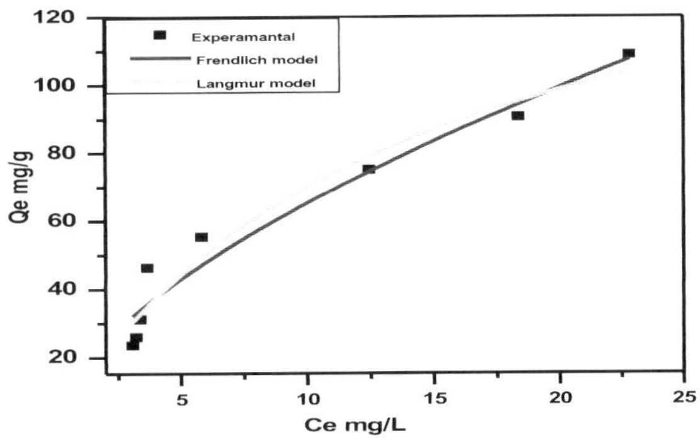

Figure 6. Adsorption models non-linear fit of adsorption **brilliant yellow** onto **pomegranate peels**, conc. = 50 mg. L^{-1}, Temp. = 25°C, and mass 0.1 g/L).

Table 1. Model Freundlich and Langmuir Factors of **brilliant yellow** adsorbed onto **pomegranate peels**(PPs)

Isotherm models	Parameters	pomegranate peels
	K_F	16.2017 ± 2.677
Freundlich	1/n	0.603 ± 0.0614
	R^2	0.98555
	q_m (mg/g)	167.147 ± 23.365
Langmuir	K_L(L/mg)	0.0718 ± 0.0192
	R^2	0.9498

4. Conclusion

pomegranate peels (PPs) are one of the agricultural wastes that have been used without any chemical treatment as effective, cheap, inexpensive and available surfaces to remove the yellow dye from water. Recycling **pomegranate peels** was studied by washing the surface loaded- dye (acid, base and distilled water). It was found that the best way to recycle and use the surface again is by washing it with CH3COOH acid . Through the results, it was found that the best weight (0.1 g) gives the removal percentage of the dye (90.68 %) at the concentration of (50 mg/L) of the dye. Also, the adsorption isotherms were studied, and it was found that it obeys the Freindelge model.

Reference

[1] Lin, S.-H. and R.-S. Juang, *Adsorption of phenol and its derivatives from water using synthetic resins and low-cost natural adsorbents: A review.* Journal of Environmental Management, 2009. **90**(3): p. 1336-1349.

[2] Aljeboree, A.M., Abbas, A.S., *Removal of Pharmaceutical (Paracetamol) by using CNT/ TiO2 Nanoparticles.* Journal of Global Pharma Technology, 2019. **11**(1): p. 199-205.

[3] Kamil, A.M., Mohammed, H.T., Alkaim, A.F.,Hussein, F.H., *Adsorption of Congo red on multiwall carbon nanotubes: Effect of operational parameters.* Journal of Chemical and Pharmaceutical Sciences, 2016. **9**(3): p. 1128-1133.

[4] Wenyan Jiang, L.Z., Xiaoming Guo, Mei Yang, Yiwen Lu, Yijun Wang, Yousen Zheng & Guangtao Wei, *Adsorption of cationic dye from water using an iron oxide/activated carbon magnetic composites prepared from sugarcane bagasse by microwave method.* Environmental Technology, 2019. **2**: p. DOI: 10.1080/09593330.

[5] Liu, M., Li, X.,Du, Y.,Han, R., *Adsorption of methyl blue from solution using walnut shell and reuse in a secondary adsorption for Congo red.* Bioresource Technology Reports, 2019. **5**: p. 238-242.

[6] Qusay K. Mojar Alshamusi, A.A.M.A., Makarim A. Mahdi, Layth S. Jasim and Aseel M. Aljeboree, *ADSORPTION OF CRYSTAL VIOLATE (CV) DYE IN AQUEOUS SOLUTIONS BY USING P(PVP-CO-AAM)/GO COMPOSITE AS (ECO-HEALTHY ADSORBATE SURFACE) : CHARACTERIZATION AND THERMODYNAMICS STUDIES.* Biochem. Cell. Arch., 2021. **12**: p. 2423-2431.

[7] Aljeboree, A.M., Alkaim A F, *Removal of Antibiotic Tetracycline (TCs) from aqueous solutions by using Titanium dioxide (TiO2) nanoparticles as an alternative material.* Journal of Physics: Conf. Series 2019. **1294** p. 052059.

[8] M. Klavarioti, D.M., D. Kassinos, *Removal of residual pharmaceuticals from aqueous systems by advanced oxidation processes.* Environ. Int., , 2009. **35** p. 402-417.

[9] Hoppen, M.I., et al., *Adsorption and desorption of acetylsalicylic acid onto activated carbon of babassu coconut mesocarp.* Journal of Environmental Chemical Engineering, 2019. **7**(1): p. 102862.

[10] Aseel M. Aljeboree, A.F.A., Ali loay, hanadi m algburi, *Photocatalytic Degradation of Textile Dye Cristal Violet Wastewater using Zinc Oxide as a Model of Pharmaceutical Threat Reductions.* Journal of Global Pharma Technology, 2019. **11**(02): p. 138-143.

[11] Hadi, Z.A., Aljeboree, A.M., Alkaim, A.F., *Adsorption of a cationic dye from aqueous solutions by using waste glass materials: Isotherm and thermodynamic studies.* International Journal of Chemical Sciences, 2014. **12**(4): p. 1273-1288.

[12] Rafid Q. Kmal, A.M.M., Aseel M. Aljeboree, Layth S. Jasim, Ayad F. Alkaim, *Synthesis and Identification of Azo Disperse Dye Derived from 4-Aminoantipyrine and Their Applications to Determine Some Drugs.* International Journal of Drug Delivery Technology. 2021;11(3):. 2021. **11**(3): p. 1040-1044.

[13] Xiong W, T., J, Zhaohui Z., Guangming Z, Yaoyu W, Dongbo S, Peipei X, Rui Z, Chen C, *Adsorption of phosphate from aqueous solution using iron-zirconium modified activated carbon nanofiber: Performance and mechanism.* Journal of Colloid and Interface Science, 2017. **493**: p. 17-23.

[14] T. Okuda, Y.K., R. Nagao, N. Yamashita, H. Tanaka, S. Tanaka, S. Fujii, C. Konishi, I. Houwa, *Removal efficiency of 66 pharmaceuticals during wastewater treatment process in Japan.* Water Science and Technology, 2008. **57**: p. 65.

[15] Gupta, V.K. and Suhas, *Application of low-cost adsorbents for dye removal â€" A review.* Journal of Environmental Management, 2009. **90**(8): p. 2313-2342.

[16] Aljeboree, A.M., Alkaim ,A. F.,Abdulrazzak , F. H. ,Abbas , A. S.,Alshirifi ,A. N., *Spectrophotometric Determination of Pharmaceutical by Oxidative Coupling of 4-Aminoantipyrine: A Short Review.* Journal of Engineering and Applied Sciences, 2019. **14**(2): p. 5561-5569.

[17] Baskaralingam, P., Pulikesi,M. ,Elango, D. ,Ramamurthi,V. ,Sivanesan ,S, *Adsorption of acid dye onto organobentonite.* J. Hazard. Mater., 2006. **128**: p. 138-144.

[18] Aseel M. Aljeboree, A.F.A., *Role of plant wastes as an ecofriendly for pollutants (Crystal Violet dye) removal from Aqueous Solutions* Plant Archives 2019 **19**(2): p. 902-905.

[19] Faiq F Karam, F.H.H., SJ Baqir, AF Alkaim, *Optimal conditions for treatment of contaminated waters with anthracene by Fenton processes in close system reactor.* Journal of Chemical and Pharmaceutical Sciences, 2016. **9**(3): p. 1111-1115.

[20] B.K. Nandi, A.G., M.K. Purkait, *Adsorption characteristics of brilliant green dye on kaolin.* J. Hazard. Mater. , 2009. **161**: p. 387-395.

[21] Yahya A. Faleh , N.D.R., *Removal of Metformin hydrochloride from Aqueous Solutions by using Carboxymethyl cellulose-g-poly(acrylic acid-co-acrylamide) Hydrogel: Adsorption and Thermodynamic Studies.* IOP Conf. Series: Earth and Environmental Science 2021. **790**: p. 012062.

[22] Aljeboree A.M. , A.A.N., Alkaim A.F., *Removal of pharmaceutical amoxicillin drug by using (Cnt) decorated clay/ fe2 o3 micro/nanocomposite as effective adsorbent: Process optimization for ultrasound-assisted adsorption.* International Journal of Pharmaceutical Research, 2019. **11**(4): p. 80-86.

[23] Thitame , P.V. and S.R. Shukla, *Adsorptive removal of reactive dyes from aqueous solution using activated carbon synthesized from waste biomass materials.* Int. J. Environ. Sci. Technol. , 2016. **13**: p. 561-570.

[24] Aljeboree A.M., A.A.N., Alkaim A.F., *Activated carbon (as a waste plant sources)-clay micro/nanocomposite as effective adsorbent: Process optimization for ultrasound-assisted adsorption removal of amoxicillin drug.* Plant Archives, 2019. **19**: p. 915 - 919.

[25] Adeyemo, A.A., I.O. Adeoye, and O.S. Bello, *Adsorption of dyes using different types of clay: a review.* Appl Water Sci,, 2017. **7**: p. 543-568.

[26] Layth S J and Aljeboree, M.A., *Removal of Heavy Metals by Using Chitosan/ Poly (Acryl Amide-Acrylic Acid)Hydrogels: Characterization and Kinetic Study* . NeuroQuantology, 2021. **19**(2): p. 31-37.

[27] Peng, G., Zhang, Menghan,Deng, Shubo,Shan, Danan,He, Qiang,Yu, Gang, *Adsorption and catalytic oxidation of pharmaceuticals by nitrogen-doped reduced graphene oxide/Fe3O4 nanocomposite.* Chemical Engineering Journal, 2017. **341**: p. 361-370.

[28] Ghaedi, M., Sadeghian, B., Pebdani, A.A., Sahraei, R., Daneshfar, A., Duran, C., *Kinetics, thermodynamics and equilibrium evaluation of direct yellow 12 removal by adsorption onto silver nanoparticles loaded activated carbon.* Chem. Eng. J., 2012. **187**: p. 133-141.

[29] Ewaid, S.H., Abed, S.A., 2017. Water quality index for Al-Gharraf river, southern Iraq. Egypt. J. Aquatic Res. 43 (2), 117–122. http://dx.doi.org/10.1016/j. ejar.201703001.

[30] Ewaid, S.H.; Abed, S.A.; Al-Ansari, N. Crop Water Requirements and Irrigation Schedules for Some Major Crops in Southern Iraq. Water 2019, 11, 756.

[31] Ewaid, S.H.; Abed, S.A.; Al-Ansari, N. Water Footprint of Wheat in Iraq. Water 2019, 11, 535.

[32] Ewaid, S.H.; Abed, S.A.; Al-Ansari, N. Assessment of Main Cereal Crop Trade Impacts on Water and Land Security in Iraq. Agronomy 2020, 10, 98.

[33] Ewaid, S.H.; Abed, S.A.; Al-Ansari, N.; Salih, R.M. Development and Evaluation of a Water Quality Index for the Iraqi Rivers. Hydrology 2020, 7, 67.

[34] Salam Hussein Ewaid et al 2020 J. Phys.: Conf. Ser. 1664 012143.

[35] Ahmed Sabah Al-Jasimee et al 2020 J. Phys.: Conf. Ser. 1664 012141.

[36] Salam Hussein Ewaid et al 2021 IOP Conf. Ser.: Earth Environ. Sci. 722 012008

[37] Salam Hussein Ewaid et al 2021 IOP Conf. Ser.: Earth Environ. Sci. 790 012075

[38] Ahmed Alaa Kandoh et al 2021 IOP Conf. Ser.: Earth Environ. Sci. 790 012073.

[39] Ewaid, S.H., Al-Farhani, B.F., Abed, S.A. & Al-Ansari, N. (2020). Modeling of trihalomethane compounds formation in Baghdad water supply network. Scientific Review Engineering and Environmental Sciences, 29 (2), 136-144. doi: 10.22630/PNIKS.2020.29.2.12.

[40] Fabryanty, R., Valencia, C., Edi Soetaredjo, F., Shella Permatasari Santoso,J. N. P., Yi-Hsu Ju, Suryadi Ismadji. A. K., *Removal of crystal violet dye by adsorption using bentonite – alginate composite.* Journal of Environmental Chemical Engineering, 2017. **5**: p. 5677–5687.

Highly adsorbent surface from watermelon peels : as non-conventional low-cost sorbent ; Equilibrium and Recycle study

Aseel M. Aljeboree[1], Ghadeer S. Hamid[2], Ayat A. Katham[3], Mohammed Mustafa Abdulkadhm[4], Holya A. Lafta[5], Aiman Mohammed Baqir Al-Dhalimy[6,7], Ayad F. Alkaim[8] and Salwan Ali Abed [9]

[1,2,3,8]Department of Chemistry, College of Sciences for Girls, University of Babylon, Hilla, Iraq

[4]Al-Manara College For Medical Sciences/Misan/Iraq

[5]Al-Nisour University College/Baghdad/Iraq

[6]College of Nursing, Altoosi University College, Najaf, Iraq.

[7]The Islamic University, Najaf, Iraq.

9 College of Science, University of Al-Qadisiyah, Iraq

Abstract. In this research, it was approved to use available environmentally friendly and low cost absorbent materials such as un-activated watermelon peels (WP) as an ideal surface to remove methylene blue (MB) dye from aqueous solution. That watermelon peels un- activated, the ability to be recycled and used again, by washing it (water, acid and base). Using several techniques to estimate the chemical-physical properties of a surface before and after the adsorption process, such as (FTIR) and (FESEM). By increasing the surface weight (WP), the percentage of removal of methylene blue dye increases, on the contrary, the adsorption efficiency decreases. The isotherm model for adsorption was based on the use of two types of Fruendlich model and Langmuir model and through the value of (R2) it was found that it obeys the Fruendlich model.

Keyword: Adsorption , Watermelon peels (WP), Removal , Isotherm model , Methylen blue dye .

1. Introduction

Dyes can have a very clear effect on organisms in water contaminated with dyes, depending on the concentration of the dye and the time of exposure to the dye. In addition, one of the most important problems of dyes causes skin infections, burns in the eyes, poisoning, diarrhea, nausea and vomiting, and also leads to irregular heartbeat and shortness of breath, and also causes irritation in the digestive system. Methylene blue (MB) was selected in this study to know the ability of non-activated sorbent materials to remove dye from aqueous solutions [1-4]. MB is a basic dye that has very wide uses in industries and dyeing for fabrics and cotton [5, 6] . It is also used for medical purposes as a disinfectant. In case of contact with the eyes, it causes burns, severe irritation, and sometimes permanent injury to the eyes . Bio-sorption is the "cost-effective technique" for removal dye[7] . several lowcost agricultural wastes like peanut husk , rejected tea , rice husk ,fruit peels like orange peel , banana peel, jackfruit peel ,grapefruit peel , custard apple shell , and pomegranate peel are utilizedas sorbents for the removal of dyes. Watermelon is one of the heaviest and largest fruits, available in abundance and cheap [8-12]. Summer watermelons produce about 6-7% of the total fruit production. In watermelon, the red pulp is sweet, edible and used in salads and juices, but the outer skin of the watermelon is considered waste and is not used for any commercial purpose. Watermelon peel is a by-product of watermelon, which consists of several components like proteins, citrulline , pectin, and carotenoids [13, 14]. In this study, one of the agricultural wastes (watermelon peels) was used as an available and inexpensive absorbent material, as it was used without any chemical treatment and is not activated with acid, and these surfaces are characterized by the ability to be recycled again to remove the methylene blue dye from the aqueous solution.

2. Materials and methods

2.1. Preparation of adsorbent

The watermelon was obtained from the local market. After removing the red core, the watermelon peel (WP) was taken and washed several times with hot water to remove any soluble substance, then with distilled water, and cut into small pieces and dried in the sun for a week to remove moisture. Then grind and sift to get watermelon peel powder without any chemical treatment. It is stored in airtight glass containers and placed in an oven at a temperature (60°C) to preserve the substance from moisture.

2.2. Preparation of MB solution

A standard solution 100 mg/L was prepared by dissolving (0.1g) from MB dye in 1000ml distilled water, the calibration curve depends by preparing a series of concentrations of MB from (1-20 mg/L), through the results it was found that it obeys Lambert-Beer's law.

2.3. Batch adsorption experiments

This research is carried out to study the parameter effecting on adsorption methods. These parameter have the equilibrium time, weight of watermelon peel. Adsorption presses were utilized via using 0.1g weight of watermelon peel in to separated elementary flasks of 100ml of MB dye solution of several concentration (1-20 mg/L) and stirred at 300 rpm at 25 oC .The equilibrium time was estimation via stirring 0.1 g from adsorbents by 100ml solution of (10 mg/L) of MB dye . The parameter of watermelon peel dosage is occurred via adding several mass (0.25- 0.2g) of watermelon peel. The removal percentage of MB dye in the solutions before and after adsorption was distinguished via using UV-visible spectrophotometer. The adsorption Percentage removal and adsorption capacity can be calculate in equation 1,2 :

$$E\% = \frac{Co - Ce}{Co} \times 100 \qquad (1)$$

$$Qe = \frac{(Co - Ce)Vml}{W\ gm} \qquad (2)$$

3. Results and Discussions
3.1. FTIR

Figure 1 appear IR spectra of the adsorbent watermelon peel (WP) before and after MB adsorption was gated to study the surface characteristics of adsorbent. The re-corded FTIR spectrum of a MB-loaded watermelon peel (WP) sample explain changes in the FTIR adsorption characteristics of several functional groups which might be involved in the method of MB adsorption. note no new beak appears on the surface after the adsorption process, but there is a difference only in the intensity of adsorption and this is evidence of the occurrence of the physical adsorption process .

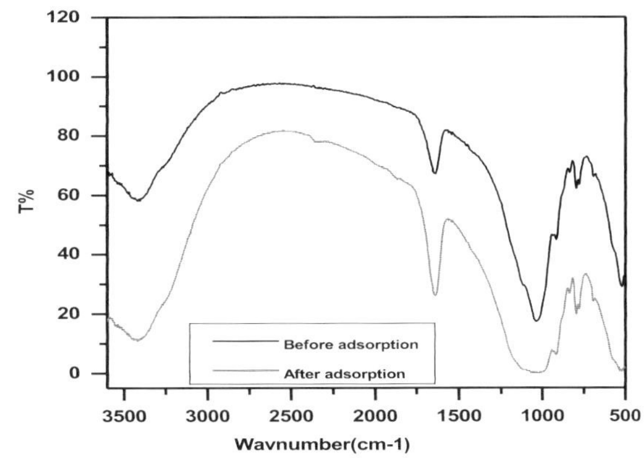

Figure 1. FT-IR spectra of surface watermelon peels (WP) before and after adsorption of dye

3.2. FESEM

The morphology of the surface prepared from watermelon peels (WP) was studied without any chemical treatment and used as an environmentally friendly surface [15]. After the adsorption process and dye loading on the surface, it was observed that agglomerate, of dye occurred on the surface and most of the active sites were filled as shown in figure 2.

Figure 2. FESEM image watermelon peels (WP) before and after Adsorption

3.3. Effect of MB concentration

The percentage removal of MB dye onto watermelon peel(WP) decreased by means of increasing primary concentration of MB dye . This may be possibly as a data of the fact that for a fixed weight of watermelon peel, the total avail-able adsorption sites remain invariable for every the concentrations checked. With increasing concentration the available adsorption sites become fewer [16, 17]. From this study concluded that the less primary MB dye concentration (1 and 10 mg/L) for MB give the high percentage removal of dye as appear in Figure 3.

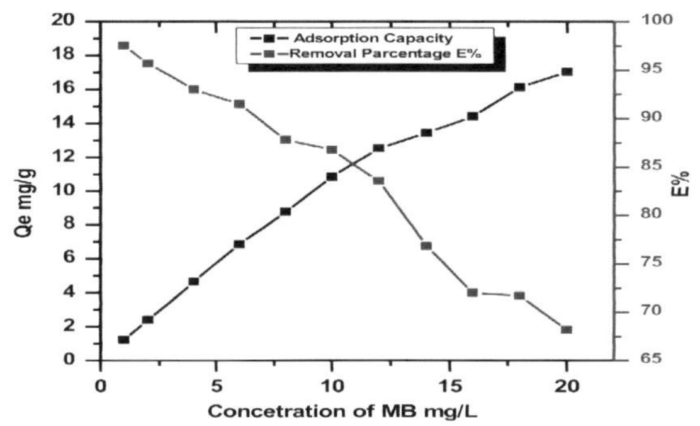

Figure 3. initial concentration of MB dye adsorption utilizing watermelon peel.

3.4. Effect of weight of watermelon peel

The percentage removal of MB increase by means of the rising weight of watermelon peel (WP) until reached at certain dosage after this dosage take place decrease in removal percentage as shown in Figure 4 .Because of the increase in watermelon peel (WP) lead to increasing in avail-able sorption sites. These data might due to the overlap of adsorption sites as a result of adsorbent particle overcrowding[18, 19] .This study appear the better weight of watermelon peel(WP) that give higher percentage removal is 0.2 g [20, 21].

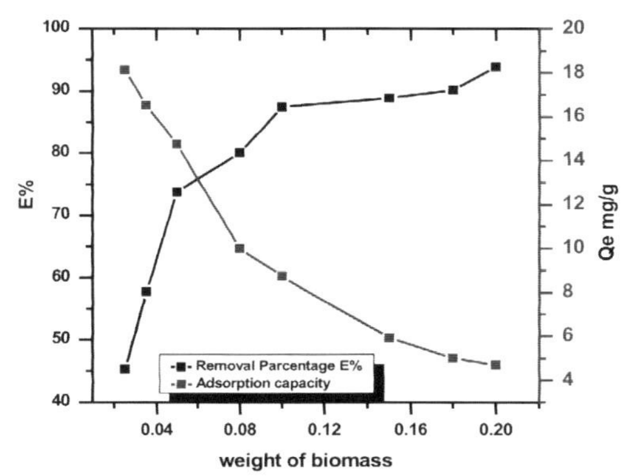

Figure 4. watermelon peel surface mass effect on MB dye optimum conditions: conc. 50mg/L , pH 6.8 and 25°C).

3.5. Study of recycle

Recycle study was perform to know the reusa-bility potential of watermelon peel(WP). recycle of MB from watermelon peel (WP) is studied via utilizing NaOH 0.1 N, HCl 0.1 N , 0.1N acetic acid and distilled water as desorbing agents. 0.1 gm of dye(MB)-loaded watermelon peel(WP) was contact with 100 mL of desorbing agents for 60 minute at 25 oC in shaker water bath at 160 rpm. The desorbed supernatant solution was subject to UV–Visible for the residual concentration of MB. It was find that HCl 0.1 N has highest desorbing efficiency of the MB followed via NaOH 0.1N, 0.1N acetic acid and distilled water as appear in Figure 5.

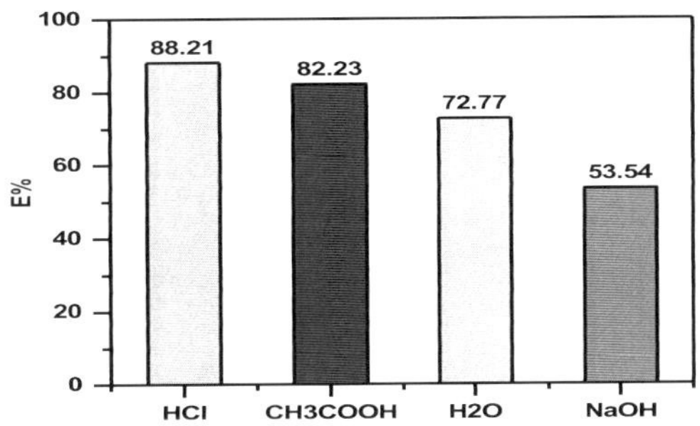

Figure 5. Comparison of desorption capacity of several Solution of the MB dye from surface of watermelon peel**(WP)** .

3.6. Adsorption Model

The equilibrium adsorption result find of the removal of MB dye at several primary dye concentrations via watermelon peel as appear in Figure 6. In order to examine the relationship among concentration of MB at equilibrium (Ce mg/L) and loading capacity (qe mg/g) equilibrium result were analyzed by Freundlich model and Langmuir model. The nonlinear form of Freundlich model calculate in equation [22, 23]:

$$Q_e = k_f\, C_e^{\frac{1}{n}} \qquad (3)$$

The Langmuir model nonlinear equation given in equation :

$$Q_e = \frac{Q_m\, K_L\, C_e}{1 + K_L\, C_e} \qquad (4)$$

The correlation coefficients (R2) and respective constants found of Freundlich model and Langmuir model as appear in Table 5. The low R2 find of langmue model suggests that this model might not be applicable to the present method. The R2 obtained of model freundlich were close to one suggesting best fit to the equilibrium result [24, 25].

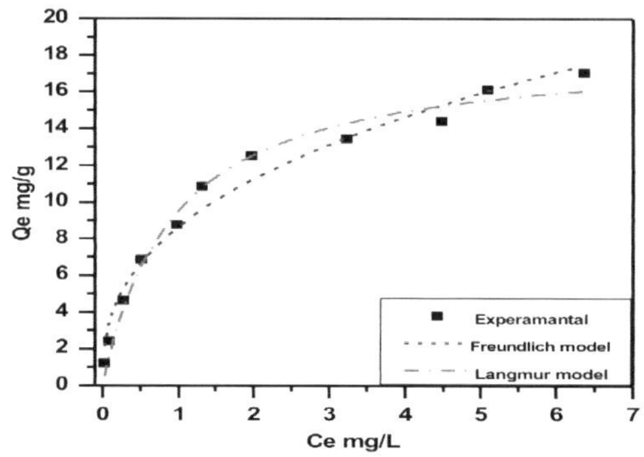

Figure 6. Adsorption models non-linear fit of adsorption **MB** on watermelon peel(WP), conc. = 50 mg. L^{-1}, Temp. = 25°C, and mass 0.1 g/L).

Table 1. Model Freundlich and Langmuir parameter **MB** adsorbed onto watermelon peel(WP)

Isotherm mode	Parameters	watermelon peel
	K_F	8.645 ± 0.334
Freundlich	1/n	0.379 ± 0.027
	R^2	0.9754
	q_m (mg/g)	18.397 ± 0.709
Langmuir	K_L(L/mg)	1/078 ± 0.1428
	R^2	0.91911

4. Conclusion

In this study, one of the types of agricultural waste (watermelon peel(WP)) was used as an effective adsorbent in removing a MB dye from aqueous solution. The surface of watermelon peels is characterize by the possibility of reactivating it and using a second time after washing (distilled water, acid and a base). It was found that the best washing process was the use of acid HCl give the best removal percentage (88.21%) comparative with base or water . also study model Isotherm found obey model freundlich depended of correlation coefficients (R2= 0.9758) .

Reference

[1] Santos, S.C.R. and R.A.R. Boaventura, *Adsorption of cationic and anionic azo dyes on sepiolite clay: equilibrium and kinetic studies in batch mode.* J. Environ. Chem. Eng., 2016. **4**: p. 1473-1483.

[2] Faiq F Karam, F.H.H., SJ Baqir, AF Alkaim, *Optimal conditions for treatment of contaminated waters with anthracene by Fenton processes in close system reactor.* Journal of Chemical and Pharmaceutical Sciences, 2016. **9**(3): p. 1111-1115.

[3] Aljeboree, A.M., Abbas, A.S., *Removal of Pharmaceutical (Paracetamol) by using CNT/ TiO2 Nanoparticles.* Journal of Global Pharma Technology, 2019. **11**(1): p. 199-205.
[4] Thitame , P.V. and S.R. Shukla, *Adsorptive removal of reactive dyes from aqueous solution using activated carbon synthesized from waste biomass materials.* Int. J. Environ. Sci. Technol. , 2016. **13**: p. 561-570.
[5] Liu, M., Li, X.,Du, Y.,Han, R., *Adsorption of methyl blue from solution using walnut shell and reuse in a secondary adsorption for Congo red.* Bioresource Technology Reports, 2019. **5**: p. 238-242.
[6] Aljeboree, A.M., Alkaim ,A. F.,Abdulrazzak , F. H. ,Abbas , A. S.,Alshirifi ,A. N., *Spectrophotometric Determination of Pharmaceutical by Oxidative Coupling of 4-Aminoantipyrine: A Short Review.* Journal of Engineering and Applied Sciences, 2019. **14**(2): p. 5561-5569.
[7] Aljeboree, A.M., Alkaim A F, *Removal of Antibiotic Tetracycline (TCs) from aqueous solutions by using Titanium dioxide (TiO2) nanoparticles as an alternative material.* Journal of Physics: Conf. Series 2019. **1294** p. 052059.
[8] Hoppen, M.I., et al., *Adsorption and desorption of acetylsalicylic acid onto activated carbon of babassu coconut mesocarp.* Journal of Environmental Chemical Engineering, 2019. **7**(1): p. 102862.
[9] Wenyan Jiang, L.Z., Xiaoming Guo, Mei Yang, Yiwen Lu, Yijun Wang, Yousen Zheng & Guangtao Wei, *Adsorption of cationic dye from water using an iron oxide/activated carbon magnetic composites prepared from sugarcane bagasse by microwave method.* Environmental Technology, 2019. **2**: p. DOI: 10.1080/09593330.
[10] Qusay K. Mojar Alshamusi, A.A.M.A., Makarim A. Mahdi, Layth S. Jasim and Aseel M. Aljeboree, *ADSORPTION OF CRYSTAL VIOLATE (CV) DYE IN AQUEOUS SOLUTIONS BY USING P(PVP-CO-AAM)/GO COMPOSITE AS (ECO-HEALTHY ADSORBATE SURFACE) : CHARACTERIZATION AND THERMODYNAMICS STUDIES.* Biochem. Cell. Arch., 2021. **12**: p. 2423-2431.
[11] Rafid Q. Kmal, A.M.M., Aseel M. Aljeboree, Layth S. Jasim, Ayad F. Alkaim, *Synthesis and Identification of Azo Disperse Dye Derived from 4-Aminoantipyrine and Their Applications to Determine Some Drugs.* International Journal of Drug Delivery Technology. 2021;11(3):. 2021. **11**(3): p. 1040-1044.
[12] Aseel M. Aljeboree, A.F.A., Ali loay, hanadi m algburi, *Photocatalytic Degradation of Textile Dye Cristal Violet Wastewater using Zinc Oxide as a Model of Pharmaceutical Threat Reductions.* Journal of Global Pharma Technology, 2019. **11**(02): p. 138-143.
[13] Aseel M Aljeboree , I.J.S., Layth S. Jasim, Ayad F. Alkaim, *Introducing a novel chemical method of treatment for dye removal: Removal of Maxillon Blue and Direct Yellow from aqueous solution.* Caspian Journal of Environmental Sciences, 2021. **19**(4): p. 673-684.
[14] Aseel M. Aljeboree, A.F.A., *Role of plant wastes as an ecofriendly for pollutants (Crystal Violet dye) removal from Aqueous Solutions* Plant Archives 2019 **19**(2): p. 902-905.
[15] Sartape, A.S., et al., *Removal of malachite green dye from aqueous solution with adsorption technique using Limonia acidissima (wood apple) shell as low cost adsorbent.* Arabian Journal of Chemistry, 2017. **10**: p. S3229-S3238.
[16] Xiong W, T., J, Zhaohui Z., Guangming Z, Yaoyu W, Dongbo S, Peipei X, Rui Z, Chen C, *Adsorption of phosphate from aqueous solution using iron-zirconium modified activated carbon nanofiber: Performance and mechanism.* Journal of Colloid and Interface Science, 2017. **493**: p. 17-23.
[17] Aljeboree A.M. , A.A.N., Alkaim A.F., *Removal of pharmaceutical amoxicillin drug by using (Cnt) decorated clay/ fe2 o3 micro/nanocomposite as effective adsorbent: Process optimization for ultrasound-assisted adsorption.* International Journal of Pharmaceutical Research, 2019. **11**(4): p. 80-86.
[18] Layth S J and Aljeboree, M.A., *Removal of Heavy Metals by Using Chitosan/ Poly (Acryl Amide-Acrylic Acid)Hydrogels: Characterization and Kinetic Study* . NeuroQuantology, 2021. **19**(2): p. 31-37.

[19] Hadi, Z.A., Aljeboree, A.M., Alkaim, A.F., *Adsorption of a cationic dye from aqueous solutions by using waste glass materials: Isotherm and thermodynamic studies.* International Journal of Chemical Sciences, 2014. **12**(4): p. 1273-1288.
[20] Peng, G., Zhang, Menghan,Deng, Shubo,Shan, Danan,He, Qiang,Yu, Gang, *Adsorption and catalytic oxidation of pharmaceuticals by nitrogen-doped reduced graphene oxide/Fe3O4 nanocomposite.* Chemical Engineering Journal, 2017. **341**: p. 361-370.
[21] Aljeboree A.M., A.A.N., Alkaim A.F., *Activated carbon (as a waste plant sources)-clay micro/nanocomposite as effective adsorbent: Process optimization for ultrasound-assisted adsorption removal of amoxicillin drug.* Plant Archives, 2019. **19**: p. 915 - 919.
[22] Ali T. Bader, Z.-A.-m., Aseel M. Aljeboree ,Ayad F. Alkaim, *REMOVAL OF METHYL VIOLET (MV) FROM AQUEOUS SOLUTIONS BY ADSORPTION USING ACTIVATED CARBON FROM PINE HUSKS (PLANT WASTE SOURCES).* Plant Archives 2019 **19**(2): p. 898-901.
[23] Fabryanty, R., et al., *Removal of crystal violet dye by adsorption using bentonite – alginate composite.* Journal of Environmental Chemical Engineering, 2017. **5**: p. 5677–5687.
[24] Ewaid, S.H., Abed, S.A., 2017. Water quality index for Al-Gharraf river, southern Iraq. Egypt. J. Aquatic Res. 43 (2), 117–122. http://dx.doi.org/10.1016/j. ejar.201703001.
[25] Ewaid, S.H.; Abed, S.A.; Al-Ansari, N. Crop Water Requirements and Irrigation Schedules for Some Major Crops in Southern Iraq. Water 2019, 11, 756.
[26] Ewaid, S.H.; Abed, S.A.; Al-Ansari, N. Water Footprint of Wheat in Iraq. Water 2019, 11, 535.
[27] Ewaid, S.H.; Abed, S.A.; Al-Ansari, N. Assessment of Main Cereal Crop Trade Impacts on Water and Land Security in Iraq. Agronomy 2020, 10, 98.
[28] Ewaid, S.H.; Abed, S.A.; Al-Ansari, N.; Salih, R.M. Development and Evaluation of a Water Quality Index for the Iraqi Rivers. Hydrology 2020, 7, 67.
[29] Ewaid, S.H., Al-Farhani, B.F., Abed, S.A. & Al-Ansari, N. (2020). Modeling of trihalomethane compounds formation in Baghdad water supply network. Scientific Review Engineering and Environmental Sciences, 29 (2), 136-144. doi: 10.22630/PNIKS.2020.29.2.12.
[30] Salam Hussein Ewaid et al 2021 IOP Conf. Ser.: Earth Environ. Sci. 722 012008
[31] Salam Hussein Ewaid et al 2021 IOP Conf. Ser.: Earth Environ. Sci. 790 012075
[32] Adeyemo, A.A., I.O. Adeoye, and O.S. Bello, *Adsorption of dyes using different types of clay: a review.* Appl Water Sci,, 2017. **7**: p. 543-568.
[33] Kamil, A.M., Mohammed, H.T., Alkaim, A.F., Hussein, F.H., *Adsorption of Congo red on multiwall carbon nanotubes: Effect of operational parameters.* Journal of Chemical and Pharmaceutical Sciences, 2016. **9**(3): p. 1128-1133.

Biosorption by Environmental, Natural and Acid-Activated Orange Peels as Low-Cost Aadsorbent: Optimization of Disperse Blue 183 as a Model

Ahmed Taifi[1], Ola Kamal A. Alkadir[2], Athraa A. Oda[3], Aseel M. Aljeboree[4], Abdul Latif Al Bayaa[5], Ayad F. Alkaim[6] and Salwan Ali Abed[7]

[1]Al-Manara College For Medical Sciences/Iraq
[2]Al-Nisour University College/Baghdad/Iraq
[3,4, 6] Department of Chemistry, College of Sciences for Girls, University of Babylon, Hilla, Iraq
[5]The Islamic University, Najaf, Iraq
[7] College of Science, University of Al-Qadisiyah, Iraq

Abstract. In this research, agricultural waste was used as effective, environmentally friendly and inexpensive adsorbents one type of agricultural waste such as orange peels (NOP) to remove Disperse Blue 183 dye from its aqueous solution in this study using three kind Natural orange peels (NOP) , Activated Carbon Orange Peels (ACOP) and Acid- Activated Carbon Orange Peels (AACOP) . Preparation of acid-activation Orange Peel (AACOP) using 0.1N phosphoric acid ,where the adsorption efficiency of NOP, ACOP and AACOP synthesis were 120.22, 125.67 and 155.12 mg/g and removal percentage E% (80.21% ,92.34% ,98.22%), respectively. The (FTIR) and (FESEM) technique was used to determine the surface characteristics before and after the adsorption process. Two types of adsorption isotherms were studied, the Freindlich and Langmuir model. The equilibrium data showed that it obeys the Freindlich model where the value of (R^2= 0.9777) was compared with the Langmuir model (R^2= 0.9533). Parameters were derived Thermodynamic including (ΔG°), (ΔH°), and (ΔS°) are found to be spontaneous and endothermic in nature.

Keyword: Activated carbon, Orange peels, Agricultural waste, Disperse Blue 183 , Removal , Isotherm model , Thermodynamic, Spontaneous , Endothermic.

1. Introduction

Water is a basic and natural resource for life. drinking water is used mainly in our daily life and is also used in industries and agricultural activities, and this, in turn, leads to water pollution with toxic hazardous materials that cause many diseases to humans [1-4]. One of the most important toxic pollutants in water is dyes because of their toxicity and high decomposition in water, which is a source of danger to humans, plants and animals. Therefore, dyes must be disposed of from drinking water, river water and sewage water using several important techniques, including precipitation and membrane filtration , photo-degradation , photo-catalytic , electrochemical , flocculation and adsorption [5-8]. The adsorption technology is very important in treating drinking water from dyes and organic pollutants compared to other techniques in terms of simplicity, cost, ease of operation, and sensitivity to toxic substances [9]. Also, the adsorption technology depends on the use of materials that are easy to prepare, inexpensive, and available to remove dyes [10-13]. For bio-sorption (ie the use of adsorbents derived from biomass) characterized by its efficiency, high selectivity, ease of preparation, cost and effective removal of dyes in high concentrations of wastewater. One of the most important cheap materials that are used as absorbent materials to remove dyes such as orange peels, lemon peels, watermelon peels, coconut husks and peels Pistachio, corn pulp, date stone and rice husks were used as effective sorbents In this study, bio-sorption (orange peels) was used as an effective, inexpensive and environmentally friendly sorbent material, which was prepared with three types of natural orange peels, activated carbon orange peels and activated-acid orange peels , where the absorbent materials are highly effective in removing dyes from water.

2. Experimental part

2.1. Biosorbent preparation

Orange peels were obtained after squeezing the orange, the pulp was removed, washed with distilled water and dried at room temperature. orange peels activated using phosphoric acid (0.1N) with stirring for one hour, then washed with distilled water several times and dried in oven at temperature 70 °C during 24 hrs . After that, it is calcinations at temperature (300°C) for one hour to be converted to Ash. The activated-acid orange peels (AACOP) are crushed and sieved by sieve to obtain the powder used in the experiment. Note: The adsorbent surface was prepared from natural orange peels (NOP), activated carbon orange peels (ACOP) and Acid -activated orange peels (AACOP) .

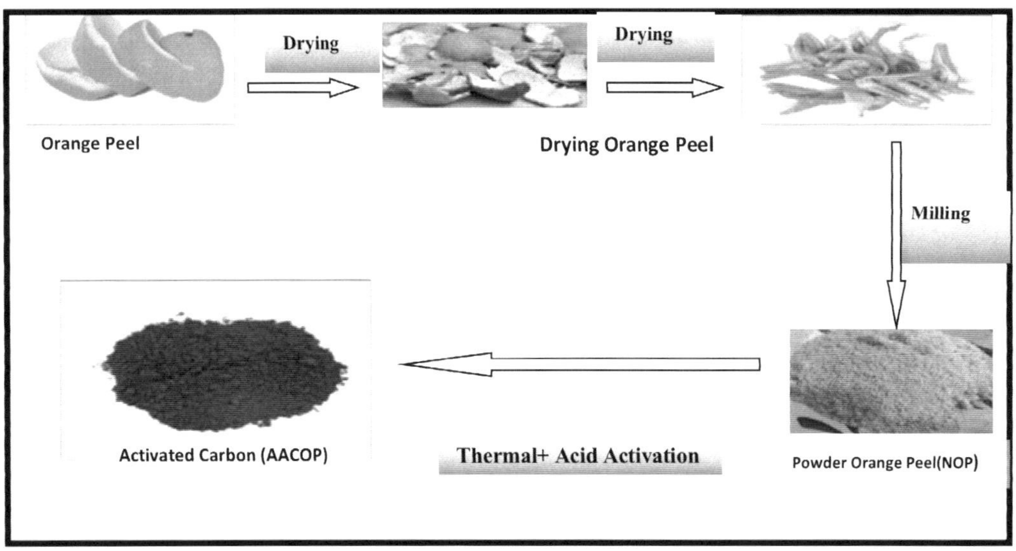

Schema 1. Preparation of Activated carbon from AACOP

2.2. Materials

Disperse Blue 183 called azo dyes ,The supplier from textile factory Al-Hilla- Iraq ,chemical formula of $C_{20}H_{21}BrN_6O_3$,molecular weight of 473.32 g mol^{-1}, and best maximum wavelength of 550 nm. chemical structures appear in Figure 1.

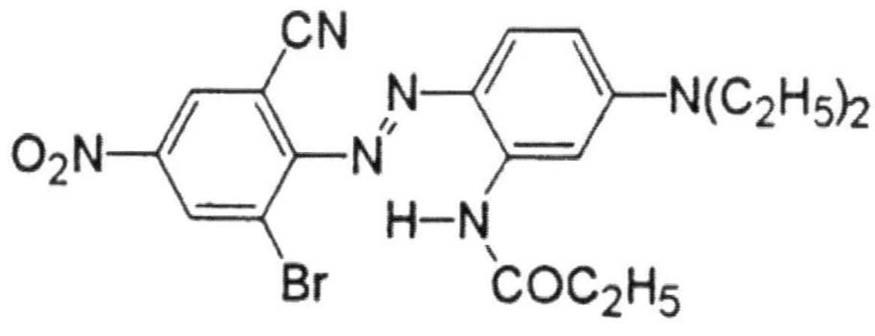

Figure 1. Chemical Structures Disperse Blue 183

2.3. Adsorption experiments

The removal capacity of orange peels based bio-sorbent for Disperse Blue 183 was investigated via usual batch adsorption experiments. Stander solutions of Disperse Blue 183 were prepared via dissolving 1.0 g of Disperse Blue 183 in 1000 ml distilled water. Solution of pH adjusted by 0.1 N HCl or NaOH. Taking 100 mL of Solution of dye put in Erlenmeyer flasks 100ml at pH 6.7,study different parameter like equilibrium time, solution of temperature, weight of adsorbent , and initial dye concentration. Adsorption experiments were conducted at 25 °C and an agitation rate of 130 rpm on a shaking shaker water bath . After the contact time, the adsorbent was separated from aqueous solutions via centrifugation at 6000 rpm for 10 minute. The residual concentration of dye in each aliquot was estimation via utilizing spectrophotometer UV-visible . The percentage removal and adsorption capacity of Disperse Blue 183 estimation by equation 1,2:

$$\textbf{Removal Percentage E\%} = \frac{\textbf{Co} - \textbf{Ce}}{\textbf{Co}} \times \textbf{100} \qquad (1)$$

$$\textbf{Adsorption capacity Qe} = \frac{(\textbf{Co} - \textbf{Ce})\textbf{Vml}}{\textbf{W gm}} \qquad (2)$$

3. Results and Discussions

3.1. Characterization of Adsorbents

3.1.1. FTIR

The adsorption on AACOP was depending of the porosity and functional groups of the chemical reactivity at the AACOP surface. As shown in Figure2. appear the appearance of a strong absorption band at the range 3430–3500 cm-1 indicating that there was over-lapping among the band O-H and the band C-H in 2922 cm-1. It was also found that no new peak appeared, only a difference in the intensity of absorption, evidence of the occurrence of the physical type due to weak bonds[14, 15].

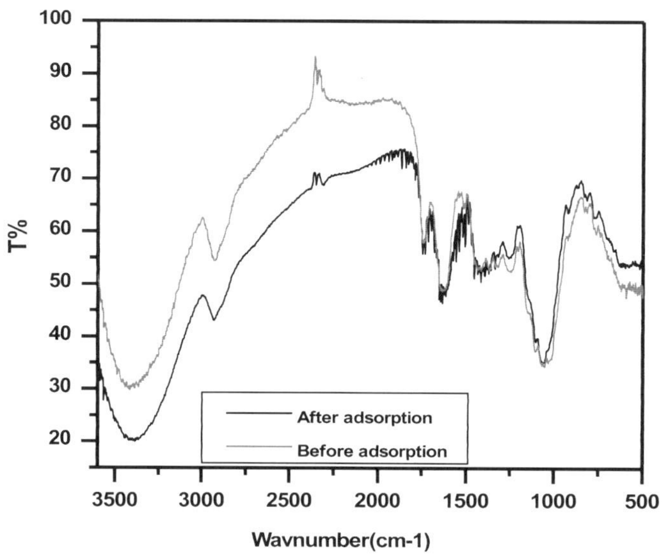

Figure 2. FTIR spectra to determination Disperse Blue 183 onto AACOP before and after adsorption

3.1.2. FESEM

The properties of the surface acid -activated carbon were studied by (FESEM) technique, where the surface before the adsorption process had the appearance of granules on the surface of the activated carbon, after the adsorption process and loading the dye on the surface we notice the occurrence of a random gathering process and swelling of the granules as a result of the occurrence of the adsorption process[16] as shown in figure 3

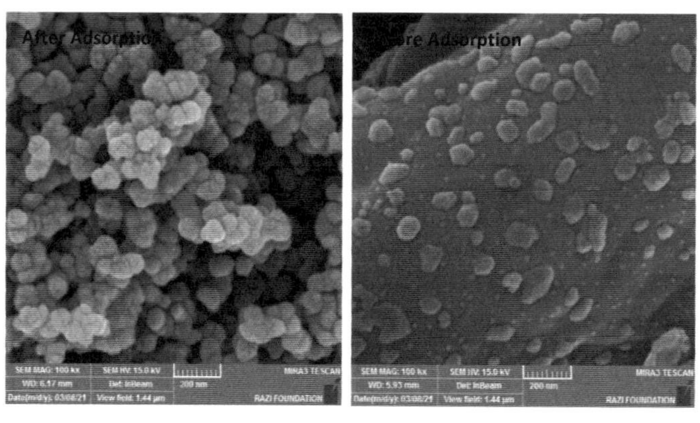

Figure 3. FESEM image of acid -activated carbon(AAOCP) after and before Adsorption

3.2. Effect of Initial Concentration of Disperse Blue 183 dye

Figure 4 shows the effect of Disperse Blue 183 dye concentration on the (adsorbent as AACOP,ACOP, NOP) under the optimum experimental conditions, including surface weight , pH solution and temperature[17]. Where different concentrations of the dye were taken (10-100mg/L) , it was found that by increasing the concentration of the dye from (10-100 mg/L) the adsorption efficiency increases from (150.1 to 19.22mg/g), (128.12 to 18.43mg/g),(120.3 to 15.34 mg/g) for each (adsorbent as AACOP,ACOP, NOP) On the contrary, the removal percentage decreases with increasing dye concentration from (99.77 to 70.12 %) , (94.44 to 52.54 %) ,(80.21 to 43.45 %) for each of the (adsorbent as AACOP,ACOP, NOP), respectively [18-20].

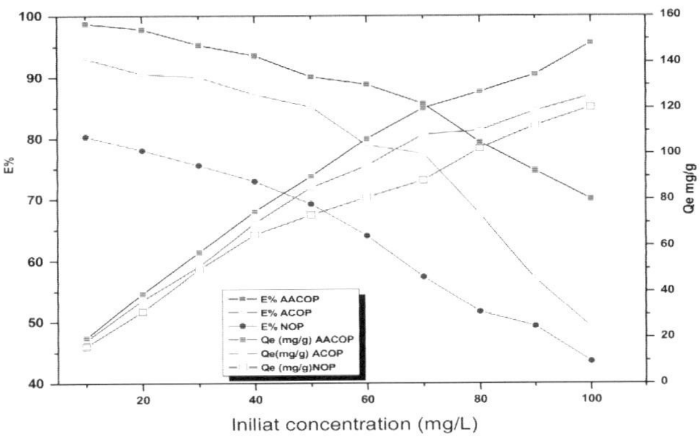

Figure 4. effect of initial concentration of Disperse Blue 183 dye onto adsorbent as AACOP,ACOP,NOP) .

3.3. Effect of adsorbent dose

The adsorption efficiency is greatly influenced via (adsorbent as AACOP,ACOP,NOP) in removal of Disperse Blue 183 from aqueous solution. The quantity of Disperse Blue 183 adsorbed via AACOP was study as a function of weight of adsorbent (0.025 -0.2 g/100 mL) at 50 mg L^{-1} concentration of dye for one hr as appear in Figure 5, the removal percentage E% of Disperse Blue 183 increased from (58.12% to 99.71 %), (44.21% to 91.11%) and (38.11% to 88.32 %) by way of increase of weight adsorbent from 0.025 to 0.2 g/100 mL [21]. while the adsorption capacity Qe of Disperse Blue 183 decrease when weight of adsorbent increase from (88.56 to 40.11 mg/g) ,(72.34 to 40.01 mg/g) and (62.22 to 40.01mg/g) onto(adsorbent as AACOP,ACOP,NOP) at the same order . The best removal of Disperse Blue 183 from aqueous solution was found at dose of 0.08 g / 100 ml, where the removal percentage E% was 93.7 %,that is because the increase in surface weight increases the number of effective sites available for adsorption, through the results shown in Figure 5 it was found that at higher doses of adsorbents an increase Slow in the percentage removal. This is because almost all the active sites are filled and equilibrium is reached [22, 23].

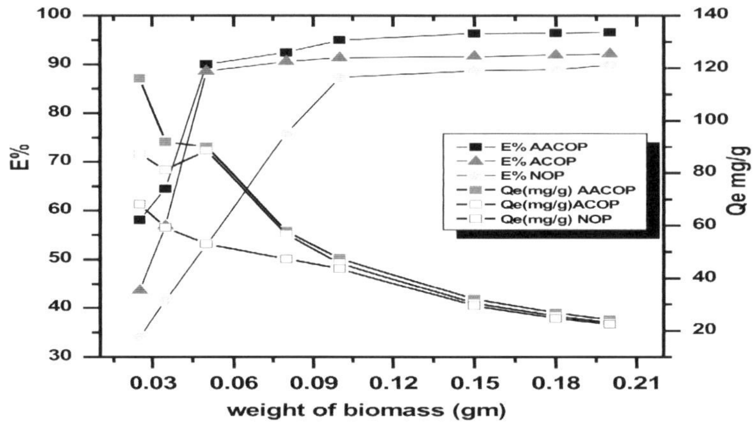

Figure 5. Effect of mass of NOP, ACOP, AACOP on the adsorption capacity and percentage removal of Disperse Blue 183 dye (Exp. Condition: equilibrium time 1 hr., Temp. = 25°C)

3.4. Effect of pH

In this study, the effect of solution pH (2–10) on the sorption of Disperse Blue 183 via AACOP,ACOP and NOP at 50mg/L concentration of Disperse Blue 183 and adsorbent doses (0.08 g/100 mL). As appear in Figure 6, the adsorption capacity (Qe) increased from (60.12 to 95.22)mg/g , (58.12 to 85.22) mg/g and (51.11 to 84.78) mg /g onto AACOP . ACOP and NOP at the same order by way of increase solution pH from 2 to 10 [17]. The data give details best adsorption capacity of Disperse Blue 183 at basic pH could be due to increase in the negatively charged moieties on the surface of the adsorbent, that an increase in electrostatic interaction between the dye molecule and surface binding legends is responsible for an initial fast rate of dye adsorption [15, 24].

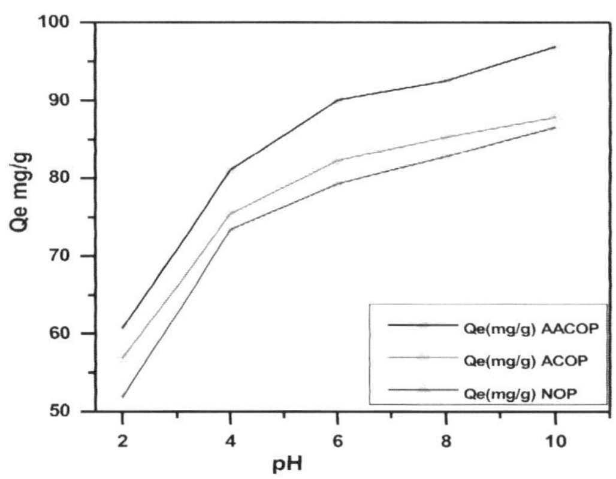

Figure 6. Effect of pH solution on the removal percentage and adsorption efficiency on to NOP,ACOP and AACOP (Exp. Condition: equlibruim time 1 hr., Temp. = 25°C, quantity of biomass 0.08 g)

3.5. Adsorption comparison (agricultural waste)

Comparing the adsorption of three surfaces (NOP), (ACOP) and (AACOP) to remove Disperse Blue 183 dye from its aqueous solution. At the optimum conditions for an experiment, the temperature (25°C) surface weight (0.08g/100ml), pH 6.7 and equilibrium time one hr .Through the results shown in the figure 7 , the surface (AACOP) gave a high percentage of dye removal compared to the surface (ACOP) (NOP) [25]. because activated surfaces, whether with acid or base, increase the active sites on the surface compared to non-activated surfaces [26].

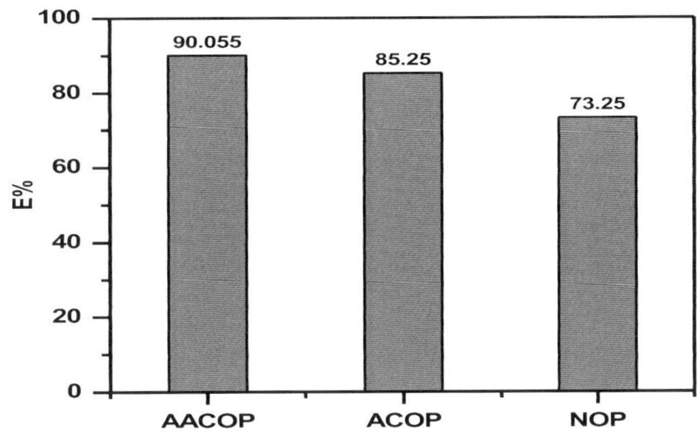

Figure 7. Comparing the adsorption of three surfaces (NOP), (ACOP) and (AACOP) to remove Disperse Blue 183 dye .

3.6. Adsorption Model

To explain the sorption capacity of the bio-sorbent, the equilibrium experimental results were studied to confirm compliance with the isothermal model equations of Freundlich and Langmuir .

Isotherm Freundlich model is an empirical equation based on adsorption onto surface heterogeneous. The equation is usually calculate in Equation (3)

$$Q_e = k_f \, C_e^{\frac{1}{n}} \qquad (3)$$

Isotherm Langmuir model is useful for the adsorption of a solute from solution as mono-layer adsorption on a surface having number finite of identical sites[27]. The isotherm is setup on various fundamental assumptions: (i) the sorption happen at set sites homogeneous in the adsorbent; (ii) once a dye molucles of site occupies ; (iii) the adsorbent (at equilibrium) has limited ability for the adsorbate ; (iv) whole of sites are congruous and energetically identical . The equation non-linear of isotherm Langmuir model in equation (4)

$$Q_e = \frac{Q_m \, K_L \, C_e}{1 + K_L \, C_e} \qquad (4)$$

The values of K_F and (R2) correlation coefficient find from isotherm Freundlich model as shown in (Table 1). The adsorption of Disperse Blue 183 best fitted to the isotherm Freundlich model with the higher correlation coefficient of >0.977 (Table 1). base on the comparison of the isotherm Langmuir and isotherm Freundlich and the data obtained [23, 28, 29], the model Freundlich has exhibit a best fit to the adsorption result than the model Langmuir as appear in Figure 8 .

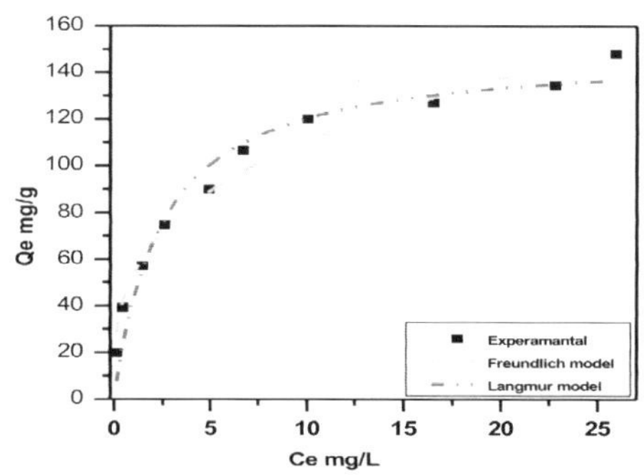

Figure 8. Several adsorption models nonlinear fit of adsorption Disperse Blue 183 on AAOP, conc. = 50 mg. L^{-1}, Temp. = 25°C, and mass of AACOP 0.08 g/L).

Table 1. Isotherms Freundlich and Langmuir model parameter Disperse Blue 183 adsorbed onto AACOP

Isotherm mode	Parameters	AACOP
	K_F	54.048 ± 2.959
Freundlich	1/n	0.3103 ± 0.0207
	R^2	0.9777
	q_m (mg/g)	149.344± 7.559
Langmuir	K_L(L/mg)	0.4122 ± 0.0845
	R^2	0.9533

3.7. Thermodynamic Parameters

Thermodynamic factors like free energy change (ΔG) (KJ/mol), enthalpy (ΔH) (KJ/mol) and entropy (ΔS) (J/K/mol), were estimated utilizing in equations:

$$\Delta G^\circ = -RT \ln K \qquad (5)$$

According to thermo-dynamic law, the Van't Hoff equation calculate via equation (6)

$$\Delta G^\circ = \Delta H^\circ - T\Delta^\circ S \qquad (6)$$

Eq. (7) above can be further transformed as follows [3]:

$$\ln k = -\Delta G^\circ/RT = -\Delta H^\circ/RT + \Delta S^\circ/R \qquad (7)$$

The values of ΔH positive indicate endothermic nature of adsorption. The values of ΔG negative indicated that the adsorption of Disperse Blue 183 on to AACOP is favor-able and a spontaneous[30] . The values of ΔS positive indicate the increased disorder and randomness at the solid solution interface of Disperse Blue 183 by the adsorbent as appear in Table 2. The raise in the capacity of the AAOP to remove Disperse Blue 183 at higher temperatures was due to activation of the adsorbent surface there by enlargement of the pore size as shown in Figure 9 [31, 32].

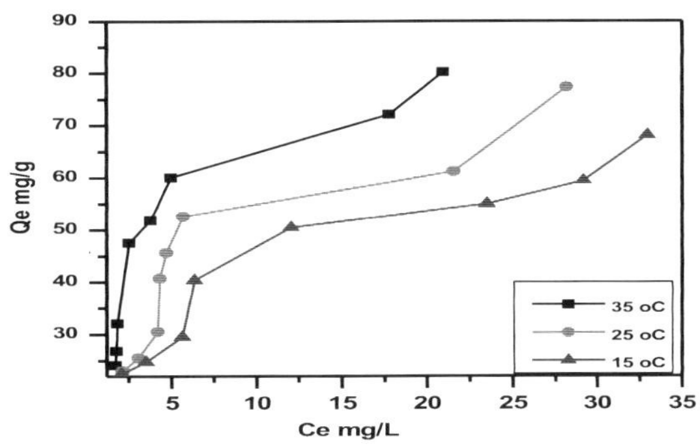

Figure 9. Adsorption isotherms of Disperse Blue 183 onto AACOP at several temperatures.

Table 2. Thermodynamic parameter for Disperse Blue 183 adsorption on to AACOP

ΔH ($KJ.mol^{-1}$)	ΔG ($kJ.mol^{-1}$)	Equilibrium constant	ΔS ($J.mol^{-1}. K^{-1}$)
4.12	-4.946	3.047	29.44

3.8. Conclusion

1-The Orange Peels adsorbent appear significant removal of Disperse Blue 183 from aqueous solutions, and best removal found at 0.08g/100 ml .

2-The surface (AACOP) gave a high percentage of dye removal compared to the surface (AOP) (NOP).

3-Isotherm Freundlich model fitted very well by the adsorption equilibrium result and therefore, multi-layer adsorption of Disperse Blue 183 molecules taken place over bio-sorbent.

4-The thermodynamic parameter appear that the adsorption method of Disperse Blue 183 was feasible, endothermic and spontaneous in nature. therefore, best removal could be find at higher temperature.

Reference

[1] Gouamid M., O.M.R., and Bensaci M. B., *Adsorption Equilibrium, Kinetics and Thermodynamics of Methylene Blue from Aqueous Solutions using Date Palm Leaves.* Energy Procedia, 2013. **36**: p. 898-907.

[2] Nasseh, N., Barikbin, Behnam,Taghavi, Lobat,Nasseri, Mohammad Ali, *Adsorption of metronidazole antibiotic using a new magnetic nanocomposite from simulated wastewater (isotherm, kinetic and thermodynamic studies).* Composites Part B: Engineering, 2019. **159**: p. 146-156.

[3] Hadi, Z.A., Aljeboree, A.M., Alkaim, A.F., *Adsorption of a cationic dye from aqueous solutions by using waste glass materials: Isotherm and thermodynamic studies.* International Journal of Chemical Sciences, 2014. **12**(4): p. 1273-1288.

[4] Xiong W, T., J, Zhaohui Z., Guangming Z, Yaoyu W, Dongbo S, Peipei X, Rui Z, Chen C, *Adsorption of phosphate from aqueous solution using iron-zirconium modified activated carbon nanofiber: Performance and mechanism.* Journal of Colloid and Interface Science, 2017. **493**: p. 17-23.

[5] Martins, A.C., et al., *Removal of tetracycline by NaOH-activated carbon produced from macadamia nut shells: Kinetic and equilibrium studies.* Chemical Engineering Journal. **260**: p. 291-299.

[6] Li, J., K. Zhang, and H. Zhang, *Adsorption of antibiotics on microplastics.* Environmental Pollution. **237**: p. 460-467.

[7] Aseel M. Aljeboree, A.F.A., Ali loay, hanadi m algburi, *Photocatalytic Degradation of Textile Dye Cristal Violet Wastewater using Zinc Oxide as a Model of Pharmaceutical Threat Reductions.* Journal of Global Pharma Technology, 2019. **11**(02): p. 138-143.

[8] Aljeboree, A.M., *Adsorption and removal of pharmaceutical riboflavin (RF) by rice husks activated carbon.* International Journal of Pharmaceutical Research 2019. **11**(2): p. 255-261.

[9] Aljeboree, A.M., Abbas, A. S., *Removal of Pharmaceutical (Paracetamol) by using CNT/ TiO2 Nanoparticles.* Journal of Global Pharma Technology, 2019. **11**(01): p. 199-205.

[10] Kerkez-Kuyumcu, İ.z., İ.a.S. Bayazit, and M.A. Salam, *Antibiotic amoxicillin removal from aqueous solution using magnetically modified graphene nanoplatelets.* Journal of Industrial and Engineering Chemistry. **36**: p. 198-205.

[11] Liu, M., Li, X.,Du, Y.,Han, R., *Adsorption of methyl blue from solution using walnut shell and reuse in a secondary adsorption for Congo red.* Bioresource Technology Reports, 2019. **5**: p. 238-242.

[12] Aljeboree, A.M., Alkaim A F, *Removal of Antibiotic Tetracycline (TCs) from aqueous solutions by using Titanium dioxide (TiO2) nanoparticles as an alternative material.* Journal of Physics: Conf. Series 2019. **1294** p. 052059.

[13] Aseel M. Aljeboree, N.D.R., Layth S. Jasim,Ayad F. Alkaim, *Removal of Pharmaceutical by Adsorption Process; Study Effect of Various Parameters and Determination of Pharmaceutical by Oxidative Coupling Reaction.* International Journal of Drug Delivery Technology, 2021. **11**(3): p. 912-916.

[14] P. Xu, G.M.Z., D.L. Huang, C. Lai, M.H. Zhao, Z. Wei, N.J. Li, C. Huang, G.X. Xie., *Adsorption of Pb(II) by iron oxide nanoparticles immobilized Phanerochaete chrysosporium: equilibrium, kinetic, thermodynamic and mechanisms analysis.* Chem. Eng. J., 2012. **203**: p. 423-431.

[15] Aljeboree A.M., A.A.N., Alkaim A.F, *Removal of pharmaceutical amoxicillin drug by using (Cnt) decorated clay/ fe2 o3 micro/nanocomposite as effective adsorbent: Process optimization for ultrasound-assisted adsorption.* International Journal of Pharmaceutical Research, 2019. **11**(4): p. 80-86.

[16] Kausar, A., et al., *Dyes adsorption using clay and modified clay: A review.* Journal of Molecular Liquids. **256**: p. 395-407.

[17] Layth S J and Aljeboree, M.A., *Removal of Heavy Metals by Using Chitosan/ Poly (Acryl Amide-Acrylic Acid)Hydrogels: Characterization and Kinetic Study* . NeuroQuantology, 2021. **19**(2): p. 31-37.

[18] Aseel M Aljeboree, A.N.A., Ayad F. Alkaim, *ACTIVATED CARBON (AS A WASTE PLANT SOURCES)–CLAY MICRO/NANOCOMPOSITE AS EFFECTIVE ADSORBENT: PROCESS OPTIMIZATION FOR ULTRASOUND-ASSISTED ADSORPTION REMOVAL OF AMOXICILLIN DRUG.* Plant Archives 2019 **19**(2): p. 915-919.

[19] Yahya A. Faleh , N.D.R., *Removal of Metformin hydrochloride from Aqueous Solutions by using Carboxymethyl cellulose-g-poly(acrylic acid-co-acrylamide) Hydrogel: Adsorption and Thermodynamic Studies.* IOP Conf. Series: Earth and Environmental Science 2021. **790**: p. 012062.

[20] Jafari, K., M. Heidari, and O. Rahmanian, *Wastewater treatment for Amoxicillin removal using magnetic adsorbent synthesized by ultrasound process.* Ultrasonics Sonochemistry. **45**: p. 248-256.

[21] Aljeboree A.M. , A.A.N., Alkaim A.F., *Removal of pharmaceutical amoxicillin drug by using (Cnt) decorated clay/ fe2 o3 micro/nanocomposite as effective adsorbent: Process optimization for ultrasound-assisted adsorption.* International Journal of Pharmaceutical Research, 2019. **11**(4): p. 80-86.

[22] Dousova, B., et al., *Removal of arsenate and antimonate by acid-treated Fe-rich clays.* p. 440-448.

[23] Aseel M. Aljeboree, A.F.A., *Role of plant wastes as an ecofriendly for pollutants (Crystal Violet dye) removal from Aqueous Solutions* Plant Archives 2019 **19**(2): p. 902-905.

[24] Chen, Y., et al., *Occurrence, removal and environmental risk assessment of pharmaceuticals and personal care products in rural wastewater treatment wetlands.* Science of The Total Environment. **566-567**: p. 1660-1669.

[25] Powers, H.J., *Riboflavin (vitamin B-2) and health.* Am. J. Clin. Nutr., 2003. **77**(6): p. 1352-1360.

[26] Gamoudi, S. and E. Srasra, *Adsorption of organic dyes by HDPy+-modified clay: effect of molecular structure on the adsorption.* Journal of Molecular Structure, 2019.

[27] del Mar Orta, M., Martn, Julia,Medina-Carrasco, Santiago,Santos, Juan Luis,Aparicio, Irene,Alonso, Esteban, *Adsorption of propranolol onto montmorillonite: Kinetic, isotherm and pH studies.* Applied Clay Science, 2019. **173**: p. 107-114.

[28] Chayid, M.A., Ahmed, Muthanna J., *Amoxicillin adsorption on microwave prepared activated carbon from Arundo donax Linn: Isotherms, kinetics, and thermodynamics studies.* Journal of Environmental Chemical Engineering. **3**(3): p. 1592-1601.

[29] Mohammed A. Jawad, A.J.K., Nadher D. Radia, *Role of Sodium Alginate-g-poly (Acrylic acid-fumaric acid) Hydrogel for Removal of Pharmaceutical Paracetamol from Aqueous Solutions by Adsorption* International Journal of Pharmaceutical Quality Assurance, 2021. **12**(3): p. 202-205.

[30] Cherdchoo, W., S. Nithettham, and J. Charoenpanich, *Removal of Cr(VI) from synthetic wastewater by adsorption onto coffee ground and mixed waste tea.* Chemosphere, 2019. **221**: p. 758-767.
[31] Ewaid, S.H., Abed, S.A., 2017. Water quality index for Al-Gharraf river, southern Iraq. Egypt. J. Aquatic Res. 43 (2), 117–122. http://dx.doi.org/10.1016/j. ejar.201703001.
[32] Ewaid, S.H.; Abed, S.A.; Al-Ansari, N. Crop Water Requirements and Irrigation Schedules for Some Major Crops in Southern Iraq. Water 2019, 11, 756.
[33] Ewaid, S.H.; Abed, S.A.; Al-Ansari, N. Water Footprint of Wheat in Iraq. Water 2019, 11, 535.
[34] Ewaid, S.H.; Abed, S.A.; Al-Ansari, N. Assessment of Main Cereal Crop Trade Impacts on Water and Land Security in Iraq. Agronomy 2020, 10, 98.
[35] Ewaid, S.H.; Abed, S.A.; Al-Ansari, N.; Salih, R.M. Development and Evaluation of a Water Quality Index for the Iraqi Rivers. Hydrology 2020, 7, 67.
[36] Ewaid, S.H., Al-Farhani, B.F., Abed, S.A. & Al-Ansari, N. (2020). Modeling of trihalomethane compounds formation in Baghdad water supply network. Scientific Review Engineering and Environmental Sciences, 29 (2), 136-144. doi: 10.22630/PNIKS.2020.29.2.12.
[37] Salam Hussein Ewaid et al 2020 J. Phys.: Conf. Ser. 1664 012143.
[38] Ahmed Sabah Al-Jasimee et al 2020 J. Phys.: Conf. Ser. 1664 012141.
[39] Mudhafar A. Salim and Salwan Ali Abed 2021 IOP Conf. Ser.: Earth Environ. Sci. 735 012021
[40] Salam Hussein Ewaid et al 2021 IOP Conf. Ser.: Earth Environ. Sci. 722 012008
[41] Salam Hussein Ewaid et al 2021 IOP Conf. Ser.: Earth Environ. Sci. 790 012075
[42] Ahmed Alaa Kandoh and Salwan Ali Abed 2021 IOP Conf. Ser.: Earth Environ. Sci. 790 012034
[43] Ahmed Alaa Kandoh et al 2021 IOP Conf. Ser.: Earth Environ. Sci. 790 012073.
[44] Carazo, E., Borrego-Snchez, Ana, Snchez-Espejo, Rita, Garca-Villn, Ftma, Cerezo, Pilar, Aguzzi, Carola, Viseras, Csar, *Kinetic and thermodynamic assessment on isoniazid/montmorillonite adsorption.* Applied Clay Science. **165**: p. 82-90.
[45] Yiming Li , S.W., Yong Zhang, Ruiming Han ,WeiWei, *Enhanced tetracycline adsorption onto hydroxyapatite by Fe(III) incorporation.* Journal of Molecular Liquids 2017. **247** p. 171-181.

Environmental Removal of Reactive Blue 49 Dye From Aqueous Solution by (Lemon peels as activated carbon): a Model of Low Cost agricultural waste

Ahmed Taifi[1] , Ola Kamal A. Alkadir[2] , Aseel M. Aljeboree[3], Abdul Latif Al Bayaa[4], Ayad F. Alkaim[5] and Salwan Ali Abed [6]

[1]Al-Manara College For Medical Sciences/Iraq
[2]Al-Nisour University College/Baghdad/Iraq
[3,5]Department of Chemistry, College of Sciences for Girls, University of Babylon, Hilla, Iraq
[4]The Islamic University, Najaf, Iraq
[6] College of Science, University of Al-Qadisiyah, Iraq

Abstract. The removal of reactive blue 49 dye on a bio-sorbent material for Natural lemon peels(NLP) and Acid-activated carbon for lemon peels(AACLP) from aqueous solutions was studied. Preparation of activated carbon for lemon peels(AACLP) by Calcinations at temperature (300°C) for one hour and chemical activation by phosphoric acid . Study different parameter such as effect of contact time , adsorbent dose , pH solution ,solution of temperature and Adsorption comparison (agricultural waste). The prepared AACLP was characterized utilizing FT-IR, FESEM technique . Equilibrium result were Isotherm utilizing Isotherm Langmuir and isotherm Freundlich. The result obey Isotherm Langmuir good correlations (R^2=0.854). Thermodynamic parameter studies have shown that the reaction is spontaneous, endothermic and convenient in nature.

Keyword: Adsorption, Bio-sorption, Lemon peels , Activated carbon , Isotherm, Reactive blue 49 dye , Thermodynamic .

1. Introduction

Textile industries use water in very large quantities in their operations, and therefore the amount of pollutants that come out of these factories is very large and accumulate in the drinking water of rivers and lakes [1-3]. The polluted water resulting from the textile industries contains many toxic and dangerous pollutants such as dyes, drugs, organic and inorganic compounds, heavy metals and others; Dyes are considered one of the most dangerous toxic pollutants due to their presence in very high concentrations in water, their high solubility in water, their ability to interact with other compounds, and their being more toxic and difficult to remove using traditional methods[4-8]. Agricultural wastes and other waste materials, like coconut husk ,cocoa pod husk, peanut hulls, peanut husk ,cashew nut shell, calcined egg shell, ,periwinkle shell, garlic peel, Apricot stone, sugarcane bagasse, corncobs ,jujuba seed, sesame stalk, sugar cane bagasse, bagasse pith, olive waste cake, cotton sticks, sunflower apple wastes, Prosopis africana seed hulls ,Bengal gram seed peel, Cedrus deodara sawdust ,oil palm shell ,oil palm fruit waste, jatropha seed coat, natural fibers of kenaf ,castor bean cake, eucalyptus leaves powder ,maize stem tissue and lignite have been used as cheap alternatives for activated carbon preparation as well as low cost adsorbents in the uptake of dyes[9-11] . In this study targeted at investigating the capacity of lemon peels (NLP) and activated carbon lemon peels (AACLP) prepared via physico-chemical activation utilizing phosphoric acid as activating agent for the removal of reactive blue 49 from aqueous solution and too estimation utmost the best isotherm model , thermodynamic .

2. Experimental part

2.1. Preparation of Acid-Activated carbon from Lemon peels

Lemon peels were obtained after squeezing the lemons, the pulp was removed, washed with distilled water and dried at room temperature. After that, it is calcinations at temperature (300°C) for one hour to be converted to Ash. Lemon peels activated using phosphoric acid (0.1N) with stirring for one hour, then washed with distilled water several times and dried in oven at temperature 102 oC during 24 hrs . The activated-acid lemon peels (AACLP) are crushed and sieved by sieve to obtain the powder used in the experiment. Note: The adsorbent surface was prepared from natural lemon peels (NLP), and activated -acid lemon peels (AACLP) .

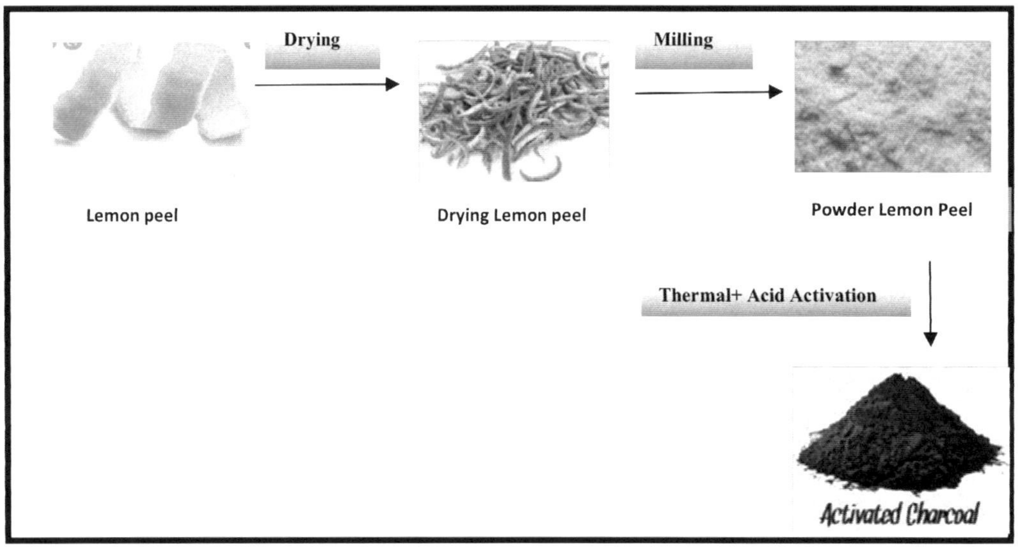

Schema 1. Preparation of Activated carbon from Lemon peels

2.2. Adsorbate

A reactive blue 49 dye was obtained from Al-Hilla textile factory. A solution of dye blue color very solubility in water, molecular weight (562) g / mole and a wavelength (585 nm). The structure of the dye is as shown in the Figure 1. The standard solution is prepared by dissolving 1 gm of the reactive blue 49 dye in 1000ml of distilled water to obtain a concentration 1000 mg/L. The calibration curve determined by preparing a series of concentrations from the original dye, where a graph between absorbance and concentration was found to obey Lambert-Beer law .

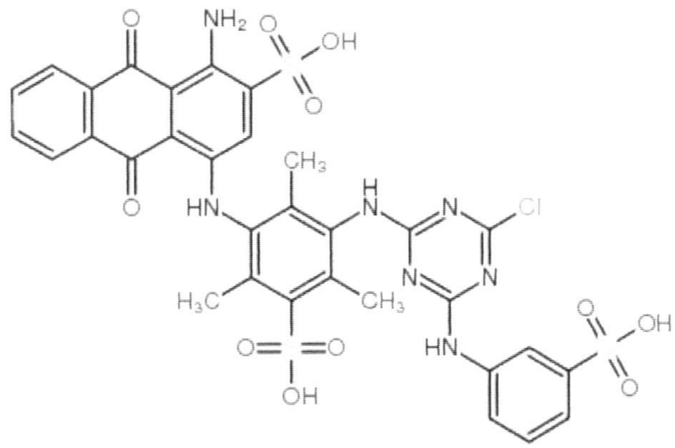

Figure 1. The chemical structure of reactive blue 49

2.3. Batch adsorption experiments

This research is carried out to study the parameter effecting onto adsorption method . These parameter like effect equilibrium time, pH solution , weight of adsorbent, solution temperature. Experimental method was carry out with two adsorbent (NLP) and (AACLP) and compare their adsorption removal capacity . Batch adsorption methods were used by using knowing weight of adsorbent into separated conical flasks with (100ml) of reactive blue 49 solution of several dye concentration (10-100 mg/L) and stirred at 300 rpm at three temperature (15°,25,35)) and solution pH (2-10).The contact time was estimation via stirring 0.05 g from adsorbents(AACLP) with 100 ml solution of 50mg/L of dye. Solution pH was adjusted by 0.1M HCl and NaOH. The parameter of weight adsorbent is occurred via addition several mass (0.02- 0.12). The removal percentage removal of dye solutions before and after adsorption was distinguished via using UV-V spectrophotometer. The adsorption percentage removal and adsorption capacity determine by equation (1), (2) at the same order .

$$\boldsymbol{Removal\ Percentage\ E\%} = \frac{\boldsymbol{Co - Ce}}{\boldsymbol{Co}} \times \mathbf{100} \qquad (1)$$

$$\boldsymbol{Adsorption\ capacity\ Qe} = \frac{\boldsymbol{(Co - Ce)Vml}}{\boldsymbol{W\ gm}} \qquad (2)$$

3. Results and Discussions

3.1. FTIR

An FTIR spectra of AACLP before and after dye-loading on the surface appear that some peaks were shifted (Figure 2). In the spectrum of AACLP before adsorption there is a strong peak at 3439 cm−1 representing the −OH stretching and the peak at 2922 cm−1 indicates the presence of -CH2 stretching . Thus the spectrum of AACLP after dye- loading we note that no new beak appears on the surface after the adsorption process, but there is a difference only in the intensity of adsorption and this is evidence of the occurrence of the physical adsorption process [12, 13].

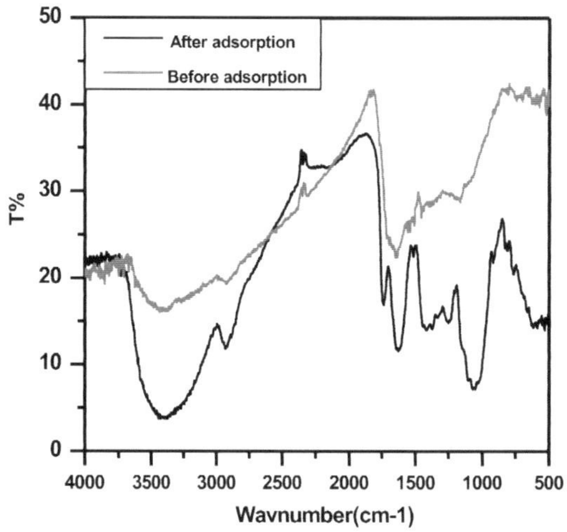

Figure 2. FT-IR image of AACLP before and after adsorption of reactive blue 49.

3.2. FESEM

Figure 3 shows a technique SEM for the surface of activated carbon before and after the adsorption process, where we note the image of the surface before adsorption contains active sites that are not loaded with dye, but after the adsorption process, we notice the appearance of agglomerations on the surface, a clear sign of loading the dye on the surface

Figure 3. SEM image of Activated carbon (AACLP) before and after adsorption

3.3. Effect of adsorbent dose

Removal percentage E% of reactive blue 49 was studied by means of varying the adsorbent dosage. The percentage of adsorption increased with increase in dosage of adsorbent. However, amount of dye adsorbed per unit mass of adsorbent decreased with increased in adsorbent dose from 0.02 to 0.12 g as appear in Figure 4. As amount of adsorbent increases, number of active sides available for adsorption also increases therefore % removal also increases. Since all active sides may not be available during adsorption due to overlapping between the active sides themselves and thus amount adsorbed mg/g of adsorbent decreases [14, 15]. Therefore, the adsorption of dye increased with the sorbent dosage and reached an equilibrium value after certain sorbent dosage (0.1 - 0.12 gm) for the adsorbent.

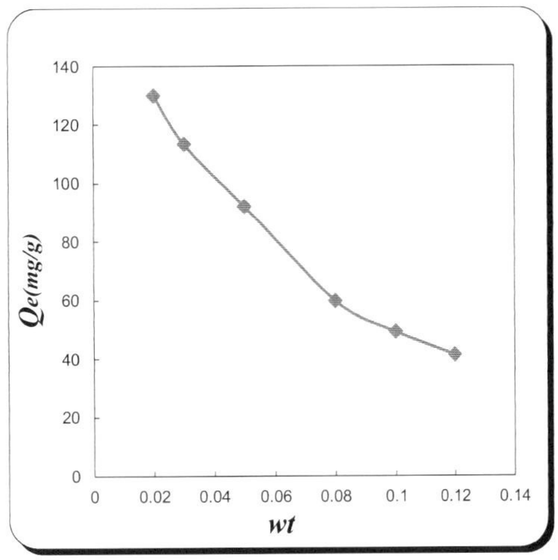

Figure 4. Effect of mass of AACLP on the adsorption capacity of reactive blue 49 dye (Exp. Condition: equilibrium time 1 hr., Temp. = 20°C)

3.4. Adsorption comparison (agricultural waste)

Comparing the adsorption of two surfaces (NLP) and (AACLP) to remove reactive blue 49 from its aqueous solution. At the optimum conditions for an experiment, the temperature (20oC) surface weight (0.05g/100ml), pH 6.5 and equilibrium time one hr .Through the results shown in the figure 5, the surface (AACLP) gave a high percentage of dye removal compared to the surface (NLP) [16, 17] .because activated surfaces, whether with acid or base, increase the active sites on the surface compared to non-activated surfaces.

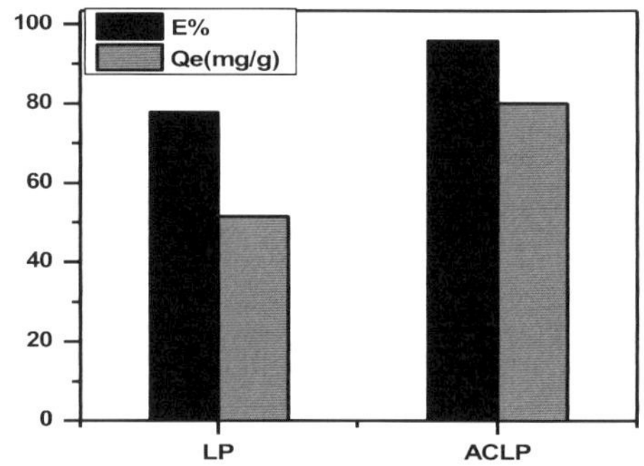

Figure 5. Comparing the adsorption of two surfaces (LP) and (AALP) to remove reactive blue 49 dye .

3.5. pH solution

Effect of the pH solution onto bio-sorbent surface and ionization/dissociation of the molecules dye . The effect of pH on bio-sorption of reactive blue 49 was studied in the series about 2–10 as appear in Figure 6. The quantity of reactive blue 49 bio-sorbed decreased by means of the increase in solution pH from 2 - 10 keeping the bio-sorbent amount and primary concentration of dye (0.05g/100 mL and 50 mg/L).

The lowest dye bio-sorption happen at pH 10 and the best bio-sorption occur at pH 2. At the pH acidic, the number of positively charged sites onto bio-sorbent surface increased which favored the bio-sorption of dye due to electrostatic attraction. furthermore, the decrease in the bio-sorption of reactive blue 49 by raise value of pH was too due to the competition among dye and excess ions OH in the solution .The best adsorption capacity pH 2, and it was used in further studies [18-20] .

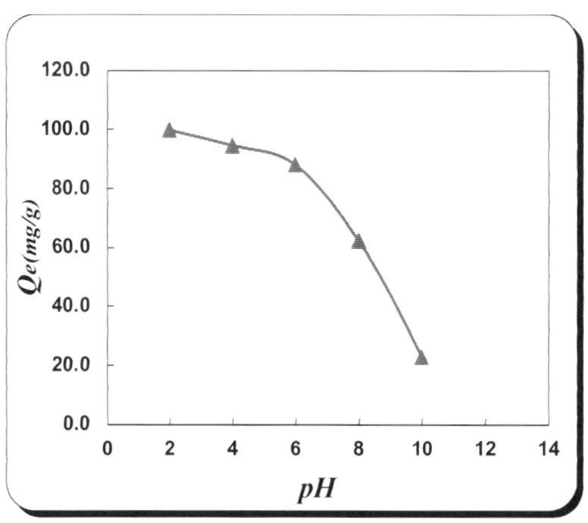

Figure 6. Effect of pH solution on the adsorption efficiency on to AACLP (Exp. Condition: equlibruim time 1 hr., Temp. = 20°C, quantity of AACLP 0.05g).

3.6. Bio-sorption Isotherm

The analysis of the Isotherm result via fitting them to two isotherm is an significant step to get the suitable isotherm that can be utilized for design purpose. In this study , two isotherm Freundlich, and Langmuir in their nonlinear forms were applied to the equilibrium result of reactive blue 49 sorption by AACLP as appear in Figure 7.

Isotherm Freundlich is an empirical equation based on adsorption onto surface heterogeneous. The equation is usually calculate in Equation (3)

$$Q_e = k_f \, C_e^{\frac{1}{n}} \qquad (3)$$

The equation non-linear of isotherm Langmuir model in equation (4)

$$Q_e = \frac{Q_m \, K_L \, C_e}{1 + K_L \, C_e} \qquad (4)$$

The parameter of isotherm Freundlich and Langmuir for reactive blue 49 and the values of (R2)correlation coefficient, for the Isotherm result are the most excellent for isotherm Freundlich appear in Figute 7 [21-23].

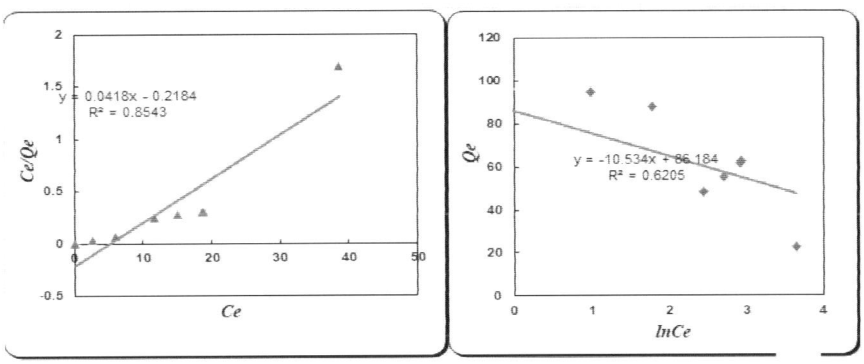

Figure 7. Adsorption models linear fit of adsorption reactive blue49 on AACLP, conc. = 50 mg. L^{-1}, Temp. = 20°C, and mass 0.05 gm).

3.7. Thermodynamic studies

Thermodynamic parameter were carry out to estimate the changes free energy (ΔG°), entropy (ΔS°) and enthalpy (ΔH°), in the solution temperature series (15-35 oC) as appear in figure 8 and the thermodynamic parameter were calculate via utilized equations :

$$\Delta G° = -RT \ln K \qquad (5)$$

According to thermo-dynamic law, the Van't Hoff equation calculate via equation (6)

$$\Delta G° = \Delta H° - T\Delta°S \qquad (6)$$

Eq. (7) above can be further transformed as follows :

$$\ln k = -\Delta G°/RT = -\Delta H°/RT + \Delta S°/R \qquad (7)$$

The thermodynamic study find of the sorption of reactive blue 49 as shown in Table 1 and figure 9. The values demonstrate spontaneous and favorable sorption processes. The values of enthalpy is negative (endothermic). also, the values of free energy is negative (spontaneous nature of the method) . the values of the entropy is negative are not very large and correspond to a decrease in the degree of freedom of the sorbed species[17, 24] .

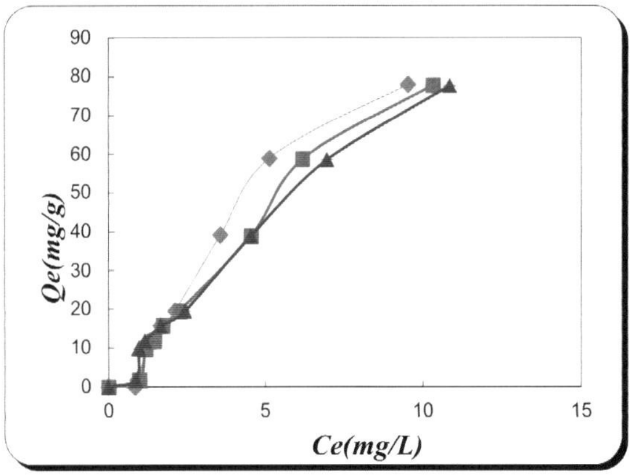

Figure 8. Adsorption isotherms of reactive blue 49 dye onto AACLP at several temperatures.

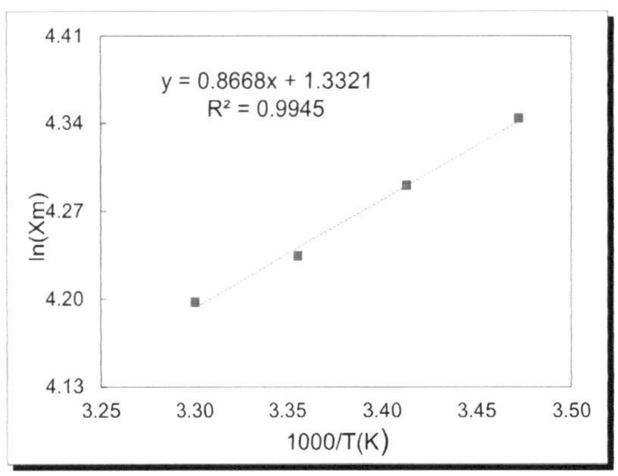

Figure 9. Plot of ln Xm against reciprocal absolute temperature for the reactive blue 49 dye absorption onto AACLP

Table 1. Thermodynamic parameter for reactive blue 49 dye adsorption on to AACLP

Thermodynamic parameter			
ΔH(kJ/mol.)	**ΔG(kJ/mol.)**	**ΔS(kJ/K.mol.)**	**Equilibrium Constant(K)**
7.20	-3.09	34.56	139.39

4. Conclusion

In this study, agricultural waste (lemon peels) was used as an environmentally friendly, inexpensive and highly efficient surface in removing the dye from its aqueous solution. The best weight of acid activated carbon from lemon peels (0.05 g) to remove the reactive blue 49 dye, as it was given to remove (95,12%) . The surface (AACLP) gave a high percentage of dye removal compared to the surface (NLP).A thermodynamic study found that the reaction is spontaneous and endothermic .The reaction obeys the Langmuir isotherm depending on the value of (R2= 0.854).

Reference

[1] Hameed, K.S., Muthirulan, P., Meenakshi, S.M., Adsorption of chromotrope dye onto activated carbons obtained from the seeds of various plants: Equilibrium and kinetics studies. Arabian J. Chem, 2017. 10: p. S2225-S2233.

[2] Hadi, Z.A., Aljeboree, A.M., Alkaim, A.F., Adsorption of a cationic dye from aqueous solutions by using waste glass materials: Isotherm and thermodynamic studies. International Journal of Chemical Sciences, 2014. 12(4): p. 1273-1288.

[3] Yongde Liu, Y.C., Yahui Shi, Dongjin Wan, Jing Chen, Shuhu Xiao, Adsorption of toxic dye Eosin Y from aqueous solution by clay/carbon composite derived from spent bleaching earth. Journal of Hazardous Materials,, 2019. 1: p. 4656-5544.

[4] Alqaragully , M.B., Removal of Textile Dyes (Maxilon Blue, and Methyl Orange) by Date Stones Activated Carbon International Journal of Advanced Research in Chemical Science, 2014. 1(1): p. 48-59

[5] Barragan, B.E., Costa, C., Carmen Marquez, M., Biodegradation of azo dyes by bacteria inoculated on solid media. . Dyes Pigments 2007. 75: p. 73-81.

[6] Aljeboree, A.M., Abbas, A. S., Removal of Pharmaceutical (Paracetamol) by using CNT/ TiO2 Nanoparticles. Journal of Global Pharma Technology, 2019. 11(01): p. 199-205.

[7] Aljeboree, A.M., Adsorption and removal of pharmaceutical riboflavin (RF) by rice husks activated carbon. International Journal of Pharmaceutical Research 2019. 11(2): p. 255-261.

[8] O. Oribayo, O.O.O., A. S. Akinyanju, K.O. Omoloja and S.O. Williams Binary Adsorption of Textile Dyes onto Zwitterionic Adsorbent Coating: Performance Study Nigerian Journal of Technology 2020. 39(4): p. 1076 - 1084

[9] Aljeboree, A.M., Alkaim A F, Removal of Antibiotic Tetracycline (TCs) from aqueous solutions by using Titanium dioxide (TiO2) nanoparticles as an alternative material. Journal of Physics: Conf. Series 2019. 1294 p. 052059.

[10] Aseel M. Aljeboree, A.F.A., Ali loay, hanadi m algburi, Photocatalytic Degradation of Textile Dye Cristal Violet Wastewater using Zinc Oxide as a Model of Pharmaceutical Threat Reductions. Journal of Global Pharma Technology, 2019. 11(02): p. 138-143.

[11] Saeed U J, A.A.K., Saad M, Iftikhar A, Guohua S, Cheng-Meng C, Rashid A, Removal of azo dye from aqueous solution by a low-cost activated carbon prepared from coal: adsorption kinetics, isotherms study, and DFT simulation. Environmental Science and Pollution Research, 2020. 1: p. https://doi.org/10.1007/s11356-020-11344-4.

[12] Yahya A. Faleh , N.D.R., Removal of Metformin hydrochloride from Aqueous Solutions by using Carboxymethyl cellulose-g-poly(acrylic acid-co-acrylamide) Hydrogel: Adsorption and Thermodynamic Studies. IOP Conf. Series: Earth and Environmental Science 2021. 790: p. 012062.

[13] Tariq J. AlMusawi, N.M., Mahmoud Taghavi, Samaneh Mohebi, Davoud Balarak, Activated carbon derived from Azolla fliculoides fern: a high-adsorption-capacity adsorbent for residual ampicillin in pharmaceutical wastewater. Biomass Conversion and Biorefnery 2021. 1: p. https://doi.org/10.1007/s13399-021-01962-4.

[14] Yan, L.-g., Qin, Li-lu, Adsorption of acid dyes from aqueous solution by CTMAB modified bentonite: Kinetic and isotherm modeling. Journal of Molecular Liquids. 211: p. 1074-1081.

[15] Aljeboree A.M., A.A.N., Alkaim A.F, Removal of pharmaceutical amoxicillin drug by using (Cnt) decorated clay/ fe2 o3 micro/nanocomposite as effective adsorbent: Process optimization for ultrasound-assisted adsorption. International Journal of Pharmaceutical Research, 2019. 11(4): p. 80-86.

[16] Farabi Temel , M., SezenKucukcongar, Removal of methylene blue from aqueous solutions by silica gel supported calix[4]arene cage: Investigation of adsorption properties. European Polymer Journal, 2020. 125(15): p. 109540.

[17] Tsai, W.T., Chang, C.Y., Lin, M.C., Chien, S.F., Sun, H.F., Hsieh, and M.F., Adsorption of acid dye onto activated carbons prepared from agricultural waste bagasse by ZnCl2 activation. Chemosphere, 2001. 45: p. 51-58.

[18] Mu, B., Wang, Aiqin, Adsorption of dyes onto palygorskite and its composites: A review. Journal of Environmental Chemical Engineering. 4(1): p. 1274-1294.

[19] Layth S J and Aljeboree, M.A., Removal of Heavy Metals by Using Chitosan/ Poly (Acryl Amide-Acrylic Acid) Hydrogels: Characterization and Kinetic Study . NeuroQuantology, 2021. 19(2): p. 31-37.

[20] Kausar, A., et al., Dyes adsorption using clay and modified clay: A review. Journal of Molecular Liquids. 256: p. 395-407.

[21] Choy, K.K.H., J.F. Porter, and G. McKay, Langmuir Isotherm Models Applied to the Multicomponent Sorption of Acid Dyes from Effluent onto Activated Carbon. J.Chem. Eng. Data. , 2000. 45 p. 575-584.

[22] Ewaid, S.H., Abed, S.A., 2017. Water quality index for Al-Gharraf river, southern Iraq. Egypt. J. Aquatic Res. 43 (2), 117–122. http://dx.doi.org/10.1016/j. ejar.201703001.

[23] Ewaid, S.H.; Abed, S.A.; Al-Ansari, N. Crop Water Requirements and Irrigation Schedules for Some Major Crops in Southern Iraq. Water 2019, 11, 756.

[24] Ewaid, S.H.; Abed, S.A.; Al-Ansari, N. Water Footprint of Wheat in Iraq. Water 2019, 11, 535.

[25] Ewaid, S.H.; Abed, S.A.; Al-Ansari, N. Assessment of Main Cereal Crop Trade Impacts on Water and Land Security in Iraq. Agronomy 2020, 10, 98.

[26] Ewaid, S.H.; Abed, S.A.; Al-Ansari, N.; Salih, R.M. Development and Evaluation of a Water Quality Index for the Iraqi Rivers. Hydrology 2020, 7, 67.

[27] Ewaid, S.H., Al-Farhani, B.F., Abed, S.A. & Al-Ansari, N. (2020). Modeling of trihalomethane compounds formation in Baghdad water supply network. Scientific Review Engineering and Environmental Sciences, 29 (2), 136-144. doi: 10.22630/PNIKS.2020.29.2.12.

[28] Salam Hussein Ewaid et al 2021 IOP Conf. Ser.: Earth Environ. Sci. 722 012008

[29] Salam Hussein Ewaid et al 2021 IOP Conf. Ser.: Earth Environ. Sci. 790 012075

[30] Shahul Hameed, K., P. Muthirulan, and M. Meenakshi Sundaram, Adsorption of chromotrope dye onto activated carbons obtained from the seeds of various plants: Equilibrium and kinetics studies. Arabian Journal of Chemistry. 10: p. S2225-S2233.

[31] Aseel M. Aljeboree, A.F.A., Role of plant wastes as an ecofriendly for pollutants (Crystal Violet dye) removal from Aqueous Solutions Plant Archives 2019 19(2): p. 902-905.

[32] Li, Q., Yue, Q., Su, Y., Gao, B., Sun, H., Equilibrium, thermodynamics and process design to minimize adsorbent amount for the adsorption of acid dyes onto cationic polymer-loaded bentonite. . Chem. Eng. J. , 2010. 158: p. 489-497.

Environmental, Biochemical and Hematological study in rats exposed to Cadmium Chloride in Drinking Water and the role of Vitamin E and C

Arjwan A Alsudani

College of Science, University of Al-Qadisiyah, Iraq

Abstract. The present study was carried out to determine the synergistic role of vitamin E and vitamin C in reducing the toxic effects of cadmium chloride on some blood parameters (RBC count, Hb concentration, PCV, Total WBC count, Differential WBC count), as well as some biochemical parameters which include (Malondialdehyde (MDA), Aspartate aminotransferase(AST), Alanine aminotransferase (ALT), Alkaline phosphatase(ALP), Creatinin, Urea) and some antioxidants (Superoxidedismutase(SOD),Glutathion-S-transferase(GST),Catalase(CAT). Forty adult male rats were divided into five equal groups. The first group was left without any treatment (the control group), the second was given cadmium chloride (30mg/L), the third was given cadmium chloride and vitamin E simultaneously, the fourth was given cadmium chloride and vitamin C simultaneously, the fifth was given cadmium chloride, vitamin E and vitamin C simultaneously. The results showed a significant decrease ($P < 0.05$) in (RBC count, Hb concentration, PCV) and a significant increase ($P < 0.05$) in WBC count (neutrophils and eosinophils). On the other hand, the results also showed an increase ($P < 0.05$) in the concentration of (ALT, AST, ALP, Urea, Creatinine, and MDA)and a significant decrease in antioxidants(SOD, CAT, GST) in (T_1) compared with the control group. On the other hand, the results showed a noticeable improvement in the parameters studied when using vitamin E and vitamin C alone or together with cadmium chloride, and this indicates the therapeutic role played by vitamins in reducing the toxicity caused by cadmium chloride

Keywords: cadmium chloride, vitamin E, vitamin C, male albino rats.

1. Introduction

Cadmium is a toxic heavy metal that pollutes the environment and puts human and animal health at risk, it is an unnecessary trace element that is found in small quantities in the body and its loss does not affect its vital activities. It is difficult to avoid exposure to this metal due to its widespread as it is found in soil, water, air, and food as a result of its use in many industries such as the plastic industries, the manufacture of phosphate fertilizers, dyes, battery electrodes, the manufacture of metal alloys and others [1]. Increasing its level in the soil as a result of the use of phosphate fertilizers leads to an increase in its concentration in plants and thus its transmission through food to humans and animals.

The percentage of cadmium chloride increases in various parts of the body with the increase in the intensity of exposure to it as well as the increase in its concentration, as it causes an increase in the processes of oxidation of fats and the formation of free radicals that affect many important cellular components such as proteins, fats, and nucleic acids and thus cause harm to cell membranes [2] [3].

Vitamin E is a fat-soluble antioxidant compound represented by two groups: tocopherols and tocotrienols, and the most effective form are α-tocopherol, which inhibits free radicals and protects unsaturated fatty acids from oxidation and conversion to peroxides. This is an antioxidant and protects cells from oxidative stress. [4] [5] As for vitamin C, it is a water-soluble and a non-fat-soluble vitamin, and it is an important antioxidant as it plays a protective role against oxidative stress and can neutralize the type of active oxygen (Reactive Oxygen Species) ROS) and protects against oxidative damage caused by free radicals by inhibiting them as well as lipid peroxidation [6] [7]. The present experiment was designed to study the effects that the use of vitamin E and vitamin C separately or together in reducing the toxicity of cadmium chloride in male albino rats.

2. Materials and methods

2.1. Experimental animals

This study was carried out in the animal house of the College of Science, University of Al-Qadisiyah. Sixty-four male albino rats (12-14 weeks old, 180-200 g weight) were used under standard conditions of temperature, ventilation, and lighting duration. Animals were housed in sterile polypropylene cages (50 × 35 × 15) cm with a stainless steel grill surface, in sterile conditions and the animals were given feed and water for the duration of the experiment.

2.2. Experimental Design

Forty adult male Albino rats were divided randomly into five groups, each group contains 8 animals and was treated for 8 weeks as follows: control group (C_1) given drinking water only, The second treatment group (T_1)given orally cadmium chloride (30mg/L), The third group (T_2) given orally cadmium chloride(30mg/L) then given vitamin E (400mg/kg) simultaneously, The fourth group (T_3) given orally cadmium chloride and vitamin C (200 mg/kg) simultaneously, the fifth group (T_4) given orally cadmium chloride, vitamin E (400mg/kg) and vitamin C (200 mg/kg) simultaneously.

2.3. Sacrifice of Animals

All animals were sacrificed at the end of the experiment, then blood was drawn directly with medical syringes using a heart punch method, with 3 ml of each animal. 1.5 ml of it was placed in tubes containing anticoagulant to study blood parameters, which include RBC count, Hb concentration, PCV according to the method of [8]and Total WBC count, Differential WBC count according to the method of [9]. The remaining blood was tilted for 30 minutes in tubes that did not contain an anticoagulant ingredient, then the tubes were placed in a centrifuge for 15 minutes at a speed of 3000 revolutions/minute to obtain the blood serum, which was kept in marked plastic tubes, frozen at a temperature (-20 C) to measure some biochemical parameters which include (Malondialdehyde (MDA) according to the method of [10], Aspartate aminotransferase(AST), Alanine aminotransferase

(ALT) according to the method of [11], Alkaline phosphatase(ALP) [12], Creatinin, Urea) and some antioxidants (Superoxide dismutase(SOD), Glutathion-S-transferase(GST), Catalase(CAT) [13].

2.4. Statistical analysis

The SPSS program (version 21) was used to compare the rates of the researched criteria for the different groups using the results of the statistical analysis. Using one-way analysis of variance, significant differences were identified at a level of probability of 5%. The least significant differences (LSD) are used to test moral differences [14].

3. Results and discussion

3.1. Hematological parameters

The current study's findings (Table 1) revealed a significant decrease ($P < 0.05$) in the RBC count, Hb concentration, PCV in the blood of white male rats treated with cadmium chloride(T2) compared to the control group, which indicates that poisoning with this element leads to a state of anemia. These results are consistent with [15].

Cadmium chloride poisoning leads to an increase in the generation of free radicals that negatively affect red blood cells due to their effect on the unsaturated fatty acids present in the cell membranes of red blood cells and thus cause the fragility of the cells and the speed of their destruction [16]. Cadmium chloride poisoning also leads to a disturbance in the production of the hormone erythropoietin, which is responsible for regulating the production of red blood cells and their maturation in the bone marrow as a result of damage to the proximal renal tubules responsible for the production of this hormone [17].

'The results in (Table 1) indicated that the use of vitamin E and vitamin C alone or together with cadmium led to an improvement in blood parameters to reach a level close to that of the control group, especially in the (T_5). The reason for the improvement may be due to the role of vitamin E in inhibiting free radicals and protecting unsaturated fatty acids from oxidation and conversion to peroxides. This is an antioxidant and protects the cell from oxidative stress. It is necessary to work synergistically with vitamin C as a free radical scavenging force by eliminating the chain reaction of radicals and protecting the cell. of poisoning [18].

Table 1. Effect of vitamin E and vitamin C in reducing the toxicity of cadmium chloride on some parameters of blood in (male albino rats).

Standards / Groups	RBC count **($x10^{12}$/L)**	Hb concentration (g/100ml)	PCV **(%)**	Total WBC count **($x10^{9}$/L)**
C_1	3.50±0.002 a	9.23±0.012 a	31.24±0.012 a	11.10±0.083 c
T_1	1.16±0.002 c	3.85±0.232 d	10.35±0.006 d	20.35±0.011 a
T_2	3.01±0.004 b	8.83±0.020 c	27.41±0.047 c	13.09±0.078 b
T_3	3..37±0.013 a	8.99±0.020 b	29.30±0.014 b	11.129±0.069

T_4	3.49±0.014 a	9.19±0.013ab	30.15±0.002ab	c 11.00±0.012c
L.S.D	0.209	0.085	1.246	0.211

The numbers represent mean ± standard error

At the level of significance (P <0.05), different letters reveal significant differences between groups.

The results showed (Table 1) a significant increase(P <0.05) in the number of total white blood cells, neutrophils, and eosinophils (T_1), offset by a decrease in the number of lymphocytes, and no significant difference was observed in the rate of the number of monocytes compared with the control group and other groups (Table 2). These results are in agreement with [19] [20].

The reason for the increase in the number of white blood cells maybe because it is a defensive means for inflammatory conditions that occur in the body, especially the liver and kidneys, as the bone marrow produces additional cells due to oxidative stress, thus increasing the total number of white blood cells [21].

On the other hand, the reason for the decrease in lymphocytes can be explained by the effect of cadmium chloride on the immune system and the inhibition of cellular immunity due to immune suppression [19].

On the other hand, the results showed that the animals that were dosed with vitamin E with cadmium chloride had the number of white cells within their normal rates, especially in the fifth treatment group, which indicates the positive synergistic role of vitamin E with vitamin C to minimize the quantity of reactive oxygen species (ROS) and thus the oxidative stress induced by cadmium chloride toxicity [20].

Table 2. Effect of vitamin E and vitamin C in reducing the toxicity of cadmium chloride on some parameters of blood in (male albino rats).

Standards Groups	neutrophil (%)	acidophilus (%)	lymphocytes (%)	monocytes (%)
C1	**34.12±0.031 c**	**6.81±0.021 c**	**55.19±0.011 c**	**4.56±0.083 a**
T1	**50.16±0.020a**	**10.84±0.102 a**	**20.00±0.006 d**	**4.46±0.011 a**
T2	**35.02±0.007 c**	**8.25±0.050 b**	**59.09±0.047 a**	**4.33±0.078 a**
T3	**36.42±0.120 b**	**7.53±0.010 c**	**57.01±0.014 b**	**4.58±0.069 a**
T4	**34.01±0.160 c**	**6.92±0.013 c**	**55.05±0.002c**	**4.71±0.012 a**
L.S.D	**0.918**	**0.435**	**0.017**	**0.130**

The numbers represent mean ± standard error

At the level of significance (P <0.05), different letters reveal significant differences between groups.

3.2. Biochemical changes

The results of the study confirmed that the liver is one of the organs most affected by cadmium chloride poisoning because the toxic substance travels through the blood and reaches the liver directly after being absorbed in the alimentary canal. The liver is responsible for representing the nutrients coming to it, including toxic substances, which negatively affects the vital activities of the liver [22] [23]

Measuring the concentration of enzymes(ALT, AST, ALP) in the serum is used as a measure to find out the pathological changes that occur in the liver[24]. The results indicated (Table 3) a significant increase ($P < 0.05$)in the concentration of ALT, AST, ALP in (T1) compared with the control group. The preceding observations could be attributable to cadmium chloride-induced liver injury, which causes widespread enzyme liberation into the bloodstream [15] [24].

Table 3. Effect of vitamin E and vitamin C in reducing the toxicity of cadmium chloride on some biochemical parameters in (male albino rats).

Standards / Groups	ALT (IU/L)	AST (IU/L)	ALP (IU/L)	Urea (mg/dl)	Creatinine (mg/dl)
C_1	1.89±1.00 c b	1.32±0.15 b	1.75±0.013 b	0.7±0.15c	0.95±0.75 b
T_1	3.50±1.07 a	2.96±1.07 a	3.99±0.006 a	2.9±0.09 a	3.51±0.95 a
T_2	1.87±0.04 b	1.25±0.77 b	1.67±0.047 b	1.1±0.01bc	1.11±0.66 b
T_3	1.93±0.08 b	1.59±0.52 b	1.70±0.004 b	0.9±0.05c	0.91±0.31 b
T_4	1.91±0-02 b	1.66±0.55 b	1.74±0.012 b	0.7±0.08c	0.99±0.11b
L.S.D	0.28	0.52	0.33	0.31	0.20

The numbers represent mean ± standard error.

At the level of significance ($P < 0.05$), different letters reveal significant differences between groups.

On the other hand, the results showed an improvement in the concentration of enzymes ALT, AST, ALP(Table3) for the groupsT_2, T_3, T_4, and it was at a level close to that of the control group, especially the groupT_4. This confirms the preventive and curative role of vitamin E and vitamin C in curbing the activity of free radicals resulting from cadmium chloride poisoning, and thus protecting the liver [25]

The results showed in (Table3) that there was a significant increase($P < 0.05$) in the level of Urea and Creatinine in the group of animals treated with cadmium chloride (T_1) compared with the control group and the other groups. This agrees with [20]. As cadmium causes nephrotoxicity, which leads to the accumulation of many metabolites in the tissues and thus damages the kidneys [26]. While the groups showed a significant decrease in the level of Urea and Creatinine compared with (T_1), the results were close to the control group, especially in the (T_4). The reason for the improvement is that vitamins are among the antioxidants that play a role in reducing oxidative damage [27].

The current study's findings (Table 4) showed a significant increase ($P<0.05$) in the concentration of (MDA) and a significant decrease ($P<0.05$) in the concentration of (SOD, GST, CAT) in (T_1) compared with the control group and the other groups.

Table 4. Effect of vitamin E and vitamin C in reducing the toxicity of cadmium chloride on some biochemical parameters in (male albino rats).

Standards / Groups	**MDA)mol/L(µ**	**SOD (U/ ml)**	**Catalase (U/ ml)**	**GST mol/L) (µ**
C_1	1.29±0.002 c	1.90±0.010a	0.89±0.013 a	3.10±0.083 a
T_1	4.16±0.005 a	0.85±0.232 b	0.35±0.006 d	1.35±0.011 b
T_2	1.64±0.007 b	1.85±0.020 a	0.82±0.047 c	2.89±0.078 a
T_3	1.42±0.120 bc	1.89±0.020 a	0.84±0.014 cb	2.99±0.069 a
T_4	1.30±0.012 c	1.92±0.013a	0.88±0.002 ab	3.01±0.012 a
L.S.D	0.216	0.075	0.047	0.230

The numbers represent mean ± standard error.

At the level of significance ($P < 0.05$), different letters reveal significant differences between groups.

The increase in MDA concentration may result from free radical targeting of fatty acids present in cell membranes and lead to peroxidation of unsaturated fatty acids and consequently decreased fluidity of cell membranes and loss of function and lead to cell death [28]. On the other hand, cadmium chloride is a cause of free radical production, which reduces the antioxidant's efficiency, as a result of the imbalance between free radicals and antioxidants, the oxidative state increases. [29].

While the results showed a significant improvement in the concentration of (MDA, SOD, GST, CAT) in (T_2, T_3, T_4) and were similar to the control group. This may be because vitamins are among the antioxidants that play a role in improving metabolic processes, protecting cells from free radicals, and reducing oxidative stress [30].

4. Conclusion

The current study showed the synergistic role of vitamin E with vitamin E, which has the greatest effect in reducing the toxicity of cadmium chloride on hematological, biochemical parameters, and antioxidants.

References

[1] Matović, V., Buha, A., Đukić-Ćosić, D., & Bulat, Z. 2015. Insight into the oxidative stress induced by lead and/or cadmium in blood, liver and kidneys. Food and Chemical Toxicology, 78, 130-140.".

[2] Matović, V., Buha, A., Bulat, Z., & Đukić-Ćosić, D. 2011. Cadmium toxicity revisited: focus on oxidative stress induction and interactions with zinc and magnesium. Arhiv za higijenu rada i toksikologiju, 62(1), 65-75.".

[3] Rani, A., Kumar, A., Lal, A., & Pant, M. 2014. Cellular mechanisms of cadmium-induced toxicity: a review. International journal of environmental health research, 24(4), 378-399.".

[4] Duncan, K.R. & Suzuki, Y.J. 2017. Review vitamin E nicotinate Journal Antioxidants, 6(20): 1-14.".

[5] Kalappurayil, T.M. & Joseph, B.P. 2017. A review of pharmacognostical studies on Moringa oleifera Lam. flowers. Pharmacogn J., 9(1): 1-7.".

[6] Chakraborthy, A.; Ramani, P.; Sherlin, H.J.; Premkumar, P. & Natesan, A. 2014. Antioxidant and pro-oxidant activity of Vitamin C in oral environment. Indian Journal of Dental Reserch, 25(4): 499-504.".

[7] Ahmed, K.S. ;Banik, R. ; Hossain, M.H. & Jahan, I.A. 2016. Vitamin C (L-ascorbic Acid) content in different parts of Moringa oleifera grown in Bangladesh. American Chemical Science Journal, 11(1): 1-6.".

[8] McCann, M. F., & Cole, L. P. 1980. Laparoscopy and minilaparotomy: Two major advances in female sterilization. Studies in Family Planning, 11(4), 119-127.".

[9] Bain, Barbara J., et al. 2016. Dacie and Lewis Practical Haematology: Expert Consult: Online and Print. Elsevier Health Sciences,.".

[10] Guidet, B. R., & Shah, S. V. 1989. In vivo generation of hydrogen peroxide by rat kidney".

[11] Asvall, J. E. 1960. "Transaminase activity after experimental hypoxia in rabbits." Scandinavian Journal of Clinical and Laboratory Investigation 12.2: 239-246.".

[12] MacDonald, D. J., et al. 1980. "Enzyme-linked immunoassay for placental lactogen in human serum." Clinical Chemistry 26.6: 745-749.".

[13] Aebi, H. 1974. Catalase. In Methods of Enzymatic Analysis (Second Edition), Volume 2 (pp. 673-684).".

[14] "Tamhane, A.C., Dunlop, D.D., 2000. Statistics and Data Analysis from Elementary to Intermediate. Upper Saddle River, USA.".

[15] Andjelkovic, M., Buha Djordjevic, A., Antonijevic, E., Antonijevic, B., Stanic, M., Kotur-Stevuljevic, J., ... & Bulat, Z. 2019. Toxic effect of acute cadmium and lead exposure in rat blood, liver, and kidney. International journal of environmental resear".

[16] Mladenovic´, J.; Ognjanovic´, B.; Dordevic´, N.; Matic´, M.; Kneževic´, V.; Štajn, A.; Saicˇic´, Z. . 2014. Protective effects of oestradiol against cadmium-induced changes in blood parameters and oxidative damage in rats. Arh. Hig Rada Toksikol. 65, 37–".

[17] Djokic, J.; Ninkov, M.; Mirkov, I.; Popov Aleksandrov, A.; Zolotarevski, L.; Kataranovski, D.; Kataranovski, M. 2014. Differential effects of cadmium administration on peripheral blood granulocytes in rats. Environ. Toxicol. Pharmacol. 37, 210–219.".

[18] Zaidi, S. K. R., & Banu, N. 2004. Antioxidant potential of vitamins A, E and C in modulating oxidative stress in rat brain. Clinica Chimica Acta, 340(1-2), 229-233.".

[19] Wershana, Karima Z. 2001. "Cadmium induced toxicity on pregnant mice and their offspring: protection by magnesium or vitamin E." Sciences 1.4: 179-186.".

[20] "Alkaladi, A., El-Deen, N. A. N., Afifi, M., & Zinadah, O. A. A. 2015. Hematological and biochemical investigations on the effect of vitamin E and C on Oreochromis niloticus exposed to zinc oxide nanoparticles. Saudi journal of biological sciences, 22(5),".

[21] "Maqbool, M., et al. 2011"Optimisation of laboratory procedures for isolating human peripheral blood derived neutrophils." Med J Malaysia 66.4: 296-299.".

[22] "Eroschenko, Victor P., and Mariano SH Di Fiore. DiFiore's 2013. atlas of histology with functional correlations. Lippincott Williams & Wilkins,.".

[23] "Habeebu, Sultan S., et al. 2000. "Metallothionein-null mice are more sensitive than wild-type mice to liver injury induced by repeated exposure to cadmium." Toxicological Sciences 55.1 : 223-232.".

[24] "Zhu, H., Jia, Y., Cao, H., Meng, F., & Liu, X. 2014. Biochemical and histopathological effects of subchronic oral exposure of rats to a mixture of five toxic elements. Food and chemical toxicology, 71, 166-175.".

[25] Harabawy, A. S., & Mosleh, Y. Y. 2014. The role of vitamins A, C, E and selenium as antioxidants against genotoxicity and cytotoxicity of cadmium, copper, lead and zinc on erythrocytes of Nile tilapia, Oreochromis niloticus. Ecotoxicology and Environmenta".

[26] Genchi, G., Sinicropi, M. S., Lauria, G., Carocci, A., & Catalano, A. 2020. The effects of cadmium toxicity. International journal of environmental research and public health, 17(11), 3782.".

[27] Adi, P. J., Burra, S. P., Vataparti, A. R., & Matcha, B. 2016. Calcium, zinc and vitamin E ameliorate cadmium-induced renal oxidative damage in albino Wistar rats. Toxicology reports, 3, 591-597.".

[28] Demirkol, O., Zhang, X., & Ercal, N. 2012. Oxidative effects of Tartrazine (CAS No. 1934-21-0) and New Coccin (CAS No. 2611-82-7) azo dyes on CHO cells. Journal für Verbraucherschutz und Lebensmittelsicherheit, 7(3), 229-236.".

[29] Bassam F. ALfarhani et al 2021 IOP Conf. Ser.: Earth Environ. Sci. 790 012037

[30] Bassam F. ALfarhani et al 2021 IOP Conf. Ser.: Earth Environ. Sci. 790 012040

[31] Alfarhani, B.F., Hammza, R.A. & Alzaidy, A.S. Potential effect of solvent and slit width on some properties of room temperature fluorescence of hydroxy polycyclic aromatic hydrocarbons. Chem. Pap. 75, 3915–3920 (2021). https://doi.org/10.1007/s11696-021-01602.

[32] Bassam Faron Alfarhani et al 2019 J. Phys.: Conf. Ser. 1294 052045

[33] Bassam Alfarhani, Maha Al-Tameemi, Hector C. Goicoechea, Fernando Barbosa, Andres D. Campiglia, Direct analysis of benzo[a]pyrene metabolites with strong overlapping in both the spectral and lifetime domains, Microchemical Journal, 137 2018, 51-61,

[34] Poli, V., Madduru, R., Aparna, Y., Kandukuri, V., & Motireddy, S. R. 2022. Amelioration of Cadmium-Induced Oxidative Damage in Wistar Rats by Vitamin C, Zinc and N-Acetylcysteine. Medical Sciences, 10(1), 7.".

[35] Kara, H., Cevik, A., Konar, V., Dayangac, A., & Servi, K. 2008. Effects of selenium with vitamin E and melatonin on cadmium-induced oxidative damage in rat liver and kidneys. Biological trace element research, 125(3), 236-244."

Environmental Study on Landfill Suitability Modeling Using Geographic Information Systems and Multicriteria Decision Analysis: A Case Study in Afak Qadhaa

Aqeel Thamer and Salwan Ali Abed

College of Science, University of Al-Qadisiyah, Iraq

Abstract. In this study, the appropriate locations for the landfill space in Afak Qadhaa determined by integrating geographic information system (GIS) and analysis decision (MCDA), additionally by implementing these methods that lead to integrated environmental management, which is required in order to take into account all components and processes in the environment. For this purpose, layers were introduced. Digital maps including settlement (industrial areas, urban center and villages), wetlands, main roads, digital elevation model, airports, land use (plants mansion campgrounds, water and sewage treatment plants, agricultural area, gardens and preparing buildings) and surface water (streams and main irrigation channels) are implemented The relative weight of the criteria with the geographic information system, which revolved around the formulation of criteria for the planning of waste dumps, giving them the appropriate ranks of each criteria and their weight, then building the model and coming up with a map that represents the most suitable sites for establishing healthy dumps in the Afak Qadhaa.

Keywords: Landfill Suitability, Geographic Information Systems

1. Introduction

The capacity to manage massive amounts of geographic data from a variety of sources distinguishes it as a useful tool, a geographic information system (GIS) is excellent for performing exploratory research on the placement of a landfill. It successfully saves, retrieves, analyzes, and displays data based on criteria that the user has specified [1]. If you're utilizing MCDA to aid decision-makers, it can be tough to make sense of enormous amounts of complex data at once. The method's assumption is that choice issues should be broken down into smaller, more understandable components, with each component being assessed individually before the parts are logically integrated together. [2]. Due to the ease with which GIS data can be edited, as well as the fact that MCDA can produce consistent rankings of potential landfill locations based on a range of characteristics, GIS may prove to be a beneficial tool in the search for the optimal landfill placement. processing the digital maps that were utilized to show the findings of this study.

2. Study Area:

Afak Qadhaa is a city in Iraq that is located in the Southeast part of the Diwaniyah Governorate. And away from the city of Diwaniyah, about 25 northeast, and from Baghdad, the capital, about 170 km to the south. (Figure 1with a population of approximately 165,000 people. It is located approximately between latitude 31°56'0.24"N, and longitude 45°24'51.41"E longitude. The Afak Qadhaa covers a total area of 3669 kilometers squared. The population is growing at a rate of 6% per year.

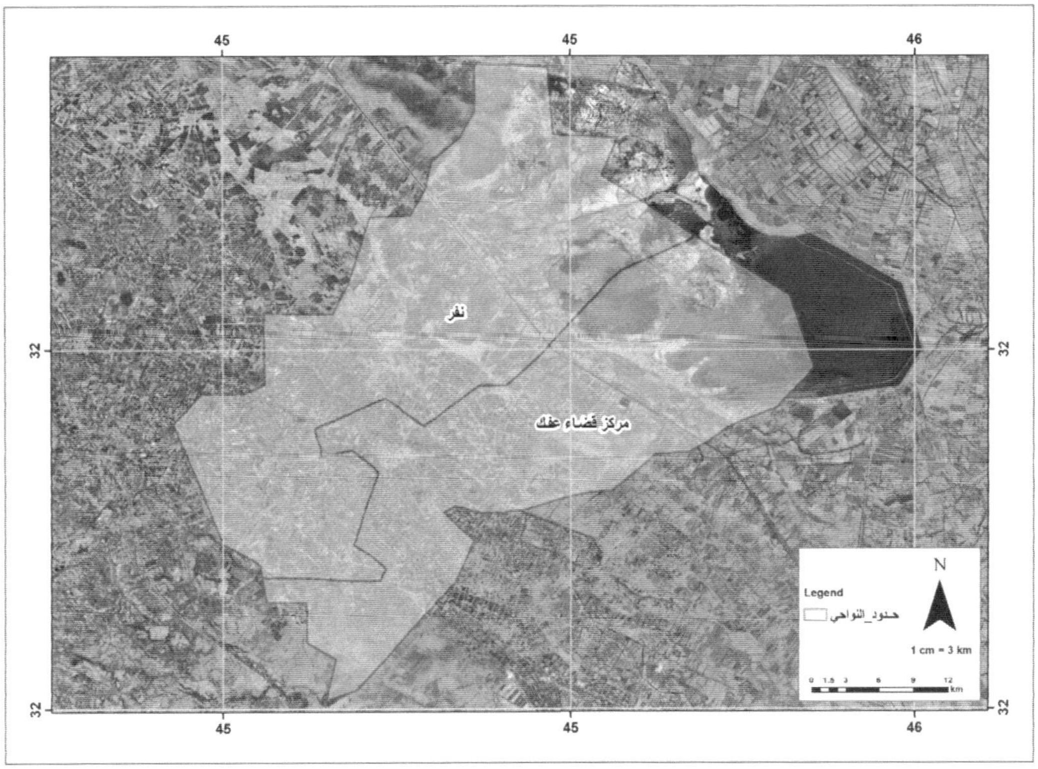

Figure 1. Map Afak Qadhaa

3. Materials and Methods

The use of geographical information systems (GIS) is very important when planning and deciding on the best landfill sites during that time. This is because the best landfill sites are chosen based on a wide range of factors and variables, which include economic, social, geological, and geomorphological criteria, as well as environmental, climate, and public acceptance criteria. In order for researchers to get accurate results and plan for the future, they need to lessen the importance of sensitive or potentially dangerous criteria that could have a negative impact on the environment and humans if the chosen criteria are met. They also need to align them with standards and conditions that have a minimal or no negative impact. There were many GIS tools that could have helped the researchers balance variables and assign one of the criteria more weight than the rest. At this point, they used weighted values to assign actors in the success of the best locations and the greater weight of the other elements, and there were many Weighted Overlay functions that went unnoticed.

3.1. Raster

All maps have been transformed from vector to raster format in order to make the process more efficient and effective. This is due to the fact that modeling methods make working with Raster data more convenient. The way cells are arranged in addition, the Raster form represents a generalization of real-world events., which is why modeling is so important!

3.2. Reclassify

These steps are useful in the rearrangement and distribution of cells; making it easier handle, in addition to being used as a standard in the form, and so it falls into a specific category into categories criteria, and is awarded the most Fit cells rank 10 According to the Common Scale, the most fit cells are assigned a score of 10, and the least fit cells are assigned a rank of 1., which leads to reclassification of each new class criterion It is automatically added to the program interface.

3.3. Weighted Overlay:

Geographic information systems (GIS) are important for planning and determining the best landfill sites, which are characterized by the application of standards, in addition to the ability to link all factors and variables in choosing the best landfill sites, which vary in nature between geo-economic, geological and environmental standards [3]. Other criteria include general acceptance criteria, climate criteria and other environmental standards. It is possible for researchers to obtain accurate results and properly plan for their future research by decreasing the importance of sensitive or potentially hazardous criteria that could pose a threat to the environment or humans if the selected criteria are met, and by matching them to criteria and conditions that have a negligible or non-existent negative environmental impact or humans. Absolutely. The use of weighted values was resorted to at this stage of the study in order to assign the actors in the success of the best sites and the largest weight of other elements, and there were numerous GIS tools available to calculate the effect of each variable. The weights of the criteria, as well as the large number of weighted overlay functions that go undetected, are at the heart of the problem, according to the researchers [4].

3.4. Buffer

In this case, it measures the size or extent of a campus service's impact on a specific point on the surface of the surrounding area's surrounding area. This is accomplished by entering a specific distance in accordance with the planning parameters; however, these spaces are not taken into consideration during the analysis and modeling processes [5]. This will have ramifications for not only the landmarks we intend to designate for them, but also for the relocation of landfills within this buffer zone. It will also have implications for other aspects of the planning process, such as the exclusion zone.

3.5. Elevation Digital Model

It's a representation of the terrain's appearance. that is generated sequentially from numerical data, and it is also known as a digital elevation model DEM [6]. Landfills are expensive, so it is preferable to locate in areas with limited space. When considering a longer period of time, it is possible that the same drop of more than 25% (excluding land) and a tendency to 5 percent will be acceptable for a landfill installation, as shown in Figure (2).

IOP Conf. Series: Earth and Environmental Science 1029 (2022) 012012 doi:10.1088/1755-1315/1029/1/012012

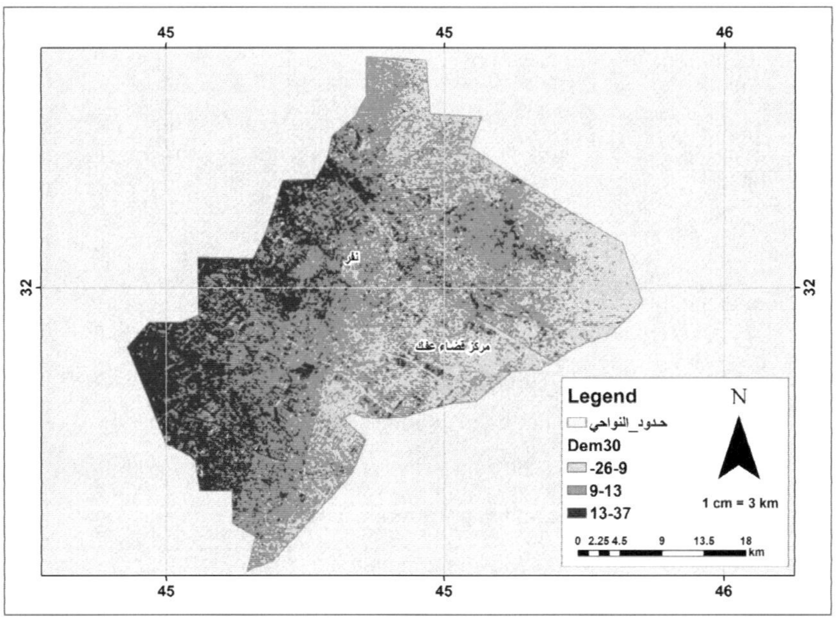

Figure 2. Map for Dem in

3.6. Land use

The use land map is obtained from the Ministry of Municipalities and then imported into the Geographic Information System (GIS), where it was recorded and then digitized, among other things. Table No. 1 is added to the map and a database is created and attached to it (1). As a starting point,

Table 1. types of land use and them classify

Type and land	symbols	RANGE
Area Organized Industrial	AOI	0
lands Agricultural	AG	0
Centers Urban	CU	0
Village	V	0
River	R	0
Forest	E	5
land Unoccupied	LU	10

convert the map into a grid system (Raster) and then work with the scores of relevance so that use of the rank 10 represents the highest suitability for the landfill site, and use of the rank 1 represents the lowest suitability for the landfill site, as shown in Figure 3.

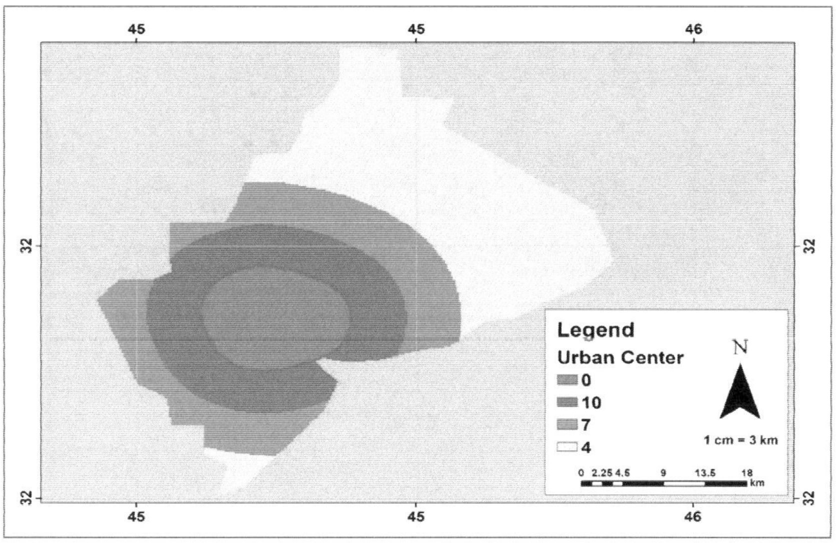

Figure 3. Map for Urban Center in Afak Qadhaa State after re-classified

3.7. cRoads

According to the reclassification and ranking system (1 - 10), the data with a rank of 10 is the most favorable for landfills and the data with a rank of 1 is the least favorable for landfills, as illustrated in Figure (4).

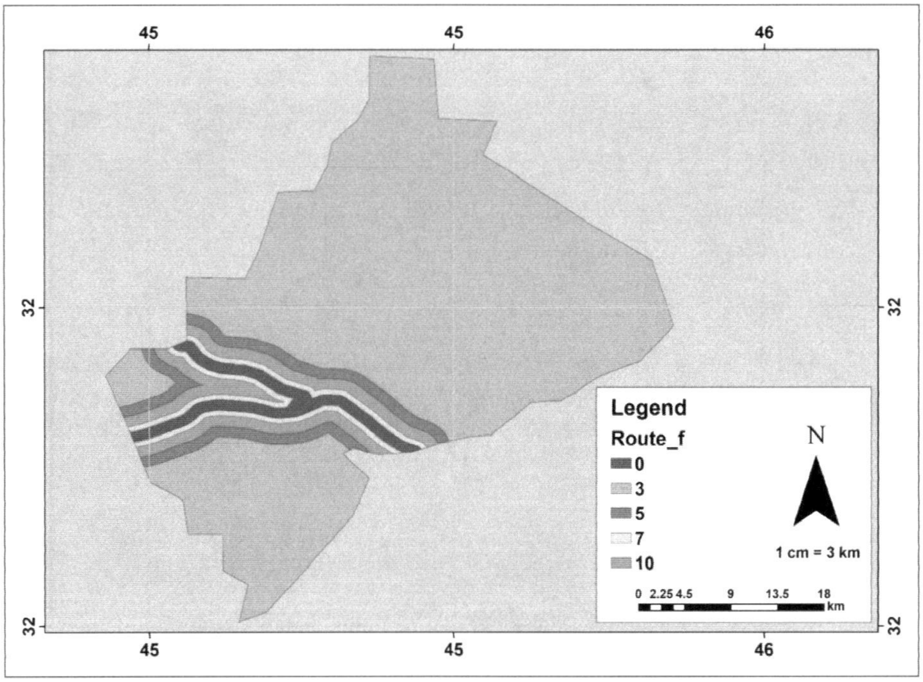

Figure 4. map for roads Afak Qadhaa after re-classified

3.8. River:

Rivers are considered one of the most important criteria and conditions to be taken into account when choosing the right place. Therefore, it is preferable that the landfill site be a thousand meters away from the river.. Reclassification and ranking of this map (5).

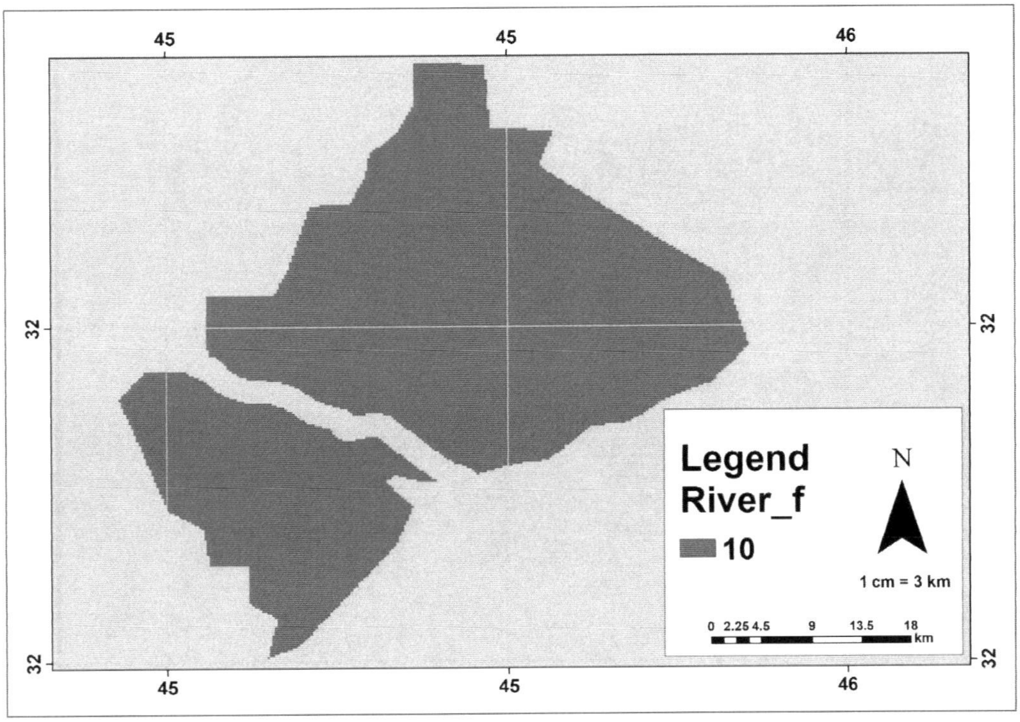

Figure 5. map for River Afak Qadhaa after re-classified

3.9. Soil Type:

In Afak Qadhaa, there are four different types of soils (Fig6). depression Basin soils A {6}, basin river soils, drained poorly B {5} silted hair and marsh soils C {9}, and river soils levee D are the soil types that fall into these categories (4). These soil types received scores of 10, 9, 7, and 5, respectively, on the soil test.

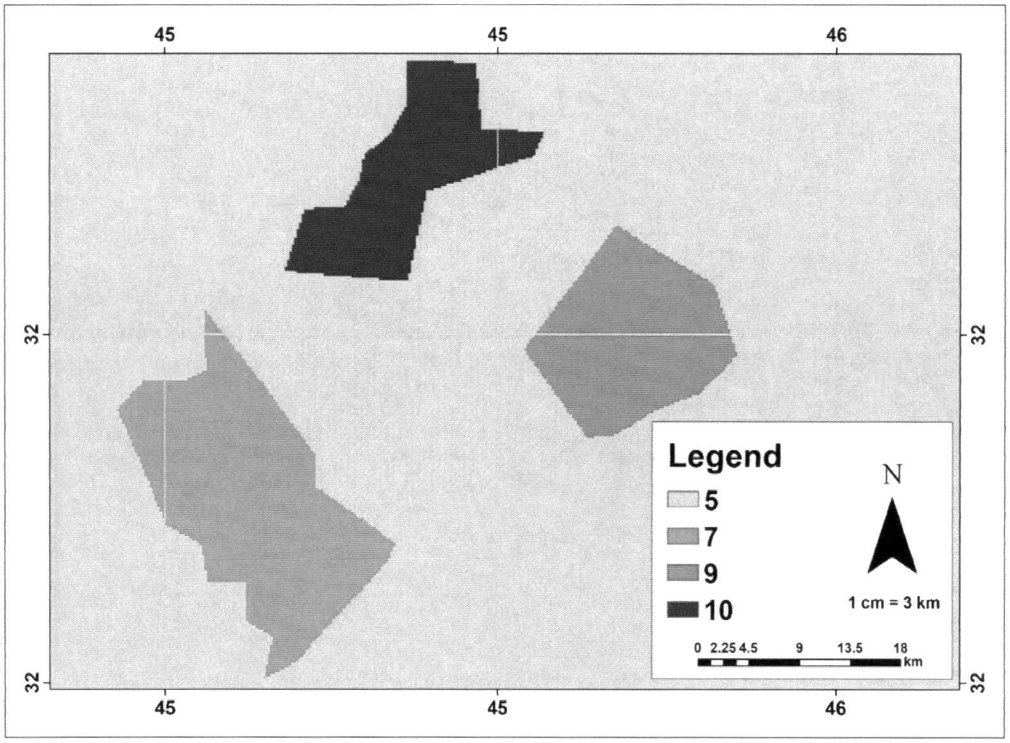

Figure 6. Map for Soil Type

3.10. Ground Water:

Groundwater has been reclassified and given ranks; the regions with the greatest amount of groundwater received the lowest rank, while the regions with the least amount of groundwater received the highest rank, as shown in Figure (7).

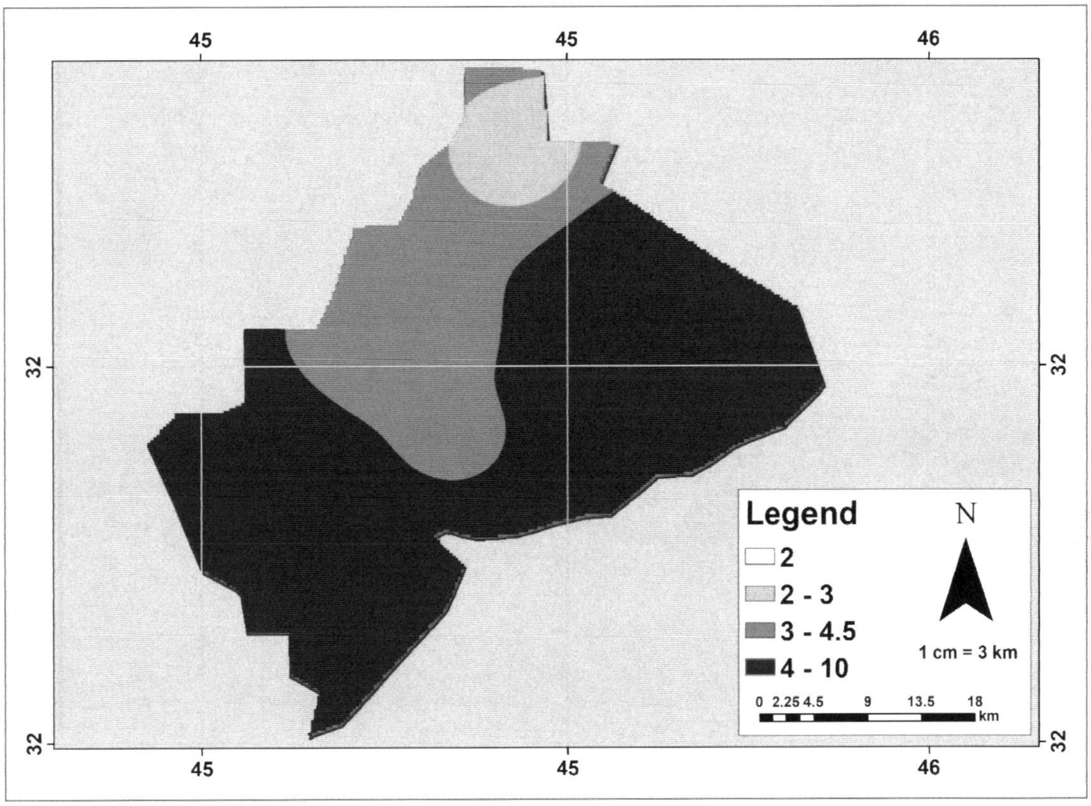

Figure 7. Map for Ground Water

4. THE RELATIVE WEIGHT FOR STANDARDS:

There are many GIS tools that can be used to determine whether a piece of land is suitable for use as a landfill site of various planning and processes, but we chose the function of (Weighted Overlay) which allows users to insert various weights provided that the total of 100%, an option that allows for greater flexibility in assessing landfill site suitability (Model Builder). We based our weights on our own personal experience, read it for numerous Arab and international references, and sought the advice of a professor supervisor during the course of developing weights for various standards. Finally, every of the standards have been assigned a weight based on their importance and influence, so that the sum of the weights equals 100% and table (2) depicts these weights.

Table 2. The weight relative of the main criteria

N.	Criteria	weight Relative %
1	Road	20
2	Valleys	20
3	Dem	10
4	Use land	17
5	Ground water	5
6	Annual mean precipitation	5
7	River	Limited

After determining the criteria and their classification, importance, and weighting, we built a model in the Arc GIS environment using Model Builder that relies on simplifying difficult problems and overlapping data and its relationship to geographical and descriptive data. As a result, the model's formulation is to simplify the problem and find a solution through (Flow Chart), Fig (8).

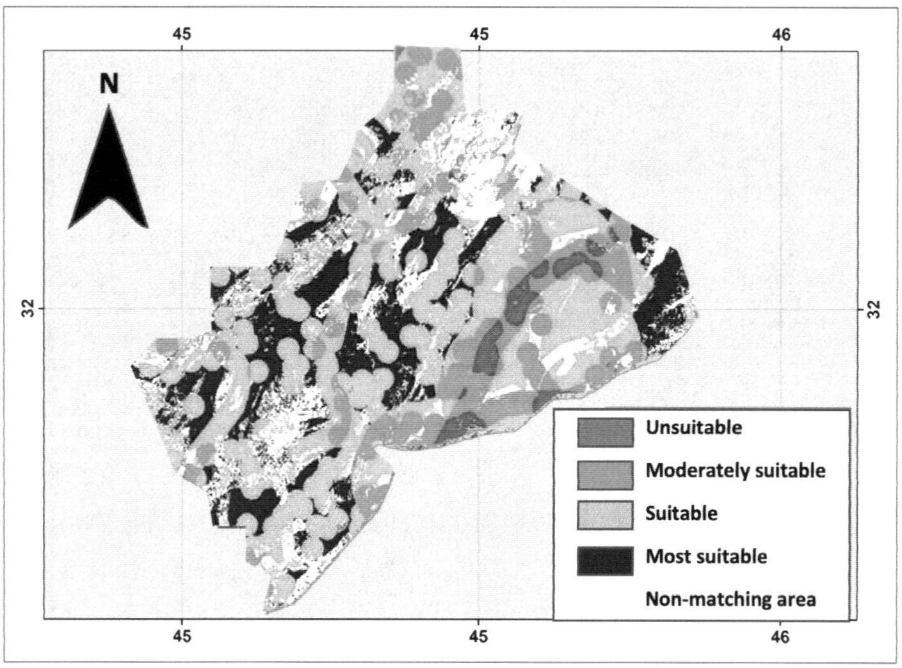

Figure 8. map Overlay of MCDM method matching and non-matching suitabile area

5. *Results and Discussions:*

The locations of landfills were determined after a succession of data layers were analyzed in detail. A new classification system was used to find the best locations, with relative weights and buffers applied to each category. The best sites were identified as follows: Urban, Village, Industrial Area, Roads, Airport, Valley, Land Use, DEM, Surface Water, Ground Water, Mean Annual Precipitation.

Urban Area: The urban area with the shortest distances between urban centers [7]and villages [2 km] and between industrial areas [250 m] has been calculated as 10 kilometers for centers urban [7] and 2 kilometers for villages and 250 m for industrial areas.

Roads: According to [8], a distance of more than (1) kilometer from major highways and roads should be avoid at all costs. Roads with the shortest possible distances on both sides for the research area are judged to be 2 kilometers on both sides.

Surface Water: A buffer zone of one kilometer on either side of a stream has been established [9].

The suitability map was converted from a raster to a polygon format, and the suitable sites were identified through the use of the selection function in the analysis tools incorporated in the Arc Toolbox software. In the study region, the areas of landfill sites that were found to be suitable were computed, see Fig (9)

6. *Conclusion:*

The following results will be obtained if the above-mentioned requirements are applied to Afak Qadhaa, for example, if the location is 10 kilometers away from the city center:

1. The dumping sites in Afak Qadhaa that are now in operation are suitable in light of the aforementioned circumstances. However, if we take into consideration the current and future increase of the population, we will discover that the most ideal spot is essentially a Al Fawara dumping ground. As for the current landfill, it is located near agricultural fields and close to residential areas

2.Because Afak Qadhaa is enormous in size, and with an eye on future expansions, we will impose special criteria on it in order to accommodate them. Consider the following example: if we choose to be 40 kilometers away from the metropolitan center, we will arrive at the dumping sites (5 and 6) depicted in figure 7. Furthermore, the Al Fawara dumping site will continue to be an appropriate location.

References:

[1] M. Z. Siddiqui, J. W. Everett, and B. E. Vieux, Landfill siting using geographic information systems: a demonstration, J. Environ. Eng., vol. 122, no. 6, pp. 515–523, 1996..

[2] J. Malczewski, Propagation of errors in multicriteria location analysis: a case study, in Multiple criteria decision making, Springer, 1997, pp. 154–155..

[3] Ewaid, S.H., Abed, S.A., 2017. Water quality index for Al-Gharraf river, southern Iraq. Egypt. J. Aquatic Res. 43 (2), 117–122. http://dx.doi.org/10.1016/j. ejar.201703001.

[4] Ewaid, S.H.; Abed, S.A.; Al-Ansari, N. Crop Water Requirements and Irrigation Schedules for Some Major Crops in Southern Iraq. Water 2019, 11, 756.

[5] Ewaid, S.H.; Abed, S.A.; Al-Ansari, N. Water Footprint of Wheat in Iraq. Water 2019, 11, 535.

[6] Ewaid, S.H.; Abed, S.A.; Al-Ansari, N. Assessment of Main Cereal Crop Trade Impacts on Water and Land Security in Iraq. Agronomy 2020, 10, 98.

[7] Salim, M. A. and Abed, S. A. (2019). The first oriental honey buzzard pernis ptilorhynchus (Temminck, 1821) in Iraq. Eco. Env. & Cons. 25 (1) ; pp. (1926-1929).

[8] Ewaid, S.H.; Abed, S.A.; Al-Ansari, N.; Salih, R.M. Development and Evaluation of a Water Quality Index for the Iraqi Rivers. Hydrology 2020, 7, 67.

[9] Ewaid, S.H., Al-Farhani, B.F., Abed, S.A. & Al-Ansari, N. (2020). Modeling of trihalomethane compounds formation in Baghdad water supply network. Scientific Review Engineering and Environmental Sciences, 29 (2), 136-144. doi: 10.22630/PNIKS.2020.29.2.12.

[10] Salam Hussein Ewaid et al 2020 J. Phys.: Conf. Ser. 1664 012143.

[11] Salam Hussein Ewaid et al 2021 IOP Conf. Ser.: Earth Environ. Sci. 722 012008

[12] Salam Hussein Ewaid et al 2021 IOP Conf. Ser.: Earth Environ. Sci. 790 012075

[13] Ahmed Alaa Kandoh et al 2021 IOP Conf. Ser.: Earth Environ. Sci. 790 012073.

[14] S. K. Yadav, GIS based approach for site selection in waste management, Int. J. Environ. Eng. Manag., vol. 4, no. 5, pp. 507–514, 2013..

[15] K. W. Holmes, O. A. Chadwick, and P. C. Kyriakidis, Error in a USGS 30-meter digital elevation model and its impact on terrain modeling, J. Hydrol., vol. 233, no. 1–4, pp. 154–173, 2000..

[16] S. Cointreau, Sanitary landfill design and siting criteria, Washington, DC World Bank, p. 6, 2004..

[17] A. R. Allen, Attenuation: A cost effective landfill strategy for developing countries, in Proceedings of 9th Congress of the International Association for Engineering Geology and the Environment, 2002, pp. 136–147..

[18] S. J. Goetz, R. K. Wright, A. J. Smith, E. Zinecker, and E. Schaub, IKONOS imagery for resource management: Tree cover, impervious surfaces, and riparian buffer analyses in the mid-Atlantic region, Remote Sens. Environ., vol. 88, no. 1–2, pp. 195–208, 2.

Environmental Influence of growth regulators and light condition on some callus traits of rosemary (Rosmarinus officinalis L.) in vitro.

Abeer Abdul Rahman Al-Saeedi[1] and Layth Sareea Al-Rekaby[2]

[1,2]College of Science- Department of Biology-University of Al-Qadisiyah-Iraq

Abstract. The study carried out in plant tissue culture laboratory at Department of Biology - College of Science - University of Al-Qadisiyah, included several experiments on the effect of different growth regulators concentration of auxins and cytokinins on the growth and characteristics of callus of *Rosmarinus officinalis L .* growing under light and dark conditions . Explant (tops and the lateral buds with a length of 0.5-1.0 cm) sterilized with ethanol solution and sodium hypochlorate, then growing on MS medium with added different concentrations of growth regulators and incubated in the growth chamber under the influence of lighting factor (light and dark) at a temperature of 25 °C, the data were taken every week and after 45 days of cultivation that . The best combination 2,4-D 1.0 mg/l with BA 0.5 mg/l that gave the largest callus weight in the light 3.75 gm, while in the dark ,the best combination 1.0 mg/l NAA and 0.5 mg/l BA gave the largest callus weight 3.60 gm, same combination which gave the highest percentage of callus emergence after 2 weeks was 100% in the dark and 95% in the light. The texture , color and shape of the callus varied as the combination

Key wards : rosemary , *Rosmarinus officinalis* L., growth regulators, light condition, callus traits, plant tissue culture and in vitro

1. Introduction

Rosmarinus officinalis L. is a medicinal plant of great economic importance that belongs to the Lamiaceae [1]. Native of the plant in southern Europe and the Mediterranean basin, then spread to all parts of the world [2] [3]. It is a perennial, aromatic, evergreen plant that has several names, including sea dew, dairy pebbles, Arabica hashish , rosemary and memory herb. [4] The ancient Greeks and Egyptians used it to activate memory [5]. It is a powerful antioxidant, anti-bacterial and anti-fungal [2], used to treat rheumatism, activates blood circulation and the nervous system, improves digestion, treats headaches and colds [6] [7]. It is widely used as a spice , and its distinctive aroma is used in the manufacture of soap and cosmetics [8], it has medicinal properties to improve memory, relieve muscle pain and stimulate hair growth [8], used as a natural food preservative that can replace industrial preservatives [9]. Antispasmodic, anticancer, antitumor, antimicrobial and anti-inflammatory [10]. Its medical importance is due to the fact that it contains essential oil, as the chemical composition of these secondary metabolites varies with different environmental conditions [11]. Recent studies have shown that plant extracts are an effective inhibitor of food pathogens [12]. Its biological properties are due to its high content of secondary metabolism such as carnosic acid, carnosol, rosmanol, rosmarinic acid [13]. Rosemary considered one of important sources of extracting phenolic compounds, as its essential oil contains many chemicals such as flavonoids, carmasol, camosic acid and volatile oils [14]. In recent years, trends have increased to use the tissue culture technique as an alternative to traditional agriculture to produce secondary metabolic materials that are available throughout the year without being restricted to a specific season, in addition to the high purity of the produced materials compared to the manufactured [15]. Callus cultures have been widely used to explore the possibility of producing useful secondary metabolites through biotechnological methods, wherein callus growth and cell differentiation are stimulated by using growth regulators and controlling environmental conditions for growth [16], and the plant is the traditional source of many chemicals that are used as medicines [17].

2. Materials and methods

Experiments conducted in plant tissue culture laboratory at Department of Biology-College of Science -University of Al-Qadisiyah for the period from October 2021 to February 2022 on rosemary plant included :

2.1. Preparation and sterilization of Explants

Explants were taken from the developing apex and lateral buds of rosemary plant with a length of 0.5-1.0 cm. The plant parts were sterilized before planting using ethyl alcohol at a concentration of 70% for 1 minute. Then the plant parts were sterilized with sodium hypochlorite following the experimental design with three concentrations (4, 8, 12) % and at intervals of 15-20 minutes, and the best result was using an 8% concentration for 15 minutes, which gave the lowest percentage of contamination. After the sterilization, the plant parts were washed by sterile distilled water 3 times for 5 minutes in in laminar flow.

2.2. Prepare MS media 2-

Murashige & Skoog media (MS pv, 1972) was used. Sucrose was added as an energy source. Growth regulators were added with three different combinations: (NAA 1.0 mg/l & BA 0.5 mg/l) , (2.4-D 1.0 mg/l & BA 0.5 mg/l) and (IAA 1.0 mg/l &BA 0.5 mg/l). Then the medium was sterilized using an autoclave for 20 minutes at 121 °C and a pressure of 1 bar, then the sterilized explant cultured on MS media in laminar flow and then placed in the growth chamber at a temperature of 25 °C under the influence of the lighting factor (light and dark)Fig 1, and the data were taken 45 days after cultivation For all experiences .

Figure 1. growth chamber

3. Results and discussion

3.1. Influence of growth regulators and light condition on callus color of rosemary plants after 45 days

Table (1) shows that different combinations of growth regulators caused significant effect on color of callus, as we see that combination of 1.0 mg/l NAA with 0.5 mg/l BA gave green callus in light and pure white callus in the dark, and treatment of 2,4 -D 1.0 mg/l with BA 0.5 mg/l in light gave a variation in the color of the callus between green and light brown, while in the dark the color of the callus varied between brown and white, combination of IAA 1.0 mg/l with BA 0.5 mg/l gave green callus Color in the light and brown callus in the dark . Fig 2.

3.2. Influence of growth regulators and light condition on the callus strength of rosemary plants after 45 days.

Table (2) shows a difference in the texture of the callus with the different combinations of growth regulators and the difference in the lighting factor. Whereas, 1.0 mg/l NAA with 0.5 mg/l BA gave a semi-solid callus in the light and a callus with a solid texture in the dark, while 2,4-D 1.0 mg/l with

BA 0.5 mg/l gave a brittle callus in the light. In the dark, solid callus, IAA 1.0 mg/l combination with BA 0.5 mg/l gave brittle callus in the light and semi-solid callus in the dark .

3.3. Influence of growth regulators and light condition on callus shape of rosemary after 45 days

Table (3) shows the different forms of callus with different growth regulators and light factor. NAA 1.0 mg/l with BA 0.5 mg/l gave a globular callus in the dark and a pear-shaped callus in the light, while the combination 2,4-D 1.0 mg/l with BA 0.5 mg/l gave a cylindrical callus in the light and a cluster callus In the dark, of IAA 1.0 mg/l with BA 0.5 mg/l gave a callus with an irregular ring shape in the light, but in the dark the callus shape varied between spherical and clustered . Fig3.

3.4. Influence of growth regulators and light condition on the period of callus emergence of rosemary plants after 45 days

Table (4) shows a difference in the periods of callus emergence with different concentrations of growth regulators under the effect of the lighting factor after 45 days of planting, where the best combination of callus emergence in a short period was NAA 1.0 mg/l with BA 0.5 mg/l, which gave a percentage of 100% After only two weeks of cultivation in the dark, in the light, it gave a percentage of 95% in the same period, while a combination 2,4-D 1.0 mg/l with BA 0.5 mg/l gave a percentage of 100% after 6 weeks in the light and dark, and a combination of IAA 1.0 mg/l With BA 0.5 mg/l it gave 90% in the light and 75% in the dark after 6 weeks of cultivation . .

3.5. Influence of growth regulators and light condition on callus weight of rosemary after 45 days

Table (5) shows a significant effect on the size of the callus by different growth regulators and the lighting factor. Where the results showed that 2,4-D 1.0 mg/l with BA 0.5 mg/l gave the highest callus weight , which reached 3.75 gm in the light, followed by 1.0 mg/l NAA and 0.5 mg/l BA reached callus weight of 3.15 gm and combination 1.0 mg/l IAA and 0.5 mg/l BA reached callus weight of 1.75 gm, while in the dark, NAA 1.0 mg/l and BA 0.5 mg/l gave the highest callus weight of 3.60 gm, followed by 2.4-D 1.0 mg/l and BA 0.5 mg/l reached callus weight of 2.50 gm, followed by IAA 1.0 mg/l and BA 0.5 mg/l reached callus weight of 1.10 gm. Based on the above result, significant increasing in the growth and characteristics of callus of Rosmary. due to different growth regulators concentration of auxins and cytokinins on growing under light and dark conditions . These increasing are typically a result of absorbing additional quantities of elements which meet's the need of the explant for the continuity of its vital activities to encourage growth .

Table 1. Influence of growth regulators and light condition on callus color of Rosmary after 45 days

growth regulators **lighting factor**	NAA + BA	2,4-D + BA	IAA + BA
Light	Green	Green + light brown	Green
Dark	pure white	Brown + white	Brown

Table 2. Influence of growth regulators and light condition on callus texture of Rosmary after 45 days

growth regulators **lighting factor**	NAA + BA	2,4-D + BA	IAA + BA
Light	semi solid	crisp	crisp
Dark	Solid	Solid	semi solid

Table 3. Influence of growth regulators and light condition on callus shape of Rosmary after 45 days

growth regulators **lighting factor**	NAA + BA	2,4-D + BA	IAA + BA
Light	pear-shaped	cylindrical shape	Irregular throat
Dark	spherical	clustered	Spherical + clustered

Table 4. Influence of growth regulators and light condition on callus emergence period of Rosmary after 45 days

and light condition on	and light condition on	and light condition on	and light condition on	and light condition on
Light	90%	50%	45%	2
	97%	90%	70%	4
	100%	100%	90%	6
Dark	100%	60%	40%	2
	100%	90%	60%	4
	100%	100%	85%	6

Table 5. Influence of growth regulators and light condition on callus weight of Rosmary after 45 days (gm)

growth regulators Lighting factor	NAA + BA	2,4-D + BA	IAA + BA
	3.15	3.75	1.10
	2.30	1.50	1.75
Light	1.55	3.30	0.60
	2.65	2.50	0.75
	0.80	1.95	1.05
	3.60	2.00	1.10
	0.60	2.50	0.90
Dark	2.50	0.85	0.75
	1.15	0.80	0.70
	0.60	1.05	0.90
R.L.S.D.	0.22	0.29	0.17

Figure 2. Influence of growth regulators on callus color of Rosmary under the influence of light factor after 45 days

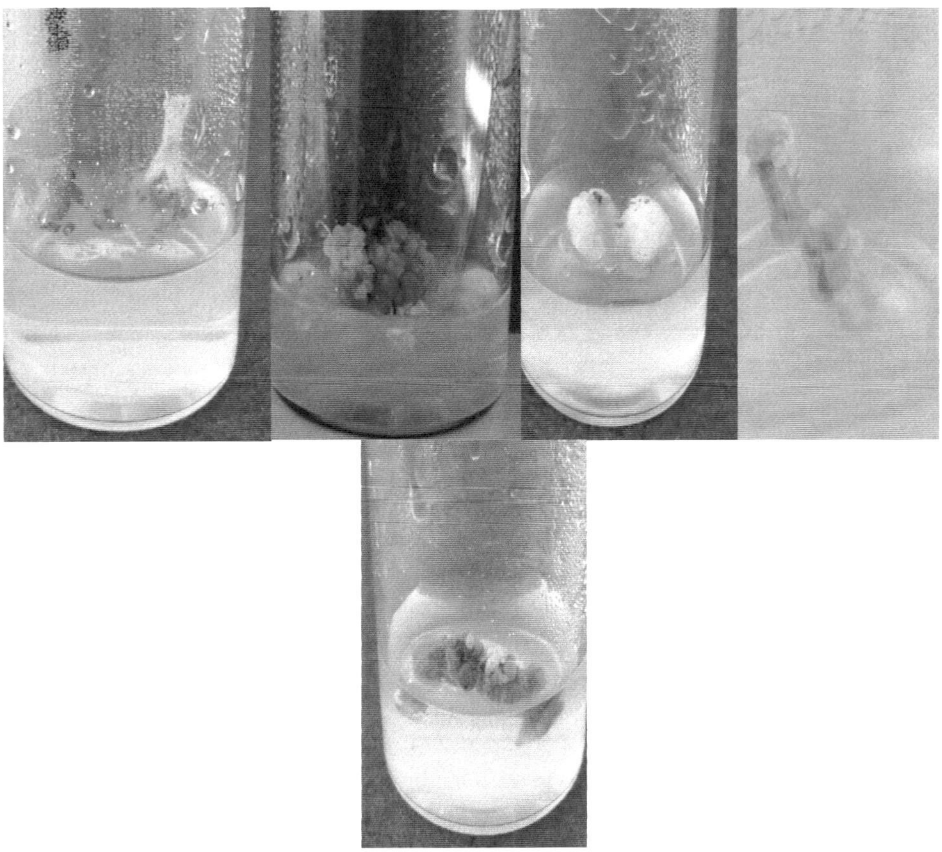

Figure 3. Influence of growth regulators on callus shape of Rosmary under the influence of light factor after 45 days

4. Conclusion

The best combination 2,4-D 1.0 mg/l with BA 0.5 mg/l gave the largest callus weight 3.75 gm in the light , while in dark the best combination NAA 1.0 mg/l with BA 0.5mg/l gave the largest callus weight 3.60 gm .

The best combination of callus emergence was NAA 1.0 mg/l and BA 0.5 mg/l, which gave the highest percentage of callus emergence after 2 weeks was 100% in the dark and 95% in the light

The texture of the callus varied as the combinations 2,4-D 1.0 mg/l with BA 0.5 mg/l, IAA 1.0 mg/l with BA 0.5 mg/l gave brittle callus, while NAA 1.0 mg/l with BA 0.5 mg/l gave semi-solid callus. In the light. in the dark, NAA gave 1.0 mg/l with BA 0.5 mg/l and 2,4-D 1.0 mg/l and BA 0.5 mg/l solid callus and IAA 1.0 mg/l with BA 0.5 mg/l semi-solid callus .

The shape of the callus differed according to the growth regulators, light and dark. combination of NAA 1.0 mg/l with BA 0.5 mg/l gave spherical callus in the dark and pear in the light, and 2,4-D 1.0 mg/l with BA 0.5 mg/l gave cylindrical callus in the light and clustered in the dark, while IAA 1.0 mg/l with BA 0.5 mg/l gave callus with irregular annular shape in the light and a contrast between spherical and clustered in the dark.

NAA 1.0 mg/l with BA 0.5 mg/l and IAA 1.0 mg/l with BA 0.5 mg/l gave green color in the light and 2,4-D 1.0 mg/l with BA 0.5 mg/l gave a contrast between green color and Light brown . in the dark NAA 1.0 mg/l with BA 0.5 mg/l gave a pure white callus, and IAA 1.0 mg/l with BA 0.5 mg/l gave brown callus, while a treatment 2,4-D 1.0 mg/l with BA 0.5 mg /l variation of the color of the callus between brown and white.

References

[1] Eva SB, Maria HT, Attila H, Csilla R and Szollosi V (2003). Antioxidant effect of various rosemary (Rosmarinus officinalis L.) clones . Acta Biologica szegediensis, 47 (1-4):111-113..

[2] Couladis M, Tzakou O, Verykokidou E(2003) . Screening of some Greek aromatic plants for antioxidant activity – J Phytother Res , p.194-196..

[3] Bradley , British herbal compendium, A handbook of scientific information on widely used plant drugs, British herbal Medicine Association, Bournemouth, 2006..

[4] Perez MB, Calderon NL, Croci CA (2007) Radiation-induced enhancement of antioxidant activity in extracts of rosemary (Rosmarinus officinalis L.). Food Chemistry 104:585-592..

[5] Marchiori, Vanderli F. 2004. Fundacao Herbarium Associacao Argentina de Fitomedicina, Rosmarinus officinalis L. Fitomedicina Herbarium,pp 34..

[6] Foster, S. and Tyler, V(1999) . The honest herbal ; A sensible guide the use of herbs and related remedies . 4th ed . Haworth Herbal Press., 321-322..

[7] Peter , K. V., .2004. Hand Book Of Herbs And Spices. CPC press LIC, North and South America. pp 215-224..

[8] Jordan MJ, Lax V, Rota MC, Loran S, Sotomayor JA (2013) Effect of bioclimatic area on the essential oil composition and antibacterial activity of Rosmarinus officinalis L. Food Control 30: 463-468..

[9] Frankel, E. N., S. W. Huang, R. Aeschbach, and E. Prior 1998. Antioxidant activity of a rosemary extract and its constituents, carnosic acid, carnosol, and rosmarinic acid, in bulk oil and oil-inwater emulsion. J. Agric. Food Chem,44: 131-135..

[10]Fernandes, L.J.; N. Zhi; CL. Aleson; perez; A.J.A. and V. Kur. 2005. Antioxidant and antibacterial activitiec of natural extracts, application in beef meat balls . Meat Sci., 69: 371-380..

[11]Macias-Alonso, M.; Sancen, S.R.; Marrero,J2019.G.Carnosic acid and its derivatives: Diterpenes of biological interest . Biomed . J. Sci. Tech. Res. 16, 12172-12173..

[12]Macias-Alonso, M.; San Andres, L.; Cordova-Guerrero, I.; Estolano-Cobian, A,; Diaz-Rubio,L,;Marrero, J.G2019. Inhibition of squalene synthase of rat liver by abietane diterpenes derivatives . Nat . Prod. Res..

[13]Carovic-Stanko, k,; Petek,M,; Grdisa, M.; Pintar , J.; Bedekovic, D.; Herak Custic, M.; Satovic, Z.2016 ,.Medicinal plants of the family Lamiaceae as functional foods-A review. Czech J. Food Sci. 34, 377-390..

[14]Okamura, N., H. Haraguchi, K. Hashmoto and A. Yagi. 1994. Flavonoids in Rosmarinus officinalis L. leaves. Phytochem.37:1463-1466..

[15]Vanisree, M.; L. Chen-Yue; S.-F. Lo; S. M. Nalawade; C. Y. Lin and H.-S. Tasay . 2004 . Studies on the production of some important secondary metabolites from medicinal plant by plant tissue culture. Bot. Bul. Acad. Sin., 45:1-22..

[16]Nagesh, K.S., C. Shanthamma and T. Pullaiah. 2010. Somatic embryogenesis and plant regeneration from callus cultures of Curculigo orchioides Gaertn. Indian J. Biotechnol., 9: 408-413..

[17]Leena, T. and Jaindra, N. T.(2003). Role of biotechnology in medicinal plants. Tropical Journal of Pharma. Res., 2(2): 243-253..

Phase Change Material of (100-200) C: Review and Taxonomy

Mushtaq A. Al-Furaiji, A.V. Baranenko,V.Y. Zakharova, and O. S. Malinina

Faculty of Energy and Ecotechnology, ITMO University, 49 Kronverksky prospect, St. Petersburg, 197101, Russian ***Federation***

Abstract. Phase Change Material (PCM) is the material where its properties changed with the temperature change. If the application is heating, the process is called Latent Thermal Energy (LTE) and it can be stored. In practical application especially for the 100 C to 200 C, where the material is converted from solid to liquid, the energy management can be efficient and the energy storage is applicable. When the material is melted, the energy can be stored in heating operation and the inverse operation which means that the material can lose the energy stored when cooling or when stored in constant temperature. To overcome this, using PCM in medium temperatures (100-200 C) becomes important in many fields such as greenhouses, solar panels, thermal management in concrete, and so on. This paper discusses the recent developments in PCM such as using nanomaterials that keep the temperature for more times and encapsulation to enhance the heating operation. The taxonomy for the use of PCM is proposed in this paper based on the material requirements and types that can be used in this temperature range, and the measurement techniques in the physical form of material and thermal conductivity enhancement. A case study of 10 materials used in the required range is also discussed based on their temperature on the different life cycles of operations. At the end of this paper, a discussion of thermal energy management especially for solar applications is illustrated. Some possible future works are also proposed.

Keywords: PCM, LTE, Thermal conductivity, Nanomaterials, encapsulation,

1. Introduction

The demand for electricity consumption in the next decade will be doubled because of the increasing use of power due to the increase of industry applications and manufacturing [1]. Besides this, 80% of the energy sources come from the use of Fossil fuels which is the primary source that most industrial sectors use [2]. This is a challenge in the upcoming decade because increasing the use of fossil fuels means that increasing the CO2 emissions by 84% which affects the climate and increases the climate change problem which is an important issue today (Maduabuchi&Mgbemene 2020). In addition to this, the attention of using renewable energy has become important for researchers and industries to enhance the performance of solar panels and enhance the energy storage in batteries [3] as shown in Figure 1.

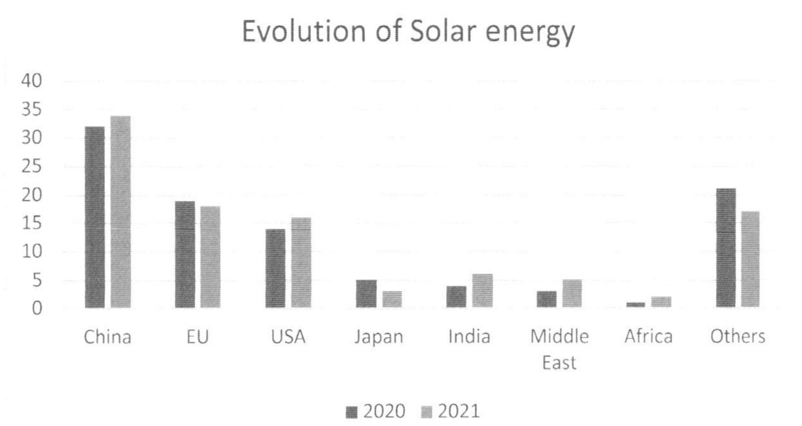

Figure 1. Evolution of solar energy worldwide

Figure 1 shows solar energy usage worldwide. it shows that the high use of solar energy is in China than in the European Union (EU) while it also shows that there is an increase in the solar energy usage from the year 2020 to the year 2021 and the total energy comes from the solar systems reaches 153.8 GW while it was 140 GW at 2020 [4].

Using Photo-Voltaic (PV) system rises another challenge of using these systems which is the time-dependent energy [5], which means that the energy comes from the solar panels and the batteries used in systems is dependent to the weather conditions [6]. This makes the attention of thinking about enhancing the energy storage ways to improve the energy system performance to satisfy stability delivering of energy [7]. Heat storage especially for applications above 100 °C is important for waste heat recovery systems where water could not be used for this purpose [8]. Because of this, the most way to store energy is the use of Latent Heat Storage (LHS) way [9]. The importance of this way is its ability to store a large amount of heat at different temperatures and its ability to gather energy from different heat sources to change its phase. Using such materials that can change their phase due to the temperature is called Phase Change Materials (PCM) [10].

One of the most important uses of the PCM is its use in thermal management applications because of its behavior of storing and releasing thermal energy during converting to another phase of operation [11]. Because of this, thermal stability is one of the things that should be taken into consideration during all the phase transaction procedures with some consideration on the enthalpy of the PCM and its temperature [12]. The reason for thinking about thermal stability is that to have proper utilization behavior of the LHS along some defined time. It is important to mention that the same amount of energy is absorbed by the environment during the melting process of the PCM material which means changing from a solid-state to a liquid state [13]

Two main parameters should be optimized which are the melting point and the heat capacity. these two parameters are very important to optimize the input/output temperature of the PCM and the heat capacity of TES according to the latent and sensible heats [14] For materials in the range of 100 °C to 200 °C, the required properties are:

- Release and absorb large amounts of energy.
- Have a fixed and determined phase change temperature.

- the stability of the PCM should be high which means that the PCM materials should be unchangeable in their physical and chemical properties and not be affected by the time of operation. This means that it is important to ensure that the PCM should not be affected by the cycles of operation and should remain stable for a long time of operation without any degradation in its energy [15]
- PCM should be safe and Non-hazardous because it is used in contact with people.
- Economical; It doesn't matter how well a substance can perform as a PCM if is prohibitively expensive.

1.1. Existing surveys

Some related works discuss the use of emerging technologies in agriculture as shown in Table I. [16]discuss the PCM used in heat storage using PCM from 120 oC to 200 oC. They evaluate sixteen materials in this temperature range. The study aims to select some materials with stable properties along any cycle of operation. They perform 0 cycles, 10 cycles, and 100 cycles of operation for each material and evaluate its performance. Their experimental results show that there are some materials in this range that have stability properties and can be used in heat storage applications using PCM. The limitation of this study is that it only represents the experimental study of these materials. There is no mention of their application and requirements.

The general review of the heating system available is presented in [17]. They review the storage methods of heating energy including sensible heat storage, Latent Heat Storage (LHS), and thermochemical energy storage. All these methods use PCM to enhance the efficiency of the system. They concentrate on PCM materials that are stable during the temperature rising above 100 oC to allow the stability of the thermal storage applications. They present the PCM types with some mention of their advantages and disadvantages. The review also discusses some methods of improving heating such as cascaded LHS, encapsulation, and shape-stabilization. Some applications of using PCM are also presented such as using PCM in buildings, power generation, and vehicular industry. There is no mention of nanomaterials as a future consideration of PCM.

[18]study the use of PCM in greenhouse applications to usefully use the thermal heating that comes from solar energy to decrease the use of fuel which is considered high in some countries. They experimented to enhance the performance of PCM used in the greenhouse. Their results say that the mean energy efficiency improved by 40% and 33% for heating systems of PCM.

[19] gives a good review about PCM materials, their usage in thermal energy, and their different properties that should be optimized to have a stable PCM. They focus on solar energy as a source of heating to decrease the use of fuel. They discuss the usage of PCM for the different temperatures to store solar energy. They study various types of PCM available at different temperatures. They conclude some considerations when choosing the proper PCM material such as its thermal and physical properties, the process of heating, and the data gathered from the operation. Six PCM materials have been discussed in this paper. They mention encapsulation as one of the techniques used in PCM performance-enhancing and mention some of the properties affecting the performance of PCMs. There is also a discussion of the economic considerations of the PCM. There is no mention of nanomaterials or general requirements for the use of PCM above 100 oC.

[20]discuss the advantages of using PCM in constant temperature heating to have high storage density for industrial applications. They discuss eleven PCM materials behavior in the temperature range 120 C to 150 ◦C that can be used for Thermo-Gravimetry (TG) and Differential Scanning Calorimetry analysis (DSC).

Table 1. comparison of some existing surveys

Ref. / Year	Interest	Remarks
(Gasia et al. 2017)	120 to 200 ◦C under partial load operating conditions	- Study 16 materials that are used as PCM in the range of 120 to 200 ◦C. - They evaluate PCM materials according to their stability when increasing cycles of operation. - There is no mention of techniques used to enhance thermal storage such as encapsulation.
(Jouhara et al. 2020)	Review of PCM materials used in several applications	- Makes a review of using such PCM materials in some applications such as buildings and vehicular industry. - They review the use of cascaded LHS, encapsulation, and shape-stabilization to improve thermal heating. - There is no mention of nanomaterials as a promising technology to improve PCM.
(Yan et al. 2020)	Focusing on enhancing the heat of PCM to be used in greenhouses.	- Focusing on decreasing the use of fuel and enhancing the heat operation of PCM. - They focus on greenhouse applications. - There is no mention of nanomaterials.
(Wani&Loharkar 2017)	Focusing on thermal and physical properties, the process of heating, and the data gathered from the operation.	- Discussing thermal and physical properties, the process of heating, and the data gathered from the operation for PCM above 100 oC. - They make a review of the economic consideration of using PCM and mention some of the properties affecting the performance of PCMs. - No mention of nanomaterials or general requirements of the use of PCM above 100 oC.
(Haillot et al. 2011)	Review of using eleven PCM materials in the range of 120 °C to 150 ◦C.	- Make a review of using 11 materials used in PCM applications. - Discuss Thermo-Gravimetry (TG) and Differential Scanning Calorimetry analysis (DSC).
Our survey	**Full review of using PCM in thermal energy storage in the range of 100 oC to 200 oC.**	**- Gives a full review of using PCM material in the LHS application.** **- Mention of recent advancement PCM materials such as nano-materials.** **- Discussing encapsulation to enhance the healing process of PCM.** **- Propose a taxonomy of the PCM used.**

1.2. This survey

To the best of our knowledge, we are the first to present the taxonomy and discussion of the use of PCM in the range of 100Co to 200 Co. Table I represents the comparison between our survey and other existing surveys which shows the novelty of our contributions compared to existing surveys. We can conclude our contributions as the following:

- We present and investigate recent developments in PCM especially for solar energy applications and thermal energy storage such as discussing the nanomaterials and encapsulation of PCM to enhance the PCM performance.
- We derive a full taxonomy for the PCM in this range of temperature based on the material types, Thermal conductivity, and PCM containers.
- At the end of the review, several recent and future trends of thermal energy management especially for solar energy applications have been mentioned and discussed such as greenhouses, concrete design, and solar energy applications.

The structure of this survey is divided into section II, discussing recent developments in PCM such as nanomaterials, thermal conductivity, leakage elimination, and encapsulation of PCMs. A taxonomy for the PCM in the discussed range on different parameters is presented in section III. Section IV mentions the recent and future trends applications in PCM. Finally, the conclusion and future recommendations are given in section V.

2. Recent Developments in PCM

This section presents two recent developments in PCM. We discuss these advances in detail as:

2.1. Nanomaterials

To enhance the performance of the PCM as a material of heat storage used widely in thermal applications such as solar applications, nanomaterials were added to the PCM to enhance its heat transfer and to enhance its stability during charging and discharging operation. Nanomaterials added are solid material in nano-sizes and should be a conductor to enhance heat transfer and to be used as a storage device. The studies in this field reveal that employing nanotechnology is an efficient method for heat transfer improvement. [21]says that adding nanomaterials to the PCM allows reducing the time required for the melting process which means that reducing the time required for charging and discharging operations when using as a storage material. For choosing the suitable nano-material, it is important to employ it with solid-state nano-particles with some consideration of the Rayleigh and Darcy numbers as two of the most important operating parameters. [22] propose Nano-enhanced PCM (NePCM) materials at high temperatures. They recommend using Al2O3, copper, and carbon as efficient nanomaterials in addition to paraffin PCM.

2.2. Encapsulation of PCMs

One of the past, present, and future technology to enhance the performance of PCM to save energy especially for solar energy is the encapsulation process before using the PCM to mitigate the major drawbacks such as limited thermal conductivity and leakage that prevent the PCMs from wide application in desired areas [23]. Encapsulation for the PCM material used in 100 oC to 200 oC aims to protect the PCM from outside conditions such as the environmental effect on the PCM and to enhance the thermal transfer by increasing the PCM size especially the surface size [24]. Recently, the encapsulation has two techniques to be used which are:

- The use of multifunctional materials is used to enhance the thermal properties of the PCM. It is almost a shell encapsulation.
- Efficient use of the wasting heat energy comes from the production of capsules. This will lead to having a passive system for thermal organization and control. This also led to the use of on-demand energy storage.

Presently, nano-encapsulation for nano-materials is used where capsules with sizes <1 μm are used [25] [26]propose a Layer-by-Layer (LbL) encapsulation technique to stabilize the PCM size during the change of heat and the change of operation cycles. There are two main materials used in encapsulation efficiently which are Polyelectrolytes Poly and Poly Sodium Salt (PSS). These two materials are used with the help of using Bovine Serum Albumin (BSA) and/or Sodium Dodecyl Sulfate (SDS) to construct the encapsulation shell to allow stability in thermal energy along with any cycles of operation.

3. PCM Proposed Taxonomy

A proposed taxonomy that is presented in Figure 2 is illustrated in this section using four main issues which are materials, Thermal conductivity, requirements, and measurements techniques.

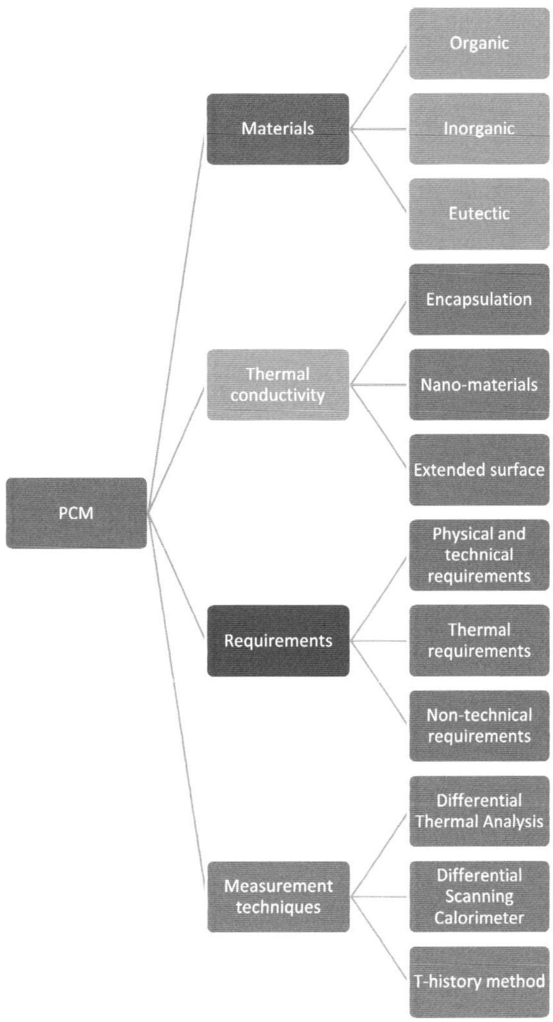

Figure 2. PCM proposed taxonomy

3.1. Materials

Table 2 shows the advantages and disadvantages of the three types of PCM materials which are organic, inorganic, and eutectic. All these types of materials are used in the temperature range 100 °C to 200 oC as mentioned in [27] [28] [29]. Organic materials are the most commonly used materials for high-temperature applications because of their works without phase segregation and consequent degradation of their latent heat of fusion. Inorganic materials are further classified as salt hydrate and metallics. This PCM is also good to be used because their heats of fusion do not degrade with cycling as shown in Table 3.

In some cases, a PCM application requires reducing the heating temperature without affecting the properties of the material used and without affecting the chemical compound [30] In this case, Eutectic PCMs are used. It can be defined as a combination of two materials especially materials that do not interact with each other to perform new properties of the PCM and a new operation. It is important to say that one of the two combined materials should have low melting point than the other one [31]. There are three combination types of the two materials which can be organic-organic, organic-inorganic, and inorganic-inorganic materials as in Molten salt. It is the wide used PCM used in the high temperature and middle one ranged from 120°C to 1000°C [32]. It is also used widely in solar panels application in the high-temperature range from 100 to 200°C. this material is useful to be used in solar energy applications because of its properties which are 1880 kg/m3 density, low vapor pressure compared with other materials, high transferring heat with low chemical properties [33]. A combination between NaNO3 and KNO3 is the basic mixture of the molten salt used in solar applications with a mixture percentage around 6:4 respectively [34]. HTEC is another molten salt Eutectic material that consists of three PCMs combined as NaNO3, KNO3, and NaNO2. However, besides all the advantages that Eutectic materials give, its challenge of not having an optimal material to be combined is a drawback of using this type and limiting its use in thermal applications.

Table 2. comparison of material types

Material	Advantages	Disadvantages
Organic	• No corrosiveness. • Low undercooling effect. • Stability of chemical and thermal properties.	• Low enthalpy. • Low thermal conductivity. • Inflammability
In-organic	• High enthalpy. • Subcooling effect.	• Corrosion. • The effect of Separation and segregation of phase. • Very low thermal stability.
Eutectic	• Sharp melting point. • Properties can be tailored to match specific requirements. • High volumetric thermal storage density.	• Limited data on thermo-physical properties for many combinations High cost.

Table III shows some of the work of the [35] which makes some measurements on sixteen materials used in the range 120 oC to 200 °C which is accepted with our review. They perform the work on different cycles of operation to evaluate the stability of these materials in this range of temperature. The fixed temperature for all cycles means the stabile material to be used in thermal storage. They start measurements by no cycle of operation, Benzamide gives the lowest temperature which is 128 oC while the Potassium thiocyanate and Hydroquinone give the highest temperature reaches 172 oC. While these two highest temperature materials give no stability of temperature after 10 and 100 cycles of operation. This variation of temperature affects the thermal storage process especially for solar energy panels and storage batteries.

There are two materials under the experiment that shows a decrease in their temperature properties after 100 cycles. These two materials are sebacic acid and D-mannitol. Their temperature reaches 122 oC and 155 oC respectively compared to their initial temperature which is 132 and 166 respectively. The conclusion here is that in some materials type, there is no need for increasing cycles of operation because the temperature will decrease. However, benzamide, HDPE, dimethyl terephthalate, adipic acid, potassium thiocyanate, and benzamide give a stable temperature range during any number of operation cycles.

Table 3. materials temperature comparison for different operation cycles

Material	Temperature C		
	Cycle 0	Cycle 10	Cycle 100
Benzamide	128	127	126
HDPE	128	129	128
Sebacic acid	132	129	122
Phthalic anhydride	130	131	131
Dimethyl terephthalate	141	140	140
D-mannitol	166	164	155
Adipic acid	151	150	150
Potassium thiocyanate	172	175	176
Hydroquinone	172	170	166
Benzanilide	162	162	163

3.2. Thermal conductivity

One of the challenges of PCM in thermal energy storage applications is that its thermal conductivity is low [31]. Besides this, changing the PCM state from solid to liquid requires decreasing the heat transfer and faces an increase in thermal resistance. [33]. These two challenges lead to an imbalance in a thermal transfer that leads to an imbalance in the thermal storage process and gives fast storage process faster than the thermal energy recovery. However, if there is a time constraint and needing a balance in the time required for storage and the time required for thermal recovery, these two challenges affect the energy efficiency of the PCM [34]. Because of this, to overcome this, some methods are used such as:

- Using nanomaterials with PCMs in case of having high thermal conductivity. Some of these materials are carbon and metal.
- Using graphite to extend the PCM properties.
- Using Encapsulation.

3.3. Requirements

Figure 3 shows the PCM requirements in the range of 100 oC to 200 oC. The main thing to be into consideration is that the matching between the transition and operation temperature is essential in the PCM operation. The second thing is that the requirement of the very high latent heat to reduce the physical size of the store. It is important also that the thermal conductivity should be also high which enhances the charging and discharging operations of the storage. The PCM properties can be classified into three main categories which are physical and technical requirements, thermal requirements, and non-technical requirements which include safety, recyclable, and the market requirements such as cost and availability.

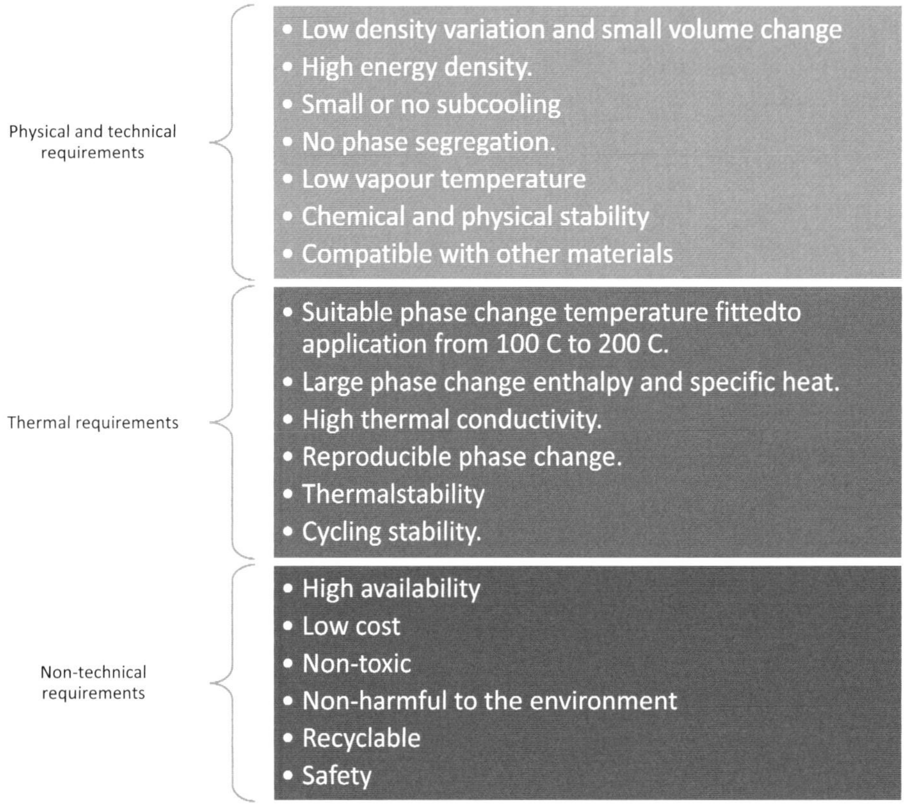

Figure 3. PCM requirements

3.4. Measurement techniques

There are three main measurement techniques used for thermo-physical properties which are:

- Differential Thermal Analysis (DTA).
- Differential Scanning Calorimeter (DSC).
- T-history method.

There is a matching between the heating rate required for the sample materials and the reference materials in DSC and DTA techniques. The heat record in DSC is obtained and proportional to the temperature difference in heat flow between the two materials. Alumina (Al2O3) is the best material that can be used as a reference material. Measuring the latent heat is performed by calculating the area under the curve and the tangent to the slope of the curve at the highest peak value is the melting temperature calculation [35]. In the T-history method, water as reference material is used to record the temperature of the melting point of the PCM as a function of time [36].

4. Thermal Energy Storage Using Molten Salt

The need for thermal energy storage in solar application is important especially on the morning of days where the solar irradiation is not high. Most of the solar applications at home are required in the morning such as water heating and cooking [18]. Because of this, the solar system in the early morning depends on the thermal storage system used which is in usual batteries. In the middle of days, where the solar irradiation reaches its high value, there is some surplus energy that can be used as charging of the system

batteries. The result of this is energy stored in PCM that can be used in other times where solar irradiation is not sufficient to be used lonely [37]. Molten salt is the best cheap and high operating temperature PCM material that can be used in solar applications efficiently.
Another application of solar thermal energy is its use in heat thermal conditioning especially on cold days and winter nights. For this, using PCM with a domestic solar water heater gives high performance with good heating conditions than the traditional solar water heater. The operation of using this material is to store the solar thermal energy during sunny hours and releases the thermal energy at night [38].

5. Conclusion

PCMs that can be used in 100 to 200 C temperatures have been briefly mentioned and studied in this review. The main aim is to propose a taxonomy that represents the PCM materials used in this range of temperature with a mention of the requirements, measurement techniques, and thermal conductivity. Two advanced improvements of the PCM are also discussed which are the use of nanomaterials and the use of encapsulation to improve the healing process. A proposed taxonomy of the most related topics to PCM was proposed. A discussion of the use of PCM in thermal management is also mentioned as an application of using PCM materials in the discussed range of temperature. In the future, reviewing the use of nanomaterials in PCM can make further discussion in this field, in addition, to discussing trends techniques in the capsulation process to enhance the PCM performance.

References

[1] Maldonado, J. M., Fullana-Puig, M., Martín, M., Solé, A., Fernández, Á. G., De Gracia, A. & Cabeza, L. F. 2018. Phase Change Material Selection for Thermal Energy Storage at High Temperature Range between 210 C and 270 C. Energies 11(4): 861..

[2] Dzikevics, M., Veidenbergs, I. & Valančius, K. 2020. Sensitivity Analysis of Packed Bed Phase Change Material Thermal Storage for Domestic Solar Thermal System. Environmental and climate technologies 24(1): 378-391..

[3] Maduabuchi, C. C. & Mgbemene, C. A. 2020. Numerical Study of a Phase Change Material Integrated Solar Thermoelectric Generator. Journal of Electronic Materials 49(10): 5917-5936..

[4] Nie, B., Palacios, A., Zou, B., Liu, J., Zhang, T. & Li, Y. 2020. Review on Phase Change Materials for Cold Thermal Energy Storage Applications. Renewable and sustainable energy reviews 134(110340.).

[5] Zhang, Y., Xie, P., Huang, Y., Liao, C. & Zhao, D. 2021. Evolution of Solar Photovoltaic Policies and Industry in China. IOP Conference Series: Earth and Environmental Science, hlm. 022050..

[6] Li, Z., Lu, Y., Huang, R., Wang, L., Jiang, R., Yu, X. & Yu, X. 2020. Parametric Study on Melting Process of a Shell-and-Tube Latent Thermal Energy Storage under Fluctuating Thermal Conditions. Applied Thermal Engineering 180(115898)..

[7] Wang, K., Qin, Z., Tong, W. & Ji, C. 2020. Thermal Energy Storage for Solar Energy Utilization: Fundamentals and Applications. Resources, Challenges and Applications 415..

[8] Nazir, H., Batool, M., Osorio, F. J. B., Isaza-Ruiz, M., Xu, X., Vignarooban, K., Phelan, P. & Kannan, A. M. 2019. Recent Developments in Phase Change Materials for Energy..

[9] Lin, Y., Jia, Y., Alva, G. & Fang, G. 2018. Review on Thermal Conductivity Enhancement, Thermal Properties and Applications of Phase Change Materials in Thermal Energy Storage. Renewable and sustainable energy reviews 82(2730-2742..

[10] Wei, G., Wang, G., Xu, C., Ju, X., Xing, L., Du, X. & Yang, Y. 2018. Selection Principles and Thermophysical Properties of High Temperature Phase Change Materials for Thermal Energy Storage: A Review. Renewable and sustainable energy reviews 81(1771.

[11] Haillot, D., Bauer, T., Kröner, U. & Tamme, R. 2011. Thermal Analysis of Phase Change Materials in the Temperature Range 120–150 C. Thermochimica acta 513(1-2): 49-59..

[12] Ji, C., Qin, Z., Low, Z., Dubey, S., Choo, F. H. & Duan, F. 2018. Non-Uniform Heat Transfer Suppression to Enhance Pcm Melting by Angled Fins. Applied Thermal Engineering 129(269-279..

[13] Senobar, H., Aramesh, M. & Shabani, B. 2020. Nanoparticles and Metal Foams for Heat Transfer Enhancement of Phase Change Materials: A Comparative Experimental Study. Journal of Energy Storage 32(101911..

[14] Taftanazia, Y., Neumanna, H., Burgera, D., Hagelsteina, G., El-Rubb, Z. A. & Gschwandera, S. 2017. Stability Study of Phase Change Material between 100 C and 200 C. Bachelorarbeit, Amman, Jordan.

[15] Zsembinszki, G., Fernández, A. G. & Cabeza, L. F. 2020. Selection of the Appropriate Phase Change Material for Two Innovative Compact Energy Storage Systems in Residential Buildings. Applied Sciences 10(6): 2116..

[16] Kumaresan, V., Chandrasekaran, P., Nanda, M., Maini, A. & Velraj, R. 2013. Role of Pcm Based Nanofluids for Energy Efficient Cool Thermal Storage System. International Journal of Refrigeration 36(6): 1641-1647..

[17] Gasia, J., Martin, M., Solé, A., Barreneche, C. & Cabeza, L. F. 2017. Phase Change Material Selection for Thermal Processes Working under Partial Load Operating Conditions in the Temperature Range between 120 and 200° C. Applied Sciences 7(7): 722..

[18] Jouhara, H., Żabnieńska-Góra, A., Khordehgah, N., Ahmad, D. & Lipinski, T. 2020. Latent Thermal Energy Storage Technologies and Applications: A Review. International Journal of Thermofluids 5(100039)..

[19] Yan, S.-R., Fazilati, M. A., Samani, N., Ghasemi, H. R., Toghraie, D., Nguyen, Q. & Karimipour, A. 2020. Energy Efficiency Optimization of the Waste Heat Recovery System with Embedded Phase Change Materials in Greenhouses: A Thermo-Economic-Environmen.

[20] Wani, C. & Loharkar, P. K. 2017. A Review of Phase Change Materials as an Alternative for Solar Thermal Energy Storage. Materials Today: Proceedings 4(9): 10264-10267..

[21] Nazari, M. A., Maleki, A., Assad, M. E. H., Rosen, M. A., Haghighi, A., Sharabaty, H. & Chen, L. 2021. A Review of Nanomaterial Incorporated Phase Change Materials for Solar Thermal Energy Storage. Solar Energy.

[22] Tofani, K. & Tiari, S. 2021. Nano-Enhanced Phase Change Materials.

[23] Al-Yasiri, Q. & Szabó, M. 2020. Incorporation of Phase Change Materials into Building Envelope for Thermal Comfort and Energy Saving: A Comprehensive Analysis. Journal of Building engineering 102122..

[24] Shchukina, E., Graham, M., Zheng, Z. & Shchukin, D. 2018. Nanoencapsulation of Phase Change Materials for Advanced Thermal Energy Storage Systems. Chemical Society Reviews 47(11): 4156-4175.in Latent Heat Thermal Energy Storage Systems: A Review. E.

[25] Kiliç, B. & İpek, O. 2020. Thermodynamic Analysis of Absorption Cooling System with Libr-Al2o3/Water Nanofluid Using Solar Energy. Thermal Science 00): 340-340..

[26] Yi, Q., Sukhorokov, G. B., Ma, J., Yang, X. & Gu, Z. 2015. Encapsulation of Phase Change Materials Using Layer-by-Layer Assembled Polyelectrolytes. International Journal of Polymer Science 2015.

[27] Senthil, R., Patel, A., Rao, R. & Ganeriwal, S. 2020. Melting Behavior of Phase Change Material in a Solar Vertical Thermal Energy Storage with Variable Length Fins Added on the Heat Transfer Tube Surfaces. International Journal of Renewable Energy D.

[28] Dindi, A., Ferber, N. L., Gloss, D., Rilby, E. & Calvet, N. 2020. Compatibility of an Aluminium-Silicon Metal Alloy-Based Phase Change Material with Coated Stainless-Steel Containers. Journal of Energy Storage 32(101961..

[29] Liu, S., Han, L., Xie, S., Jia, Y., Sun, J., Jing, Y. & Zhang, Q. 2017. A Novel Medium-Temperature Form-Stable Phase Change Material Based on Dicarboxylic Acid Eutectic Mixture/Expanded Graphite Composites. Solar Energy 143(22-30..

[30] Li, Z., Lu, Y., Huang, R., Chang, J., Yu, X., Jiang, R., Yu, X. & Roskilly, A. P. 2020. Applications and Technological Challenges for Heat Recovery, Storage and Utilisation with Latent Thermal Energy Storage. Applied Energy 116277..

[31] Jacob, R., Liu, M., Sun, Y., Belusko, M. & Bruno, F. 2019. Characterisation of Promising Phase Change Materials for High Temperature Thermal Energy Storage. Journal of Energy Storage 24(100801..

[32] Muñoz-Sánchez, B., Iparraguirre-Torres, I., Madina-Arrese, V., Izagirre-Etxeberria, U., Unzurrunzaga Iturbe, A. & García-Romero, A. 2015. Encapsulated High Temperature Pcm as Active Filler Material in a Thermocline-Based Thermal Storage System. Energ.

[33] Naghavi, M., Metselaar, H., Ang, B., Zamiri, G., Esmailzadeh, A. & Nasiri-Tabrizi, B. 2021. A Critical Assessment on Synergistic Improvement in Pcm Based Thermal Batteries. Renewable and sustainable energy reviews 135(110259..

[34] Nayak, A. O., Gowtham, M., Vinod, R. & Ramkumar, G. 2011. Analysis of Pcm Material in Thermal Energy Storage System. International Journal of Environmental Science and Development 2(6):.

[35] Agyenim, F., Eames, P. & Smyth, M. 2011. Experimental Study on the Melting and Solidification Behaviour of a Medium Temperature Phase Change Storage Material (Erythritol) System Augmented with Fins to Power a Libr/H2o Absorption Cooling System. Renew.

[36] Floros, M. C., Kaller, K. L., Poopalam, K. D. & Narine, S. S. 2016. Lipid Derived Diamide Phase Change Materials for High Temperature Thermal Energy Storage. Solar Energy 139(23-28.437.Storage Applications: A Review. International Journal of Heat a.

[37] Qin, Z., Ji, C., Low, Z. & Duan, F. 2019. Angled Fin Effect on Pcm Melting for Latent Heat Storage. ICTEA: International Conference on Thermal Engineering, hlm..

[38] Peiró, G., Gasia, J., Miró, L. & Cabeza, L. F. 2015. Experimental Evaluation at Pilot Plant Scale of Multiple Pcms (Cascaded) Vs. Single Pcm Configuration for Thermal Energy Storage. Renewable energy 83(729-736..

Regeneration equations for the Rankine cycle with super-heated steam

Mushtaq A. AL-Furaiji[1,2] ,Fawzi Sh. Alnasur[3], Hayder salah AL sammarraie [4,] and Muhammed Im. Kareem[5.]

[1]Faculty of Cryogenic Engineering, ITMO University, Kronverksky prospect, 49, St. Petersburg, 197101, Russia,
[2] Department of Mechanical Engineering,College of Engineering,University of Misan, Al Amarah, 10062, Misan, Iraq.
[3]College of Science,Al-Qadisiyah University,Al-Qadisiyah, Iraq,
[4] South - Russian State Technical University (NPI) of the M.I. Platov , Department of Thermal power plants Prosveshcheniya st . , 132, Novocherkassk, Rostov Oblast, Russian Federation.
[5]Islamic Azad University Science and Research Branch, Iran.

Abstract. The paper presents a mathematical model to calculate the thermal efficiency of the steam turbines with superheating, methodology forecasting for the Rankin cycle with superheated steam using Regression equations, MathCad, SPSS, and Statistatica programs. The resulting regression equation is applied to calculate the indicator values in a given range of variation of parameters, and it is also limitedly suitable for calculation outside this range. A MathCad-based approach for calculating the predictive model was created and presented. This work presents the formulation of a problem for the method of experiment planning. A turbine unit with three input parameters that influence the value of the output parameter Y (the thermal efficiency of the Steam turbine unit) was studied through the method of experiment planning. Along with that, the first parameter (initial pressure turbine, bar) varied within: $100< x_1 <128$; the second parameter (the turbine-inlet temperature, °C) varied within: $450 < x_2 < 550$; the third parameter (the fraction of dryness) varied within: $0.79 < x_3< 0.83$.
Keywords: thermal efficiency , steam turbines , Rankine cycle, MathCad.

1. Introduction:

One of the most frequently-researched subjects in thermal energy is the power generation plant. A combined-cycle power plant is a large scale electrical power generation plant in which electricity is obtained from both gas and steam turbines. Energy is transferred in the form of heat or gaseous flow through each turbine to produce electricity [1].

The efficiency of the use of steam extraction from cogeneration turbines for the needs of heat consumption largely determines the economics of the operation of heat and power plants. It is no coincidence that in the former cogeneration based district heating was used as the main way to save organic fuel throughout the country. Cogeneration based district heating is a centralized heat supply based on combined generation of electrical and heat energy [2]. Also, considerable attention has always been payed to the development of internal heating - the use of steam turbine selections for heating feed-water and other technological in-station heat carrier flows [3].

At present, the technical and economic indicators of most CHP plants with an initial vapor pressure of 12.8 MPa are comparable to those of supercritical pressure condensation stations. The main reasons for the reduction in the efficiency of CHP are, firstly, a significant reduction in the generation of electricity by heat consumption, and secondly, significant losses in the transportation of heat carriers. At the same time, at the majority of operating CHP plants there are significant energy efficiency reserves that are related to provision of intrastation thermal loads [4] [5]. A significant part of these loads is accounted for water treatment plants (VAP), which compensate for both steam and condensate losses from the plant cycle, and delivery water from the pipelines of the heating network [6]. The main disadvantage of the technologies used at CHP for providing heat loads to water treatment plants is the almost ubiquitous use of a high grade waste heat as a heating medium, which significantly reduces the share of electricity generation by heat consumption and, consequently, the economy of the power plant. these study sunlight to improve the application technology of regenerative steam extraction from steam turbine units of a combined heat and power plant by traditional ways of using steam extraction of turbines of CHP for covering thermal loads of water treatment plants are analyzed and thermal calculation of the steam-turbine unit is made and regression equations for various parameters of a heating turbine are obtain and regression equations for the Rankin cycle with superheated steam are obtained [7] [8].

2. Methodology

The possibility of increasing the initial temperature is much less than that of the initial pressure. Therefore, an intermediate superheat of steam is introduced at a certain initial pressure. In this case (Figure 1) [9]

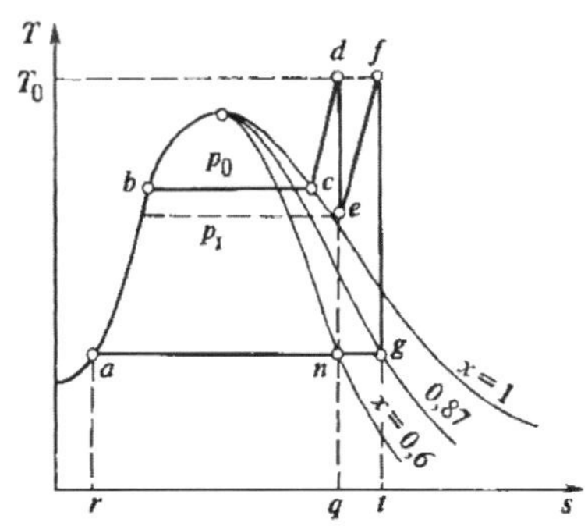

Figure 1. Ideal cycle with intermediate superheating of steam

Steam is first expanded along the adiabatic *de* to a pressure p_1 at which steam moisture does not yet arise and then is sent to the intermediate super heater of the boiler where the vapor temperature rises in the isobaric *ef* to the initial temperature. Then the vapor expands along the adiabatic *fg* to the pressure in the condenser. At the same time, the humidity of the steam at the end of the expansion process becomes acceptable.

With the correct choice of the separation pressure, it is possible to increase the economy of the turbine simultaneously. Figure 1 shows that the thermal efficiency of the additional cycle *nefgn* is greater than that of the main cycle *abcdena* without intermediate overheating of the steam. In general, the use of intermediate superheating of steam allows to obtain 5-6% of fuel economy. This complicates the design of the boiler, turbines and their operation.

Regenerative heating of feed water with steam taken from the intermediate stages of the turbine is used in all modern STPs (Figure 2).

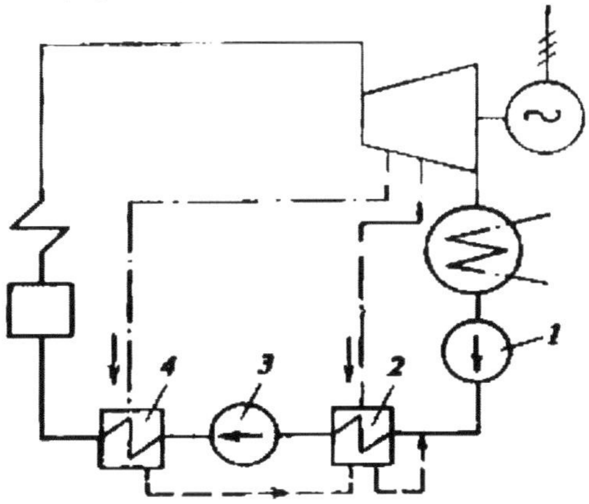

1 - Condenser pump; 2 - HPH; 3 - feeding pump; 4 - LPH

Figure 2. steam turbine units (STU) with regenerative feed water heating

In this case, the heat from the steam is supplied to the feed water. The more the feed water is heated up before being fed into the boiler, the greater the thermal efficiency of the cycle. At a certain limit, the temperature of the feed water can be brought to the saturation temperature which corresponds to the pressure of the extraction steam. However, excessive heating of feed water not only increases investment in heaters, but also worsens the use of heat from the flue gases of the boiler: as the temperature of the feed water rises, the temperature of the boiler's exhaust gases will increase and its efficiency will decrease. Therefore, the temperature of the feed water is chosen on the basis of technical and economic calculations when all the equipment of the heat and power plant is taken into account [10].

3. Mathematical modeling and results

The turbine unit at the nominal mode with the initial parameters of the turbine unit T-100 / 120-130 are as follows:

- Fresh steam pressure, P_0= 12.75 MPa;
- The enthalpy of fresh steam, h_0 = 3485 kJ / kg;
- Temperature of fresh steam, t_0 = 555 °C;
- Consumption of fresh steam, D_0 = 89 kg / s;
- feed water flow rate, D_{fw}= 130.6 kg / s;
- feed water temperature, t_{fw} = 249 °C;
- Heating load, Q_t = 75 MW;
- Coefficient of cogeneration-based district heating α = 0.5;
- Pressure at the inlet to the feed pump, P_{pum}= 0.59 MPa;

A turbine unit with three input parameters (Table 1) that influence the value of the output parameter *Y* (the thermal efficiency of the STP) was studied through the method of experiment planning. Along with that, the first parameter (initial pressure turbine, bar) varied within: $100< x_1 <128$; the second parameter (the turbine-inlet temperature, °C) varied within: $450 < x_2 < 550$; the third parameter (the fraction of dryness) varied within: $0.79 < x_3 < 0.83$.

3.1. Problem statement

An experimental study of a heat engineering device with three input parameters (Table1) that affects the value of the output parameter Y (thermal efficiency of the steam turbine) was carried out using the experiment planning method.

Table 1. Input parameters

Parameter	Pressure, (bar)	Turbine inlet temperature, (°C)	The dryness fraction of water vapor
Max	128	550	0.83
Min	100	450	0.79

To plan an experiment, it is necessary to determine the fractionally of the used replica, calculate the regression coefficients and write down the mathematical model in a normalized form and in a dimensional form. It is also necessary to conduct a complete analysis of the accuracy of the experimental results, to evaluate the significance of the obtained regression coefficients and the adequacy of the obtained mathematical model. Using the resulting model, calculate the Y value for the parameter values and evaluate the accuracy of the result.
Calculation of thermal efficiency in Mathcad, the results obtained by us using the Mathcad software are shown in Table 2

Table 2. Results of calculating the thermal efficiency

№ experience	Pressure P_1 ,bar	Temperature T_1,°C	Dryness x_2	Thermal efficiency η %
1	128	550	0.83	39.459
2	128	550	0.79	42.377
3	128	450	0.83	33.942
4	128	450	0.79	37.126
5	100	550	0.83	40.025
6	100	550	0.79	42.915
7	100	450	0.83	34.985
8	100	450	0.79	38.118

Table 3. Planning matrix and results of a randomized experiment

№ experience	*X1*	*X2*	*X3*	*Y'*	*Y''*	*Y'''*	*Y''''*	*Y'''''*
1	1	1	1	39.459	39.424	39.485	39.437	39.314

2	-1	-1	1	34.985	34.549	34.234	34.047	34.877
3	1	-1	-1	37.126	37.664	36.889	36.719	37.919
4	-1	1	-1	42.915	42.622	42.612	42.141	41.359

The number of factors k in the used replica is 3. The number of full-factor experiments is expressed by the dependence $N = 2^k = 2^3 = 8$. The number of experiments in this study is N = 4, which corresponds to one of the half-replicas.

3.2. Identifying Suspicious Results

The most suspicious results in a series of experiments are identified. The average values of the results of the experiment for the entire series of experiments are found:

$$\bar{y} = \frac{1}{n}\sum_{i=1}^{n} y_i \qquad (1)$$

Where *n* is the number of parallel experiments in the series under study.
Next, the variance and root-mean-square error are calculated for each series of experiments. The results of calculations to identify suspicious results are shown in Table 4

Table 4. Results of calculations to identify suspicious results.

Y'	Y''	Y'''	Y''''	Y'''''	$Y_{av.}$	S^2	S
39.459	39.424	39.485	39.437	39.314	39.4	0.0043	0.0656
34.985	34.549	34.234	34.047	34.877	34.5	0.1622	0.4027
37.126	37.664	36.889	36.719	37.919	37.3	0.2614	0.5113
42.915	42.622	42.612	42.141	41.359	42.3	0.3714	0.6094

3.3. Test for misses by t-criterion

The average values of the results of the experiment, variances and *root-mean-square errors* are determined without taking into account the suspicious result. The calculation results are shown in Table

Table 5. Calculation results

Y'	Y''	Y'''	Y''''	Y'''''	$Y_{av.}$	S^2	S	$S^2_{clarify}$	$S_{clarify}$	$Y_{i\ clarify}$
39.459	39.424	39.485	39.437	39.314	39.4	0.0043	0.0656	0.0043	0.0656	39.424
34.985	34.549	34.234	34.047	34.877	34.5	0.1622	0.4027	0.1622	0.4027	34.538
37.126	37.664	36.889	36.719	37.919	37.3	0.2614	0.5113	0.2614	0.5113	37.263

42.915	42.622	42.612	42.141	41.359	42.3	0.3714	0.6094	0.3714	0.6094	42.330

For each suspicious value, the criterion value$y_{i\ susp.}$ is calculated using the formula:

$$t_{\text{i-calcuated}} = \frac{y_{\text{i susp.}} - \bar{y}_{\text{i clarify}}}{\sqrt{S^2_{\text{2 clarify}}}} \qquad (2)$$

And checks with the tabular value of the Student's T-test (we take a confidence probability of 0.999).The results of the Student's T- test are summarized in Table 6.

Table 6. The results of the test according to the Student's criterion (Student's T-test)

experience 3		
$Y_{\text{i clarify}}$	S^2_{clarify}	S_{clarify}
37.263	0.2614	0.511271291
Check		
$t_{\text{i-calcuated}}$		$t_{tabular}$
1.282851845		2.776445105

The results of the check showed that the suspicion is not a miss.

3.4 Test for homogeneity of dispersions

The value of the Fisher criterion is calculated as the ratio between the largest and smallest partial variances [11]

$$F_{i\ calcul.} = \frac{S^2_{i\ max}}{S^2_{i\ min}} \qquad (3)$$

Further, according to the special tables of the Fisher *F-distribution* or using the (fdisp) function in the Excel package, the value of F_{tab} is found, which also depends on the accepted significance level q and two degrees of freedom $f_1 = n - 1$ and $f_2 = N$, if $F_{\text{расч}} > F_{\text{табл}}$ $F_{calcul} > F_{table}$ then the dispersions are non-uniform and in one of the two rows (where the variances are largest or smallest) in present data, which are either undetected misses or approaching them in their magnitude.
In this case, it is recommended to increase the number of parallel experiments in one of these rows, calculate the partial variance and repeat the analysis for heterogeneity. The calculation results are summarized in Table. 7

Table 7. Calculation results

$S^2_{\text{clarify. }max}$	$S^2_{\text{clarify. }min}$	F_{criteria}	F_{table}	**Result**
0.3714	0.0043	86.34930949	6.38823290 0	Dispersions are heterogeneous

So the experiment was successful.
The regression equation has the following form, taking into account the found coefficients:

$$Y = 77.44 - 0.028 \cdot X_1 + 0.052 \cdot X_2 - 75.781 \cdot X_3 \qquad (4)$$

3.5. Structure a three-dimensional relationship between parameters

For the calculation results presented in Table 2, we construct a surface in the STATISTICA software [12] [13] [14] [15] [16] [17] [18].

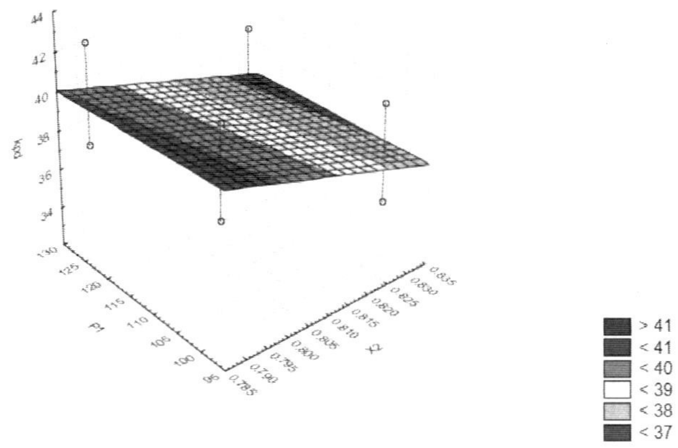

Figure 3. Surface for dependence $\eta_t = f(p_1,x_2)$

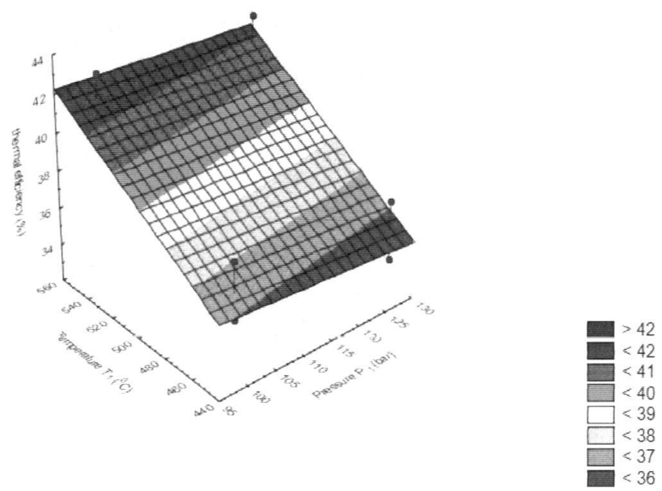

Figure 4. Surface for dependence $\eta_t = f(p_1,t_1)$

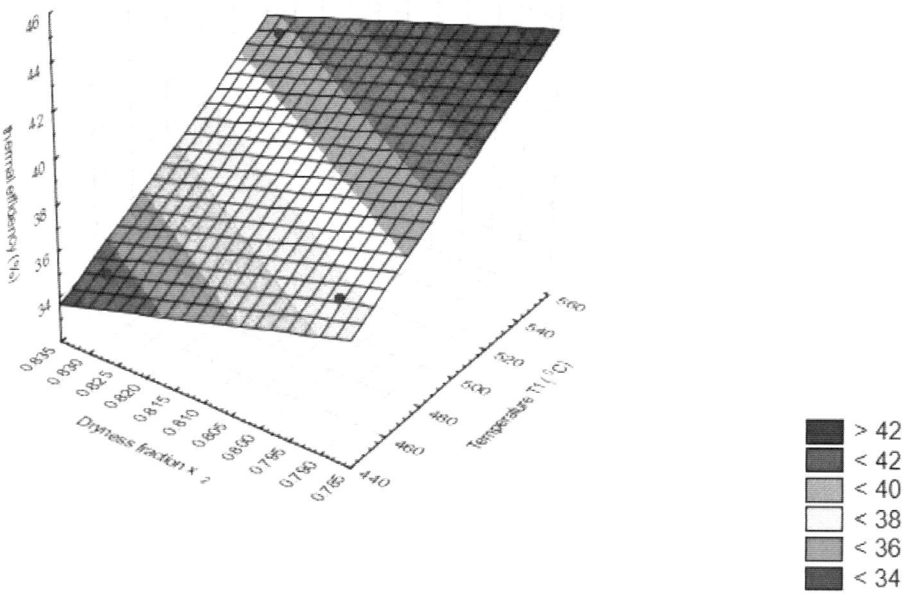

Figure 5. Surface for dependence $\eta_t = f(x_2, t_1)$

Thus, the level surfaces for Table 1 are obtained (Figures 4 - 6), and also the regression equations of the corresponding functions are derived:

$$\eta_t = 103.1962 - 75.7812\,x_2 - 0.028\,p_1\,; \quad (5)$$

$$\eta_t = 16.0572 - 0.028\,p_1 - 0.0515\,t_1\,; \quad (6)$$

$$\eta_t = 74.2449 + 0.0515\,t_1 - 75.7812\,x_2\,; \quad (7)$$

4. Conclusions

Traditional methods of using steam extractions of turbines of CHPP for covering thermal loads of water treatment plants were analyzed where the thermal calculation of the basic thermal circuit is performed at the nominal mode. The calculation showed the thermal efficiency of the turbine unit, as well as energy indicators and regression equations for various parameters of the heating turbine were obtained, that shows the maximum efficiency in the steam turbine plant with increasing fresh vapor temperature. The resulting regression equation is applied to calculate the indicator values in a given range of variation of parameters, and it is also limitedly suitable for calculation outside this range.

References

[1] Force, T. (2003). Modeling of gas turbines and steam turbines in combined cycle power plants. CIGRE Technical Brochure..

[2] N. Hasan, B. B. Arora and J. N. Rai, Performance Analysis of Combined Cycle Gas Turbine for Frequency Deviation and Temperature Control..

[3] J. N Rai, N. Hasan, B. B. Arora, R. Garai, R. K. Gupta, and R. Kapoor, Study the Effect of Temperature Control on the Performance of the Output of Combined Cycle Gas Tur-bine. International Journal of Theoretical and Applied Mechanics, 8(1), 15-23, 2013..

[4] Zamaleev M.M. Efficiency upgrading of regeneration systems of thermal steam tur bines: the author's abstract. Dis. cand. Tech.Sciences: 05.14.14. Ulyanovsk, 2008.,19 p..

[5] Trukhniy A.D. Thermal steam turbines and turbine plants. A.D. Trukhniy, B.V. Lomakin. - Moscow, MEI, 2002. 540 p..

[6] M. Hanss and S. Turrin. A fuzzy-based approach to comprehensive modeling and analy sis of systems with epistemic uncertainties. Structural Safety, 32(6):433–441, 2010. https://doi.org/10.1016/j.strusafe.2010.06.003..

[7] N.A. Harish, R. Ismail and T. Ahmad. Graphical representation of fuzzy state space of a boiler system. In Proc. of the 11th WSEAS Int. Conf. on Neural Networks, NN'10, Prceedings of the 11th WSEAS Int. Conf. on Evolutionary Computing, EC'10, Proc. of the.

[8] I.U. Khan, T. Ahmad and N. Maan. Feedback fuzzy state space modeling and optimal production planning for steam turbine of a combined cycle power generation plant. Research Journal of Applied Science, 7(2):100–107, 2012 https://doi.org/10.3923/rjasci.2012.

[9] A. Chaibakhsh and A. Ghafari. Steam turbine model. Simulation Modelling Practice and Theory, 16(9):1145-1162, 2008. https://doi.org/10.1016/j.simpat.2008.05.017.

[10] M. Hanss. Applied Fuzzy Arithmetic. An Introduction with Engineering. Springer,Berlin, 2005. https://doi.org/10.1007/b138914..

[11] Li S.Z., Jain A. (eds) Encyclopedia of Biometrics. Springer, Boston, MA. [(2009) Fisher Criterion.. https://doi.org/10.1007/978-0-387-73003-5_585].

[12] W.M. Wan Munirah, T. Ahmad, S. Ahmad and A. Ashaari. Simulation of furnace sys-tem with uncertain parameter. Malaysian Journal of Fundamental and Applied Sciences, 2015. https://doi.org/10.11113/mjfas.v11n1.334..

[13] W.M. Wan Munirah, T. Ahmad, A. Ashaari and M. Adib Abdullah. Modeling fuzzy state space of reheater system for simulation and analysis. In AIP Conference Proceed-ings, volume 1605, pp. 488–493, Malaysia, 2014. AIP. https://doi.org/10.1063/1.4887637..

[14] W.M. Wan Munirah, A. Tahir and A. Azmirul. Identifying the uncertain model parame-ter of a steam turbine system. Pertanika Journal of Science and Technology, 25(2):545–560, 2017..

[15] M. Hanss. The transformation method for the simulation and analysis of systems with uncertain parameters. Fuzzy Sets and Systems, 130(3):277-289, 2002. https://doi.org/10.1016/S0165-0114(02)00045-3.

[16] M. Hanss and O. Nehls. Enhanced parameter identification for complex biomedical models on the basis of fuzzy arithmetic. Proceedings Joint 9th IFSA World Congress and 20th NAFIPS International Conference (Cat. No. 01TH8569), volume 3, pp. 1631-1636, 2001.

[17] M. Hanss and S. Turrin. A fuzzy-based approach to comprehensive modeling and anal-ysis of systems with epistemic uncertainties. Structural Safety, 32(6):433- 441, 2010. https://doi.org/10.1016/j.strusafe.2010.06.003..

[18] A. Ordys, A.W. Pike, M.A. Johnson, R.M. Katebi and M.J. Grimble. Modelling and Simulation of Power Generation Plants. Advances in Industrial Control. Springer-Verlag, London, 1994. https://doi.org/10.1007/978-1-4471-2114-5.
[19] M. Hanss and O. Nehls. Simulation of the human glucose metabolism using fuzzy arithmetic. In PeachFuzz 2000. 19th International Conference of the North American Fuzzy Information Processing Society - NAFIPS (Cat. No.00TH8500),pp. 201{205, 2000. DOI: 10..

Determination of Polycyclic Aromatic Hydrocarbons in Selected Soils from Al-Diwaniyah governorate, Iraq-Case Study

Tania S. Kadhim and Faiq F. Karam

Department of chemistry, college of science, University of Al-Qadisiyah, Iraq

Abstract. This study includes determination of 16 (PAHs) compounds from soils of Al-Diwaniyah governorate in six polluted sites from August 2020 to May 2021, (PAHs) compounds have carcinogenic and mutagenic activity in biological systems, so their identification in soil samples is considered important. Soxhlet extraction is used to extract polycyclic aromatic hydrocarbons and concentration of these compounds in all samples was determined by using (GC). The summation of all (PAHs) compounds during ten months were (4842.666667 ppm), among all these compounds Benzo[G,H,I]Perylene recorded the maximum cumulative value of (388.8333333 ppm) whereas the minimum value was for Acenapthylene (219.9 ppm).

Keyword: contamination with polycyclic aromatic hydrocarbons, Soil contamination, Al-Diwaniyah pollution.

1. Introduction

Environmental pollution, such as water, soil and air pollution, is one of the main causes that lead to a high death rate worldwide, chemicals in the environment have effects on living organisms and these effects depend on several factors, including:

1- Kind of pollutants

2- Health status of the individual

3- Gens

4- Degree of exposure pollutants

While the types of pollutants differ according to human activities [1]. PAHs tend to remain in the soil because they are lipophilic and have high stability [2]. After these compounds are deposited on the surface soil, these compounds accumulate in (vegetables, other organisms) and are transmitted through the food chain to humans [3] and remain for longer periods in the soil due to their strong absorption, where they persist for longer periods of time [4]. Soil type and characteristics of its content of organic compounds play an important role in the uptake of PAHs into soil [5]. The increase in the number of means of transportation in urban areas leads to the release of (PAHs) through (erosion of asphalt pavements, exhausts of vehicles and other means of transport, tires), these compounds move to agricultural soils on the sides of the roads by:

1- Surface runoff

2- Mist

3- Dust

4- Airborne particles [6].

Polycyclic aromatic hydrocarbons (PAHs) constitute a public concern and become a threat to life in the environment. PAHs circulate and remain as long as possible in the environment and are the main cause of cancers and are mutagenic in human genes [7], where the (US EPA) listed (16) compounds of hydrocarbons as high-risk contamination [8]. figure.1 shows the structures of 16 (PAHs) compounds [9].

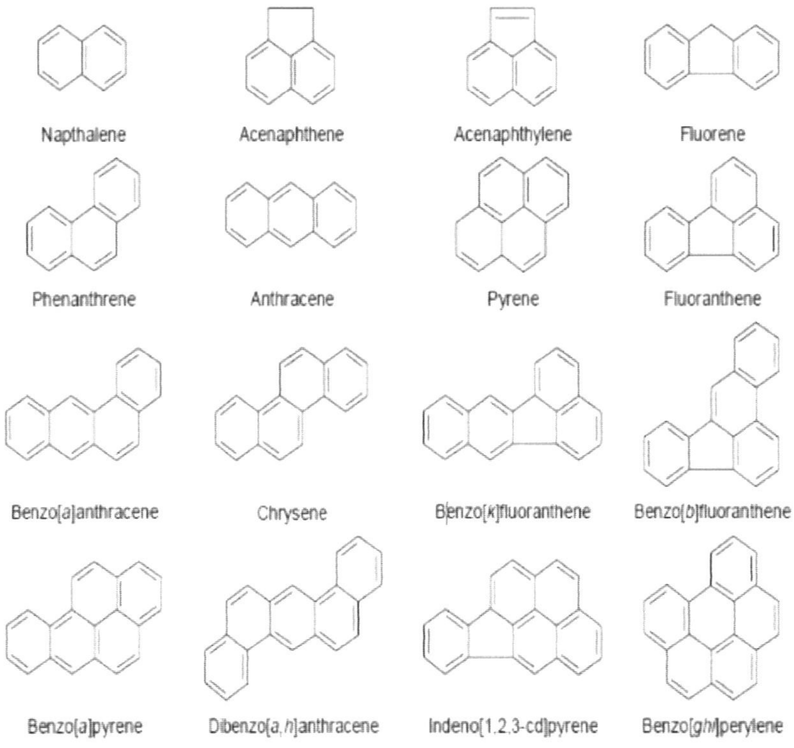

Figure 1. Structures of the 16 US-EPA PAH_s

Polycyclic aromatic hydrocarbons are characterized as compounds with high melting point, low vapor pressure and low solubility in water. The melting point and boiling point increase, but the vapor pressure decreases with the increase in molecular weight. Whereas PAHs are classified as non-polar compounds because they are highly hydrophobic compounds, PAHs don't affect specific body organs or biochemical pathways, but cause irreversible dysfunction [10]. PAHs ratio profiles are used as a way to distinguish hydrocarbons that originate from pyrolysis from hydrocarbons of petroleum origin such as: Fluoranthene/Pyrene and Phenanthrene/Anthracene, these two pairs of compounds differ in reactivity and solubility, so they were used to find out the origin of PAHs [11]. When the Flur/Pyr ratio is greater than 1 and the Phen/Ant ratio is <10, this indicates PAHs originating from thermal decomposition [12], the ratio (LMW- PAHs /HMW -PAHs) are often wont to determine the origin of PAHs, if the ratio was < (1) this means that the PAHs are of petrogenic origin [13].

The Soxhlet extraction method is done to extract PAHs from soil or sediment samples in an extraction thimble, after which the samples to be examined are extracted using suitable solvents by the reflux cycle, whereby the solvent is boiled by the heating source and therefore the vapor then passes through the bypass arm within the condenser. When the solvent reaches the highest of the tube arm, the extract and solvent are returned to the lower flask to evaporate the solvent again, then the cycle is repeated until the whole sample is extracted into the lower flask employing a temperature suitable for heating and evaporation of the used solvent [14].

The GC/FID device is used to analysis of contaminated soil samples, this device has the following features:

1- Relatively inexpensive and widely available even in small factories.

2- It allows sensitive detection of a wide range of carbon-containing compounds.

3- There is no limit to the range of (FID) and quantitative measurement is allowed because it does not lead to blocking the activity of the ion, and therefore (GC/FID) is more useful than (GC/MS) in some cases such as prediction and classification based on the quantities of certain compounds [15].

This work aimed to determination the quantities of PAHs in contaminated soil samples of Al-Diwaniyah Governorate.

2. Experimental:

Study area:

Sample containers and glasses are washed with detergent and rinsed with distilled water then with acetone. All samples in this study were collected from the topsoil once a month from six areas in Al-Diwaniyah Government in dark colored plastic bags and then kept in a cool box and then placed in the refrigerator at a temperature (4°C) until then transferred to the laboratory. Soil samples are extracted using the soxlet extraction method, where (2.5 gram) of dry soil was mixed with about (2.5 gram) of anhydrous sodium sulphate and placed during a porous extraction thimble. The solvent used in the extraction process is (100 ml) of a mixture (1:1 n-hexane, acetone) and this process is continued for (24 hours) until the sample is totally extracted into the lower flask. The solution in the lower flask (representing the solvents and PAHs extracted from the soil) is concentrated to 1 ml employing a rotary evaporator, the concentrated solution is kept in small dark colored glass containers in the refrigerator at a temperature below (4 °C) [16] until do the measurement. Figure 2 shows the map of the study area.

S1: Represents the Northern Electric Power Plant, which is located 5 km north of the Diwaniyah Governorate Center. This station operates on two types of fuel (heavy fuel oil, light diesel oil), this station contains 16 exhausts and pollutes the eastern areas with 80% according to the wind direction in the governorate. This site was chosen to check the pollution levels of PAHs.

S2: Represents the eastern electric power plant, which is located near the village of Al Hamad, 8 km east of the center of Al Diwaniyah Governorate. This station contains 14 exhausts that pollute the eastern and southern regions, especially the village of Al Hamad. This site was chosen to refer to the effect of polycyclic aromatic hydrocarbons.

S3: It represents the Diwaniyah gas power station, which is located in the second area of Al-Haffar (Al-Shamiya), 22 km from the center of Al-Diwaniyah Governorate. This station operates on natural gas fuel and contains four turbines, each turbine having an exhaust. This station pollutes the eastern and southern regions, especially Al-Abtan and Al-Haffar River.

S4: It represents the area of the industrial district that includes the auto engine oil exchange station. The soil of this station may be contaminated with polycyclic aromatic hydrocarbons, so this site was chosen to study the pollution levels of these compounds.

S5: It represents the Diwaniyah tire factory (the rubber factory), which is located 8 km from the center of Diwaniyah Governorate. This plant contains storage that contradicts the environmental and safety requirements of the chemicals and used tires, as these materials are considered environmental pollutants, may contain PAHs. Therefore, this site was chosen to watch the amount of soil contamination of the plant with these compounds.

S6: Represents the waste incineration site, which is located in the area behind the old camp (in front of the Caliphs Mosque), these wastes contain oils, greases or fats and others, which are basically the cause of the presence of polycyclic aromatic hydrocarbons, so this site was chosen to refer to the cumulative effect of these compounds.

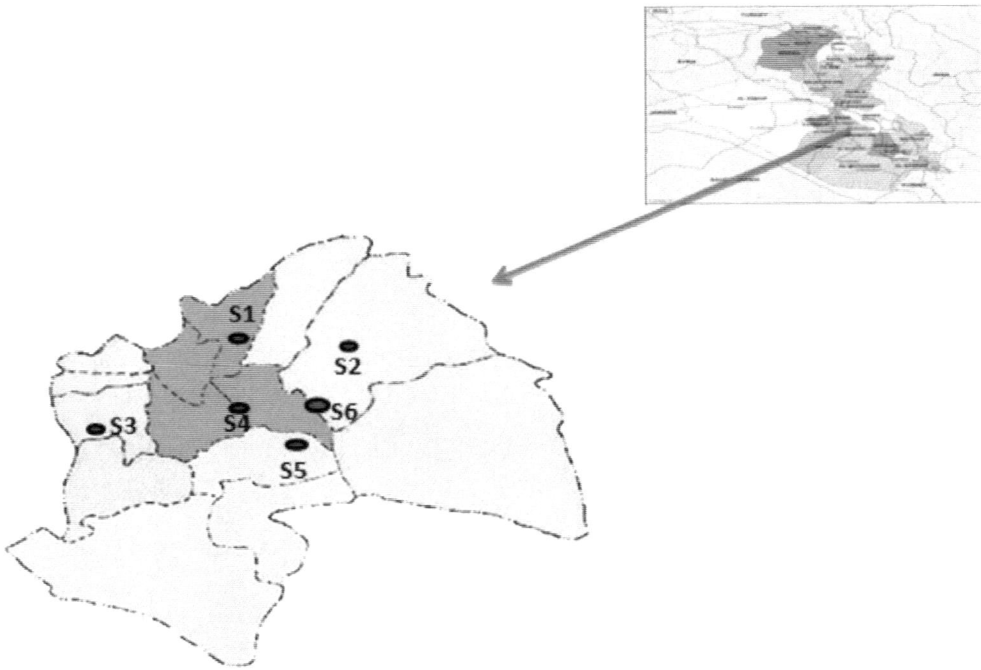

Figure 2. Map indicating the sampling points in the soil of Al-Diwaniyah Governorate

Gas chromatography technology was wont to detect (PAHs), therefore the carried was separated during a DB-5 column (30 m × 0.25 mm) covered with a (0.25 μm-thick) film of fifty (diphenyl-polydimethylsiloxane). Samples were injected at an injection temperature of (280 °C). Then the sample was kept at the column temperature initially held at (40°C) for (1 min), then it had been raised to (120°C) at a rate of (25°C/min), then the temperature was raised to (160°C) at a rate of (10°C/min). Eventually to (300 °C) at a rate of (5 °C/min), whereby the sample is kept at the ultimate temperature for (15 min), the temperature of the detector (FID) is then maintained at (330 °C). Helium gas was used as a carrier at a continuing flow of (5 ml/min).

3. Result and discussion

The concentrations of the 16 detected (PAHs) from soil of six sites located within the soil of Al-Diwaniyah are shown in table.1, as table 1 shows the average concentrations of (16 PAHs) for each site within the soil of Al-Diwaniyah.

Table 1. levels of 16 PAHs (ppm) for each site from which soil samples were collected in Al-Diwaniyah Governorate.

Site No.	S1	S2	S3	S4	S5	S6	Mean	Sum.
AcP	117.6	112.8	74.6	257.2	273.8	502.8	223.1333	1561.933
AcPy	65.4	67.4	42.6	208.8	322.8	612.4	219.9	1539.3
Ant	133.8	123	99.8	299	409.8	830.4	315.9667	2211.767
BaA	133.4	149.6	116.6	348.8	290.6	514.4	258.9	1812.3
BbFl	104.6	134.6	129.4	349.6	490	1024.4	372.1	2604.7
BkFl	124	144.4	112	271.2	439.2	475.8	261.1	1827.7
BghiP	130.6	156	124.4	372.4	456.2	1093.4	388.8333333	2721.833
BaP	105.2	136	116	202.6	329.2	621.4	251.7333	1762.133
Chr	131.2	155.6	129.8	268.2	501	1004.4	365.0333	2555.233
DbA	140.4	134	100	160.2	421.2	1092.6	341.4	2389.8
Flu	138.8	184	142.4	373.6	438	1021.2	383	2681
Fl	155.6	166	135.8	240.2	366.8	648.4	285.4667	1998.267
InP	135.8	126.4	103.4	330.2	393.6	643	288.7333	2021.133
NaP	153.8	151.2	107.4	351.8	245	577.2	264.4	1850.8
Phe	142.8	150.8	113.2	399.4	333.8	552.4	282.0667	1974.467
Pyr	164.8	173.6	83.8	317.6	438.4	867.2	340.9	2386.3
∑PAHs	2077.8	2265.4	1731.2	4750.8	6149.4	12081.4	4842.666667	33898.67

Sixteen compounds of polycyclic aromatic hydrocarbons detected in the collected soil samples (Table1), benzo[g,h,i]perylene recorded (the highest concentration) of PAHs (2721.833 ppm), in while acenaphthylene recorded (the lowest concentration) of PAHs (1539.3 ppm). Total content of PAHs was (1731.2 to 12081.4 ppm), mean concentration of total PAHs was (4842.666667 ppm).

Other studies of polluted sites recorded in some soil samples from Al-Diwaniyah Governorate, (Pyrene) recorded the highest level of PAHs (5.36 ng/g) compared with acenaphthylene, which recorded the lowest concentration of PAHs (1.11 ng/g). Total content of polycyclic aromatic hydrocarbons were (4.46 to

17.24 ng/g) [17].

In China (Beijing) urban park topsoil, recorded that pyrene had the highest concentration of PAHs (1.341 mg/kg) and acenaphethylene had the lowest concentration of PAHs (0.032 mg/kg), total hydrocarbon content between (0.066 - 6.867 mg/kg), while PAHs mean was (0.460 mg/kg) [18]. In the study conducted by Faiq *et al.* . Results showed that the rate of recovery of chrysene is (72.23% - 96.11%), while the relative standard deviation values ranged from (1.65%- 4.87%). The highest total concentration of chrysene (57.9 ppb) was recorded in the samples of Al-Diwaniyah government center, while the lowest total concentration (16.5 ppb) was recorded in Afak district [19]. Site number 6 was the most polluted place than others because this site may contain places for burning coal and petroleum products coming from residential waste, and this is one of the main sources of PAHs, and there are four categories of soil pollution with these components have been proposed:

1- If the concentration level of PAHs is (> 200 µg /kg), then the soil is not contamination.

2- Soil is considered less contaminations when the PAHs level is recorded 200-600 µg/kg.

3- Soil is considered contaminated when the level of PAHs is between (600-1000 µg/kg).

4- Soil is considered severely contaminated if the PAHs level is (<1000 µg/kg) [20].

Table No. (2): shows the classification of PAHs consistent with the amount of rings available within the soil samples, their collection, their percentage. This might be attributed to the upper rate of biodegradation of the (low rings) than the (high rings) of the PAHs of soil bacteria once they are used as energy [21]. The lowest percentage was for (6-rings) PAHs (13.7%) followed by (5- rings, 3-rings, 4-rings) while the highest percentage was for (2-rings) (53.6%). In the study carried out by Luma and her groups, the highest percentage was for (5-rings) PAHs (33.07%), followed by (4-rings, 6-rings) polycyclic aromatic hydrocarbons, while the (3-rings) PAHs is the lowest percentage (17.99%) [17]. In urban soils of Beijing, China, the classification of PAHs was (4-rings) (45.95%) > (5-rings) (27.38%) >(3-rings) (16.43%) > (6-rings) (6.38%) > (2-rings) (3.68%) [18].

Table 2. The classification of polycyclic aromatic hydrocarbons according to the amount of rings within the soil samples collected

No. of ring	Two rings	Three rings	Four rings	Five rings	Six rings
∑Mean Concentration ppm	1586.4	7939.2	8095.8	7858	4065.4
Percentage abundance	53.6%	26.8%	27.4%	26.5%	13.7%

Hites, Wagrowski (1997) recorded that the chemical composition of PAHs can effectively indicate the source of pollutants [22]. Where (oil spills) contribute to the formation of (2-rings) PAHs, as well as the combustion of (coal, biomass) mainly produces (3-rings), (4-rings) polycyclic aromatic hydrocarbons, and the (combustion of gasoline) contributes mainly to the formation of (5-rings) PAHs, Diesel combustion mainly produces (6-rings) polycyclic aromatic hydrocarbons [23].

4. Conclusion

This research paper contains important data on the concentration of PAHs in Al-Diwaniyah government soil samples. 160 samples of soil were collected from six selected sites in Al-Diwaniyah government for the period from (August 2020 - May 2021) and analyzed by GC-2010 and the results recorded significant differences in the concentrations of PAHs from (site to site) and from (month to month). Human activities contributed to the pollution for each site each month.

Acknowledgments:

The authors extend their gratitude and thanks for the support provided by the (University of Al-Qadisiyah), for its contribution in helping to complete this research.

References

[1] Dubey, B.(2010). A model for the effect of pollutant on human populationdependent on a resource withenvironmental and health policy. J. Biol. Syst. 18(03), 571—592..

[2] Zhang, J.; Fan, S.K.(2016). Influence of PAH speciation in soils on vegetative uptake of PAHs using successive extraction. J. Hazard. Mater. 320, 114–122..

[3] Bortey-Sam, N., Ikenaka, Y., Nakayama, S.M.M., Akoto, O., Beyene, Y., Baidoo, E., Ishizuka, M.(2014). Occurrence, distribution, sources and toxic potential of polycyclic aromatic hydrocarbons (PAHs) in surface soils from the Kumasi Metropolis, Ghana. Sci..

[4] Banan, S.; Khaled, E.H.; Mohamad, E.H.; Helene, B.; Farouk, J.(2018). Impact of Lebanese practices in industry,agriculture and urbanization on soil 3 toxicity. Evaluation of the Polycyclic Aromatic Hydrocarbons (PAHs)levels in soil. Chemosphere. 06, 178..

[5] Jonker MTO, Smedes F (2000). Preferential Sorption of Planar Contaminants in Sediments from Lake Ketelmeer, The Netherlands.Environ. Sci. Technol.34, 1620-1626..

[6] Kibblewhite, M. (2018). Contamination of agricultural soil by urban and peri-urban highways: an overlooked priority? Environ. Pollut. 242, 1331–1336..

[7] Zhang,J., Liu,G., Wang,R., Huang,H.(2017).Polycyclic aromatic hydrocarbons in the water SPM sediment system from the middle reaches of Huai River, China: distribution, partitioning, origin tracing and ecological risk assessment.Environ.Pollut.230,61–71..

[8] US EPA, (2014, December). Priority Pollutant List. Retrieved August 23, 2020, from https://www.epa.gov/sites/production/files/2015-09/documents/priority- pollutant-list-epa.pdf..

[9] M. Bergknut. (2006). Characterization of PAH-contaminated soils focusing on availability, chemical composition and biological effects Umea University Department of Chemistry Environmental Chemistry SE-901 87 Umea Sweden, 6..

[10] Benlahcen, K.T., Chaoui, A., Budzinski, H.,Bellocq, J. and Garrigues, P.H. (1997).Distribution and source of polycyclic aromatic hydrocarbons in some Mediterranean coastal sediment. Mar.Poll. Bull., 34: 298-305..

[11] Zhu, L., Chen., Wang, J. and Shen, H.(2004). Pollution survey of polycyclic aromatic hydrocarbons in surface water of Hazhou, China. Chemosphere, 56: 1085-1095..

[12] Doong, R. and Line, Y. (2004). Characterization and distribution of polycyclic aromatic hydrocarbon contamination in surface sediment.

[13] Zhang, W., Zhang, S., Wan, C., Yue, D., Ye, Y., Wang, X. (2008).Source diagnostics of polycy-clic aromatic hydrocarbons in urban road runoff, dust, rain and canopy throughfall. Environ. Pollut. 153 (3), 594–601..

[14] E. V. Lau, S. Gan, and H. K. Ng (2010). Extraction Techniques for Polycyclic Aromatic Hydrocarbons in Soils. Int. J. Anal. Chem., Volume 2010, Article ID 398381, 9 pages.

[15] Wang, C., Yu, D., Shi, W., Jiao, K., Wu, B., Xu, H. (2016). Application of spent mush-room (Lentinulaedodes) substrate and acclimated sewage sludge on thebioremediation of polycyclic aromatic hydrocarbon polluted soil. RSC Adv. 6,37274-37285..

[16] ZheSun, Ying Zhu,ShaojieZhuo, WeipingLiu, Eddy Y.Zeng, XilongWang, BaoshanXing,ShuTao.(2017). Occurrence of nitro- and oxy-PAHs in agricultural soils in eastern China and excess lifetime cancer risks from human exposure through soil ingestion. Environ. I.

[17] Luma Abdalalah Sagban Alabadi, Dunya A. AL-jibury, Faiq F. Karam. (2017). Evaluation of Surface Soils Contamination by Polycyclic Aromatic Hydrocarbons of Selected Samples in Al-Diwaniya City, Iraq. Journal of Global Pharma Technology; 10(9), 2017:419-42.

[18] YajingQu, Yiwei Gong, Jin Ma, Haiying Wei, Qiyuan Liu, Lingling Liu, Haiwen Wu, Shuhui Yang, Yixiang Chen.(2020).Potential sources, influencing factors, and health risks of polycyclic aromatic hydrocarbons (PAHs) in the surface soil of urban parks in Bei.

[19] FAIQ F. KARAM and MUQDAD I. KADHIM. (2014). Determination of Chrysene by Gas Chromatography in Selected Soils from Al-Diwaniya Province, Iraq. Asian Journal of Chemistry; Vol. 26, Supplementary Issue (2014), S259-S261..

[20] S. Orecchio.(2010). Assessment of polycyclic aromatic hydrocarbons (PAHs) in soil of a Natural Reserve (Isola delle Femmine) (Italy) located in front of a plant for the production of cement. J. Hazard. Mater. Volume 173, Issues 1–3, Pages 358-368..

[21] Bossert I, Bartha R (1984). Fate of hydrocarbons during oily sludge disposal in soil. Appl. Environ. Microbial. 47,

763–767..
[22] Wagrowski, D.M., Hites, R.A.(1997).Polycyclic aromatic hydrocarbon accumulationin urban, suburban, and rural vegetation. Environ. Sci. Technol. 31 (1), 279-282..
[23] Yu G G, Wang T G, Wu D P.(2007). Study on fingerprints of PAHs from the combustion of bavin and coal. Ecology and Environment, 16(2): 285-289..

Determination of Polycyclic Aromatic Hydrocarbons in Surface Water of Al-Diwaniyah River-Case Study

Tania S. Kadhim and Faiq F. Karam
Department of chemistry, college of science, University of Al-Qadisiyah, Iraq

Abstract. The current study includes the identification of PAHs compounds in seven contaminated sites from the surface water of Al-Diwaniyah river from August 2020 to May 2021. Solid phase extraction was used for extract PAHs compounds from the surface water samples by specified columns and gas chromatography with (FID) detector was used for analysis of these compounds. Significant differences appeared in the concentration of PAHs according to different spatial, temporal conditions and human activities during the study period. The aggregate of PAHs samples for seven sites during ten months were (253.1714ppm). Benzo[A]Pyrene had the lowest cumulative value (7.1428ppm), whereas anthracene compound had the highest cumulative value (21.45714ppm).

Keywords: PAHs compounds, Gas chromatography with flame ionization detector, surface water, Al-Diwaniyah river.

1. Introduction

Can be defined the environmental contamination as every human activity that lead to the retrogradation or decline of the natural environment. In 2015 was predicted that poor health resulting from contamination had caused (9 million) precocious deaths and this is three times more than the number of deaths resulting from other diseases such as (AIDS, tuberculosis, malaria) [1]. In general, environmental pollution in depressed and middle revenue countries is greater than in progressing countries, due to lack of awareness and poverty [2]. Water is essential to sustain life and must be kept free from all pollutants. In addition to human body, other living organisms need water, but in its natural form free of pollutants [3]. Agriculture plays an important part in water contamination, as farmers emptying large amount of organic materials, agricultural chemicals, drug and sediments residues into water bodies, and this water contamination has many risks to human health and production activities [4]. There are many studies that have shown that the spread of typhoid fever, cholera and other diseases is linked to water contamination [5]. PAHs compounds are organic pollutants formed from 2 or more fused aromatic rings and containing only (hydrogen and carbon) atoms [6]. PAHs originate from several sources including (incomplete combustion processes, human activities, automobile exhaust, home heating, petroleum refining) [7]. Polycyclic aromatic hydrocarbons resulting from human activities are transferred to the water through several processes including:

1- Sewage disposal
2- Runoff
3- Accidental leakage
4- Dry or wet deposits in the atmosphere
5- River transport [8]

After the PAHs compounds enter the water bodies, a small portion can resolve in the water, while the other portion may link to the pending particles in water (because of its lipophilic nature and its low solubility in water), then move with the flow of water [9]. The discovery of PAHs in the aquatic environment is challenging due to their hydrophobic nature and low solubility in addition to being found in small concentrations and ordinarily existing in complex form, to beat these difficulties, appropriate pretreatment must be undertaken to pre-concentrate these compounds [10]. Polycyclic aromatic hydrocarbons are categorized to two set according to the number of aromatic rings. These two set include:

1- High molecular weight polycyclic aromatic hydrocarbons (HMW) where these hydrocarbons contain more than four fused aromatic rings such as Chrysene (contains four rings), Benzo[A]Pyrene (contain five rings), Benzo[GHI)Perylene (contain six rings), Coronen (contain seven rings).
2- Low molecular weight polycyclic aromatic hydrocarbons (LMW): These compounds contain two rings such as naphthalene, three rings such as anthracene. In general LMW PAHs compounds volatile more than HMW PAHs compounds and found fundamentally in the gas case, while high molecular weight hydrocarbons are found in the particulate case. Because PAHs can found in the gaseous and solid case, it can move in the environment freely and has the ability to spread in the surface of water, air and soil. PAHs are large groups of constant organic contaminants that have carcinogenic properties. Thus, they cause a danger to human health and for these reasons, the US EPA has listed 16 compounds of PAHs as primary contaminants [11]. LMW PAHs compounds like anthracene, phenanthrene, and naphthalene are toxic to a large degree, but they are not carcinogenic to some organisms, while high molecular weight polycyclic aromatic hydrocarbons such as chrysene and pyrene are considered less toxicity but have

greater strength in carcinogenicity compared to low molecular weight hydrocarbons [12], table 1 shows carcinogenic and some physical properties of polycyclic aromatic hydrocarbons.

Table 1. Carcinogenic activity and some physical properties of PAHs [12]

Compound	**Molecular weight (g/mol)**	**Melting point (C^0)**	**Boiling point (C^0)**	**Aqueous solubility (µg/L) at 25 C^0**	**Carcinogenic Activity**
Naphthalene	128.00	81	200	31700	0
Fluorene	166.22	115	298	1980	0
Chrycene	228.29	254	448	2	+
Acenapthene	154.21	96.2	279	3930	0
Benzo[a] anthracene	228.3	158.4	400	14	+
Fluoranthene	202.26	107	384	260	0
Benzo[k]Fluoranthene	252.32	217	480	-	++
Indeno[1,2,3-cd]pyrene	276	161-163.5	-	-	+
Phenanthrene	178.23	99-101	340	1290	0
Pyrene	202.26	156	393	135	0
Benzo[a]Pyrene	252.3	179	495	3800	++
Acenapthylene	152.21	92	265-275	16100	0
Anthracene	178.23	216.4	340	73	0
Benzo[b]Fluoranthene	252.32	168		-	++

(0) non-carcinogenic, (+) carcinogenic to a degree of 33%, (++) carcinogenic to a degree of more than 33%

Solid phase extraction is one of the (fastest, cleanest, most economic) pre-concentration method for separating, concentrating of trace metals from water [13], and it is preferred method for extracting polycyclic aromatic hydrocarbons in water [14]. There are a lot of materials that used as adsorbents for pre-concentration of these compounds in the surface water samples such as (polyvinylidene fluoride, bamboo charcoal, porous organ clay composite, matrix-immobilized organ clay). But C18 (they are compounds based on the use of SiO_2 with a carbon chain consisting of 18 carbon atoms) is the most common adsorbent chosen to extracting of PAHs from samples of water [15]. The use of gas chromatography (GC) has several advantages including:

1- Reproducibility
2- Stability
3- Large peak capacity

Gas chromatography can handle less kind of compounds as compared to liquid chromatography (LC) [16], flame ionization detector detects ions generated through the combustion of compounds discrete by GC chromatography in the flame of an air/H_2 and its response rely on the ions created by the compound [17]. Several studies were interested in method for removal of PAHs compounds from the environment [18] [19].

The objective of this research is for determination of concentrations of polycyclic aromatic hydrocarbons compounds in the surface water of Al-Diwaniyah river.

2. Experimental:

Study area: Al-Diwaniyah River : It is stretch of Al-Hilla river in the area extending from the Al-Suniyah side in the north to the south of the city of Al-Diwaniyah about (15 km) inside the city, is a major source of drinking, for agriculture and animals, as well as fishing for the people who live around this river. There are many sources of pollution near the river such as sewage discharges, agricultural pumps, outlet pipes from drinking water treatment plants, polluted water from textile and rubber factories, as well as pollution from atmospheric deposition of PAHs and other volatile compounds. Several sites have been selected to find out the percentage of contamination with polycyclic aromatic hydrocarbons. Figure1 explain map of the pollution sites of Al-Diwaniyah River.

S1: Represents the starting point before entering Al-Diwaniyah governorate (Al-Suniyah area), which is located about (11 km) north of the center of Al-Diwaniyah city.

S2: This site represents the heavy water treatment unit, which is located in the gas-electric power plant. The water is drained into the Al-Haffar river without treatment because of this unit has not working for a long time, the number of employees of this station exceeds (600) people, in addition to draining the water of the houses adjacent to the Al-Haffar river directly into the river.

S3: Al-Thowra street bridge, which is located in the center of Al-Diwaniyah city, the source of pollution in this location is the rainwater drainage pipe located at the bottom of the bridge the right bank of the river.

S4: Represents the rainwater drainage pipe for all districts of Al-Diwaniyah city, which is located in the Rafaat neighborhood.

S5: Represents drainage station of the heavy water, which is located in the Umm Al-Khail area, behind the sabean building.

S6: Located near five Martyrs bridge, drivers use this area to wash the cars. Wash water is often mixed with oil derivatives such as lubricating oil, wax, transmission oil.

S7: Represents the river's exit point from Al-Diwaniyah city (Al-Sudair area), which is located about (12 km) south of the center of the city.

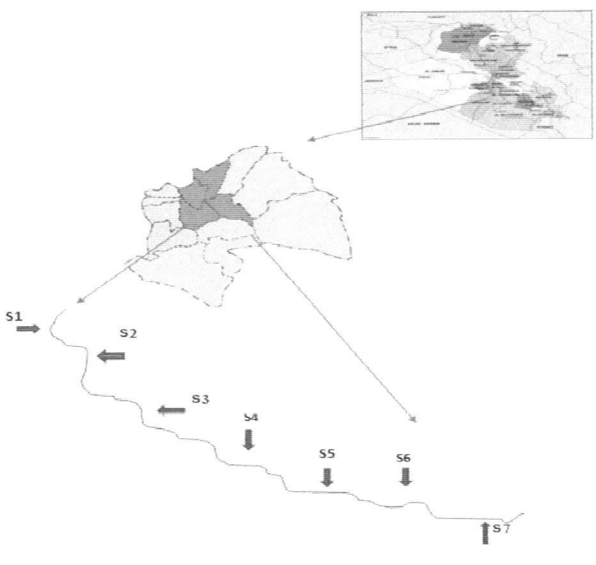

Figure 1. Map of the sampling points of Al-Diwaniyah river

Surface water samples are collected once a month, the sampling distance is about (3 meters) from the source of pollution. The water is collected in a (1 liter) amber glass container lined with foil aluminum and fitted with a screw cap to block adhesion occurs between polycyclic aromatic hydrocarbons and the container's plastic cap, containers labeled from (S1) to (S7). After collection, samples are kept in a cooler box at a temperature (4^0C) and stored in the refrigerator until transported to the laboratory for extraction. When keeping samples, it must be ensured that they are not exposed to light in order to prevent their decomposition because of these compounds are sensitive to the light, all used glass are cleaned with (hot water, detergents) and rinsed with distilled water and dried for 1 hour at a temperature of (180^{0}C). Extraction of PAHs compounds carried out by SPE using special columns (SUPELCLEAN ENVI-18 solid phase extraction Tube. bed wt. 500mg, volume 6 ml, pkg of 30 ea). The extracted of PAHs compounds eluted from column cartridge by 10 ml of dichloromethane in 2 batches each time 5 ml. The solution concentrated to 1 ml by rotary evaporator device at a temperature 45^{0}C, finally the concentrated solution of samples kept in a small dark colored glass containers in the refrigerator at a temperature below 4^{0}C until measurement, the analysis of PAHs compounds was carried out by GC-FID. Separation was carried out in a (30 m × 0.25 mm i.d.) DB-5 column (J&W Scientific, Folsom, CA) coated with (0.25 μm) thick film of (5%) diphenyl-poly dimethyl siloxane. The samples were injected (in the split mode) at an injection temperature of (280^{0}C), the column temperature was initially held at (40^{0}C) for (1 min), raised to (120^{0}C) at the rate of (25^{0}C/min), then to (160^{0}C) at the rate of (10^{0}C/min), finally to (300^{0}C) at the rate of (5^{0}C/min). Held at final temperature for (15 min), FID detector temperature was keep at (330^{0}C), helium use as a carrier gas at constant flow rate of (5ml/min).

3. Results and discussion:

Table 2 explain average of concentration of 16 PAHs compounds in the surface water of Al-Diwaniyah samples in seven study sites.

Table 2. Average concentration of 16 PAHs compounds (in part per million unit) in the surface water samples of Al-Diwaniyah river

No. of site	S1	S2	S3	S4	S5	S6	S7	Mean	Sum.
AcP	27.8	16.8	4.6	22.8	4.4	47.2	18.8	20.34286	142.4
AcPy	30.6	18	8.6	19.2	5.8	46.4	11.4	20	140
Ant	21.4	14.2	3.4	14.4	10.4	64	22.4	21.45714	150.2
BaA	0	12.6	1.2	15.2	2.6	76.8	0	15.48571	108.4
BbFl	6.6	7.4	9.2	9.8	0	71.6	0	14.94286	104.6
BkFl	20.6	10.4	0	1	3.6	43.6	12	13.02857	91.2
BghiP	0	0	0	2.4	2.4	93.8	0	14.08571	98.6
BaP	0	3.6	0	4.8	0	41.6	0	7.142857	50
Chr	4.4	10.2	11.4	4.8	7.2	66.8	5.2	15.71429	110

DbA	0	11.8	7.6	11	2.2	28.8	0.6	8.857143	62
Flu	0	9.4	8	7.2	0	100.4	0	17.85714	125
Fl	0	4	0	8.8	4.2	53	0	10	70
InP	2.4	6.8	0	2	0	72	3.6	12.4	86.8
NaP	12.4	7.4	0	4.6	0	77.6	6	15.42857	108
Phe	7.8	6.6	5.4	8.2	2.6	86.2	3.6	17.2	120.4
Pyr	0	4.8	0	0	0	73.8	0	11.22857	78.6
∑PAHs	134	144	59.4	136.2	45.4	1043.6	83.6	235.1714	1646.2

Anthracene is the highest concentration of PAHs compounds (150.2 ppm), while Benzo[A]Pyrene has less concentration (50 ppm). The total PAHs content ranged from (45.4 ppm) to (1043.6 ppm) with mean concentration (235.1714 ppm), the results in this study are compared with previous studies. In the study carried out by Faiq and his group, the highest concentration was for phenanthrene (1177.20 ng/ml^{-1}) [20]. In the study that carried out by Hamid and his group show that the highest concentration was for naphthalene (1.13 $ng.l^{-1}$) and total content of PAHs compounds was from (7.24 $ng.l^{-1}$) to (17.39 $ng.l^{-1}$) [21], while in the study carried out by Chunhui Wang and his group show that the highest concentration was for naphthalene (10.677 ng/l) and the lowest concentration was for Benzo[A]Pyrene (1 ng/l), the total content of PAHs compounds was in the range from (4.076 ng/l) to (29.455 ng/l) with mean concentration (17.212 ng/l) [22]. The results of the study carried out by Zhao et al. (2015) show that the total content of PAHs compounds was in the range from (54.8 ng/l) to (2.589 ng/l) and the mean concentration (1.375 ng/l) [23]. Site No.6 was the most pollutant site than other because of the drivers use this area to wash the cars and this wash water maybe mixed with oil derivatives. Pollution of water with PAHs compounds is related to human activities which represents main source of PAHs compounds compared to the natural sources [24], and the process of atmospheric precipitation in the form of particulate matter resulting from burning coal and fuel also is the main source of these compounds in the water, in addition to rain and fumes, pollutants it carries with it [25]. A lot of previous studies prove that the concentration of PAHs compounds are different according to human activities in the sites and weather of sampling whether it was in water or soil samples [26] [27] [28] [29] [30].

4. Conclusion:

16 PAHs compounds have been identified in the seven study sites of the surface water of Al-Diwaniyah

river. The extraction of these compounds carried out by SPE extraction, while the analysis by GC chromatography device with FID detector. The results show that the site number 6 was most pollutant with polycyclic aromatic hydrocarbons compounds than other sites according to human activities.

Acknowledgements:
The authors are gratefully acknowledge the support provided by the University of Al-Qadisiyah without which this, research would not have been possible.

References

[1] Landrigan, P.J., Fuller, R., Acosta, N.J.R., Adeyi, O., Arnold, R., Basu, N. (2017). The Lancet Commissionon pollution and health. The Lancet .391 (10119), 462-512..
[2] Muralikrishna,I.V., Manickam,V.(2017). Analytical methods for monitoring environmental pollution. Environmental Management.Butterworth Heinemann, Elsevier, pp. 495-570..
[3] A.K. Pathak,(2013) Water pollution and treatment, Int. J. Environ. Eng. Manag. 3 (4): 191–198..
[4] UNEP. 2016. A snapshot of the world's water quality: towards a global assessment. Nairobi, United Nations Environment Programme (UNEP)..
[5] I.J. Gambo, Y.A. Yusuf, G. Ahmed, S.A. Garba, A.M. Idris, K. Abdulkarim,(2018). Causes and health effects of water pollution in domestic water sources in Hadeija metropolis, Nigeria, using statistical modeling, Nigerian Res. J. Chem. Sci. 4: 101–112..
[6] L. Yan, X. Li, J. Chen, X. Wang, J. Du and L. Ma, (2012). Source and deposition of polycyclic aromatic hydrocarbons to Shanghai, China, Journal of Environmental Science, 24(1): 116-123..
[7] Liu, Y., Zarfi, C., Basu, N.B., Cirpka, O.A. (2019). Turnover and legacy of sediment-associated PAH in a baseflow- dominated river. Sci. Total Environ. 671, 754-764..
[8] Kanzari, F., Syakti, A.D., Asia, L., Malleret, L., Piram, A., Mille, G., Doumenq, P. (2014). Distributions and sources of persistent organic pollutants(aliphatic hydrocarbons, PAHs, PCBs and pesticides) in surface sediments of an industrialized urban riv.
[9] Yuan, Z.J., Wang, R.W.,Da, C.N. (2014). Polycyclic aromatic hydrocarbons in sediments from old Yellow River Estuary, China: Occurrence, Sources, characterization and correlation with the relocation history of the yellow River. Ecotox. Environ. Saf. 109, 1.
[10] X. Yang, Y. Yin, Y. Zong, T. Wan, X. Liao (2019). Magnetic nanocomposite as sorbent for magnetic solid phase extraction coupled with high performance liquid chromatography for determination of polycyclic aromatic hydrocarbons, Microchem. J. 145: 26–34..
[11] US EPA, (2014, December). Priority Pollutant List. Retrieved August 23, 2020, from https://www.epa.gov/sites/production/files/2015-09/documents/priority- pollutant-list-epa.pdf..
[12] Walters ,R.W. and Luthy ,R.G. (1984). Equilibrium adsorption of polycyclic aromatic hydrocarbons from water onto activated carbon, Environ, Sci. Technol. Vol. 18, No. 6-395..
[13] I. S.Trujillo, E.V. Alonso, A.G.de Torres andJ.M. C.Pavon (2012). Development of a solid phase extraction method for the multielement determination of trace metals in natural waters including sea-water by FI-ICP-MS, Microchemical Journal,101:87-94..
[14] P. Poppa, C.Bauer, M.Moder andA.Paschke (2000). Determination of polycyclic aromatic hydrocarbons in waste water by off-line coupling of solid-phase microextraction with column liquid chromatography, Journal of Chromatography A,897,153–159..
[15] P. Oleszczuk, and S. Baran, (2004). Application of solid-phase extraction to determination of polycyclic aromatic hydrocarbons in sewage sludge extracts,Journal of Hazardous Material B, 113:237-245..
[16] Kadri, T., Rouissi, T., KaurBrar, S., Cledon, M., Sarma, S., Verma, M. (2017). Biodeg-radation of polycyclic aromatic hydrocarbons (PAHs) by fungal enzymes. J. Environ. Sci. 51, 52-74..
[17] Wang, C., Li, Y., Tan, H., Zhang, A., Xie, Y., Wu, B., Xu, H. (2019). A novel microbeconsortium, nano-visible light photocatalyst and microcapsule system todegrade PAHs. Chem. Eng. J. 359, 1065-1074..
[18] Faiq F. Karam, Falah H. Hussein, Sadiq J. Baqir and Ayad F. Alkaim. Optimal conditions for treatment of

contaminated waters with anthracene by Fenton processes in close system reactor. JCPS Volume 9 Issue 3, 1111-1115..

[19] Faiq F. Karam, Falah H. Hussein, Sadiq J. Baqir, Ahmed F. Halbus, Ralf Dillert, and Detelf Bahnemann. Photocatalytic Degradation of Anthracene in Closed System Reactor. International Journal of Photoenergy Volume 2014, Article ID 503825, 6 pages..

[20] Faiq. F. Karam, Falah H. Hussein , Sadiq J. Baqir and Fadhil M. Abid, (2012). Determination of Polycyclic Aromatic Hydrocarbons in Surface Water of Shatt Al-Hilla River,. Asian Journal of Chemistry; Vol. 24, No. 12, 5589-5591..

[21] Hamid T. Al-Saad, Dhfar. Dh. Al-Khion, Ebtihal S. Majeed, Abas A. Hantoush, Salah M. Saleh, Abdulzahra A. Alhello, (2016). Environmental Assessment of Polycyclic Aromatic Hydrocarbons pollutants in the waters of the side branches of the Shatt Al-Arab.

[22] Chunhui Wang, Shenglu Zhou, Shaohua Wu, Jing Song, YaxingShi,Baojie Li, Hao Chen.(2017). Surface water polycyclic aromatic hydrocarbons (PAH) in urban areas of Nanjing, China,. Water Science &Technolog. 76 (8): 2150–2157..

[23] Zhao, X., Qiu, H., Zhao, Y., Shen, J., Chen, Z. & Chen, J. (2015) Distribution of polycyclic aromatic hydrocarbons in surface water from the upper reach of the Yellow River,Northwestern China. Environ. Sci. Pollut. Res. 22,6950–6956..

[24] Yan, L., Li, X., Chen, J., Wang, X., Du, J. and Ma, L. (2012). Source and deposition of polycyclic aromatic hydrocarbons to Shanghai, China. J. Environ. Sci., 24(1):116-123..

[25] Kucklick, J. R. and Bidleman, T. F. (1994). Organic contaminants in Winyah bay, South Carolina 1: pesticides and polycyclic aromatic hydrocarbons in subsurface and microlayer water. Mar. Envi. Res., 37: 63 –78..

[26] FALAH H. HUSSEIN, FAIQ F. KARAM and SADIQ J. BAQIR. (2014). Monitoring of Polycyclic Aromatic Hydrocarbons in Surface Water of Shatt Al-Hilla River. Asian Journal of Chemistry; Vol. 26, No. 9 (2014), 2768-2772..

[27] FAIQ F. KARAM and MUQDAD I. KADHIM.(2014). Determination of Chrysene by Gas Chromatography in Selected Soils from Al-Diwaniya Province, Iraq. Asian Journal of Chemistry; Vol. 26, Supplementary Issue (2014), S259-S261..

[28] FALAH. H. HUSSEIN, FAIQ F. KARAM and SADIQ J. BAQIR. (2014). Determination of Benzo[a]anthracene in Surface Water of Shatt Al-Hilla River by Solid Phase Extraction and High Performance Liquid Chromatography. Asian Journal of Chemistry; Vol. 26, No. 9 (201.

[29] Luma Abdalalah Sagban Alabadi, Dunya A. AL-jibury, Faiq F. Karam. (2017). Evaluation of Surface Soils Contamination by Polycyclic Aromatic Hydrocarbons of Selected Samples in Al-Diwaniya City, Iraq. Journal of Global Pharma Technology; 10(9), 2017:419-42.

[30] FALAH. H. HUSSEIN, FAIQ F. KARAM and SADIQ J. BAQIR. (2014). Determination of Benzo(a)pyrene in Surface Water of Shatt Al-Hilla River, Iraq. Asian Journal of Chemistry; Vol. 26, No. 24 (2014), 8349-8352..

The inhibitory effect of ginger and lemongrass plants extracts on the growth of some fungi associated with stored yellow corn grains

Ali A Alsudani[1] and Ghaidaa Raheem Lateef Al-Awsi[2]
[1]Environmental Research and Pollution Prevention Unit, College of Science, University of Al-Qadisiyah, Diwaniyah, Iraq
[2]Department of Radiological Techniques, Al-Mustaqbal University College, Babylon, Iraq

Abstract. The current study included testing the effect of the aqueous and alcoholic extracts of the rhizomes of the ginger plant (*Zengiber officinal*) and the leaves of the lemongrass plant (*Cymbopogon citratus*) on the germination of local yellow corn grains stored in some local markets in the city of Al-Diwaniyah and studying the effect of four different concentrations of these extracts (10, 20, 30, and 40) mg/ml in the radial growth of some fungi isolated from yellow corn grains. The results showed that there were significant differences in the germination ratios of yellow corn grains for the different concentrations of the extracts in comparison with the control treatment, and there were no significant differences in the germination rates of yellow corn grains at a concentration of 40 mg/ml for the different treatments with the treatment of the fungicide Raxil at a probability level of 5 %. It was found that all the extracts had a significant inhibitory effect on the growth of the fungi tested on Potato Dextrose Agar medium by comparison with the control treatment, and these fungi are: *Aspergillus niger*, *Ulocladium* sp., *Rhizopus stolonifer* and *Fusarium graminearum*, and the alcoholic extracts of ginger and lemongrass extracts were higher to inhibition of the radial growth of fungi tested in the different treatments, the inhibition ratios were between 73.33-87.12 % in the alcoholic extract treatments for ginger and between 75.15-86.98 % in the alcoholic extract treatments for the lemongrass plant, and the two concentrations were 30 and 40 mg/ml for the alcoholic extracts of the ginger plant and lemongrass has a significant effect similar to that of the fungicide Raxil for all the tested fungi, while it was found that the concentration of 40 mg/ml of the aqueous extracts of ginger and lemongrass was the only concentration that produced a significant effect similar to that of the fungicide Raxil for all tested fungi.

Keywords: Yellow corn, ginger, lemongrass, Raxil

1. Introduction

The yellow corn crop (Zea mays L.) is an important economic field crop on a global and local level for its use as the primary source of oils used in various agricultural industries, as well as its entry into the animal feed industry as basic food for poultry and other animals. Yellow corn grains are infected with various types of fungi such as *Aspergillus* spp, *Alternaria* spp, *Penicillium* spp, *Fusarium* spp, *Rhizopus* spp, *Trichoderma* spp, and others during harvesting, transportation, and storage [1]. These fungi greatly affect the vitality of the yellow corn grain and reduce their germination rates and thus cause great economic losses by reducing agricultural production when using such grains in agriculture, some fungal species can produce some toxins, such as Aflatoxins, Ochratoxins, Zearalenone, and others [2]. These toxins are among the most dangerous agricultural pollutants because of their cumulative carcinogenic effects and the ability to degrade various vital tissues, especially the liver and kidneys [3]. The percentage of apparent damage caused by the fungi to the yield of yellow corn was 50.8%. It was also found that some fungal species that are transmitted with the yellow corn grains to the field when planted, such as *Fusarium* spp and other fungi, may cause important diseases on the plant in the field such disease as rotting corn stalks [4]. Therefore, the importance of how to protect humans and animals from the damage caused by these toxins and by various means, the most important of which is controlling production and storage conditions to reduce the growth of contaminated fungi by using some different chemical pesticides, and because the majority of these pesticides have many side effects, in addition to being pollutants to the environment, they are toxic and carcinogenic to humans and animals if grains are used as a food source. [5]. It also has the property of accumulating in soil particles, resulting in the death of a significant number of vital microorganisms in the soil, as well as the potential for leaking into various water sources and adverse effects on them [6]. Therefore, it has become necessary to find environmentally safe alternatives to such pesticides, so several studies have focused on the use of plant extracts to control agricultural pests, including various fungi that lead to a significant reduction in plant production [7]. So, this study came to investigate the effect of the aqueous and alcoholic extracts of the rhizomes of the ginger plant (*Zengiber officinal*) and the leaves of the lemongrass plant (*Cymbopogon citratus*) on the germination of local yellow corn grains stored in some local markets in the city of Al-Diwaniyah.

2. **Material and Methods**

2.1 Yellow corn grains collection

The yellow corn grains used in this study were obtained from local markets in Al-Diwaniyah city, with three random samples (1 kg) collected from each site.

2.2 Preparation of plant extracts

The hot aqueous extract of ginger and lemongrass was prepared separately, depending on the method mentioned in [8]. and the same steps were followed in preparing the alcoholic extract but using ethyl alcohol at a concentration of 70% instead of distilled water.

2.3 Prepare the stock solution

For each variety of plant studied, the stock solution was made by dissolving 10 g of dry extract in 100 ml of sterile distilled water to achieve a concentration of 100 mg/ml.

2.4 Isolation and identification of fungi

The corn grains were divided into two groups; The first group was superficially sterilized using sodium hypochlorite solution at a concentration of 1% for 3 minutes, then washed with sterile distilled water and left to dry. As for the second group, it was washed with sterile distilled water only and left to dry. Then, the grains were planted in plastic Petri dishes containing Potato Dextrose Agar (5 grains in each dish) and with three replicates for each group. The dishes were left in the incubator at a temperature of 25° C. After four days, the dishes were tested to see which fungi had grown, and then they were purified on the PDA culture medium [9]. They were diagnosed according to the approved taxonomic bases and using the taxonomic keys in [10] [11], and after their diagnosis, four fungi were selected for this study which are *Aspergillus niger*, *Ulocladium* sp., *Rhizopus stolonifer*, and *Fusarium graminearum*.

2.5 Effect of tested plant extracts on the germination of corn grains

Four concentrations of aqueous and alcoholic extracts were prepared separately (10, 20, 30, and 40) mg/ml by dilution with sterile distilled water, and the fungicide Raxil was used at a concentration of 5 mg/ml, and the control treatment included untreated corn grains with additional material, and after treating the yellow corn grains with different concentrations of extracts and pesticide for 3 minutes, they were planted in plastic Petri dishes (10 grains in each dish) containing PDA sterilized culture media and with three replications for each treatment. The dishes were incubated in the incubator at a temperature of 25° C for 7 days [12], germination percentages were calculated using the following equation:

$$\text{Germination percentage} = \frac{\text{Germinated seeds}}{\text{Total seeds}} \times 100$$

2.6 Effect of the tested plant extracts on the radial growth of the fungi

Four concentrations of the tested extracts were prepared, which are (10, 20, 30, and 40) mg/ml of sterile PDA culture medium. As for the treatment of the fungicide Raxil, it was prepared at a concentration of 5 mg/ml of PDA culture medium, then distributed in plastic Petri dishes. The control treatment included Petri dishes containing PDA only, then a piece of 7.5 mm diameter was transferred from pure cultures of the fungi under study at the age of 7 days using a cork borer and placed in the middle of the dish and the dishes were incubated at a temperature of 25 ° C with three replications for each treatment and each fungus, and the growth rate of each fungus was measured in the different treatments using the usual ruler (the average of two perpendicular diameters) after the mycelium in the control treatment reached the edge of plate [13], and the inhibition percentages were calculated using the following equation:

$$\text{Inhibition percentage} = \frac{\text{Growth in control} - \text{Growth in treatment}}{\text{Growth in control}} \times 100$$

2.7 statistical analysis

The results were statistically analyzed to determine the significant differences at the probability level of 5%, as the statistical analysis included the two-way analysis of variance, and the test for significant differences between the means using Duncan's polynomial test [14].

3. Results and Discussion

3.1 Effect of aqueous and alcoholic extracts of tested plants on the germination of corn grains

The results in Table (1) showed that the effect of alcoholic extracts of ginger and lemongrass plants was higher than the effect of aqueous extracts in increasing the germination rates of the treated grains, the germination rates of grains at the concentration of 40 mg/ml for the alcoholic ginger and lemongrass extracts were 93.33 %, while the aqueous extracts gave at the same concentration, germination rates reached 86.66 % and 80.00 %, respectively. It was also found that the concentrations used and for all tested extracts increased the germination rates of the treated grains with significant differences in comparison with the control treatment whose grain germination rates ranged between 60.00-53.33 %, and the absence of significant differences with treatments of the fungicide Raxil at a concentration of 5 mg/ml, whose germination rates ranged between 90.00-93.33 %. This confirms the effectiveness of the aqueous and alcoholic extracts of ginger and lemongrass with their different concentrations in raising the germination rates of the yellow corn grains because the ginger plant contains many active substances, the most important of which are volatile oils known for their inhibitory effects on the growth of fungi [15]. As for the lemongrass plant, the studies indicate that its different parts contain many active substances, such as the volatile oils Nirol 33.2 % and Geraniol 37.8 %, which have an inhibitory effect on the growth of different fungi [16].

These findings support previous research that demonstrated that most plant extracts improve the rates of germination of various plant grains treated with them, with lemongrass extract showing the highest germination of rice seed and being the most effective in suppressing rice seed-borne fungus [17]. In a different investigation, ginger extracts were found to be more effective than fungicides (mancozeb and benomyl) at improving calliandra seed germination and suppressing fungal infection [18]. The decrease in germination rates of untreated grains (the control treatment) may be attributed to the effect of fungi on the internal tissues of the grains and their negative effect on the embryo, or the fungi attacking the grains during the germination process, which reduces germination rates, also it was found that the competition of fungi with grains for the amount of oxygen in the growth medium is related to germination rates [19].

Table 1. Effect of aqueous and alcoholic extracts of tested plants on the germination of corn grains

Concentration (mg/ml)	Grain germination percentage (%)			
	Aqueous ginger extract	Alcoholic ginger extract	Aqueous lemongrass extract	Alcoholic lemongrass extract
10	73.33 c	80.00 c	73.33 c	80.00 c
20	80.00 b	86.66 b	73.33 c	86.66 b
30	80.00 b	86.66 b	80.00 b	86.66 b
40	86.66 a	93.33 a	90.00 a	93.33 a
Raxil 5 (mg/ml)	90.00 a	93.33 a	93.33 a	93.33 a
Control	53.33 d	60.00 d	53.33 d	53.33 d

-- The results in the table are based on three replicates on average.

- Averages with the same letters do not differ significantly among themselves in vertical comparisons according to Duncan's polynomial test at a 5% probability level.

3.2 Effect of aqueous and alcoholic extracts of tested plants on the radial growth of fungi

The results of the effect of aqueous and alcohol extracts of the tested plants on the radial growth of some fungi isolated from stored corn grans showed that it had a significant inhibitory effect on the growth of all tested fungi at a probability level of 5%, and these fungi are *Aspergillus niger* (Table 2), *Ulocladium* sp. (Table 3), *Rhizopus stolonifer* (Table 4) and *Fusarium graminearum* (Table 5), and the percentages of inhibition increase with the increase in the concentration of the extract used. The results also showed that the effect of alcoholic extracts of ginger and lemongrass was higher than the effect of aqueous extracts in inhibiting the radial growth of fungi tested in different treatments, the percentages of inhibition reached between 73.33-87.12 % in the alcoholic extract of ginger treatments and between 75.15-86.98 % in treatments of alcoholic extract of lemongrass, as for aqueous extracts, the inhibition rates were between 66.66-85.93 % in treatments of aqueous extract of ginger and between 65.93-86.52 % in treatments of aqueous extracts of lemongrass compared to the control treatments for these fungi that the inhibition percentage was 0.00 %, and the concentrations of 30 and 40 mg/ml for alcoholic extracts of ginger and lemongrass had a significant effect similar to that of the fungicide Raxil for all tested fungi, while it was found that the concentration of 40 mg/ml for aqueous extracts of ginger and lemongrass was the only concentration which had a significant effect similar to that of the fungicide Raxil for all tested fungi, and on the in spite of that, all the other concentrations caused a significant reduction in the growth rate compared to the control treatment. The antifungal activity of the aqueous and alcoholic extracts of ginger and lemongrass is because they contain many volatile oils that have an antimicrobial effect on the growth of microorganisms, including fungi [20], the difference in the extent to which fungi are affected by the plant extracts used may be due to the nature of the fungus in terms of composition, the thickness of its cell membranes, its content of fats and proteins, and its relationship to the mechanism of action of the effective compounds of those extracts, as the effect of the extracts on the fungi may be a result of distortions in their membranes and internal structures [21]. The higher effect of the alcoholic extracts of ginger and lemongrass than what they gave in the aqueous extracts may be because the alcoholic extract contains more active substances than the aqueous extract and the reason for this is the high ability of the alcoholic solvent to dissolve more than one active substance, and also the alcoholic extract contains a percentage of water, which leads to dissolving the active substances that can dissolve in water and alcohol [22].

Table 2. Effect of aqueous and alcoholic extracts of tested plants on the radial growth of *Aspergillus niger*

Concentration (mg/ml)	Aqueous ginger extract	Alcoholic ginger extract	Aqueous lemongrass extract	Alcoholic lemongrass extract
	Inhibition (%)	Inhibition (%)	Inhibition (%)	Inhibition (%)
10	**72.22 d**	**79.26 b**	**65.93 d**	**76.66 c**
20	**75.98 c**	**81.22 b**	**73.33 c**	**81.11 b**
30	**80.37 b**	**85.55 a**	**78.88 b**	**85.44 a**
40	**85.18 a**	**87.09 a**	**84.44 a**	**86.66 a**
Raxil 5 (mg/ml)	**87.04 a**	**87.41 a**	**87.11 a**	**87.18 a**
Control	**0.00**	**0.00**	**0.00**	**0.00**

-- The results in the table are based on three replicates on average.
- Averages with the same letters do not differ significantly among themselves in vertical comparisons according to Duncan's polynomial test at a 5% probability level.

Table 3. Effect of aqueous and alcoholic extracts of tested plants on the radial growth of *Ulocladium* sp.

Concentration (mg/ml)	Aqueous ginger extract	Alcoholic ginger extract	Aqueous lemongrass extract	Alcoholic lemongrass extract
	Inhibition (%)	Inhibition (%)	Inhibition (%)	Inhibition (%)
10	**66.66 d**	**73.33 c**	**62.66 d**	**75.99 c**
20	**72.22 c**	**80.74 b**	**67.30 c**	**78.41 b**
30	**78.52**	**84.44**	**74.82**	**82.96**

	b	**a**	**b**	**a**
40	**83.71**	**85.93**	**83.33**	**85.55**
	a	**a**	**a**	**a**
Raxil	**85.93**	**86.33**	**85.93**	**85.93**
5 (mg/ml)	**a**	**a**	**a**	**a**
Control	**0.00**	**0.00**	**0.00**	**0.00**

- The results in the table are based on three replicates on average.
- Averages with the same letters do not differ significantly among themselves in vertical comparisons according to Duncan's polynomial test at a 5% probability level.

Table 4. Effect of aqueous and alcoholic extracts of tested plants on the radial growth of *Rhizopus stolonifer*

Concentration (mg/ml)	Aqueous ginger extract	Alcoholic ginger extract	Aqueous lemongrass extract	Alcoholic lemongrass extract
	Inhibition (%)	Inhibition (%)	Inhibition (%)	Inhibition (%)
10	**73.33**	**81.37**	**70.00**	**78.15**
	d	**b**	**d**	**c**
20	**77.41**	**83.71**	**75.55**	**81.03**
	c	**b**	**c**	**b**
30	**81.44**	**86.66**	**80.00**	**84.82**
	b	**a**	**b**	**a**
40	**85.93**	**87.12**	**84.82**	**86.95**
	a	**a**	**a**	**a**
Raxil	**87.41**	**88.24**	**87.41**	**87.34**
5 (mg/ml)	**a**	**a**	**a**	**a**
Control	**0.00**	**0.00**	**0.00**	**0.00**

- The results in the table are based on three replicates on average.
- Averages with the same letters do not differ significantly among themselves in vertical comparisons according to Duncan's polynomial test at a 5% probability level.

Table 5. Effect of aqueous and alcoholic extracts of tested plants on the radial growth of *Fusarium graminearum*

Concentration (mg/ml)	Aqueous ginger extract	Alcoholic ginger extract	Aqueous lemongrass extract	Alcoholic lemongrass extract
	Inhibition (%)	Inhibition (%)	Inhibition (%)	Inhibition (%)
10	**68.88 d**	**74.80 c**	**66.740 d**	**75.15 c**
20	**74.44 c**	**81.25 b**	**76.30 c**	**81.77 b**
30	**81.12 b**	**84.76 a**	**82.66 b**	**85.74 a**
40	**84.93 a**	**86.33 a**	**86.52 a**	**86.98 a**
Raxil 5 (mg/ml)	**85.31 a**	**86.66 a**	**86.41 a**	**86.74 a**
Control	**0.00**	**0.00**	**0.00**	**0.00**

- The results in the table are based on three replicates on average.
- Averages with the same letters do not differ significantly among themselves in vertical comparisons according to Duncan's polynomial test at a 5% probability level.

4. Conclusion

The aqueous and alcoholic plant extracts of the tested plants had a high inhibitory effect on the growth of the fungi under study, with significant differences from the control treatment and effects similar to that of the fungicide Raxil treatment. It was discovered that the two concentrations were 30 and 40 mg/ml for the alcoholic extracts of the ginger plant and lemongrass has a significant effect similar to that of the fungicide Raxil for all the tested fungi, while it was discovered that the concentration of 40 mg/ml of aqueous extracts of ginger and lemongrass produced a significant effect comparable to that of the fungicide Raxil for all tested fungi.

References

[1] García-Lara, S. and Serna-Saldivar, S.O., 2019. Corn history and culture. Corn, pp.1-18..

[2] Ji, F., He, D., Olaniran, A.O., Mokoena, M.P., Xu, J. and Shi, J., 2019. Occurrence, toxicity, production and detection of Fusarium mycotoxin: A review. Food Production, Processing and Nutrition, 1(1), pp.1-14..

[3] Prescott, L.M., Harley, J.P. and Klein, D.A., Microbiology.2002. Dubuque, IA: McGraw-Hill Higher Education..

[4] Khokhar, M.K., Hooda, K.S., Sharma, S.S. and Singh, V., 2014. Post flowering stalk rot complex of maize-present status and future prospects. Maydica, 59(3), pp.226-242..

[5] Gyawali, R. and Ibrahim, S.A., 2014. Natural products as antimicrobial agents. Food control, 46, pp.412-429..

[6] Singh, N.S., Sharma, R., Parween, T. and Patanjali, P.K., 2018. Pesticide contamination and human health risk factor. In Modern age environmental problems and their remediation (pp. 49-68). Springer, Cham..

[7] Al-douri, N.A., 2000. A survey of medicinal plants and their traditional uses in Iraq. Pharmaceutical biology, 38(1), pp.74-79..

[8] Habourne, J.B.C., 1998. Phytochemical Methods: A Guide To Modern Techniques of Plants Analysis..

[9] Sauer, D.B. and Burroughs.

[10]Domsch, K.H., Gams, W. and Anderson, T.H., 2007. Compendium of soil fungi, 2nd taxonomically revised edition by W. Gams. IHW, Eching..

[11]Hunter, B.B. and Barnett, H.L., 2019. Deuteromycetes (fungi imperfecti). In CRC handbook of microbiology (pp. 448-476). CRC Press..

[12]Marcos Filho, J., 2015. Seed vigor testing: an overview of the past, present and future perspective. Scientia agricola, 72, pp.363-374..

[13]Yu, D., Wang, J., Shao, X., Xu, F. and Wang, H., 2015. Antifungal modes of action of tea tree oil and its two characteristic components against Botrytis cinerea. Journal of applied microbiology, 119(5), pp.1253-1262..

[14]McDonald, J.H., 2014. Handbook of Biological Statistics 3rd ed Sparky House Publishing. Baltimore, MD: Available http://www. biostathandbook. com/transformation. html..

[15]Hussein, K.A. and Joo, J.H., 2018. Antifungal activity and chemical composition of ginger essential oil against ginseng pathogenic fungi. Curr. Res. Environ. Appl. Mycol, 8, pp.194-203..

[16]Boukhatem, M.N., Ferhat, M.A., Kameli, A., Saidi, F. and Kebir, H.T., 2014. Lemon grass (Cymbopogon citratus) essential oil as a potent anti-inflammatory and antifungal drugs. Libyan Journal of Medicine, 9(1)..

[17]Rahman, M.Z., Ali, M.A., Hossain, M.S., Monsur, M.A. and Zaman, M.Z., 2013. Effect of lemon grass (Cymbopogen citratus) extract on seed health of rice. Bangladesh Rice Journal, 17(1-2), pp.105-108..

[18]Suharti, T., Nugraheni, Y.M.M.A., Suita, E. and Sumarni, B., 2020, August. Effect of plant extracts and chemical fungicide on viability and percentage of seed-borne fungal infection on calliandra (Calliandra callothyrsus) seed. In IOP Conference Series: E.

[19]Tian, Y., Guan, B., Zhou, D., Yu, J., Li, G. and Lou, Y., 2014. Responses of seed germination, seedling growth, and seed yield traits to seed pretreatment in maize (Zea mays L.). The Scientific World Journal, 2014..

[20]Liu, C.X., 2021. Overview on development of ASEAN traditional and herbal medicines. Chinese Herbal Medicines, 13(4), pp.441-450..

[21]Rella, A., Farnoud, A.M. and Del Poeta, M., 2016. Plasma membrane lipids and their role in fungal virulence. Progress in lipid research, 61, pp.63-72..

[22]Abarca-Vargas, R. and Petricevich, V.L., 2018. Bougainvillea genus: A review on phytochemistry, pharmacology, and toxicology. Evidence-Based Complementary and Alternative Medicine..

The First record of Greater Painted-snipe *Rostratula benghalensis* (Linnaeus, 1758) for Iraq

Mudhafar A. Salim[1], Salwan Ali Abed[2] and Saifuldeen M. Alsaffah[3]

[1] Arab Regional Centre for World Heritage, Manama, Kingdom of Bahrain
[2] College of Science, University of Al-Qadisiyah, Diwaniyah, Iraq
[3] Iraq Organization for Conservation of Nature (IOCN), Iraq

Abstarct. Greater Painted-snipe (or Greater Painted Snipe) *Rostratula benghalensis* is medium-sized wader that inhibits the wetland habitats in Africa and south and southeast Asia. Despite of its English name, Greater Painted-snipe does not belong to the true *Gallinago* taxa. The female in this species is the colourful mate while the male brownish with pale spots in general and smaller in size. This species is of least concern (LC) conservations status on the IUCN Red List. During one of the field surveys carried out by the Iraq Organization for Conservation of Nature (IOCN) in Iraq, the team was able to find this species in a meadow wetland of low grass close to River Euphrates bank, northwest of Basra Governorate which represents the first observation of this species in Iraq. By providing the documented observation and detailed description of this species, this paper confirms the first record of Greater Painted-snipe (or Greater Painted Snipe) *Rostratula benghalensis* as an addition to the Iraqi birds checklist.

Keywords: Greater Painted Snipe, *Rostratula benghalensis*, Iraq, New bird record.

1. Introduction

Greater Painted-snipe (or Greater Painted Snipe) *Rostratula benghalensis* [1]is a medium-sized, coloured (especially the female) wader (23-26 cm), belongs to Rostratulidae Family. Despite of the English name of the species, it is not included within the true Snipe species *Gallinago*. It was formerly considered conspecific with Australian *R. australis* but is now generally considered monotypic [2]. On the contrast with the majority of bird species, the female of this species is larger and more colourful than the male. It has the distinctive white eye-patch that contrasts the bark reddish-brown head and neck. The upper-parts of the body are dark bronze-green with fine blackish barring. The wings are dark grey with whitish and golden colours. The under-parts and under-tail coverts are white. There is a whitish, V-shaped stripe between the chest and the front of the wing goes back along the base of the wings. The bill is of fleshy colour, and the relatively long legs are pale olive grey. The juvenile resembles the colours of the male [3] [4] [2]. Painted Snipe is least concern (LC) on the Redlist, with estimated population of 31,000-100,0000 [5].

This species inhibits the freshwater marshes' margins, especially those that include muddy patches, vegetated edges of lakes and the rivers, and other types of seasonal wetlands and flooded areas [4].

Mainly, the Greater Painted-snipe *Rostratula benghalensis* is African and South-Asian bird species. Its distribution is in sub-Saharan Africa and Madagascar (uncommon in southern Africa), the Delta of the Nile in Egypt; and its distribution in Asia extends from northeast parts of China, south-eastern Russia, and as south as southeast Asia region. Some references consider this species as nomadic that affected by water availability and precipitation levels and it has northern migration in Africa to avoid the dry seasons. [6] [3] [2].

Within the Middle East, Greater Painted-snipe *Rostratula benghalensis* has bred in Israel and Saudi Arabia [7]. The species has been observed in Iran, Oman, and the UAE where it is considered as vagrant [4]; and it has been listed as rare/accidental to Jordan [8].

2. Methodology

As part of the comprehensive surveys scheme that targets different habitats and areas in Iraq adopted and carried out by the Iraqi Organization for Conservation of Nature (IOCN), a survey was planned and conducted targeted the eastern parts of Thi-Qar and the north-western parts of Basra within the area of Chibayish and Mdeina districts, 70km northwest of Basra city. The surveyed areas were to the north of River Euphrates, and to the west of Al-'Ezz river. The surveyed area consists of various types of habitats (mainly aquatic and semi-aquatic) that include: Riparian habitat, marsh and marsh-margins, in addition to some terrestrial areas and orchards of date-palm. The microhabitat where the bird was observed at consists of riparian meadow with medium-sizes and low grass with few patches of shallow water and mud. The area is rich with the green plant cover and water-resources (mainly small canals originated from Euphrates (Figure -1).

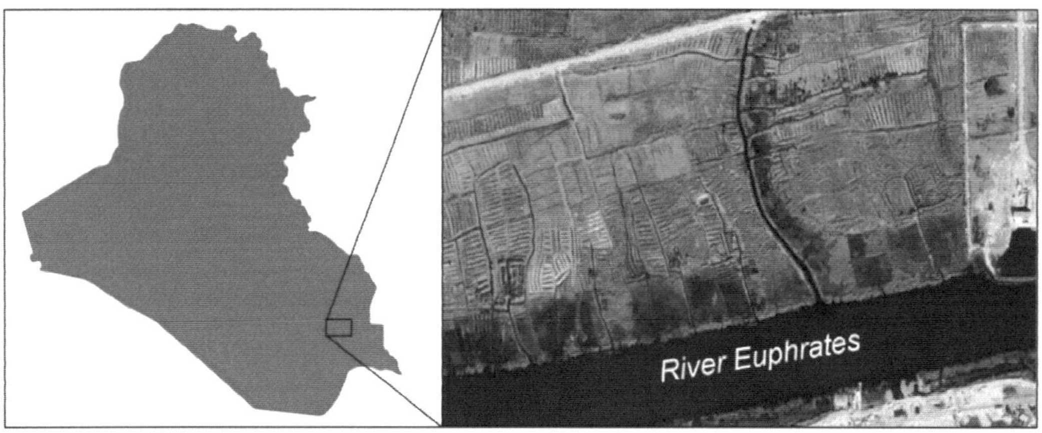

Figure 1. Map of Iraq shows part of the location of the survey made by the IOCN team close to River Euphrates, southern Iraq. Sat image is from Google Earth.

The survey in this area was carries out at the 28th of Sept 2021, where the weather was sunny and the wind was quiet. Three members of IOCN were contributed in this survey. The binoculars (8X30 & 12X45) and the professional cameras (Canon and Nikon) with telephotos (50-500mm &, 24-2000mm, 24-3000mm respectively) were used to observe the birds in the targeted areas within the survey stripe where the first record of Greater Painted Snipe was made.

3. The observation of the first Painted-snipe *Rostratula benghalensis* in Iraq

While birding in the wetlands of Thi-Qar and Basra southern Iraq by the IOCN team at the 28th of Sept 2021, the team decided to make a survey points along the northern part of Euphrates within the area of northwest of Mdeina town, north-western Basra. This stripe is well-known of its dense aquatic and semi-aquatic vegetation of reed-beds and riparian grass and other low to medium height plants. Such habitats used to host considerable content of birds and other fauna species. Adopting the same methodology to cover as wider area as possible by the team, splitting the team members was the team's strategy adopted to survey this area.

While watching the waders and other passerines close to the northern bank of Euphrates, between Chibayish-Mdeina paved road and the river, a small, brownish wader was seen foraging among the low grass not far from a small canal fed by Euphrates (*called Nahr Thoni*), within the coordinates (30.95749°, 47.24773°). Based on its foraging habits and the type of habitat where the bird was, the first impression on the observed bird was a Common Snipe *Gallinago gallinago* which is common wader in this area at this time of the year [4] [9] [10]. After closer observation by binoculars, and based on the observed key features, specifically the distinctively shorter bill (in comparison with Common Snipe *Gallinago gallinago*), whitish belly, the whitish eye-stripe, spotting pattern, the whitish stripes beside the chest, and its behaviour that seemed to be differs from the Common Snipe or even other *Gallinago* birds; it became obvious that this bird was the male Greater Painted-snipe *Rostratula benghalensis*. The bird was relatively shy, and have flew for short distance once it was approached by the observer. (Figure - 2).

When fled, it was quite obvious that the bird didn't fly as fast as a *Gallinago;* and after landing in closer spot of same habitat (approximately 40m), it started in skulking out trying to hide beyond the short grass, but it flew further after the observers tried to approach it closer where its wings' beats were comparatively slower than *Gallinago* genus, and its legs were trailed under its body; in addition, the whitish stripe along the breast sides and the wing-base was very clear when watching by binoculars. After the bird have flown beyond the intensive reed that's close to the river, it wasn't seen after that despite that the team has continued searching the entire area.

Figure 2. A male Greater Painted-snipe *Rostratula benghalensis* feeding at the grassy ground (left); the location and habitat where the 1st Greater Painted-snipe was found, north Euphrates, northwest of Mdeina city (right), the 28th of Sept, 2021. ©IOCN.

After consulting the Field Guide to the Birds of the Middle East [4], and discussing the key features that the team have clearly seen during the direct observation of the bird, it was very clear that the observed individual was the male Greater Painted-snipe *Rostratula benghalensis*. Based on the clear observation made by the team members, and in reference to the literature of ornithology in Iraq [11] [12] [9] [10] being not listed as Iraqi bird species, this observation represent the first confirmed record of Greater Painted-snipe *Rostratula benghalensis* for Iraq, and adds this bird species as an addition to the Checklist of the Birds of Iraq.

4. Acknowledgement

The authors are very grateful for the continuous support of the Iraqi Organization for Conservation of Nature (IOCN) and the College of Science at the University of Al-Qadisiyah for facilitating and covering the field surveys in Iraq. We are also grateful for the generous help of the locals during conducting the surveys.

References

[1] Linnæus, C. 1758. Systema naturæ per regna tria naturæ, secundum classes, ordines, genera, species, cum characteribus, differentiis, synonymis, locis. Tomus I. Editio decima, reformata. - pp. [1-4], 1-824. Holmiæ. (Salvius)..

[2] Kirwan, G. M. 2016. Greater Painted-snipe Rostratula benghalensis. In Del Hoyo, J., Elliott, A., Sargatal, J., Christie, D. A. & de Juana, E. (eds.) Handbook of the Birds of the World Alive. Barcelona: Lynx Edicions. www.hbw.com/node/53745 ..

[3] Marchant, S. & Higgins, P.J. (editors). 1993. Handbook of Australian, New Zealand & Antarctic Birds. Volume 2, Raptors to lapwings. Melbourne, Oxford University Press. Pages 648-649, 658-666..

[4] Porter, R. & S. Aspinall. 2010. Birds of the Middle East. Christopher Helm, London..

[5] BirdLife International. 2022. IUCN Red List for birds. Downloaded from http://www.birdlife.org (2022)..

[6] Brown, L.H., E.K. Upon4, and K. Newman. 1982. The birds of Africa. Academic Press. London, U.K..

[7] Sapir, N. 2003. Six New Breeding Bird Species In Israel During 1995–2002. Israel. JOURNAL OF ZOOLOGY, Vol. 49, 2003, pp. 203–218..

[8] Avibase (Birds Checklists of the World). 2021. https://avibase.bsc-eoc.org/checklist.jsp?region=JO&list=howardmoore.

[9] Salim, M.A., R. Porter, P. Schiermacker-Hansen, S. Christensen & S. Al- Jbour. 2006. [Field guide to the birds of Iraq]. NI/BirdLife International, Baghdad. [In Arabic]..

[10] Salim M.A., Al-Shekly OF, Majeed KA, Porter RF. 2012. An annotated checklist of the birds of Iraq. Sandgrouse. 2012; 34:4-43..

[11] Allouse, B. 1953. The Avifauna of Iraq. Iraq Natural History Museum, Baghdad..

[12] Allouse, B. 1962. [Birds of Iraq]. Vol. II. Al-Rabita Press, Baghdad. [In Arabic].

The First Confirmed Record Of European Pied Flycatcher *Ficedula Hypoleuca* For Iraq

Salwan Ali Abed [1], Mudhafar A. Salim [2], and Zaman Salman Harbi [3]

[1] College of Science, University of Al-Qadisiyah, P.O.Box.1895 , Iraq
[2] Arab Regional Centre for World Heritage, Manama, Kingdom of Bahrain
[3] Iraqi Organization for Conservation of Nature (IOCN), Iraq

Abstract. While conducting intensive field surveys in different locations of various habitats in Iraq arranged by the Arab Regional Centre for World Heritage, Manama, Kingdom of Bahrain and the Iraqi Organization for Conservation of Nature (IOCN), some considerable observations on birds and their habitats in various locations were made. This included adding new bird-species to the Avifaunal of Iraq, in addition to new observations of confirmed breeding of some bird species. A new Passerine species, European Pied Flycatcher *Ficedula hypoleuca,* was added to the Iraqi bird checklist after it was clearly observed and photographed during one of IOCN's surveys in Zakho mountains, Kurdistan Region, northern Iraq. This observation was documented and the habitat was described as well. The current paper documents the first confirmed record of this species for the bird-list of Iraq.

Keywords: *Ficedula hypoleuca*, European Pied Flycatcher, Iraq, Iraqi Organization for Conservation of Nature (IOCN), Kurdistan Region.

1. Introduction

Various field visits that targeted different areas within Iraq were arranged by the Iraqi Organization for Conservation of Nature (IOCN) in order to discover the status of the different habitats in Iraq recently and contribute in their prioritization and protection through designation of Protected Areas network. These surveys have contributed to better understanding of the current status of the species and the different habitats in Iraq. The current observation of the European Pied Flycatcher *Ficedula hypoleuca* was part of field surveys conducted by three of IOCN's team in a mountainous areas few kilometers to the east of Zakho town, at the extreme north of Iraq. IOCN team has visited and surveyed this area more than one time as it demonstrated remarkable richness in biodiversity in general and in avifauna particularly. The results of the intensive surveys in this area have ended in discovering new bird species to Iraq that was not listed previously as an Iraqi bird species [1] [2] [3]. Discovering two new bird species at this area within Zakho district, represented by the current species in addition to Crested Honey Buzzard *pernis ptilorhynchus* [4], and adding them to the Iraqi avifauna checklist, indicates clearly to the biological importance of this area and the necessity of shading more scientific light and consideration on the biodiversity components of this area.

The European (some references name it 'Eurasian') Pied Flycatcher *Ficedula hypoleuca* is a small Passerine from the Muscicapidae family, one of the four species of Western Palearctic Pied (black-and-white) flycatchers that breeds in most of Europe and western Asia with 'Least Concern' conservation status [5] [6] It is one of the active migratory bird species that migrates from its breeding northern grounds through the Middle East, and north Africa, to central Africa reaching west Africa [7]and as south as Kenya [8]. European Pied Flycatcher is one of the variable bird species in phenotype in Europe and there is a relationship between the geographic and spatial variation in phenotypes with the distribution of Collared Flycatcher [9].

Within the Middle East, the European Pied Flycatcher *Ficedula hypoleuca* is a passage migrant to the northern and western parts. The map below illustrates the distribution of the European Pied Flycatcher in the Middle East countries where it has been considered as winter visitor to Turkey, northern Iran, Syria, Jordan, and east of the Mediterranean [10] [2] [11].

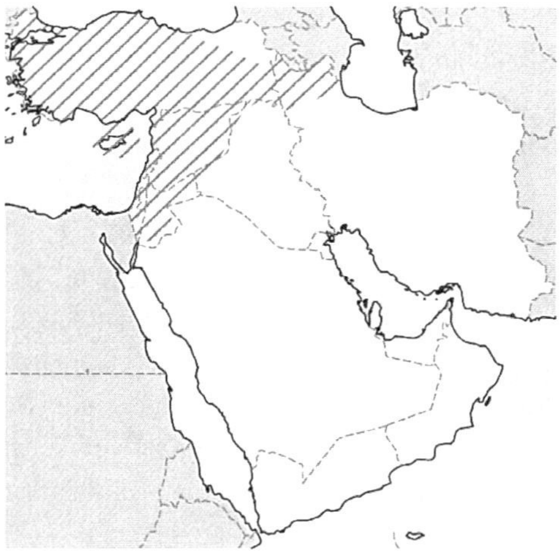

Figure 1. Map of the Middle East shows the passage area of the European Pied Flycatcher. Source: Courtesy, Birds of the Middle East [2].

2. Methodology

At the 31st of March 2021, during large-scope survey made by the IOCN that extended over various areas and habitats in Iraq, the team was able to survey different areas in Dahuk, Kurdistan Region of Iraq, few kilometers to the east of Zakho, at the coordinates (N 37.14737, E 42.8459). The site (called Tawke) consists relatively deep and medium-sized valleys and few villages and some agricultural fields wherever the land become flat. The valley consists of wet grassy bed with areas with dense bushes and trees in addition to rocky slopes. The valley also includes scattered trees and some patches of reed-bed. In order to getting better coverage, the team used the spreading technique surveying the different parts and habitats of the valley where different bird species have been observed by the team (see the list below).

The map below (Fig. 2) shows Dahuk district in northern Iraq and part of the area surveyed by the IOCN team as part of Tawke plain and valleys. IOCN team has surveyed this area more than one time during more than one season due to its richness in biodiversity – specifically in birds.

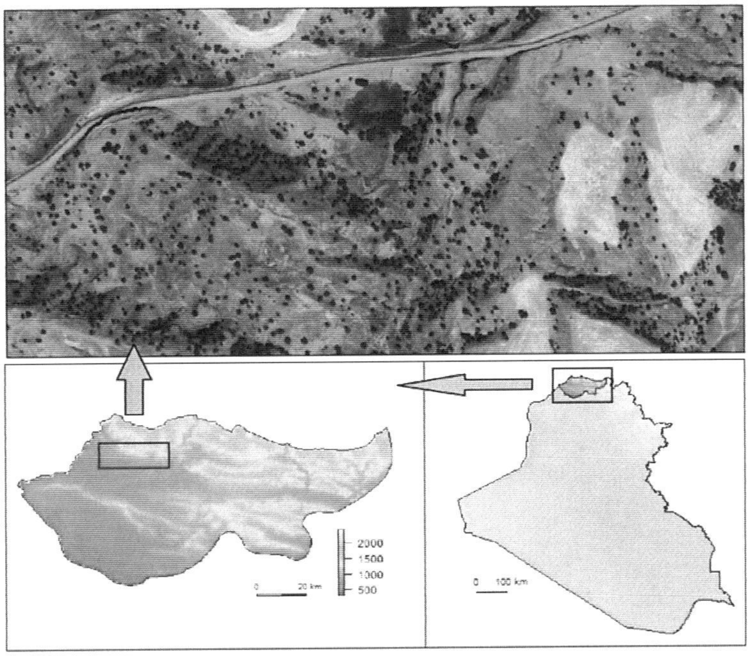

Figure 2. Map of Iraq (below, right) shows the locations where the surveys were made in northwest of Dahuk district (below, left). The upper satellite image shows the exact area where the first Pied Flycatcher *Ficedula hypoleuca* was observed by IOCN team.

3. Results and Discussion

While surveying one of the bottom of the valleys in middle Tawke by one of IOCN team members, two other members of the team were carefully searching for birds along each side of the valley's slopes; a small, white-and-black bird suddenly flew from a small tree at the bottom, close to the reed-bed's edge. The bird continued its flying to a group of Oak trees heading to the middle of Oak tree at the eastern side of the valley where the team was able to spot it again at that tip and followed it quietly to check it by binoculars and to take some more and closer photos. After few minutes of observation, it became obvious that the targeted bird was the European Pied Flycatcher *Ficedula hypoleuca*. The team was quite sure of the identification of the bird due to the key feature that it showed that were different from the other similar species: Collared Flycatcher *Ficedula albicollis*, and Semi-collared Flycatcher *Ficedula semitorquata*.

The observed individual was an adult male, and was moving deeply inside and at the tip of the tree as the team follow it; nevertheless, we were able to notice clear details that were quite enough for identifying this bird as European Pied Flycatcher *Ficedula hypoleuca*. The hindneck was black, and the white projection at the neck-sides ends under the ear-coverts and does not exceed it. The small white spot at the forehead was divided into two smaller white spots. The white streak of the base of the primaries that's obvious in the Collared Flycatcher *Ficedula albicollis*, and Semi-collared Flycatcher *Ficedula semitorquata* is absent in this individual, which is key sign to separate this species from Collared and Semi-collared flycatchers.

Figure 3. Left – Pied Flycatcher that was observed by the IOCN team, east of Zakho. Right – landscape of the habitat where the first Pied Flycatcher has been observed by IOCN team. ©IOCN.

The first comprehensive checklist of the avifauna of Iraq made by [1], did not list the European (or Eurasian) Pied Flycatcher *Ficedula hypoleuca* among the birds species of Iraq [1]; however, in his updated book "Birds of Iraq", Allouse has indicated to the possibility of the occurrence of this species in northern Iraq as a spring visitor indicating to a mid-April observation made by Weigold, H. in Urfa (or Sanliurfa), southern Turkey [12], and this remained the most accepted assumption regarding this species until publishing the Annotated Checklist of the Birds of Iraq at 2012. In the provisional checklist of the birds of Iraq that was published at 2010, the European Pied Flycatcher was not listed among the Flycatcher species in this paper [2].
In the updated checklist that was written by [3]where thorough revisions of all of the Iraqi bird species were made, only Semi-collared Flycatcher *Ficedula semitorquata* was listed as an Iraqi species out of the rest of the black-and-white flycatcher species based on the historical reliable observations [3]. Based on the detailed status of the black-and-white flycatchers in Iraq mentioned above, and on the clear evidence of the occurrence of European Pied Flycatcher *Ficedula hypoleuca* in Zakho mountains, in Kurdistan Region northern Iraq, this paper confirms the first record of the European Pied Flycatcher *Ficedula hypoleuca* in Iraq and adds it to the Iraqi checklist.

4. Other Bird Species Observed at the Same Area

In order to provide more details regarding the circumstances of the current observation, the authors would like to list the other bird species that have been observed during surveying the valley (around the observation location of Pied Flycatcher, where the following bird species were observed as well:

1. Eurasian Sparrowhawk
2. Common Buzzard
3. Long-legged Buzzard
4. Eurasian Scops Owl
5. Great Tit
6. Crested Lark
7. Wood Lark
8. Eurasian Blackbird

9. Magpie
10. Common Chiffchaff
11. Lesser Whitethroat
12. Menetries's Warbler
13. Common Redstart
14. Yellow Wagtail
15. Common Chaffinch
16. Corn Bunting

Acknowledgement

We are very grateful for the continuous support of the Iraqi Organization for Conservation of Nature (IOCN) and College of Science, University of Al-Qadisiyah for making the field surveys possible in different locations and habitats in Iraq.

References

[1] Allouse, B.E. (1953) The Avifauna of Iraq. Publication. No. 3.Iraq. Natural History Museum. Baghdad, Iraq: Al-Rabita Press..

[2] Porter R.F., Aspinall R.S. (2010) Birds of the Middle East. Christopher Helm, London..

[3] Salim M.A., Al-SheikhlyOF, Majeed K.A., Porter, R.E. (2012) An annotated checklist of the birds of Iraq. Sandgrouse. 2012; 34:4-43.14..

[4] Salim, M.A.; and Abed, S.A. (2019) The First Oriental Honey Buzzard pernis ptilorhynchus (Temminck, 1821) in Iraq. Eco. Env. & Cons. 25 (4) : 2019; pp. (1926-1929). Copyright@ EM International. ISSN 0971–765X..

[5] BirdLife International (2021) Species factsheet: Ficedula hypoleuca. Downloaded from http://www.birdlife.org..

[6] IUCN. The IUCN Red List of Threatened Species. Version 2018-2. Available at, 2018, www.iucnredlist.org. (Accessed: 15 November 2018).9..

[7] BirdLife International. (2018) European Pied Flycatcher (Ficedula hypoleuca). The IUCN Red List of Threatened Species 2018: e.T22709308A131952521. https://dx.doi.org/10.2305/IUCN.UK.2018-2.RLTS.T22709308A131952521.en..

[8] Hentze, N., and Bradley J. (2018) A new record of the Pied Flycatcher Ficedula hypoleuca from Kakamega Forest, western Kenya. Scopus 38(2): 19–20, July 2018. (Received 22 March 2018)..

[9] Lehtonen, P. K., Laaksonen, T., Artemyev, A. V., Belskii, E. , Christiaan Both, Stanislav Bureš, Andrey V. Bushuev. (2009) Geographic patterns of genetic differentiation and plumage colour variation are different in the pied flycatcher Ficedula hypoleuca.

[10]Kirwan, G.M., Martins, R.P., Eken G., & Davidson, P. (1999) A checklist of the birds of Turkey. Sandgrouse Supplement issue 1:1-32..

[11]Khaleghizadeh A., Roselaar K., Scott D.A., Tohidifar M., Mlíkovský J., Blair M. & Kvartalnov P. (2017) Birds of Iran: Annotated Checklist of the Species and Subspecies. Iranian Research Institute of Plant Protection..

[12]Allouse, B.E. (1962) Birds of Iraq (Volume 3). [Arabic]. Baghdad, Iraq: Al-Rabita Press..

[13]Porter, RF, M Salim, K Ararat & O Fadhel. (2010) A provisional checklist of the birds of Iraq. Marsh Bulletin 5(1): 56–95..

[14]Snow, D.W., Perrins, C.M. (1998) The Birds of the Western Palearctic Concise Edition, 1998, ISBN 0-19-854099-X.18..

Bio-Environmental preparation of Selenium Nanoparticle using *Klebsiella Pneumonia* and their Biomedical Activity

Soroor Mohammed Hadi[1] and Nawfal Hussein Aldujaili[2]

[1,2] Kufa University, faculty of science, Department of biology, Iraq

Abstract. The aim of this study was to study the biomedical activity (biofilm by fluoresence microscope) and Characterization of Selenium NPs synthesized by *klebsiella pneumonia*. Different types of bacterial strains *(*S1-S35*) aureus,* were screened for biosynthesis of selenium NPs. Isolate *S*6 was selected as efficient isolate based on color change and antibacterial activity against the pathogenic bacteria, and it was diagnosed as *klebsiella pneumonia* depending on the morphology, microscopic examination and VITEK2 compact system. Characterization of bio-selenium NPs was performed using: UV-visible spectroscopy was used to characterize the synthesis of selenium NPs by *Klebsiella pneumonia*, with an absorption peak at 280 nm.The size and structure of the bioavailable selenium NPs were measured using an XRD occurring at 20°-26°. AFM analysis showed that the mean diameter of the bio-selenium NPs was 47. 18. Antibacterial activities have been studied for bio-selenium against pathogenic bacteria (B. cepacia, S. aureus, A. baumanni and E. coli).Biogenic Selenium expressed high significant antibiofilm activity with increasing concentration of Selenium NPs. DPPH reducing activity of nanoparticles which increased with increase concentration of biogenic Selenium NPs.

Keywords: fluorescence, *Klebsiella pneumonia and* 16sRNA

1. Introduction

Many distinct bacteria species produce nanomaterials, each with its own set of features. The bulk of these events include metal nanoparticles binding to bacterial biomass outside of the cell.For NPs, there are biological and physicochemical characterization methodologies available. Antibacterial activity, genotoxicity, and antibiofilm activity are all part of biological characterization. Physicochemical methods, on the other hand, include spectrophotometry.Lorem ipsum dolor sit amet, consectetuer adipiscing elit. Selenium (Se) NPs have high photoconductivity, catalytic action in hydration and oxidation reactions and they have high piezoelectric, thermoelectric, and nonlinear optical responses (14). electron microscopy, and X-ray processing. Also, (Se) NPs have been synthesized by reducing selenium tetrachloride in the presence of ascorbic acid. An optimization study with poly (sodium 4-styrene sulfonate) produced stable and spherical narrowly size-distributed

NPs (46 nm), which are considered highly monodisperse [1]. Selenium is widely utilized in medicine for anticancer treatment and cholesterol lowering, as well as medicinal membranes, wound dressings, and controlled-release medicinal mate [21]

2. Materials and methods

2.1. Selection and identification of a reliable isolate capable of generating Selenium nanoparticles

Twenty-five isolates from various strains were cultured for 24 hours at 37°C in brain heart infusion broth. After incubation, the colloidal suspension was centrifuged for 15 minutes at 10,000 rpm to separate the precipitate, and the supernatant was added to the Selenium and acetic acid solution used in industry. The Muller Hinton agar well diffusion test was used to screen all isolates (S1-S35) for E. coli and incubated for 24 hours at 37°C.

Based on morphological and microscopic examination, biochemical identification, and VITEC, an efficient isolate (S6) was selected and identified as *Klebsiella pneumonia* based on antibacterial activity, color change, and absorption spectrum efficiency isolate (S6) was selected and identified as *Klebsiella pneumonia* based on antibacterial activity, color change, and absorption spectrum efficiency isolate (S6) was selected and identified as *Klebsiella pneumonia* based on morphological and microscopic examination, biochemical identification. [2]

2.2. Wide scale synthetization of Selenium nanoparticles by using Klebsiella pneumonia

Antibacterial activity of several strains (S1-S35) against E. coli as an indicator strain was tested, and the most effective isolate was chosen based on antibacterial activity and color change. Based on morphological and microscopic analysis, biochemical identification, and VITEC, an efficient isolate (S6) was chosen and identified as *Klebsiella pneumonia*.

Bacteria cultivated for 24 hours in brain heart infusion broth at 37°C. The colloidal suspension was centrifuged at 10,000 rpm for 15 minutes after incubation, then the precipitate was removed and the supernatant was added to a solution comprising distilled water (Selenium 2 mg), and (acetic acid, 2ml).

The cell free supernatant was made in several steps: activation of the inoculum, inoculation of the BHI broth (incubation for 24 hours at 37°C), incubation of the culture for biomass production, and centrifugation at 4200 xg for 15 minutes at 4°C to make the cell free supernatant. Pellets of microorganism cells were discarded, and the cell free supernatants were collected for use in the biosynthesis of Selenium.

The supernatant was collected by centrifuging broth culture with large scale chilled centrifuges with 5ml tubes at 4200 xg to pellet cells at the bottom of the tubes, leaving cell free supernatant.

The supernatant was then collected in a sterile flask and used to make Selenium nanoparticles. The strains were inoculated into three autoclaved flasks containing 400ml BHI broth, 4gm Selenium as a substrate, and 4ml acetic acid, which were then incubated for 24 hours at 37°C in a shaking incubator at 150 rpm. This happened a number of times. [8]

2.3. Classification of biosynthesized Selenium Nanoparticles

The physical characteristics of biosynthesis nanoparticles were characterized by UV visible, SEM, FTIR and XRD. [7]

2.4. UV-Visible Spectroscopy

The selenium nanoparticles formation and their size estimation within the composite suspension were confirmed by UV-Visible spectroscopy (shimadzu UV- visible 1800) in AL ameen center for Advanced Biotechnology and research. [10]

2.5. Scanning Electron Microscope (SEM) Analysis

Selenium nanoparticles were examined at the University of Tehran's electron microscopy unit.

2.6. Fluorescence Microscopy of Bacterial Biofilm and Image Analysis

Fluorescence is used to create images in a fluorescence microscope. Biofilm that has been dyed with fluorescent dye or tagged with fluorescent protein is exposed to light of a specified wavelength (or wavelengths) that is absorbed by the fluorophores, leading them to emit longer wavelengths of light (i.e., of a different colour than the absorbed light).

The filters and dichroic are chosen to match the fluorophore employed to designate the specimen's spectrum excitation and emission characteristics. A single fluorophore (color) distribution is photographed at a time in this manner. Multiple single-color photos must be combined to get multicolor images of various fluorophores.

Each dye has its unique emission and excitation spectra, which is important for filter selection when viewing the material under a microscope.

Acridine orange (AO) is a fluorescent dye that selectively binds to nucleic acids. It is permeable to cells. It has an excitation maximum of 502 nm and an emission maximum of 525 nm when interacting with DNA (green).

However, when interacting with RNA, the highest excitation moves to 460 nm and the maximum emission shifts to 650 nm (red).

After staining, the biofilm adhered to a surface can be dyed with acridine orange and seen at both above and below excitation wave lengths. 0.02 percent acridine orange solution is used to stain biofilm samples properly.

It's made by combining 20 mg of acridine orange powder with 100 mL milli-Q water. The dye solution should be kept at room temperature and kept in the dark.[19].

2.7. Molecular Identification

Sequencing the 16sRNA gene has used in this study for the identification and differentiation of the isolates and detecting of the phylogenetic relationship among the isolated of the bacteria.

2.8. Extraction of Bacterial Genomic DNA

An appropriate number of bacterial cell (up to $1x10^9$) was transferred to a 1.5 ml microcentrifuge tube and centrifuged at 14,000 rpm for 1 minute. the supernatant was discard.

Two hundred µl of FATG Buffer were added and resuspended the pellet by vortex. Incubated for 5 minutes at room temperature. During incubation, inverted the tube every 3 minutes. The sample mixture was transferred (including any precipitate) carefully to FABG Column, which was then washed and dried.

The dried FABG Column was placed in new 1.5 ml microcentrifuge tube, and 100µl of preheated Elution Buffer or TE were added to the membrane center of FABG Column, FABG Column for 3-5 minute or until the buffer was absorbed by the membrane then it was Centrifuged for 1 minute at 14000 rpm to elute the DNA. DNA solution was stored at -20°C. [16]

2.9. Amplification of 16sRNA.

The polymerase chain reaction (PCR) was used to amplify fragments of DNA using Maxime PCR PreMix Kit (i-Taq) (Intronbio/Korea).

The template DNA, forward and reverse primers, and RNase-free water were added to PCR tube, containing i-Taq DNA polymerase, dNTP mixture, and the reaction buffer was mixed together. [6]

Table 1. Components of Mixture PCR PreMix kit (i-Taq)

Material	*Concentration*
dNTPs (dATP, dTTP, dGTP, dCTP)	*2.5mM*
i -Taq DNA Polymerase	*5U/5µl*
Reaction buffer (10X)	*1X*

2.10. Preparations of Primer in 16sRNA Gene.

PCR assay was carried out upon molecular identification using universal primers (Table 2).These primers synthesized by (Macrogen/Korea). Primers were dissolved using sterile dd water.

The stock solution (100 pmol/µl) was prepared by adding dd H2O to the vial containing lyophilized primer while working solution of 10 pmol/µl was made by mixing 10 µl of the stock primer and 90 µl of ddH2O. the stock and working solution (10 pmol/µl) were stored in -20°C.

Table 2. 16S rRNA primer pair

Primer Name	*Sequence 5′ - 3′*	*Reference*
27F	*AGAGTTTGATCCTGGCTCA*	[5]
1492R	*GGTTACCTTGTTACGACTT*	

3. Results & Discussion

3.1. Selection of the efficient isolate that producing Selenium NPs

Thirty-five isolates from various sources were tested against E. coli as an indicator strain for Selenium Nanoparticle production. The most effective isolate was chosen based on the color shift from light to dark (Figure 1) and the biological activities of Selenium NPs in E. coli against multidrug resistant bacteria.
K. Pneumonia was the most effective isolate. Because each microbe has a unique metabolic process and enzyme activity, not all bacteria can manufacture NPs.

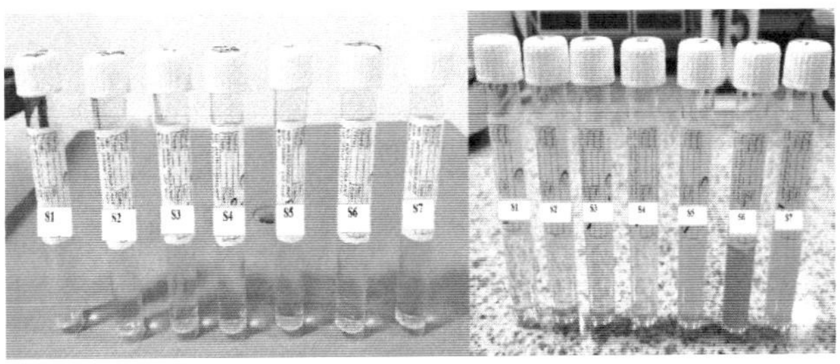

Figure 1. Selection of efficient strain S6 (*Klebsiella pneumonia*) Based on color change and antibacterial activities.

3.2. Production of Selenium NPs using K. Pneumoniae

Klebsiella pneumonia, which was used in the biosynthesis of Selenium nanoparticles, demonstrated the ability to extracellular biosynthesis by using cell free supernatant after the addition of Selenium as a substrate under previously optimized conditions. The colour changes of the reaction mixture from light to dark. after incubation in shaking at 150 rpm, as well as the colour change and antibacterial behaviour, served as an indicator for the biosynthesis of Selenium NPs by Klebsiella pneumonia. [13]

3.3. Identification of efficient isolate K. Pneumoniae

This bacterium K. Pneumoniae was used for the biosynthesis of Selenium NPs and identified based on: (i) morphological examination, which has a rod shape and is Gram-negative.
When cultured on MacConkey agar, the morphology of this bacteria's is large circular colony shape with convex elevation, mucoid and pink to red in color due to lactose fermentation, opaque, non-hemolytic (Gamma hemolysis) nonmotile and nonsporing, appeared under a microscope. (Figure 2) ;(ii) The isolate was identified by VITEK2 with BCL/ID card with 64 biochemical tests. The isolate was identified as K. Pneumoniae with probability 99% (Figure 3).

IOP Conf. Series: Earth and Environmental Science 1029 (2022) 012021 doi:10.1088/1755-1315/1029/1/012021

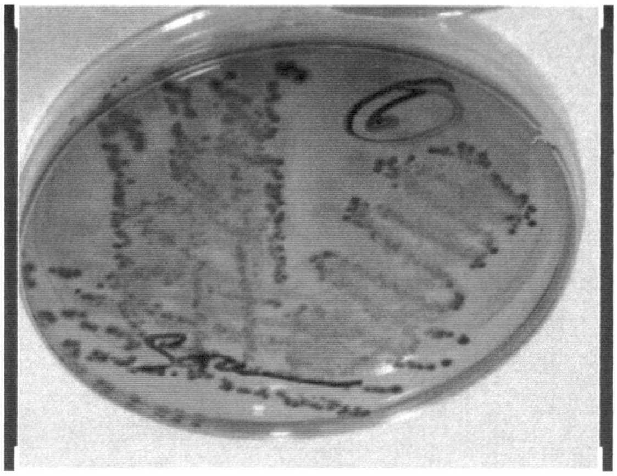

Figure 2. Morphological and microscopical examination of K. Pneumoniae on MacConkey agar.

bioMérieux Customer: Microbiology Chart Report Printed February 14, 2022 9:13:11 AM CST

Patient Name: Patient ID: 31320216
Location: Physician:
Lab ID: 31320216 Isolate Number: 1

Organism Quantity:
Selected Organism : Klebsiella pneumoniae ssp pneumoniae

Source: **Collected:**

Comments:	

Identification Information	**Analysis Time:**	3.87 hours	**Status:**	Final
Selected Organism	99% Probability **Bionumber:**	**Klebsiella pneumoniae ssp pneumoniae** 6607734753564010		
ID Analysis Messages				

Biochemical Details

2	APPA	-	3	ADO	+	4	PyrA	+	5	IARL	-	7	dCEL	+	9	BGAL	+
10	H2S	-	11	BNAG	-	12	AGLTp	-	13	dGLU	+	14	GGT	+	15	OFF	+
17	BGLU	+	18	dMAL	+	19	dMAN	+	20	dMNE	+	21	BXYL	+	22	BAlap	-
23	ProA	-	26	LIP	-	27	PLE	+	29	TyrA	+	31	URE	+	32	dSOR	+
33	SAC	+	34	dTAG	-	35	dTRE	+	36	CIT	+	37	MNT	+	39	5KG	-
40	ILATk	+	41	AGLU	-	42	SUCT	+	43	NAGA	-	44	AGAL	+	45	PHOS	+
46	GlyA	-	47	ODC	-	48	LDC	+	53	IHISa	-	56	CMT	-	57	BGUR	-
58	O129R	+	59	GGAA	-	61	IMLTa	-	62	ELLM	-	64	ILATa	-			

Figure 3. The result of Biochemical test of K. Pneumoniae isolate by VITEK2 compact system.

3.4. Sequencing the 16sRNA Gene

Sequences for K. pneumonia strains were received online and aligned to NCBI data base using Blast software, multiple aligned to each other using BioEdit software and submitted in fasta format to NCBI through Sequence software. [20]

3.5. Pairwise Alignment

K. pneumoniae strain was found to be nearest and neighbors to *Klebsiella pneumoniae* strain SGH10 with Query cover 99%, E-value 0.0, and identity 97.04%, and with Score 1631 bits (883), Expect 0.0, Identities (97%), and Gaps (1%) rooted from Enterobacteriaceae

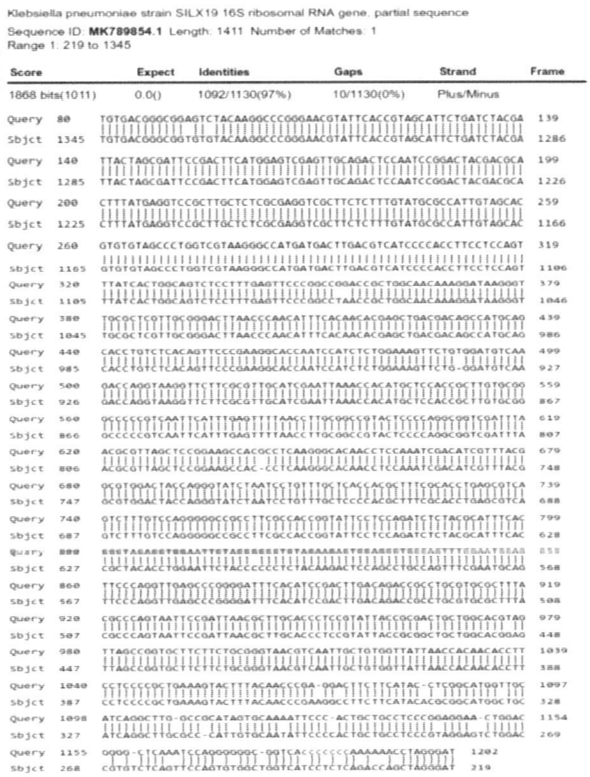

***Figure* 4.** Pair-wised alignment of partial nucleotide sequence of 16S ribosomal rRNA gene (Query) to that of Klebsiella pneumoniae strain SILX19 16S ribosomal whose sequence producing highest score (97%) of homology during BLASTN search.

3.6. Characterization of biosynthesized Selenium NPs

3.6.1. Spectroscopy in the UV-visible range

Visual inspection and measurement of the absorbance band using UV-visible spectroscopy can validate the biogenesis of nanoparticles. Figure shows the absorption spectra of nanoparticles formed in the reaction mixture, which exhibits a peak at 280 nm (3-2). The presence of nanoparticle surface plasmon resonance (SPR) is demonstrated, and a single SPR band indicates that the nanoparticles are spherical in shape.

A UV-Visible spectrophotometer was used to evaluate the synthesis of nanoparticles, which revealed a single peak at 280 nm, indicating the presence of Selenium nanoparticles in the reaction solution.

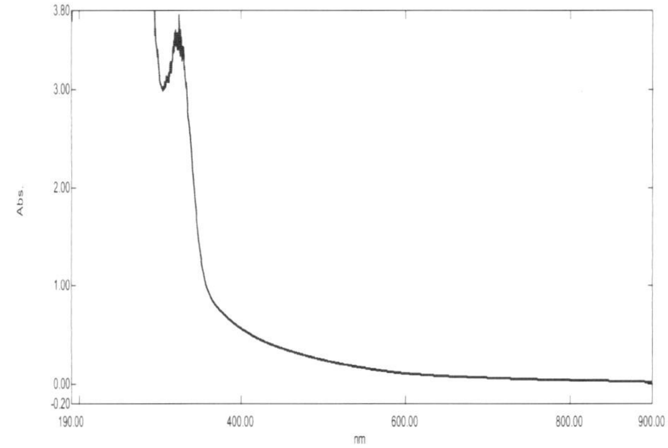

Figure 5. UV-visible spectroscopy analysis of Selenium NPS synthesis by *Klebsiella pneumonia*

3.6.2. Scanning Electron Microscopy Analysis

SEM was used to approve the morphology and size of the biogenic Selenium nanoparticles and well distributed and spherical shaped Selenium nanoparticle synthesized by Klebsiella pneumonia with a size range between (29.9 - 51.8) figure (6).

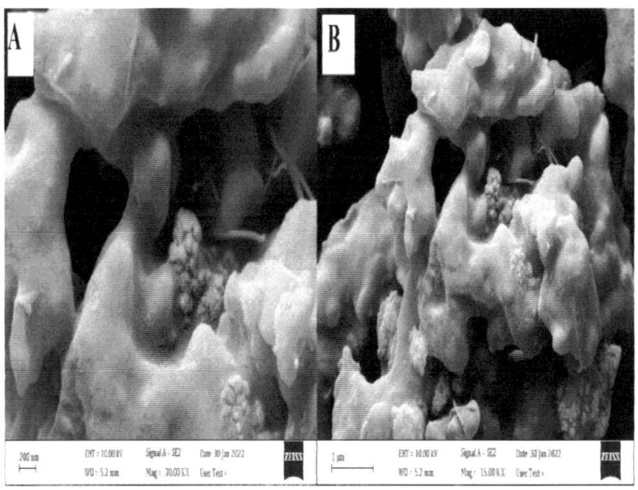

Figure 6. SEM micrograph of biogenic Cs-NPs synthesis by Klebsiella pneumonia showed Selenium NPs with spherical well dispersed with size (29-51 nm) at A: 200 nm, and B: 1 µm.

The size of the NPs ranged from 10 to 60 nanometers. The majority of NPs were the same size, with only a few particles larger than 60 nm. In comparison to prior research that employed other biological substances as reducers and stabilizers, the approach used in this study allowed for smaller Selenium NPs. [17]

3.6.3. X-ray Diffraction (XRD) Analysis

XRD was used to investigate the crystalline structure of CHNPs. The distinctive peaks of Selenium polymer discovered at 20-21 were ascribed to the polymorph's diffraction pattern. The biogenic Selenium nanoparticles XRD spectrum, which appeared at 20-26. During the manufacturing process, the strong peak showed the presence of bioorganic coupons/proteins in the nanoparticles. figure (7). [4].

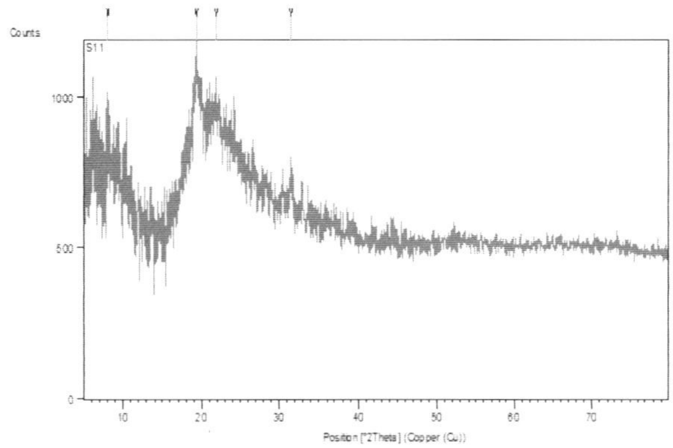

Figure 7. XRD analysis of Selenium NPs synthesis by *Klebsiella pneumonia* showing the structure and size.

The average size of the Selenium nanoparticle was determined using X-ray diffraction. The crystalline structure of CHNPs was investigated using XRD peaks of Selenium polymer observed at 20-21 was attributed to the polymorph's diffraction pattern. The XRD spectrum of CHNPs which appeared at 20-26 sharp peak indicated some bioorganic couponed/ proteins in the nanoparticles during synthesis. [3]

3.6.4. FTIR of selenium NPs

The transmittance of the prepared formulations was accomplished by FT-IR spectrophotometer, in a spectral range of 4000-400 cm-1 with 4 cm-1 resolutions. The data sets were averaged over 24 scans [15].

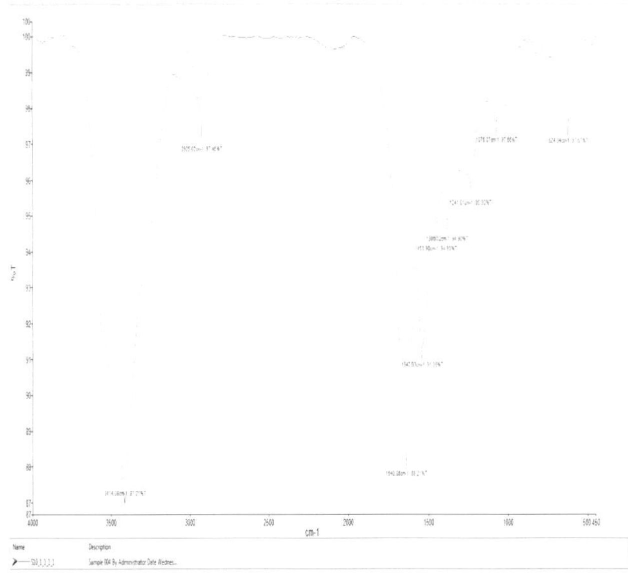

Figure 8. Fourier transform infrared (FTIR) spectroscopy of Nano biosynthesized Selenium NPs by K. pneumonia.

3.6.5. Fluorescence Microscopy of Bacterial Biofilm and Image Analysis

Biofilm that has been dyed with fluorescent dye or tagged with fluorescent protein is exposed to light of a specified wavelength that is absorbed by the fluorophores, leading them to emit longer wavelengths of light (i.e., of a different colour than the absorbed light).

The filters and dichroic are chosen to match the fluorophore employed to designate the specimen's spectrum excitation and emission characteristics.

Each dye has its unique emission and excitation spectra, which is important for filter selection when viewing the material under a microscope.

Acridine orange is a fluorescent dye that selectively binds to nucleic acids. However, when interacting with RNA, the highest excitation moves to 460 nm and the maximum emission shifts to 650 nm. After staining, the biofilm adhered to a surface can be dyed with acridine orange and seen at both above and below excitation wave lengths. 02 percent acridine orange solution is used to stain biofilm samples properly. [5]

Figure 9. Fluorescence microscope Analysis of bacterial biofilm by selenium nanoparticles.

bioMérieux Customer: Microbiology Chart Report Printed February 14, 2022 9:13:11 AM CST

Patient Name: Patient ID: 31320216
Location: Physician:
Lab ID: 31320216 Isolate Number: 1

Organism Quantity:
Selected Organism : Klebsiella pneumoniae ssp pneumoniae

Source: **Collected:**

Comments:	

Identification Information	**Analysis Time:**	3.87 hours	**Status:**	Final
Selected Organism	99% Probability	**Klebsiella pneumoniae ssp pneumoniae**		
	Bionumber:	6607734753564010		
ID Analysis Messages				

Biochemical Details

2	APPA	-	3	ADO	+	4	PyrA	+	5	IARL	-	7	dCEL	+	9	BGAL	+
10	H2S	-	11	BNAG	-	12	AGLTp	-	13	dGLU	+	14	GGT	+	15	OFF	+
17	BGLU	+	18	dMAL	+	19	dMAN	+	20	dMNE	+	21	BXYL	+	22	BAlap	-
23	ProA	-	26	LIP	-	27	PLE	+	29	TyrA	+	31	URE	+	32	dSOR	+
33	SAC	+	34	dTAG	-	35	dTRE	+	36	CIT	+	37	MNT	+	39	5KG	-
40	lLATk	+	41	AGLU	-	42	SUCT	+	43	NAGA	-	44	AGAL	+	45	PHOS	+
46	GlyA	-	47	ODC	-	48	LDC	+	53	lHISa	-	56	CMT	-	57	BGUR	-
58	O129R	+	59	GGAA	-	61	lMLTa	-	62	ELLM	-	64	lLATa	-			

Acknowledgements

The authors would like to thank Kufa University, faculty of science, Department of biology as they partially supported this study in gathering samples and analyzing them.

References

[1] Gangadoo, S., Stanley, D., Hughes, R. J., Moore, R. J., & Chapman, J. (2017). The synthesis and characterization of highly stable and reproducible selenium nanoparticles. Inorganic and Nano-Metal Chemistry, 47(11), 1568–1576.

[2] Abdel-Moneim, A.-M. E., El-Saadony, M. T., Shehata, A. M., Saad, A. M., Aldhumri, S. A., Ouda, S. M., & Mesalam, N. M. (2022). Antioxidant and antimicrobial activities of Spirulina platensis extracts and biogenic selenium nanoparticles against selected pathogenic bacteria and fungi. Saudi Journal of Biological Sciences, 29(2), 1197–1209.

[3] Abdelmigid, H. M., Hussien, N. A., Alyamani, A. A., Morsi, M. M., & AlSufyani, N. M. (2022). Green Synthesis of Zinc Oxide Nanoparticles Using Pomegranate Fruit Peel and Solid Coffee Grounds vs. Chemical Method of Synthesis, with Their Biocompatibility and Antibacterial Properties Investigation. Molecules, 27(4), 1236.

[4] Basak, M., Rahman, M. L., Ahmed, M. F., Biswas, B., & Sharmin, N. (2022). The use of X-ray diffraction peak profile analysis to determine the structural parameters of cobalt ferrite

nanoparticles using Debye-Scherrer, Williamson-Hall, Halder-Wagner and Size-strain plot: Different precipitating agent approach. Journal of Alloys and Compounds, 895, 162694.
[5] Bridier, A., & Briandet, R. (2022). Microbial Biofilms: Structural Plasticity and Emerging Properties. In Microorganisms (Vol. 10, Issue 1, p. 138). Multidisciplinary Digital Publishing Institute.
[6] Celis, A. I., Aranda-Díaz, A., Culver, R., Xue, K., Relman, D., Shi, H., & Huang, K. C. (2022). Optimization of the 16S rRNA sequencing analysis pipeline for studying in vitro communities of gut commensals. IScience, 103907.
[7] Fan, D., Li, L., Li, Z., Zhang, Y., Ma, X., Wu, L., Zhang, H., & Guo, F. (2020). Biosynthesis of selenium nanoparticles and their protective, antioxidative effects in streptozotocin induced diabetic rats. Science and Technology of Advanced Materials, 21(1), 505–514.
[8] Fesharaki, P. J., Nazari, P., Shakibaie, M., Rezaie, S., Banoee, M., Abdollahi, M., & Shahverdi, A. R. (2010). Biosynthesis of selenium nanoparticles using Klebsiella pneumoniae and their recovery by a simple sterilization process. Brazilian Journal of Microbiology, 41(2), 461–466.
[9] Huang, X., Chen, X., Chen, Q., Yu, Q., Sun, D., & Liu, J. (2016). Investigation of functional selenium nanoparticles as potent antimicrobial agents against superbugs. Acta Biomaterialia, 30, 397–407.
[10] Ingole, A. R., Thakare, S. R., Khati, N. T., Wankhade, A. V, & Burghate, D. K. (2010). Green synthesis of selenium nanoparticles under ambient condition. Chalcogenide Lett, 7(7), 485–489.
[11] Kawai, Y., Ozawa, N., Fukuda, T., Suzuki, N., & Mikata, K. (2022). Development of an efficient antimicrobial susceptibility testing method with species identification by Nanopore sequencing of 16S rRNA amplicons. Plos One, 17(2), e0262912.
[12] Loy, A., Lehner, A., Lee, N., Adamczyk, J., Meier, H., Ernst, J., Schleifer, K.-H., & Wagner, M. (2002). Oligonucleotide microarray for 16S rRNA gene-based detection of all recognized lineages of sulfate-reducing prokaryotes in the environment. Applied and Environmental Microbiology, 68(10), 5064–5081.
[13] Maiyo, F., & Singh, M. (2017). Selenium nanoparticles: Potential in cancer gene and drug delivery. Nanomedicine, 12(9), 1075–1089. https://doi.org/10.2217/nnm-2017-0024
[14] Sanchez-Rodas, D., Corns, W. T., Chen, B., & Stockwell, P. B. (2010). Atomic fluorescence spectrometry: a suitable detection technique in speciation studies for arsenic, selenium, antimony and mercury. Journal of Analytical Atomic Spectrometry, 25(7), 933–946.
[15] Shnoudeh, A. J., Qadumii, L., Zihlif, M., Al-Ameer, H. J., Salou, R. A., Jaber, A. Y., & Hamad, I. (2022). Green Synthesis of Gold, Iron and Selenium Nanoparticles Using Phytoconstituents: Preliminary Evaluation of Antioxidant and Biocompatibility Potential. Molecules, 27(4), 1334.
[16] Toyomane, K., Yokota, R., Watanabe, K., Akutsu, T., Asahi, A., & Kubota, S. (2022). Optimization of microbial DNA extraction from human skin samples for CRISPR typing. Forensic Science International: Reports, 100259.
[17] Uchaev, D. V, Uchaev, D. V, & Malinnikova, O. N. (2022). Assessing the degree of coal tectonic disturbance by multifractal analysis of scanning electron microscopy images of coal specimen surfaces. IOP Conference Series: Earth and Environmental Science, 990(1), 12062.
[18] Verma, P., & Maheshwari, S. K. (2018). Preparation of sliver and selenium nanoparticles and its characterization by dynamic light scattering and scanning electron microscopy. Journal of Microscopy and Ultrastructure, 6(4), 182.
[19] Ali A Alsudani et al 2019 J. Phys.: Conf. Ser. 1294 062099
[20] Bassam Faron Alfarhani et al 2019 J. Phys.: Conf. Ser. 1294 052045

[21] Bassam Alfarhani, Maha Al-Tameemi, Hector C. Goicoechea, Fernando Barbosa, Andres D. Campiglia, Direct analysis of benzo[a]pyrene metabolites with strong overlapping in both the spectral and lifetime domains, Microchemical Journal, 137 2018, 51-61,

[22] Wang, Z., & Xing, B. (2022). Small-molecule fluorescent probes: big future for specific bacterial labeling and infection detection. Chemical Communications, 58(2), 155–170.

[23] Zheng-Li, C., Yu, P., Guo-Sheng, W., Xu-Dong, H., Hao, F., Xu-Dong, Z., Guang-Yi, W., Zhu, S., & Jin, J. (2022). Characterization of bacterial community structure dynamics in a rat burn wound model using 16S rRNA gene sequencing. Journal of Burn Care & Research.

[24] Zhou, Y., Linden, A., & Heimgartner, H. (2000). Selenium-Containing Heterocycles from Isoselenocyanates: Synthesis of 1, 3Selenazoles from N-Phenylimidoyl Isoselenocyanates. Helvetica Chimica Acta, 83(7), 1576–1598.

Environmental Removal of some Dyes from Aqueous Solution by using highly adsorbent surface derived from Malva parviflora plant leaves: multifactors Studies

Ola Kamal A. Alkadir[1] Ahmed Taifi[2] Aseel M. Aljeboree[3], Abdul Latif Al Bayaa[4] , Salwan Ali Abed [5] and Ayad F. Alkaim [6]

[1]Al-Nisour University College/Baghdad/Iraq

[2]Al-Manara College For Medical Sciences/Iraq

[3,6]Department of Chemistry, College of Sciences for Girls, University of Babylon, Hilla, Iraq

[4]The Islamic University, Najaf, Iraq

[5] College of Science, University of Al-Qadisiyah, Iraq

Abstract. In this research, the efficiency of activated carbon prepared from the leaves of a perennial plant called Malva parviflora plant (MPP) to remove Maxillon Blue GRL and Disperse Blue 183 dyes from aqueous solution was studied. The preparation process was carried out in three ways, firstly, using the leaves of the plant without any chemical treatment (MPP), and secondly, Activated carbon without any chemical treatment (AC-MPP), third, acid-activated carbon (AAC-MPP) with several solutions such as ((H_2SO_4, HCl, H_3PO_4, CH3COOH, and NaOH). The results that were reached indicate that the surface activated with (phosphoric acid) is characterized by its high efficiency in removal the two dye from aqueous solution, where the removal percentage reaches (94.69 to 84.14 %) compared acid (94.44 to 79.96 %) for Maxillon Blue GRL and Disperse Blue 183 dyes respectively. The properties of the prepared and activated surface were determined by using FESEM technique, and the functional aggregates of the surface (AAC-MPP) were also detected using the (FTIR) technique. The role of recycling carbon prepared from the Malva parviflora plant reactivity and can be re-used again was studied .

Keywords: Dyes; Malva parviflora plant leaves; pollutants; multi-factors Studies

1. Introduction

The adsorption technique is defined as a phenomenon of the adhesion of atoms, molecules, or ions on the surface[1, 2]. The accumulation of adsorption is attributed to the molecules or atoms on the surface and is subject to unbalanced attractive forces perpendicular to the surface of the earth [3-6]. These forces possess a certain degree of un-saturation, where the adsorption process depends on the use of surfaces that are inexpensive It is highly efficient in removing dyes from its aqueous solution, such as agricultural waste, Roman peels, coconut shells, and the Leaves of some plants such as castor paper, fern leaf, fig leaf[2, 7-14].

Malva (family Malvaceae) that includes trees, shrubs and herbs is a very common herbal medicinal plant that grows in the Middle East and was used by the Romans and Greeks as an emollient and laxative. It has also been used to treat some diseases, respiratory and digestive disorders, and infections in most countries. People use Malva species as food in their food; They eat it cooked as a vegetable or raw mixed into a salad. Its flowers are used in medical treatments in dermatology , incorporated into a lotion to treat bruised and broken limbs [1, 15].

this work; depends on the use of the Malva parviflora plant as a green chemistry surface, friendly to the environment, inexpensive, and available, and also several solutions were activated and compared with the non-activated surface. And study the recycling of the surface and re-use it again to remove the dye.

2. Experimental part

2.1. Adsorbate

A Maxillon Blue GRL dye was obtained from Al-Hilla textile factory. A solution of dye blue color very solubility in water, molecular weight 456 g.mol^{-1} and a wavelength 599 nm Chemical formula: $C_{20}H_{26}N_4O_6S_2$, Disperse Blue 183 called azo dyes ,The supplier from textile factory Al-Hilla- Iraq ,chemical formula of $C_{20}H_{21}BrN_6O_3$,molecular weight of 473.32 g mol^{-1}, and best maximum wavelength of 550 nm. The chemical structures for two dyes appear in Figure 1. The standard solution is prepared by dissolving 1 gm of the dyes in 1000ml of distilled water to obtain a concentration 1000 mg/L. The calibration curve determined by preparing a series of concentrations from the original dye, where a graph between absorbance and concentration was found to obey Lambert-Beer law .

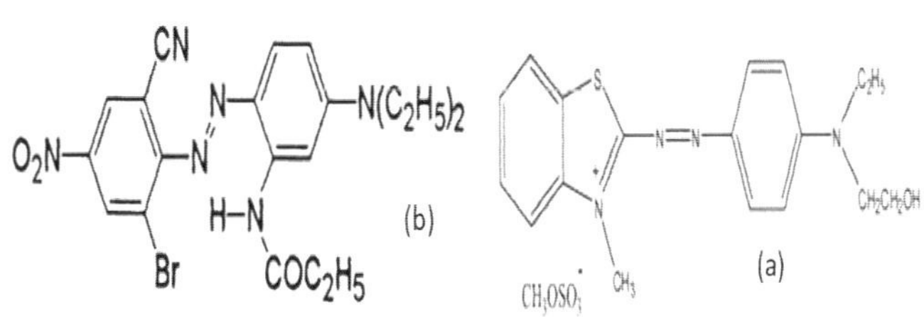

Figure 1. Chemical Structures a) Maxilon Blue GRL and b) Disperse Blue 183 dyes

2.2. Preparation Activated Carbon

Malva parviflora plan (MPP) was obtained from Hilla -Iraq. First, washed with distilled water several times to remove the suspended dust, then dried at room temperature and ground. Secondly, several solutions are activated such as ((H_2SO_4, HCl, H_3PO_4, CH_3COOH, and NaOH), With stirring and left for one hour, then washed with distilled water several times to get rid of the excess amount of acid. And dried at a temperature (60°C) for one hour . Then burn at a temperature (300 °C) for one hour to obtain acid-activated carbon powder used in the experiment.

2.3. Effect of acidic treatment

The study was carried out (MPP) with several solutions like ((H_2SO_4, HCl, H_3PO_4, CH_3COOH, and NaOH), After this all samples were prepared using as adsorbents for two dyes under study (0.05gm) dose adsorbent AACMPP of adsorbent of (75μ) average particle size mixing with (100 ml)aqueous solution of dye concentration (2, 6 and 10 mg.L^{-1})and the samples were shaking for a period of (1 hour) at a fixed temperature (20 ^{0}C) at pH 6.7, then the remaining two dyes concentration in the aqueous phase is measured spectrophotometer for the chosen wavelength.

3. Results and Discussion

3.1. Characterization

FTIR spectroscopy performed on AACMPP before and after adsorption on two dyes revealed the presence of many peaks indicating several functional groups as shown in Fig. 2. Displaying the morphology of AACMPP, FE-SEM was used. The image is shown in Figure 3, indicating that the AACMPP is composed of different pores coordinated on the surface prior to the adsorption process. Note that the surface after Maxilon Blue dye loading, all the active sites were filled in compared to the surface loaded with Disperse Blue 183, depending on the high adsorption capacity of the GRL dye. [16-18].

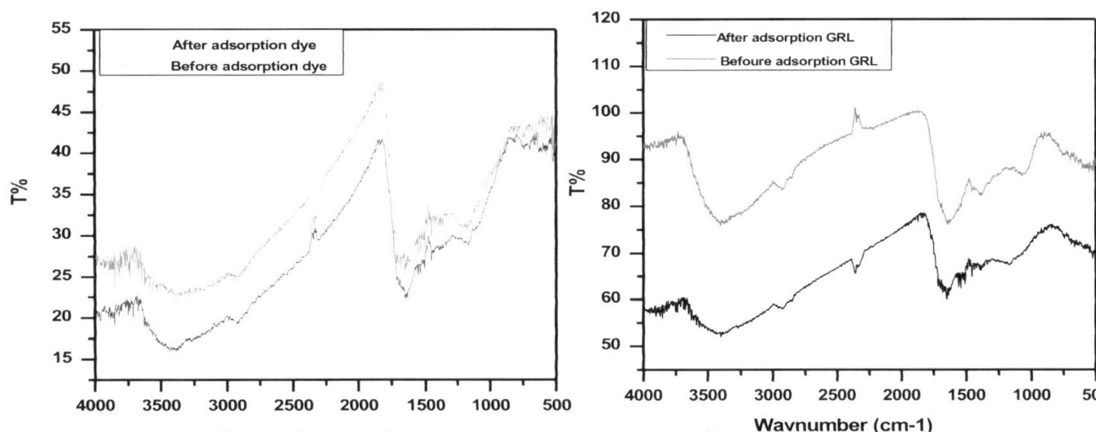

Figure 2. FT-IR spectra of AACMPP before and after adsorption of Maxilon Blue GRL and b) Disperse Blue 183 dyes .

Figure 3. FE-SEM image AACMPP before and after Adsorption **for** Maxilon Blue GRL and Disperse Blue 183 dyes

3.2. Effect of acid treatment on the adsorbent surfaces

The study of the effect of acid treatment was necessary to show the maximum adsorption. The MPP adsorbent was treated by different acids like ((H_2SO_4, HCl, H_3PO_4, CH_3COOH, and NaOH),. It was obvious from the results shown in figures 4 and 5, the best adsorption capacity when the adsorbent treated by H3PO4, this is may be due to the increased in acid acidity caused to re-activated the active sites for adsorbent surface [5, 19-21].

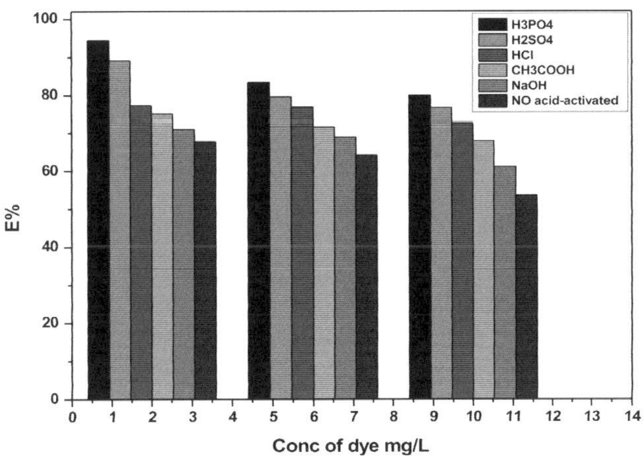

Figure 4. Effect of acid treatment onto Malva parviflora plan (MPP) to removal Disperse Blue 183 dye .

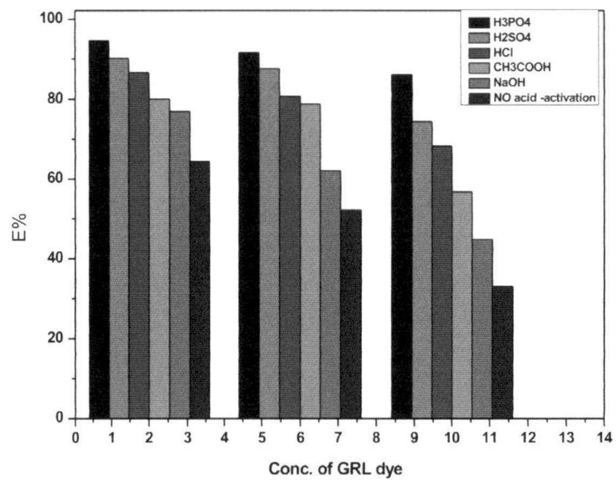

Figure 5. Effect of acid treatment onto Malva parviflora plan (MPP) to removal Maxillon Blue GRL dye .

3.3. Adsorption comparison

Comparing the adsorption of three surfaces (MPP), (ACMPP) and (A-ACMPP) to remove Maxillon Blue GRL and Disperse Blue 183 dyes from aqueous solution. At the optimum conditions for an experiment, the temperature (20°C) surface weight (0.05 g/100ml), pH 6.7 and equilibrium time one hr[22] .Through the results shown in the figure 6 and 7 , the surface (AACMPP) gave a high percentage for two dyes removal compared to the surface (ACMPP) (MPP) [23]. Because activated surfaces, whether with acid increase the active sites on the surface compared to non-activated surfaces [24, 25].

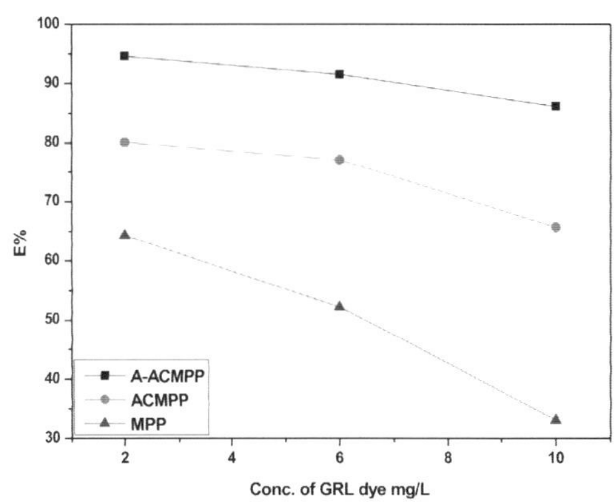

Figure 6. Comparing the adsorption of three surfaces (MPP), (ACMPP) and (A-ACMPP) to remove Maxillon Blue GRL dye.

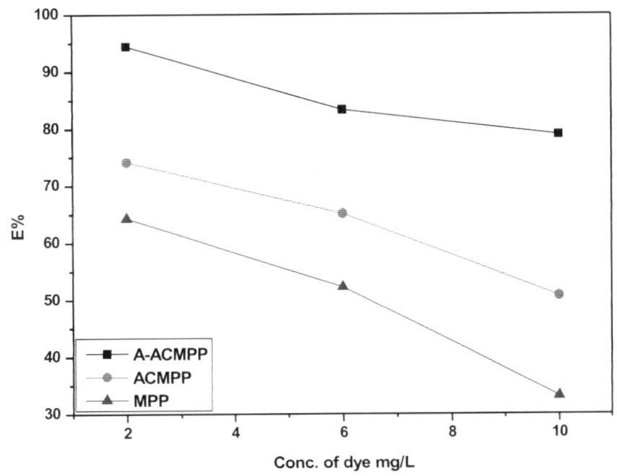

Figure7: Comparing the adsorption of three surfaces (MPP), (ACMPP) and (A-ACMPP) to remove Disperse Blue 183 dyes

4. Conclusion

Malva parviflora plan (MPP) can be utilized as an alternative bio-sorbent for two dyes removal from aqueous solutions. The advantages of this bio-sorbent are low cost, high availability, and good bio-sorption efficiency. Acid-activated carbon AAC-MPP was prepared via thermal treatment for Malva parviflora plan (MPP) in electrical furnace at 300°C in absence of air for 1 hr then treated with several solution ((H2SO4, HCl, H3PO4, CH3COOH, and NaOH). The best experimental conditions for the adsorption of Maxillon Blue GRL and Disperse Blue 183 dyes at equilibrium time 1hr. solution of pH 6.7 and 7.2 at adsorbent dose 0.05 g and concentration of two dye 100 mg/L with removal percentage (94.69 to 84.14 %) compared acid (94.44 to 79.96 %) for Maxillon Blue GRL and Disperse Blue 183 dyes onto AAC-MPP.

References

[1] Tariq J. AlMusawi, N.M., Mahmoud Taghavi, Samaneh Mohebi, Davoud Balarak, *Activated carbon derived from Azolla ficuloides fern: a high adsorption capacity adsorbent for residual ampicillin in pharmaceutical wastewater.* Biomass Conversion and Biorefnery 2021. **1**: p. https://doi.org/10.1007/s13399-021-01962-4.

[2] Hameed, K.S., Muthirulan, P., Meenakshi, S.M., *Adsorption of chromotrope dye onto activated carbons obtained from the seeds of various plants: Equilibrium and kinetics studies.* Arabian J. Chem, 2017. **10**: p. S2225-S2233.

[3] Adriano, W.S., et al., *Adsorption of amoxicillin on chitosan beads: Kinetics, equilibrium and validation of finite bath models.* Biochemical Engineering Journal, 2005. **27**(2): p. 132-137.

[4] Baskaralingam, P., Pulikesi,M. ,Elango, D. ,Ramamurthi,V. ,Sivanesan ,S, *Adsorption of acid dye onto organobentonite.* J. Hazard. Mater., 2006. **128**: p. 138-144.

[5] Layth S J and Aljeboree, M.A., *Removal of Heavy Metals by Using Chitosan/ Poly (Acryl Amide-Acrylic Acid)Hydrogels: Characterization and Kinetic Study* . NeuroQuantology, 2021. **19**(2): p. 31-37.

[6] Jafari, K., M. Heidari, and O. Rahmanian, *Wastewater treatment for Amoxicillin removal using magnetic adsorbent synthesized by ultrasound process.* Ultrasonics Sonochemistry, 2018. **45**: p. 248-256.
[7] Wared S.H.H. , R.N.D., *Synthesis and characterization of sodium alginate-g-polyacrylic acid hydrogel and its application for crystal violet dye adsorption.* International Journal of Drug Delivery Technology, 2021. **11**(2): p. 556 - 565.
[8] Mohammed A. Jawad, A.J.K., Nadher D. Radia, *Role of Sodium Alginate-g-poly (Acrylic acid-fumaric acid) Hydrogel for Removal of Pharmaceutical Paracetamol from Aqueous Solutions by Adsorption* International Journal of Pharmaceutical Quality Assurance, 2021. **12**(3): p. 202-205.
[9] Kalibantonga, P.D., *Adsorption of Heavy Metals from Solution by South African Industrial Clay.* A thesis submitted at Tshwane University of Technology, 2004.
[10] Aseel M. Aljeboree, A.F.A., Ali loay, hanadi m algburi, *Photocatalytic Degradation of Textile Dye Cristal Violet Wastewater using Zinc Oxide as a Model of Pharmaceutical Threat Reductions.* Journal of Global Pharma Technology, 2019. **11**(02): p. 138-143.
[11] Aljeboree , A.M., ALSHIRIFI, A. N., *Spectrophotometric Determination of phenylephrine hydrochloride drug in the existence of 4- Aminoantipyrine: Statistical Study.* International Journal of Pharmaceutical Research, 2018. **10**(4).
[12] Aljeboree, A.M., Alkaim A F, *Removal of Antibiotic Tetracycline (TCs) from aqueous solutions by using Titanium dioxide (TiO2) nanoparticles as an alternative material.* Journal of Physics: Conf. Series 2019. **1294** p. 052059.
[13] Divyapriya, G., R. Srinivasan, I.M. Nambi, and J. Senthilnathan, , *Highly active and stable ferrocene functionalized graphene encapsulated carbon felt array - A novel rotating disc electrode for electro-Fenton oxidation of pharmaceutical compounds.* . Electrochimica Acta,
[14] , 2018. **283**: p. 858-870.
[15] S. Tanada, M.K., N. Kawasaki, T. Sakiyama, T. Nakamura, M. Araki, T. Tamura, *Removal of phosphate by aluminum oxide hydroxide.* J. Colloid Interface Sci., 2003. **257** p. 135-140.
[16] Prahas, D., Kartika, Y., Indraswati, N., Ismadji, S., *Activated carbon from jackfruit peel waste by H3PO4 chemical activation: pore structure and surface chemistry characterization.* . Chem. Eng. J. , 2008. **140** p. 32-42
[17] L. Yan, L.Q., H. Yu, S. Li, R. Shan, B. Du, *Adsorption of acid dyes from aqueous solution by CTMAB modified bentonite: kinetic and isotherm modelling.* J. Mol. Liq., 2015. **21**: p. 1074-1081.
[18] Zaheer, Z., Al-Asfar, Aisha,Aazam, Elham Shafik, *Adsorption of methyl red on biogenic Ag@Fe nanocomposite adsorbent: Isotherms, kinetics and mechanisms.* Journal of Molecular Liquids, 2019. **283**: p. 287-298.
[19] Wenyan Jiang, L.Z., Xiaoming Guo, Mei Yang, Yiwen Lu, Yijun Wang, Yousen Zheng & Guangtao Wei, *Adsorption of cationic dye from water using an iron oxide/activated carbon magnetic composites prepared from sugarcane bagasse by microwave method.* Environmental Technology, 2019. **2**: p. DOI: 10.1080/09593330.
[20] Adejumoke A. Inyinbor, F.A.A.a.G.A.O., *Adsorption of Rhodamine B Dye from Aqueous Solution on Irvingia gabonensis Biomass: Kinetics and Thermodynamics Studies.* S. Afr. J. Chem., 2015. **68**: p. 115-125.
[21] Aseel M Aljeboree, A.N.A., Ayad F. Alkaim, *ACTIVATED CARBON (AS A WASTE PLANT SOURCES)–CLAY MICRO/NANOCOMPOSITE AS EFFECTIVE ADSORBENT: PROCESS OPTIMIZATION FOR ULTRASOUND-ASSISTED ADSORPTION REMOVAL OF AMOXICILLIN DRUG.* Plant Archives 2019 **19**(2): p. 915-919.

[22] Aseel M. Aljeboree, N.D.R., Layth S. Jasim,Ayad F. Alkaim, *Removal of Pharmaceutical by Adsorption Process; Study Effect of Various Parameters and Determination of Pharmaceutical by Oxidative Coupling Reaction.* International Journal of Drug Delivery Technology, 2021. **11**(3): p. 912-916.

[23] Yahya A. Faleh , N.D.R., *Removal of Metformin hydrochloride from Aqueous Solutions by using Carboxymethyl cellulose-g-poly(acrylic acid-co-acrylamide) Hydrogel: Adsorption and Thermodynamic Studies.* IOP Conf. Series: Earth and Environmental Science 2021. **790**: p. 012062.

[24] Salam Hussein Ewaid et al 2020 J. Phys.: Conf. Ser. 1664 012143.

[25] Ewaid, S.H., Abed, S.A., 2017. Water quality index for Al-Gharraf river, southern Iraq. Egypt. J. Aquatic Res. 43 (2), 117–122. http://dx.doi.org/10.1016/j. ejar.201703001.

[26] Ewaid, S.H.; Abed, S.A.; Al-Ansari, N. Crop Water Requirements and Irrigation Schedules for Some Major Crops in Southern Iraq. Water 2019, 11, 756.

[27] Ewaid, S.H.; Abed, S.A.; Al-Ansari, N. Water Footprint of Wheat in Iraq. Water 2019, 11, 535.

[28] Ewaid, S.H.; Abed, S.A.; Al-Ansari, N. Assessment of Main Cereal Crop Trade Impacts on Water and Land Security in Iraq. Agronomy 2020, 10, 98.

[29] Salam Hussein Ewaid et al 2021 IOP Conf. Ser.: Earth Environ. Sci. 722 012008

[30] Salam Hussein Ewaid et al 2021 IOP Conf. Ser.: Earth Environ. Sci. 790 012075

[31] Jawad, A.A.R., Aljeboree, Aseel M.,Jasim, Layth S.,Alkaim, Ayad F., *Comparative between the Syzgium Aromaticum Flower Bud (SAFB) and Activated (SAFB) for the Removal Reactive Green Dye from Aqueous Solution.* Egyptian Journal of Chemistry. **64**(10): p. 3-4.

[32] Y. Cao, G.Z., R. Zhou, C. Wang, B. Chi, Y. Wang, C. Hua, J. Qiu, Y. Jin, S.S. of R. M. γ-F.M. by D.T.H.S.N. for R. of B. Wu, *Green Synthesis of Reusable Multifunctional γ-Fe2O3/Bentonite Modified by Doped TiO2 Hollow Spherical Nanocomposite for Removal of BPA,*. Sci. Total Environ., 2019. **708**: p. 134669.

[33] Ewaid, S.H.; Abed, S.A.; Al-Ansari, N.; Salih, R.M. Development and Evaluation of a Water Quality Index for the Iraqi Rivers. Hydrology 2020, 7, 67.

[34] Aseel M. Aljeboree, A.F.A., *Role of plant wastes as an ecofriendly for pollutants (Crystal Violet dye) removal from Aqueous Solutions* Plant Archives 2019 **19**(2): p. 902-905.

Bio-Environmental preparation of chitosan nanoparticle using *Bacillus subtilis* and their biomedical activity

Zahraa Hussein Ali[1] and Nawfal Hussein Aldujaili[2]

[1]Kufa University, faculty of science, Department of biology, Iraq
[2]Kufa University, faculty of science, Department of biology, Iraq

Abstract. The aim of this study was to study the biomedical activity (antibacterial, antibiofilm, and antioxidant activity) and Characterization of Chitosan NPs synthesized by *B. subtilis*. Different types of bacterial strains(Z1-Z25) were screened for biosynthesis of chitosan NPs, Isolate (Z2) was selected as efficient isolate based on color change and antibacterial activity against the pathogenic bacteria, and it was diagnosed as *Bacillus subtilis*, depending on the morphology, microscopic examination and VITEK2 compact system. The characterization of biogenic chitosan NPs was achieved, using: UV visible spectrophotometry was used to describe the synthesis of chitosan NPs by *B. subtilis*, with an absorption peak at 280nm wavelength. SEM analysis showed the spherical, homogenous with average sizes from 29-51nm. The elemental analyses of biogenic chitosan with EDS showed 33.31% carbon, 10.24 % nitrogen, 50.55% oxygen, 5.34% phosphate and 0.55% chloride. The size and structurer of biogenic chitosan NPs was measured using XRD that appeared at 20°-26°. AFM analysis showed the average diameter of the biogenic chitosan NPs was 47. 18**.** The antibacterial activity was studied for biogenic chitosan against pathogenic bacteria *(P.aeruginosa*, *S. aureus*, *K. pneumonia*, *E.coli*, and *P. mirabilis)*.

Biogenic chitosan NPs with deferent concentration (100, 200, 400ug/ml) showed inhibition against all tested bacteria and largest inhibition zone against *S. aureus*(26mm) with concentration 400 μg/ml. Biogenic chitosan expressed high significant antibiofilm activity with increasing concentration of chitosan NPs. DPPH reducing activity of nanoparticles which increased with increase concentration of biogenic chitosan NPs.

Keywords: antibacterial , antibiofilm , antioxidant chitosan ,and *B. subtilis*

1. Introduction

Many different species of bacteria synthesize nanomaterials, each of which has number of characteristics. The majority of these reactions involve the extracellular binding of metal nanoparticles to bacterial biomass [1]. Biological and physicochemical characterization strategies for NPs are available. Biological characterization includes antimicrobial activity, genotoxicity, and antibiofilm activity. In contrast, spectrophotometry, electron microscopy, and X-ray processing are examples of physicochemical techniques.

Chitosan is a semi-synthetic polyaminosaccharide formed when chitin is N-deacetylated. Chitosan has garnered interest in a variety of medicinal, pharmaceutical, food, and environmental fields due to its low toxicity, biodegradability, and biocompatibility, as well as its bacteriostatic and mucoadhesive properties [2].

Chitosan is commonly used in medicine for a variety of purposes, including antitumor therapy and cholesterol reduction, as well as in medicinal membranes, wound dressings, and controlled-release medicinal mate [3].

The presence of NH2 groups in its structure confers on it a cationic nature, which confers biological activity on chitosan in low pH environments, specifically through association with negatively charged compounds found in bacterial cells, such as proteins, polysaccharides, and phospholipids [4].

Chitosan nanoparticles have been shown to interact with bacterial membrane proteins, intracellular proteins, and phosphate residues in DNA, interfering with cell division and ultimately killing bacteria [5].

The antibiofilm properties of chitosan were examined against *Bacillus cereus*, *Staphylococcus aureus*, *Salmonella enterica*, and *Cryptococcus neoformans* [6]. Chitosan was mixed with other substances in experiments to determine its efficacy in preventing and treating biofilm formation [7].

2. Materials and methods

Selection and identification of efficient isolate that producing chitosan NPs

Twenty-five of isolates from different strains were grown in brain heart infusion broth for 24 hours at 37C0. After incubation, the colloidal suspension was centrifuged at 10,000 rpm for 15 minutes, the precipitate was separated, and the supernatant was applied to the industrial chitosan and acetic acid solution. All isolates (Z1-Z25) were screened against *E. coli* using the Muller Hinton agar well diffusion assay and incubated for 24 hours at 37Co. An efficient isolate (Z2) was selected based on antimicrobial activity, color change, and absorption spectrum

efficient isolate(Z2) was selected and identified as *B. subtilis* based on morphological and microscopic examination, biochemical identification and VITEC

Large scale production of chitosan nanoparticles using *B. subtilis*

Screening of different strains (Z1-Z25) for antibacterial activity against *E. coli* as indicator strain and Selection the efficient isolate based on antimicrobial activity, and color change

efficient isolate(Z2) was selected and identified as *B. subtilis* based on morphological and microscopic examination, biochemical identification and VITEC.

Bacteria grown in brain heart infusion broth at 370C for 24h. After incubation, the colloidal suspension was centrifuged at 10,000 rpm for 15 minutes, then the precipitate was removed and the supernatant was added to the solution containing distilled water, (chitosan 2 mg), and (acetic acid, 2ml). The cell free supernatant was prepared through multiple steps: the activation of the inoculum, inoculation of the BHI broth (24 hours at 37°C of incubation), incubation of the culture for biomass production and centrifugation at 4200 xg for 15 minute at 4°C to prepare the cell free supernatant, pellets of cells of microorganism were discarded and the cell free supernatants were collected for use in the biosynthesis of chitosan nanoparticles.

The supernatant was harvested by centrifuging broth culture using large scale cooled centrifuged with the capacity of 50ml tubes at 4200xg to the pellet cells at the bottom of the tubes leaving cell free supernatant. The supernatant then collected in a sterile flask for use in the production of chitosan nanoparticles. The strains were inoculated into autoclaved three flasks: each contained 400ml of BHI broth, 4gm of chitosan as a substrate, and 4ml acetic acid then were incubated for 24h at 370C in a shaking incubator at 150 rpm. This was repeated in different times

Characterization of biosynthesized chitosan Nanoparticles

The physical characteristics of biosynthesis nanoparticles were characterized by XRD ,UV visible and AFM.

UV-Visible Spectroscopy

Detection of The chitosan nanoparticles formation was estimated with
UV-Visible spectroscopy (Shimadzu UV-visible1800 spectrophotometer) in research laboratory of Imam Ali Holy Shrine.

Scanning Electron Microscope (SEM) Analysis

Chitosan nanoparticles were examined at the University of Tehran's electron microscopy unit

Energy Dispersive X-Ray Spectroscopy (EDS) Analysis

At the University of Tehran, energy dispersive x-ray spectroscopy (EDS) was used to perform an elemental analysis of the Chitosan NPs

Atomic force microscope

AFM was used to estimate the size of chitosan nanoparticale in university of Tehran

X-ray Diffraction (XRD) Analysis

The X-ray diffraction is used for characterization of chitosan nanoparticle in Tehran.

Antibacterial activity of chitosan NPs

The antibacterial activity of chitosan was tested with different concentration against multidrug resistance bacteria *P. aeruginosa, S. aureus , K. pneumonia , E. coli and P. mirabilis* isolated from burn infections, bacteria using the agar diffusion method. With 100 μl of bacterial suspension, the agar plate was inoculated. Bores (6 mm in diameter) were then produced using a sterile borer and filled with 150 μL of chitosan . After that, First, the petri dishes were kept at 4 ° C for 2 h, then incubated at 37 ° C for 24 antibacterial activities was evaluated and the values were provided as means of triplicate analysis by measuring the growth inhibition zone diameter in millimetres.

Anti-biofilm activity of chitosan NPs

Congo red agar is a generally prepared medium Comprising of a broth brain heart infusion (37 g/l) supplemented with sucrose (50 g/l),Congo red (0.8 g/l) and agar No 1 (10 g/l) We prepared a concentrated Congo red stain aqueous solution which was then autoclaved for 15 minutes at 121oC, the autoclaved BHI agar with sucrose was added. With the isolated uropathogens, prepared CRA plates were vaccinated and aerobically incubated for 24 hours at 37 °C. The emergence of dry black crystalline colonies on the Congo red agar plates indicated the production of biofilms, while the colonies of non-producing biofilms remained red or pink [8].

Antioxidant Activity of Chitosan Nanoparticles

The DPPH solution (0.006 % w/v) was prepared in 95 percent methanol. Freshly prepared DPPH solution was placed in test tubes, and chitosan NPs (100, 200, 300, 400 ug/ml) were applied to each test tube until the final volume was 2 ml, and discoloration was calculated at 517 nm after 30 minutes in the dark incubation (UV visible spectrophotometer). Measurements were implemented at least in triplicate.

DPPH solution was used as a control which contained the same volume (without chitosan NPs) and 95% methanol was used as the blank. The percentage of DPPH free radical scavenging was calculated using the following equation: DPPH scavenging impact (%) = (Ao –A1) / Ao X 100 where Ao was the absorbance of the control and A1 was the absorbance in the presence of the chitosan. The real absorption decrease caused by test compounds was compared with the positive controls [9].

3. Results & Discussion:

Selection the efficient isolate that producing chitosan NPs

Twenty- five isolates from deferent strains (Z1-Z25) were screened based on the ability of bacteria for chitosan NPs biosynthesis depending on color change from light to dark , UV-Visible Spectroscopy and the biological activities of the chitosan NPs against *E. coli* to select the efficient strains. The efficient isolate Z2 (figure3-1). Z2 was identified as *Bacillus subtillas* based on morphological, biochemical identification and VITEK2 compact system

Figure (3-1): Selection of efficient strain Z2 (*B. Subtilis*) Based on color change and antimicrobial activities

Different microorganisms can synthesize inorganic materials both intracellularly and extracellularly. The microbial cell has an outstanding ion transfer mechanism in the intracellular approach. The negatively charged bacterial cell wall absorbs positively charged metal ions due to electrostatic activity [10] . Furthermore, enzymes in the bacterial cell wall reduce the metal ions to their respective nanoparticles. In contrast, the microbial cell secretes reductases used in the bioreduction of metal ions into the corresponding NPs in the extracellular method [11].

Large scale production of chitosan nanoparticles using *B. subtilis*

B. subtilis, which was used in the biosynthesis of chitosan nanoparticles, demonstrated the ability to extracellular biosynthesis by using cell free supernatant after the addition of chitosan as a substrate under previously optimized conditions. The color changes of the reaction mixture from light to dark. after incubation in shaking at 150 rpm, as well as the color change and antibacterial behavior, served as an indicator for the biosynthesis of chitosan NPs by *B. subtilis*

Chitosan nanoparticles can be prepared using many methods such as ionic gelation, complex coacervation, emulsion cross-linking , spray drying and biological method(Yien et al., 2012). The use of biological systems for production is an emerging area and the scientists have been viewing with interest microorganisms as probable eco-friendly nano-factories [12] . The biological method was applied due to its reproducibility and flexibility of implementation. This method is highly controllable: it allows for a certain degree of control of the produced particles size and surface charge [13].

Characterization of biosynthesized chitosan NPs

UV-visible Spectroscopy

The biosynthesis of nanoparticles can be confirmed by visual observation and measuring the absorbance band using UV-visible spectroscopy. The absorption spectrum of nanoparticles produced in the reaction mixture has a peak at 280nm figure (3-2). This demonstrates the existence of nanoparticle surface plasmon resonance (SPR), and a single SPR band shows that the nanoparticles are spherical in form.

The production of nanoparticles was investigated using a UV-Visible spectrophotometer, which indicated a single peak at 280 nm, suggesting the presence of chitosan nanoparticles in the reaction solution. According to a report performed by [14].

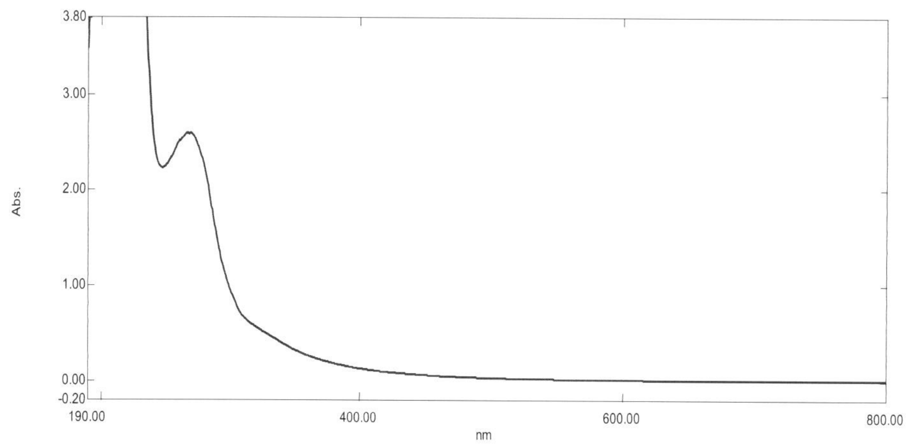

Figure (3-2) UV-visible spectroscopy analysis of chitosan NPS synthesis by *B. Subtilis*

Scanning Electron Microscopy Analysis

SEM was used to approve the morphology and size of the biogenic chitosan nanoparticles and well distributed and spherical shaped chitosan nanoparticle synthesized by *Bacillus subtilis* with a size range between (29 .9 - 51.8) figure (3-3).

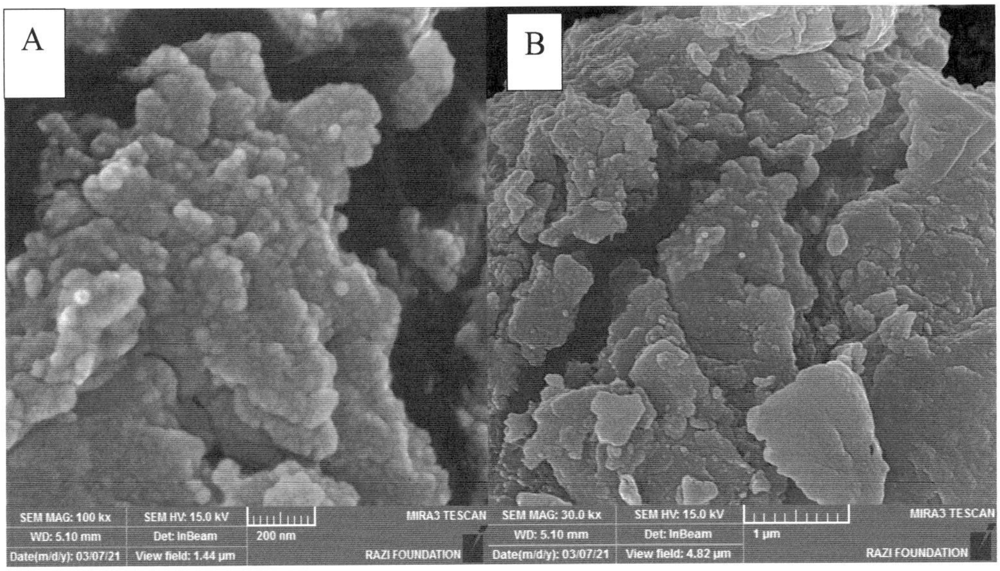

Figure (3-3) SEM micrograph of biogenic Cs-NPs synthesis by *B. Subtilis* showed chitosan NPs with spherical well dispersed with size (29-51 nm) at A: 200 nm, and B: 1 um

The size distribution of NPs was ranging from 10 to 60 nm. Most NPs were in the same size while very few particles were above 60nm. The method used in the present work allowed obtaining chitosan NPs with smaller size, in comparison with different studies using other biological compounds as reducer and stabilizer agents

EDS analysis of biogenic chitosan NPs

Energy Dispersive Spectroscopy (point and mapping analysis) was used to quantify the presence of chitosan nanoparticles by observing the optical absorption peaks of elements. Elemental analysis exhibited the percentage of elements constituent of CHNPs synthesis by *B. Subtilis* was 33.31% carbon, 10.24 % nitrogen, 50.55% oxygen, 5.34% phosphate and 0.55% chloride(figure (3-4) and table(3-1)

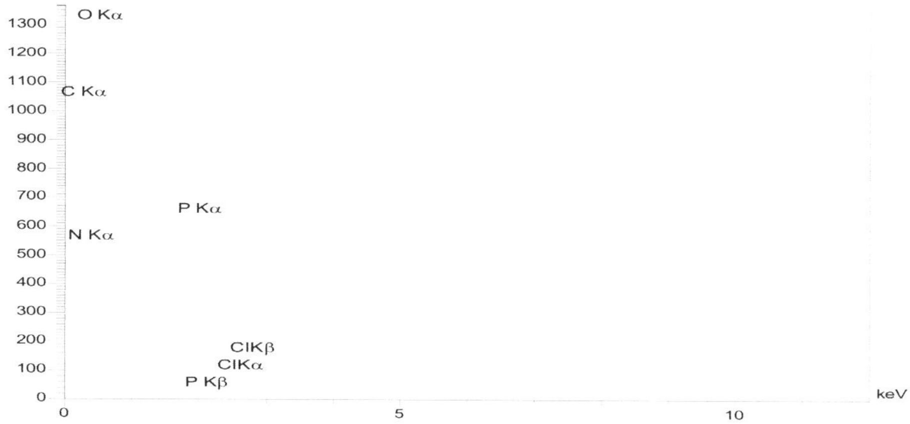

Figure (3-4) EDS analysis of chitosan NPs by synthesis using *B.Subtilis*

The energy dispersive spectroscopy analysis recognized components, allowing for quantitative and qualitative analysis of the carbon, oxygen, nitrogen, phosphate, and chloride constituents in the formulations produced.

Table (3-1) elemental analysis of chitosan nanoparticle synthesis by *B. Subtilis*

Elt	Line	Int	Error	K	Kr	W%	A%	ZAF
C	Ka	80.6	49.2781	0.3945	0.1350	33.31	40.48	0.4052
N	Ka	8.3	50.0926	0.0572	0.0196	10.24	10.67	0.1909
O	Ka	170.5	50.9072	0.4088	0.1399	50.55	46.11	0.2767
P	Ka	134.1	4.3776	0.1263	0.0432	5.34	2.51	0.8101
Cl	Ka	11.2	2.1428	0.0132	0.0045	0.55	0.23	0.8159
				1.0000	0.3422	100.00	100.00	

Atomic Force Microscope AFM

AFM image of chitosan nanoparticle synthesis by *Bacillus subtilis,* gave information of the morphology of biogenic chitosan NPs . The average diameter of chitosan NPs biosynthesis from *B. subtilis* was 47.18 figure (3-5) which showed three dimension images.

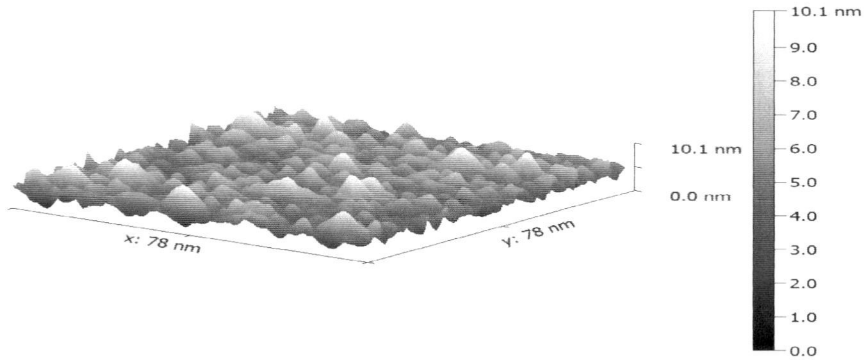

Figure (3-5) AFM analysis of biogenic chitosan synthesis by *B. Subtilis*

AFM extraordinary resolution allows for precise three-dimensional visualization of molecular structures, as well as atomic-scale strategies. The procedure for preparing samples for AFM is straightforward. Because samples can be viewed under near-physiological conditions, AFM can record the critical procedures of molecules, organelles, and other structures in living cells in real time [15].

X-ray Diffraction (XRD) Analysis

The crystalline structure of CHNPs was studied using XRD. The diffraction pattern of the polymorph was assigned to the characteristic peaks of chitosan polymer found at $2\Theta=21^{o}$. The XRD spectrum of biogenic chitosan nanoparticles which seemed at 20° -26°. The sharp peak indicated the presence of bioorganic coupons/proteins in the nanoparticles during the synthesis process. figure (3-6).

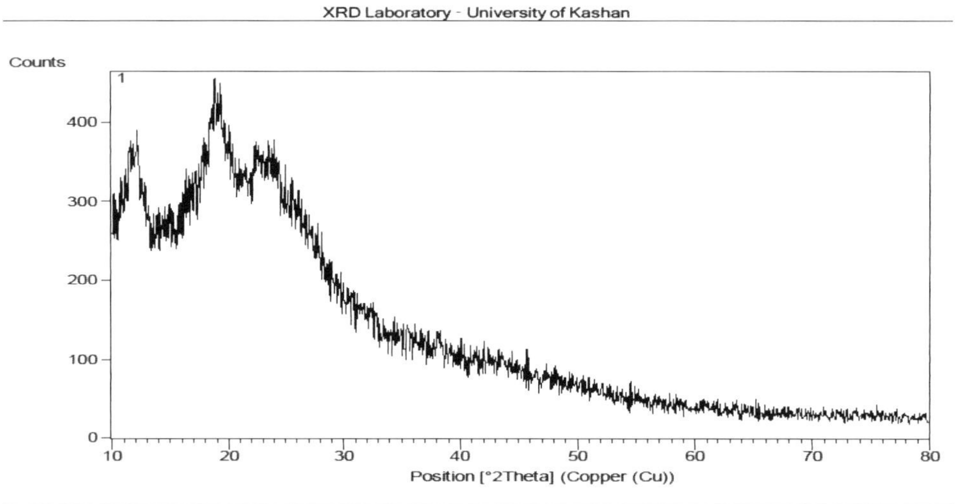

figure (3-6)

The average size of the chitosan nanoparticle was determined using X-ray diffraction. The crystalline structure of CHNPs was investigated using XRD peaks of chitosan polymer observed at 2Θ=21° was attributed to the polymorph's diffraction pattern. The XRD spectrum of CHNPs which appeared at 20° -26°. Sharp peak indicated some bioorganic couponed/ proteins in the nanoparticles during synthesis

Antibacterial activity of chitosan NPS

Biogenic chitosan nanoparticles synthesized by *B. subtilis* have been evaluated for their antibacterial activity against some multi drug resistance bacterial pathogens(MDR).The agar well diffusion method was used for detecting the antibacterial activity of biogenic chitosan nanoparticles. Chitosan NPS with different concentration (100, 200, 400 ug/ml) showed showed inhibition activity against all tested bacteria and the highest inhibition zone of Chitosan NPs observed in Gram- Positive bacteria was (26mm) in *S. aureus* with concentration (400 ug/ml) , while the lower inhibition zone in Gram –Negative bacteria was (18mm)in *E. coli* with same concentration . This inhibitory effect increased with increasing concentrations of chitosan NPS as in figure (3-7), and table (3-2). It was very similar to other reports that the Chitosan NPs could inhibit the growth of bacteria at high concentration [16].

The results showed that nanoparticles have the ability to inhibit the bacterial growth of both gram positive and gram negative. The gram positive bacteria were with higher sensitivity than gram negative bacteria to biogenic chitosan NPs. This agree with [17] reported GP bacteria are more susceptible to chitosan's antibacterial effect than GN bacteria. As a result, interpreting bacterial exposure to chitosan is difficult

Because of the rising threat of multidrug resistance (MDR), the scientific community has made it their mission to find an effective alternative to replace existing antibiotics that have become resistant. Areas an effective new entry-drug supplement to antibiotics Nanoparticles are now seen as a promising solution to antibiotics and seem to have a high potential to overcome the bacterial MDR problem [18]

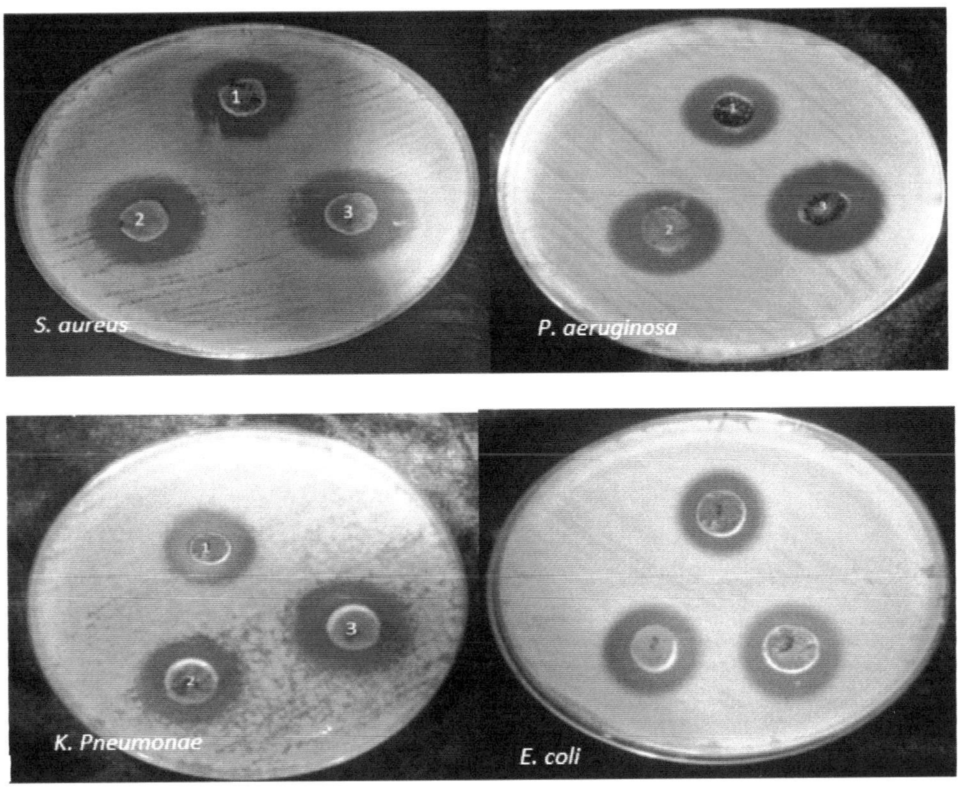

Figure (3-7) anti-bacterial activity of biogenic Chitosan nanoparticles against some pathogenic bacteria

Table (3-2) inhibition zone of chitosan nanoparticles synthesis by *B. subtili*

Bacteria	Inhibition zone (mm)		
	1(100 ug/ml)	2 (200 ug/ml)	3(400 ug/ml)
S. aureus	18	20	26
P. aeruginosa	17	19	23
K. Pneumonaei	16	18	22
E. coli	14	16	18
P. mirabilis	15	17	20

Anti-biofilm activity of chitosan

Biofilm is the most important feature of bacteria that enhance the binding of bacteria to surfaces of surgical instruments and prostheses. congo red method used for qualitative assessment of biofilm formation the result recorded as change the color of colony on the congo red agar Figure (3-8). Chitosan NPs was tested with 3 concertation (100,200,400) for antibiofilm activity against five strains of bacteria including (*p.aeruginosa*, *S. aureus*, *K. pneumonia*, *E.coli* and *P. mirabilis*) . The formation of biofilm in all isolates was prevented by chitosan NPs.The anti-biofilm activity increased with increased concentration of biogenic Chitosan NPs figure (3-8).

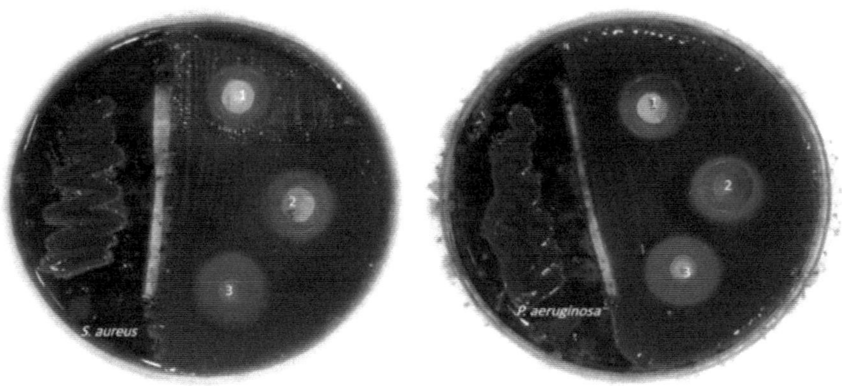

Figure (3-8) anti-biofilm activity of biogenic Chitosan NPs against S. *aureus*, and *P. aeruginosa*.

Biofilms are complex bacterial populations that resist the action of antibiotics and the human immune system . Due to the lack of effective antibiofilm antibiotics. Nanoparticles were used to resolve this problem. One potentially important candidate treatment uses NPs to show antibiofilm activity. However, in most cases antibiotics fail to eradicate these cells, as the concentration of antibiotics required to eradicate bacteria in biofilms is much higher than MIC of planktonic cells [19].

Antioxidant Activity of Biogenic Chitosan Nanoparticles

The 1-Diphenyl-2- picrylhydrazyl (DPPH), free radicals' scavenging assay was used to detect the antioxidant ability of the NPs biosynthesized from *B. Subtilis,* in vitro by reducing DPPH free radicles . After adding the nanoparticles (100, 200, 300 ,400 ug /ml) to DPPH solution , the absorbance was measured at 517 nm after 30 minutes. The results revealed the ability of nanoparticles to scavenge DPPH free radicals, indicated by observing the color change

DPPH reducing activity of nanoparticles which increased with the increase in concentration of biogenic chitosan NPs. It was 28% in 100ug /ml , 30% in 200ug/ ml, 42% in 300ug /ml, and 54in 400ug/ml, for chitosan NPs figure (3-9).

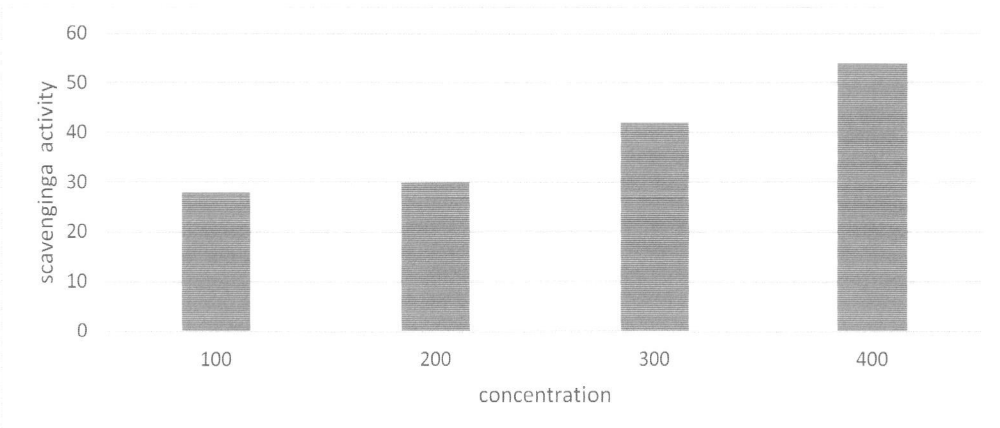

Figure (3-9) antioxidant activity of chitosan NPs by DPPH assay

The DPPH scavenging activity of NPs increased as their concentration increased, suggesting that the percentage of inhibition of DPPH increased as the concentration of chitosan NPs increased, indicating that DPPH showed more inhibition due to increased electron donation and agrees.

Antioxidants work not only by scavenging free radicals, but also by inhibiting free radical formation [20] .

4. Conclusion

- *B. Subtilis* has ability to synthesis of Chitosan NPs through biological techniques and the CHNPs showed optimal properties AFM, XRD, SEM, and EDS.
- biogenic CHNPs showed a great inhibition activity against pathogenic bacteria (*P. aeruginosa*, *S. aereus*, *K. Pneumonae,* and *E. Coli*)at concentrations 400ug/ml, and the largest inhibition was against *S. aureus.*
- Chitosan NPs showed anti-biofilm activity against pathogenic bacteria MDR (*P. aeruginosa*, and *S. aereus*)
- Biogenic Chitosan NPs exhibited antioxidant activity .

5. Recommendations

- Testing a larger number of bacteria to find out their ability to produce chitosan nanoparticles.
- Study the antimicrobial activity of chitosan NPs against MDR bacteria from different clinical sources.
- Optimal properties test to determine perfect conditions such as pH, temperature, etc. for the production of chitosan NPs.
- Toxicity study of chitosan NPs.
- study the anticancer activity of chitosan NPs.

References

[1] Manivasagan, P., & Kim, S. (2015). Biosynthesis of nanoparticles using marine algae: A review. Marine Algae Extracts: Processes, Products, and Applications, 295–304..

[2] Savin, C.-L., Popa, M., Delaite, C., Costuleanu, M., Costin, D., & Peptu, C. A. (2019). Chitosan grafted-poly (ethylene glycol) methacrylate nanoparticles as carrier for controlled release of bevacizumab. Materials Science and Engineering: C, 98, 843–860..

[3] Linden, J. C., Stoner, R. J., Knutson, K. W., & Gardner-Hughes, C. A. (2000). Organic disease control elicitors. Agro Food Industry Hi-Tech, 11(5), 32–34..

[4] Ang, W., Wang, S., Zhang, L., & Guo, Y. (2017). An overview of the protective effects of chitosan and acetylated chitosan oligosaccharides against neuronal disorders. Marine Drugs, 15(4). https://doi.org/10.3390/md15040089.

[5] Xu, X.-H. N., Brownlow, W. J., Kyriacou, S. V, Wan, Q., & Viola, J. J. (2004). Real-time probing of membrane transport in living microbial cells using single nanoparticle optics and living cell imaging. Biochemistry, 43(32), 10400–10413..

[6] Orgaz, B., Lobete, M. M., Puga, C. H., & San Jose, C. (2011). Effectiveness of chitosan against mature biofilms formed by food related bacteria. International Journal of Molecular Sciences, 12(1), 817–828..

[7] Mu, H., Zhang, A., Zhang, L., Niu, H., & Duan, J. (2014). Inhibitory effects of chitosan in combination with antibiotics on Listeria monocytogenes biofilm. Food Control, 38, 215–220..

[8] Ruchi, T., Sujata, B., & Anuradha, D. (2015). Comparison of phenotypic methods for the detection of biofilm production in uro-pathogens in a tertiary care hospital in India. Int J Curr Microbiol App Sci, 4(9), 840–849..

[9] Goyal, A. K., Middha, S. K., & Sen, A. (2010). Evaluation of the DPPH radical scavenging activity, total phenols and antioxidant activities in Indian wild Bambusa vulgaris Vittata methanolic leaf extract. Journal of Natural Pharmaceuticals, 1(1)..

[10] Bhardwaj, K., Sharma, A., Tejwan, N., Bhardwaj, S., Bhardwaj, P., Nepovimova, E., Shami, A., Kalia, A., Kumar, A., & Abd-Elsalam, K. A. (2020). Pleurotus Macrofungi-Assisted Nanoparticle Synthesis and Its Potential Applications: A Review. Journal of Fungi.

[11] Ovais, M., Khalil, A. T., Ayaz, M., Ahmad, I., Nethi, S. K., & Mukherjee, S. (2018). Biosynthesis of metal nanoparticles via microbial enzymes: a mechanistic approach. International Journal of Molecular Sciences, 19(12), 4100..

[12] Iravani, S. (2011). Green synthesis of metal nanoparticles using plants. Green Chemistry, 13(10), 2638–2650..

[13] Yien, L., Zin, N. M., Sarwar, A., & Katas, H. (2012). Antifungal activity of chitosan nanoparticles and correlation with their physical properties. International Journal of Biomaterials, 2012..

[14] Gabriel, J. S., Gonzaga, V. A. M., Poli, A. L., & Schmitt, C. C. (2017). Photochemical synthesis of silver nanoparticles on chitosans/montmorillonite nanocomposite films and antibacterial activity. Carbohydrate Polymers, 171, 202–210..

[15] Deng, X., Xiong, F., Li, X., Xiang, B., Li, Z., Wu, X., Guo, C., Li, X., Li, Y., & Li, G. (2018). Application of atomic force microscopy in cancer research. Journal of Nanobiotechnology, 16(1), 1–15..

[16] Liu, N., Chen, X.-G., Park, H.-J., Liu, C.-G., Liu, C.-S., Meng, X.-H., & Yu, L.-J. (2006). Effect of MW and concentration of chitosan on antibacterial activity of Escherichia coli. Carbohydrate Polymers, 64(1), 60–65..

[17] Alishahi, A. (2012). Chitosan: a bioactive polysaccharide in marine-based foods. The Complex World of Polysaccharides, 409–428..

[18] Franci, G., Falanga, A., Galdiero, S., Palomba, L., Rai, M., Morelli, G., & Galdiero, M. (2015). Silver nanoparticles as potential antibacterial agents. Molecules, 20(5), 8856–8874..

[19] Roy, R., Tiwari, M., Donelli, G., & Tiwari, V. (2018). Strategies for combating bacterial biofilms: A focus on anti-biofilm agents and their mechanisms of action. Virulence, 9(1), 522–554..

[20] Huang, T.-W., Ho, Y.-C., Tsai, T.-N., Tseng, C.-L., Lin, C., & Mi, F.-L. (2020). Enhancement of the permeability and activities of epigallocatechin gallate by quaternary ammonium chitosan/fucoidan nanoparticles. Carbohydrate Polymers, 242, 116312..

Environmental Effects Of Monosodium Glutamate On (NF-κB) Levels In The Male Reproductive System Of Rats

Alaa Mohammad Hasson Al-Husseini[1], Leena Adeeb Mehdi Al-Waely[2], Ahmed Abdel Ameer Kazem[3],Nabeel Rahi Mashkoor[4]

[1] Department Of Biology, College of Science, University of Al-Qadisiyah, Iraq.
[2] Department Animal Production, Faculty of Agriculture, University of Kufa, Iraq.
[3] Engineering unit, University of Al-Qadisiyah, Iraq.
[4]Scientific Affairs , University of Al-Qadisiyah, Iraq.

Abstract. Monosodium glutamate (MSG) has been identified as a food additive that Adversely affects MSG use on the reproductive efficiency by several mechanisms, this study was carried out to determine its role in stimulating the immune response in the reproductive system by estimating levels of NF-κB in serum and of its expression in reproductive organs. Thirty sex males rats were used in this study each included 12 male rats. Male rats A and B are receiving 60 and 120 mg/kg for 28 days respectively male rats of the control group were left without treatment for the period of the experiment. The levels of Nuclear factor-kappa NF-κB in serum and tissue fluid of testes and epididymis were measured. histological changes in addition to detection NF-κB expression were studying in testes and epididymis. Results of the present study recorded the toxic effect of MSG as it caused an elevation in the levels of NF-κB in serum, testicular, and epididymis tissue fluid. histological alternation in the reproductive organs was observed to represent detachment and vacuolation of the seminal epithelium, degeneration of spermatogenesis edema in a lumen in testicular tissues. Epididymal sections appeared to sever necrosis, degeneration in the epithelial layer and stereocilia, and the lumen of the epididymis. Also there was an increasing density of NF-κB protein immunoreactivity in both testis and epididymis.

Keywords: Monosodium Glutamate, NF-κB, Male, Reproductive, testis, epididymis.

1. Introduction

Monosodium glutamate (MSG) is a subset of glutamate and among the food additives commonly used in our everyday diet [1], it is in many foods, including vegetables, beef, tuna, and milk, which is a necessary but non-essential amino acid and plays a key role in metabolic processes [2]. MSG is created from sodium, glutamate, and water. It is used as a flavor enhancer for foods. Excessive consumption of MSG will result in symptoms such as fatigue, headaches, muscle spasms, dizziness, numbness, and redness [3]. . The preclinical research Confirmed that repeated ingestion of can often induces oxidative stress, asthma, diabetes, and obesity. MSG consumption also causes some, toxicities such as genotoxicity, neurotoxic, renal toxicity, hepatotoxicity, and reproductive toxicity e-ects [4].MSG can elicit male reproductive impairment by different mechanisms As studies indicated that it causes oxidative damage, histopathological alterations, hormonal disruption, and diminished sperm quality [5]. Fernandes et al. [6] proved that the Subcutaneous injection of male rats with MSG at a level dose of 4 mg/kg for 120 days leads to decline levels of FSH and testosterone and a decrease in sperm counts in addition to a reduction of reproductive organs weights.

The reproductive system is one of the targets of ROS due to the presence of adipose tissue within those organs [7]. Many studies have proved that MSG administration induced oxidative stress in testicular tissues as lead to an increment of malondialdehyde (MDA) and decrement in the activity of the antioxidant enzyme [8]. Excessive free radicals generation triggered by MSG caused in sperm dysfunction due to richness of sperm membrane with unsaturated fatty acid and low concentration of cytoplasmic antioxidants Which makes spermatozoa and testis more sensitive and vulnerable to oxidative stress damage .its resulted in damage of sperm's DNA and impaired sperm vitality [9]. the high generation of ROS has been noted in the seminal fluid of Patients with asthenozoospermia in addition to the destruction of sperm membrane caused by MSG [2].
ROS is one of the causes of infertility and leads to inflammation [10]. The essential mediator of inflammation is nuclear factor κB (NF-κB) and the production of NF-κB affects cellular functions and cell death [11]. NF-κB as the transcription factor plays an important role in the proliferation and apoptosis of germ cells and the results of Delfino and walker [12] proved the role of NF-B in spermatogenesis. It has been recorded that intraperitoneal injection of males rats with MSG induced NF-κB and TNF-α activity in the gastrointestinal and central nervous systems, indicate that MSG contribution in the inflammatory response [13]. as a result of the routine use of these preservatives, research and clinical studies on the toxic effects that may result from these preservatives have been published. This study aimed to discover some of the adverse effects that result from using MSG as a preservative agent that can cause male productive dysfunction and its role in the inflammatory response process by determining the expression of NF-κB.

2. Materials and Methods

MSG Organic bulk food E621 halal extra-fine 50mesh MSG Monosodium Glutamate was used in this study, which was obtained from Henan Sinowin Chemical Industry Co., Ltd.

2.1. Experimental Animal

Thirty-six adult male Wistar rats were used in this study, they were purchased from the animal house of veterinary medicine college /Al-Qadisiyah, the rats were placed in plastic standard cages at 25°C, they were categorized randomly into experimental and control. It was provided with water and food freely throughout the trial period. The treated groups were received MSG at60 mg/kg body weight [3] and120 mg/kg body weight [14]. The animals were categorized randomly into three equal groups as follow :
Group A: included (12) male rats receive daily 60 mg/kg via oral gavage technique for 28 days.
Group B: included (12) male rats receive daily 120 mg/kg via oral gavage technique for 28 days.
Group C: included (12) male rats who were left without treatment for the duration of the experiment (28 days).
After the end of the experimental period, the animals were anesthetized utilizing chloroform. Blood samples were obtained directly from the heart approximately (2 ml) per rat, then, blood samples we placed in the gel tubes and centrifuged to isolate sera for measuring the levels of NF-κB. testis and epididymis removed for determining NF-κB concentrations, histological alternation, and study of immune staining of NF-κB.

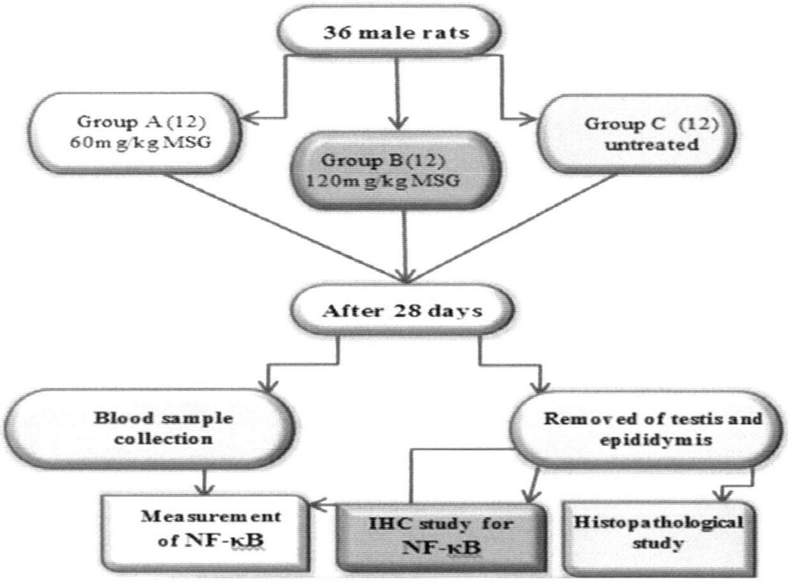

Figure 1. Explain experimental design

2.2. Estimation of NF-κB levels.

To determine NF-κB concentrations in serum and studied tissues, the ELISA technique has been depended on. Rat NF-κB p65 (Nuclear Factor Kappa B p65) ELISA Kit supplied by (MYBIOSOURCE.com) Detection Range was (78.13-5000pg/mL)and Sensitivity (46.88pg/mL). The described protocol steps were followed by the kit's manufacturer.

The testicular and epididymal plasma samples were prepared using BPS (PH 7.4). The tissue samples were homogenized in the ice using a homogenizer, next centrifuged via cooling centrifuge for 30 minutes at 10 000 × rpm/ minute. The supernatant represents tissues plasma, where NF-κB concentrations were measured using ELISA kits

2.3. Histological sections preparation

To assess the possible histological changes that may be caused by MSG treatment in the tissues of the testes and epididymis of the treated rat's males, routine tissue sections were prepared with hematoxylin and eosin stain depending on the Preparation steps Described by Steven and Bancroft [15]

2.4. Immunohistochemical sections preparation

The Immunohistochemical sections were prepared to detect the localization of NF-κB in the testicular and epididymis tissues, The described IHC staining protocol for paraffin sections has been adopted within the website of the company www.abcam.com/technical. Anti-NF-kB p65 antibody as primary antibody and horseradish peroxidase (HRP) as secondary antibody have been used.

2.5. Statistical data analysis

Data of NF-κB levels of serum and tissue fluid of testis and epididymis have been statistically analyzed by using Graphpad prism 7. ANOVA One –Way analysis was applied to detected present significant differences. also , Tukey's multiple comparisons test was performed to find the differences among the groups. Linear regression between the serum levels of NF-κB and its levels in the histological fluid of testis and epididymis was performed by finding Pearson Correlation Coefficient.

3. Result

3.1. NF-κB Levels Assay

The results of NF-κB analysis shown in figure (2) indicated a significant increase (P<0.0001) in circulation NF-κB levels in the male rats that treated daily with MSG at two concentration (60 and 120 mg/kg) in comparison with a control group and this increase was more in the group B which given 120 mg/kg of MSG as shown in (figure 2 A), Testicular NF-κB recorded high concentration (P<0.0001)in both groups A and B that treated with (60 an 120 mg/kg)respectively compared to a control group .while there was no significant difference in the testicular NF-κB levels between A and B (figure 2 B).on another hand, the results of the epididymal NF-κB investigated that treated with MGS at level dose 120 mg/kg in the male rats of group B lead to significant increment (p<0.0001) in comparison with group A and C, As the treatment with MSG at concentration 60 mg/kg did not cause any significant difference in comparison with a control group (figure 2 C).

the association results of NF-κB protein levels indicated to presence significant positive relation between serum and testicular NF-κB protein (figure 3) in the groupss A (r=0.9608 ,p<0.0001) and group B (r= 0.6301,p<0.281) .the correlation results of NF-κB protein levels ,also confirmed presence positive association between serum and epididymal NF-κB in two concentration 60 and 120 0f MSG in the group A (=0.842 ,p<0.0025) and group B (r=0.6861 ,p<0.0137) (figure 4)

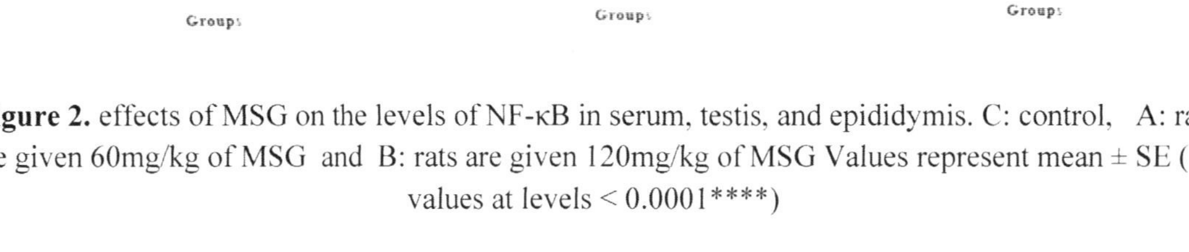

Figure 2. effects of MSG on the levels of NF-κB in serum, testis, and epididymis. C: control, A: rats are given 60mg/kg of MSG and B: rats are given 120mg/kg of MSG Values represent mean ± SE (p-values at levels < 0.0001****)

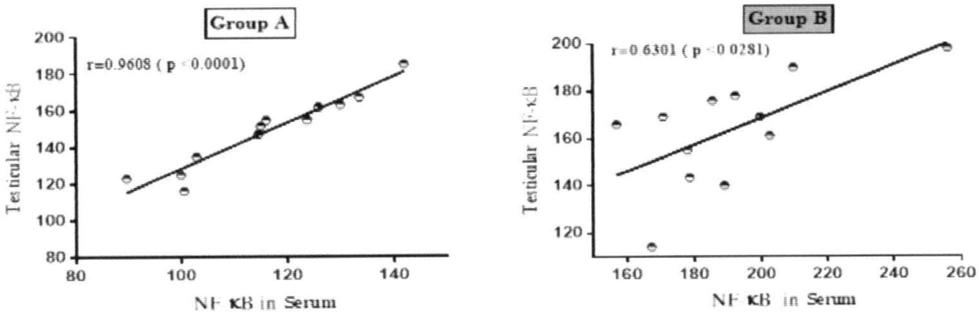

Figure 3. association between levels of NF-κB in serum and testicular tissues. (A) a group that was treated with MSG at level 60mg/kg (B) a group that treated with MSG at level 60mg/kg. (r values represent correlation coefficient, p represent level values.

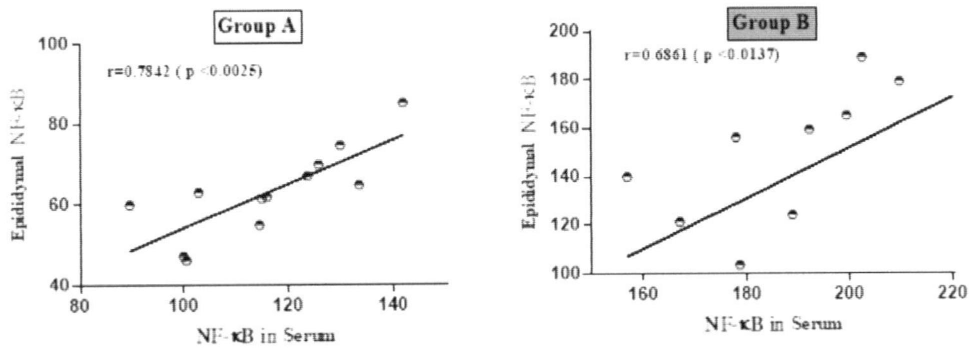

Figure 4. association between levels of NF-κB in serum and epididymis tissues. (A) a group that was treated with MSG at level 60mg/kg (B) a group that treated with MSG at level 60mg/kg. (r values represent correlation coefficient, p represent level values.

3.2. Histopathological Examination

The histological section of the present study shown in (figure 5), proved the presence of histological alternation in the testis and epididymis tissues of groups A and B which were treated with MSG compared to the untreated males. Control testis showed normal structure represent with normal structures of seminiferous tubules with normal stages of the spermatogenesis process and normal Leydig cells groups in interstitial space (figure 5 A) while testis sections of the treated groups with MSG at dose 60 mg/kg(figure 5 B) demonstrated presence vacuolation in the seminal epithelium in addition to a few sperms with edema in the lumen and there was Congestion in the blood vessel. Whereas, the treatment of male rats with MSG at concentration 120 mg/kg caused more severe damage in the testis tissues (figure 5 C) .they were uncompact seminiferous tubules with dilated lining, detachment of the seminal epithelium, degeneration of spermatogenesis very clear, and the number of Leydig cells was fewer in the testicular interstitial space compare to group A and C . sections of the epididymis tissues shown in figure (6) reveal normal epididymal tubules in the control groups, they have a columnar epithelial lining with long stereocilia inward to the lumen and the epididymal lumen was full with sperms (figure (6A). In the rat's epididymis that given 60 mg/kg of MSG in group A investigated a few sperm in some epididymal tubules with slight degeneration of epithelial layer and there was simple congestion, all these changes are shown in (figure 6B). Epididymal sections of male rats that orally receive 120 mg/kg in group B appeared to sever necrosis, degeneration in the epithelial layer and stereocilia and the lumen of epididymis tube (L) was empty from sperm (figure 6C).

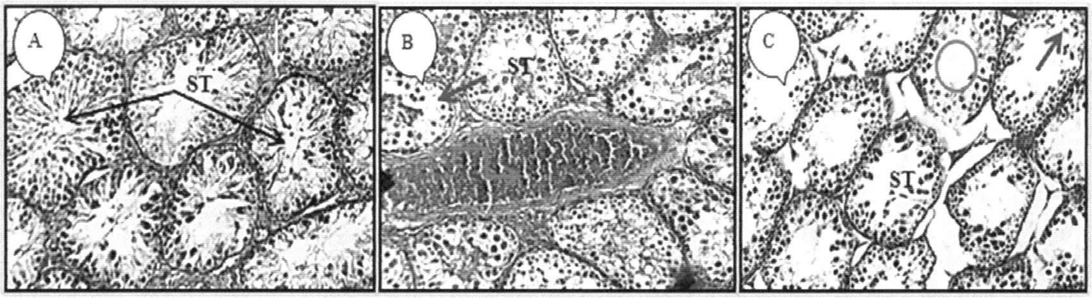

Figure 5. show the effects of MSG on testicular tissues of males rats. A represents the Control group, B represents A group which treated with 60 mg/kg of MSG and C represents B group which treated with 120 mg/kg of MSG. control group appeared normal structure of tubules (ST) seminiferous (black arrows) with normal Leydig cells (yellow arrow). Treated groups show the Sloughing of the epithelial layer (blue arrow), congestion (red arrow), and lumen empty of sperm (red circle), E&H(100X).

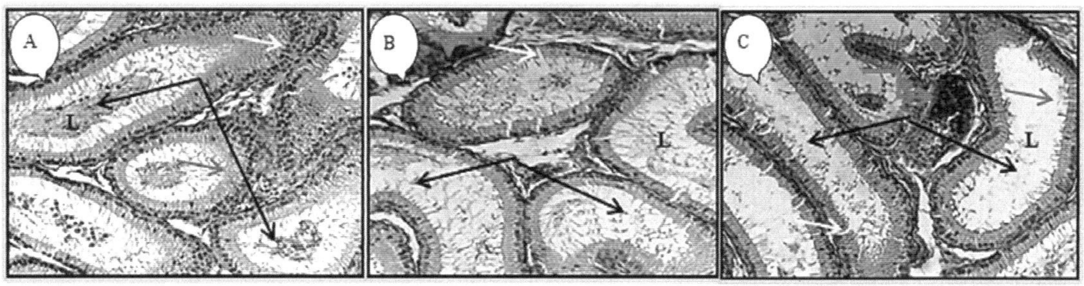

Figure 6. show the effects of MSG on epididymal tissues of males rats. (A) represent the Control group, (B) represent A group which treated with 60 mg/kg of MSG, and (C) represent B group which treated with 120 mg/kg of MSG. section of the control group showed normal structure of epididymal tubules consist of one layer of columnar epithelium cells (yellow arrows) which have stereocilia (green arrow)and expanding lining (L) fill with sperm seminiferous (black arrows) .in B and C, congestion (red arrows) degeneration of columnar epithelium cells and their stereocilia, in (C), the epididymal lumen was empty from sperm. ,E&H(100X).

3.3. Immunohistochemical Examination

the immunostaining examination study has been investigated positive immunoreactivity for NF-κB in the testicular and epididymal tissues of males rats that treated with MSG for 28 days. In the male rats of a control group, there was no immune reactively of NF-κB in the testis and epididymis sections (figure 7A and 8A). While animals of group A that received 60 mg/kg of MSG showed weak immune reactive for NF-κB protein in Leydig cells while noted strong immune staining of NF-κB protein in the germinal epithelial layer of seminiferous tubules (figure 7 B).In group B that received 120 mg/kg of MSG revealed strong and high-density immunoreactivity of NF-κB protein in the germinal layer of seminiferous tubules and interstitial space including Leydig cells(figure 7 C). NF-κB protein in the epididymis tissues shows moderate immune staining in the epithelial layer of epididymal tubules of group A (figure 8B). while immune staining in the epithelial layer of epididymal tubules of group A was strong and high density (figure 8C)

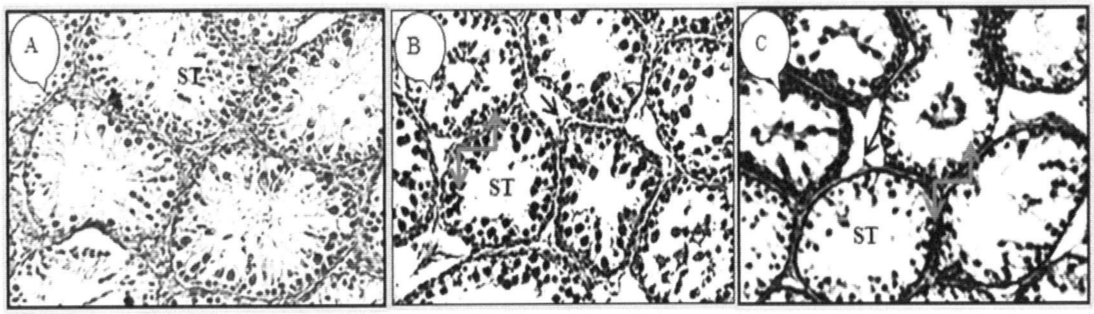

Figure 7. immunoreactivity of NF-κB in the testicular tissues of males rats that received orally MSG. (A) represent Control group, (B) represent A group which treated with 60 mg/kg of MSG, and (C) represent B group which treated with 120 mg/kg of MSG. in control (image A)there was no immunoreactivity of NF-κB protein in testis, in treated groups showed weak immunoreactivity of NF-κB protein in the Leydig cells(black arrows) of group A, while showed strong immunoreactivity NF-κB protein in spermatogenic cells (red arrows) in an image (B) in group B there was strong positive immunoreactivity of NF-κB protein in the germinal epithelium and Leydig cells (C). IHC staining (100X).

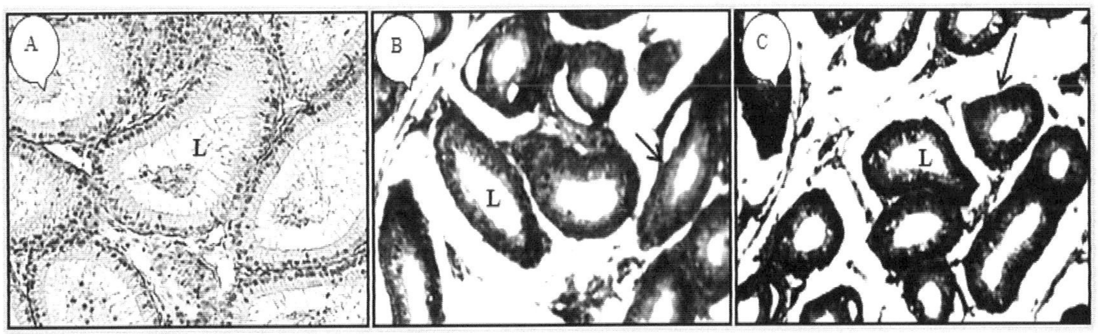

Figure 8. immunoreactivity of NF-κB in the testicular tissues of males rats that received orally MSG. (A) represent Control group, (B) represent A group which treated with 60 mg/kg of MSG, and (C) represent B group which treated with 120 mg/kg of MSG. in control (A)there was no immunoreactivity of NF-κB protein in epididymis tissue. in treated groups, the epithelial lining layer showed moderate immunoreactivity in group A and high density and strong in B (black arrows) in an image (B and C) respectively. No immune staining in the lumen of epididymal tubules (L). IHC staining (100X).

4. Discussion

The increased daily consumption of food additives causes damage to the body's systems, as it affects the liver, the nervous system, and the reproductive system through induction of oxidative stress, which negatively affects the tissues of the body. [16] [17] [18]. One of the principal processes of oxidative harm is lipid peroxidation, which has a key role in the toxicity of the biosystem [19]. It is known that the administration of MSG induces oxidative stress by producing ROS [20]. Hamza and AL-Harbi [3] investigated that treated with MSG caused elevation of MDA in rat's testicular tissues this indicate to presence of oxidative stress. ROS has been traditionally regarded as particular messengers for signal-induced NF- κB stimulation. NF-κB is known to be an important regulator of the stress and immune responses [21]. MSG caused in elevation of Tumor necrosis factor alpha (TNF-α) and activation of NF- κB this indicate to that MSG stimulate of inflammatory responses [22]. in this study we proved elevation of the levels of NF- κB in the serum ,epididymal and testicular tissues of male rats that treated with MSG , This indicates that glutamate stimulated the immune response in treated males rats . In the study of Lei et al [22] noted that inflammatory response and intestinal NF- κB activation are stimulated after increasing of intracellular glutamate concentrations Which suggests the oxidative effect of glutamine may be through activation of a NF- κB. NF- κB a transcription factor is a main regulator of the stress and immune responses [23].

Many studies investigated the effects of oral ingestion of MSG on male fertility indicators such as serum testosterone concentrations, sperm count, in addition to morphological and histological alteration in testicular tissues. Abd-Elaziz and Ashoush [24]showed histological Alteration in testicular tissues and Seminiferous tubules retrogradation In male albino rats that treated orally with 350 and 500 mg/kg for 90 days

Our results show histopathological change of male reproductive organs in the rats received 120 mg/kg MSG . high concentrations of glutamate maybe lead to hyper stimulate of the glutamate receptors which localized in spermatogenic cells, this causes induce many pathological alternations in testicular tissues Which were clear in the current study as it led to Epithelial detachment and azoospermia in the seminiferous lumen, in addition to the apparent damage to the interstitial tissue, which is reflected in androgen synthesis. Then damage of androgenic dependent organs as epididymis, these are results in agreement with Jubaidi et al [25]. One of the possible actions of MSG effects on rat's testicular tissue may be related to oxidative damages in the germ cells. Many studies proved that elevation of MDA contents produce from lipid peroxidase in animals treated with MSG [26]

Aitken et al [27] have shown that the main cause of testicular tissues damage is a high level of lipid peroxidation in the testis which lead to membrane Oxidative damage of DNA sperm and, defect in the development and motility of spermatozoa growth. this indicates that the toxic effects of MSG on male fertility might belong to the increasing in free radicals production in the reproductive organs.

In current study showed that MSG caused many histopathological alternations such as edema, and fewer spermatozoa in the seminiferous lumen It may be correlated with oxidative effects caused by MSG on testicular tissues and especially in the sperm's membrane, many studies indicated that treatment with MSG leads to Oxidative damage by a generation of reactive oxygen species ROS.

Degeneration of germinal epithelium in the male rats that treated with MSG, Its causes may be due to the decline of testosterone which responsible for the spermatogenesis process stops, as Previous studies have shown another mechanism by which MSG impaired the spermatogenesis. as known the Glutamate receptors are found in many tissues like the liver, kidneys, endocrine system testis, ovary, and hypothalamus [28].On another hand, Takarada [29] proved that glutamate transporters and receptors are located in rat testes This may be explained the mechanisms by which glutamate affects directly by its association with the transporters and receptor on the germinal epithelial cells that lining seminiferous tubules.

immunoreactivity of NF-κB investigated increase protein expression in the epididymis and testicular tissues Which indicates the inflammatory effect of MSG by stimulation of NF-κB. Activation of NF- κB in rat's testis increases by TNF-α, Delfino, and Walker [30] investigated that NF- κB protein expression in Sertoli cells induced by TNF-α release by spermatid and this lead to the alternation of spermatogenesis.NF-KB is an important candidate for the regulation of male germ cell apoptosis because it has a functional role in the proliferation and apoptosis of cells this indicates the contribution of NF- κB in the mammalian spermatogenesis [31]

5. Conclusions

This study confirmed, that the uses of MSG as a food additive at 120 mg/kg for 28 days has toxic effects by inducing an inflammatory response in reproductive organs in males which can be reflected in male efficiency, this was demonstrated by increasing levels of NF- κB in the treated testis and epididymis tissues as a key regulatory inflammatory response in addition to histological change represented very clear damage of reproductive organs and degeneration of spermatogenesis.

References

[1] Walker, R., Lupien ,J.R. The safety evaluation of monosodium glutamate. J Nutr 2000; 130: 1049S-1052S..

[2] Bera, T.K., Kar, S.K., Yadav, P.K., Mukherjee, P.; Yadav, S., Joshi, B. E_ects of monosodium glutamate on human health: A systematic review. World J. Pharm. Sci. 2017; 5: 139–144..

[3] Hamza, R.Z., and AL-Harbi, M.S. Monosodium glutamate induced testicular toxicity and the possible ameliorative role of vitamin E or selenium in male rats. Toxicol. Report .2014, 1: 1037–1045. [CrossRef]..

[4] Kazmi, Z.; Fatima, I.; Perveen, S.; Malik, S.S. Monosodium glutamate: Review on clinical reports. Int. J.Food Prop. 2017, 20 (2): 1807–1815. [CrossRef].

[5] Omowumi ,T. K., Damilare E. R., Abolanle, A. A. K., Tomilola D. O ., and Oluyomi S. A. Monosodium Glutamate (MSG)-Induced Male Reproductive Dysfunction: A Mini Review Toxics 2020, 8(7): 1-7..

[6] Fernandes, G.S.A.; Arena, A.C.; Campos, K.E.; Volpato, G.T.; Anselmo-Franci, J.A.; Damasceno, D.C.;Kempinas,W.G. Glutamate-induced obesity leads to decreased sperm reserves and acceleration of transit time in the epididymis of adult male rats. Reprod. Bio.

[7] Hanipah, E.N.A.; Yahya, N.J.; Ajik, E.M.; Yuso_, N.A.; and Taib, I.S. Monosodium Glutamate Induced Oxidative Stress in Accessory Reproductive Organs of Male Sprague-Dawley Rats. J. Sains Kesihat. Malays. 2018, 16: 67–73. [CrossRef]..

[8] El-Sawy, H.B.I.; Soliman, M.M.; El-Shazly, S.A.; and Ali, H.A. Protective e_ects of camel milk and vitamin E against monosodium glutamate induced biochemical and testicular dysfunctions. Prog. Nutr. 2018, 20: 76–85..

[9] Khaled, F.A.; Yousef, M.I.; Kamel, K.I. The protective role of propolis against the reproductive toxicity of mono-sodium glutamine in male rabbits. Int. J. Chem. Stud. 2016, 4: 4–9..

[10] Kong, Z.L.; Johnson, A.; Ko, F.C.; He, J.L.; Cheng, S.C. E_ect of cistanche tubulosa extracts on male reproductive function in streptozotocin(-)nicotinamide-induced diabetic rats. Nutrients 2018, 10: 1562.[CrossRef] [PubMed]..

[11] Lawrence ,T. The Nuclear Factor NF-κB Pathway in Inflammation.Cold Spring Harb Perspect Biol. 2009,1(6): a001651.

[12] Delfino F, Walker WH: Hormonal regulation of the NF-_B signaling pathway. Mol Cell Endocrinol .1999, 157:1–9..

[13] Lei Xu, Jie Sun, Ran Lu, Qing Ji, Jian-Guo Xu Effect of glutamate on inflammatory responses of intestine and brain after focal cerebral ischemiaWorld J Gastroenterol 2005;11(5):733-736.

[14] Brosnan JT, Drewnowski A, Friedman MI. Is there a relationship between dietary MSG and obesity in animals or humans? Amino Acids. 2014, 46(9):2075–2078..

[15] Bancroft, J.D and Stevens, A. Theory and practic of histological techniques. Churchill living ston, New York. 1982, 76(2): 351 – 356..

[16] Mehdi ,L.A. and HASSON ,A. M. Estimate levels of the liver enzymes and serotonin immune localization in the brain of rats administrated orally with sodium benzoate. International J. .Pharma. Res . 2020 12 (4): 2751-2760.

[17] A,-Husseini,A.M.H. The effect of induced oxidative stress by Short and long –terms Exposure to potassium sorbate (E202) as a food additive on the Female Reproductive System of Wistar Rats. International J. .Pharma. Res . 2020 ,12 (2): 1407-1422..

[18] Hasson ,A.M. Majhwol E.M. and Almuoswi,H.J valuation of Some Kidney Functions of Rates Treated with Sodium Benzoate. Annals of R.S.C.B. 2021,25(1): 4859 – 4866..

[19] Ongjanovic, L.B.I. , Markovic, L. Dor.d-evic, S.D. Trbojevic , N.Z. tajn,, I.S.A. Saicic,L.Z.S. Cadmium-induced lipid peroxidation and changes in antioxidant defense system in the rat testes: protective role of coenzyme Q10 and Vitamin E, Reprod. Toxic.

[20] Fábio,R. Seiva, F. Gustavo, L. Chuffa,A. Braga,C.P. Paulo,J. Amorim, A. Fernandes., A.A.H..Quercetin ameliorates glucose and lipid metabolism and improves antioxidant status in postnatally monosodium glutamate-induc metabolic alterations, Food Chem..

[21] Pahl ,HL: Activators and target genes of Rel/NF-_B transcription factors.Oncogene 1999, 18:6853–6866.

[22] Xu, L.; Sun,J.; Lu,R., Ji,Q.;and Xu .JG .Effect of glutamate on inflammatory responses of intestine and brain after focal cerebral ischemia. World J Gastroenterol 2005;11(5):733-736..

[23] Obidoa,A.C. O. Ezeokonkwo, C.A. Ezeanyika, L.U. Ejikeme,P.M. Hepatotoxic effects of low dose oral administration of monosodium glutamate in male albino rats, Afr. J. Biotechnol. 8(2009) 3031–3035.

[24] Abd-Elaziz, A.M.S.; Ashoush, I.S. E_ect of monosodium glutamate administration on the reproductive performance in male male albino rats. Egypy J. Basic Appl. Physiol. 2007, 6: 101–110..

[25] Jubaidi,F.F., Jubaidi,R.D., Noor,M.M., Taib,I.S.,and Budin,S.B.(2019).Monosodium glutamate daily oral supplementation: study of its effects on male reproductive system on rat model.Systems Biology In Reproductive Medicine. 2019, 65(3): 194–204.

[26] Kianifard,D., Ehsani,A., Daneshgar,P,Z. Effect of monosodium glutamate on testicular tissue of paclitaxel-treated mice: An experimental study. Int J Reprod Biomed. 2019 , 17(11): 819–830.

[27] Aitken,R.J. Clarkson,J. Fisher,S. Generation of reactive oxygen species, lipid peroxidation and human sperm function, Biol. Reprod. 40 (1989) 183–197..

[28] Gill, S. and Pulido, O. Glutamate Receptors in Peripheral Tissue Excitatory Transmission Outside the CNS, Kulwer Academic/Plenum Publisher, New York, 2005.

[29] Takarada,T. Hinoi, E. Balcar, V. Taniura,H. Yoneda,Y. Possible expression of functional glutamate transporters in the rat testis, J. Endocrinol. 181 (2004) 233–244..

[30] Bassam Faron Alfarhani et al 2019 J. Phys.: Conf. Ser. 1294 052045

[31] Bassam Alfarhani, Maha Al-Tameemi, Hector C. Goicoechea, Fernando Barbosa, Andres D. Campiglia, Direct analysis of benzo[a]pyrene metabolites with strong overlapping in both the spectral and lifetime domains, Microchemical Journal, 137 2018, 51-61,

[32] B. Alfarhani, M. Al-tameemi, A.V. Schenone, H.C. Goicoechea, F. Barbosa, A.D. Campiglia. Microchem. J., 2016 129 83–89. https://doi.org/10.1016/j.microc.2016.06.010.

[33] Delfino F, Walker WH: Stage-specific nuclear expression of NF-_B in mammalian testis. Mol Endocrinol 1998, 12:1696–1707.

[34] Wang CY, Mayo MW, Baldwin AS: TNF- and cancer therapy-induced apoptosis: potentiation by inhibition of NF-κB. Science 1996, 274:784–787.

The Agricultural Water Footprint of Al-Qadisiyah Governorate, Southern Iraq

Zainab Wisam Wedaa [1], Salwan Ali Abed [2], Salam Hussein Ewaid[3]

[1] College of Science, University of Al-Qadisiyah, Diwaniyah, P.O.Box. 1895, Iraq.
[3] Technical College of Shatra, Southern Technical University, Basra, Iraq.

Abstract. Rapid urbanization, pollution, and increasing water consumption together with climate change necessitated to process of more effective measurement, management, and decision mechanisms on regional water resources. The concept of water footprint (WF) is a parameter that has been introduced to the scientific literature in recent years similar to the ecological and carbon footprints. The WF of any field or product refers to the total volume of water resources that are processed or contaminated directly or indirectly during the production process. The current work is the first study assessing and discussing the agricultural water footprint of an Iraqi governorate by analyzing blue and green WFs of agricultural production in Qadisiyah governorate, southern Iraq for 2010-2020. Recently developed WF methodology has been used. The blue and green evapotranspiration amounts were estimated by the crop water requirement (CWR) option in CROPWAT 8.0 software. The statistical data including meteorological data, rainfall statistics, local crop coefficients, cultivation area, crop production amounts and animal statistics data have been utilized. The average annual agricultural WF of Qadisiyah governorate for the 10 years between 2010-2020 was determined to be 1,315,201,621 Mm^3/yr. The largest water-consuming sector is crop production (54%). Cereal and feed crops are the main component of water consumption. The rice crop followed by wheat is the primary crop production comprising about 44% of the total WF and require water supplied from rivers. Vegetable production has only 14% of the crops WF. The green WF was only 15% of the crop production WF. The largest share of water used for animal production is related to broiler chickens (44%) and 37% for dairy cattle. The study area is fertile land for crop production. However, limited water resources and scarcity of the region restrict the agricultural activities. The sustainability of freshwater resources of the governorate could be provided by reducing the WF and blue water contents. This study is expected to contribute to the national authorities to develop more accurate irrigation water management policies.

Keywords: Iraq, water footprint, Qadisiyah, water management, irrigation, environment.

1. Introduction

Water is the most important natural resource used in all sectors. Major civilizations were founded around important rivers such as the Tigris, Euphrates, and the Nile. Today, water resources have an extremely important position in terms of the supply of water for drinking, irrigation, industry, and sustainable development in energy production [1, 2]. However, the increase in water consumption have begun to threaten the existing water resources [3]. In this context, the number of international studies that call for the used water more efficiently and to provide more effective water management in terms of quality and quantity has increased rapidly in recent years [4, 5]. Especially in this period when the effects of global climate change are intense, estimation, modeling, correct planning, and management of regional water capacities and consumption are of great importance [6].

The importance of the topic can be understood when we know that the world's freshwater potential equals 2.5% of the total amount of water and that only 0.3% of the total water resources are suitable for human use on a global scale [7, 8]. Despite the increase in water demand and pollution, limited or decreasing resources have increased the importance of national and regional water management efforts [9, 10]. The concept of water management encompasses the whole range of studies on the use of water resources in a more effective and planned way, increasing the efficiency in distribution and operation, as well as controlling and interfering with water quality parameters [11, 12].

One of the most important international strategies on water management, that water is not a commercial product and that it is a heritage that must be defended, protected, and used properly, and that all nations can reveal the ecological status of their water bodies and official obligations are imposed [13,14].

Similar to carbon and ecological footprints, the concept of water footprint (WF) is one of the topics that have been introduced into the scientific literature in recent years. Regional WF accounts contribute to water management efforts to identify the impact of human factors on local water resources and to ensure the sustainability of the natural ecosystem [15].

The concept of water footprint is based on the studies of Hoekstra *et al.* [16] Chapagain and Hoekstra [17, 18], Hoekstra and Hung [19], Mekonnen and Hoekstra [20]. Using the WF method, the total WF of any product, process, or area can be calculated. The WF of any product is defined as the total volume of water used, consumed, processed, or polluted during the production process of that product. The WF in any area refers to the total water treated or polluted to a certain spatial scale. Conventional water use calculations only take into account the surface and groundwater consumed [21]. However, in a globalizing world, this approach is no longer recommended and needs to be approached from a holistic perspective [22]. The WFf method differs from conventional water usage accounts in that it represents the total water volume as well as the type and extent of water used.

In WF theory, the type of water used is examined in three main categories: blue, green, and gray. The blue WF includes the total volume of surface and groundwater used; the green WF represents the total volume of rainwater used [23]. Evapotranspiration (ET) and water transfer to other regions and basins are also embedded in the blue and green WF components. Therefore, if an artificial or human process is mentioned, the higher the blue water footprint, the more surface, and groundwater are used, which will lead to an increase in water stress problems [15]. If the green water footprint of any product is high, it means that its production is provided with rainwater contributions [24]. As a sustainability parameter, it is better that the green WF of agricultural products, in particular, is expected to be relatively higher than the blue WF [25]. Finally, the gray WF refers to the volume of clean water required for any product to make throughout the production process or following the environmental standards of pollutants in the same area [26]. The total WF is the sum of the blue, green, and gray components. The gray WF is low, especially in areas with low industrial capacity. Therefore, the calculation of blue and green WFs is sufficient in terms of reflecting the total WF of any region [17, 27].

Due to a large number of data to be used in WF studies, the complexity of the analyses to be performed, and the concept of WF being a relatively new parameter, studies on this subject are relatively inadequate [28]. One of the first country-based WF studies was conducted by Mekonnen and Hoekstra [29], in this study, WFs of many countries due to production and consumption were calculated at relatively low resolution. Similarly, in the study conducted by Chapagain and Hoekstra [17], WFs on a country basis was calculated at a wider resolution and the concept of international virtual water transfer was examined in detail. Again, the WF of agricultural products with different spatial resolutions was analyzed by Mekonnen and Hoekstra [30, 31]. Since agricultural products represent a significant proportion of the regional WF calculations, there is a wide variety of studies [32, 33,34]. However, because the agricultural WF depends entirely on the meteorological and climatological effects of the region, regional calculations come to the fore [35, 36, 37, 38]. Half of the world's population lives in cities with a total settlement area of 2%. This ratio is expected to increase to 70% in 2050 [39]. Since population density is a parameter that directly affects the use of water resources [40, 41], domestic WF analyzes are important for more regular and efficient use of water resources [42]. In this context, when the population parameters of Qadisiyah governorate are examined, a significant increase has been observed in the last years. In addition to the arid climate of the region, the increase in population and agricultural production capacity caused a large part of the water that needs to be met from blue water. However, the decline in water capacity across the governorate has been reported in many studies [2, 5, 43]. Therefore, a comparative analysis of water-using sectors and processes throughout the governorate will support the governorate's water management and planning efforts.

The main purpose of the present study is to analyze the agricultural WF of Qadisiyah governorate based on production. In this context, blue and green WFs of agricultural and livestock sectors were calculated. Gray WF has not been studied due to the limited water pollution data [44]. The present study has been conducted for the years 2010-2020 and uses relatively recent data. On the other hand, when calculating the agricultural WF, a wide range of crops cultivated in the region was used and all the districts of Qadisiyah were evaluated together. According to the authors' knowledge, the present study is the first provincial WF analyzes in Iraq. However, several studies model the WFs of different countries or basins globally, and it is possible to make some inferences about the WF of our country or the study area from these studies. Chapagain and Hoekstra's study [17] is one of the first analyzes on this subject, however, in this study, the inputs (crop characteristics, evapotranspiration amounts, crop production statistics) used to model WFs are not high resolution and average data of countries between 1997-2001 are used. On the other hand, in the studies conducted by Mekonnen and Hoekstra [31, 45], high-resolution spherical WF maps were produced in light of the estimated information obtained from international databases. A study conducted by Mekonnen and Hoekstra developed by the basic data was analyzed taking WF of production and consumption in many countries including Iraq [21, 29].

The main advantages of this study over the above studies are; working on a specific area, using current data produced by local authorities, and choosing a broad period (10 years). Accordingly, it was paid attention to the use of data developed by local authorities such as measured or calculated spatial precipitation averages, crop coefficients, crop production, animal husbandry water consumption statistics for the study area. In the literature, the number of governorate-scale WF studies is quite limited compared to the basin and country studies. In the study conducted by Vanham and Bidoglio [46], the WF of the Milan governorate was analyzed, and the amount and proportions of the total water used in domestic, industrial, and agricultural production and consumption were found to be significantly dependent on the outside. Hoff *et al.* [42] discussed the details of water imported by the governorates of Berlin, Delhi, and Lagos in the context of virtual water trade.

In addition, for different governorates and cities; industrial water footprint and optimum use of water [47], WF of the energy sector in large cities [48], the impact of urbanization on WF [49-51], WF of regions with low urbanization rate and rich water resources [52, 53], urban-based agricultural [54], domestic [55], groundwater footprint [56], gray and green WF [57] and direct water use [58] WFs were discussed in the literature.

2. Data and Method

2.1. Study Area

The governorate of Qadisiyah is located in the south of Iraq, the governorate area is 8521 km2, the population is 1,291,048 M, and the capital city is Diwaniya. It borders the governorates of Muthanna, Najaf, Wassit, Babil, and Dhi-Qar. It has four districts: Diwaniya, Afaq, Al-Shamiya, and al Al-Hamza. The Euphrates and its main branch, Shamiya, run through the area (Iraqi CSO, 2019), Figure 1.

The climate in the governorate is semi-arid, continental subtropical, winter and spring rainfall occur from November to April, the average annual rainfall is about 100-120 mm, winter is cold, average high temperatures are 15°C in January to 48°C in (July-August). Average low temperatures are 6°C in January to 27°C in July, a significant quantity of water is lost by evapotranspiration (Bishay, 2003; Al-Ansari and Knutsson, 2011). The regional economy is based primarily on government employment, agricultural activities, and very limited industries (Iraqi CSO, 2019).

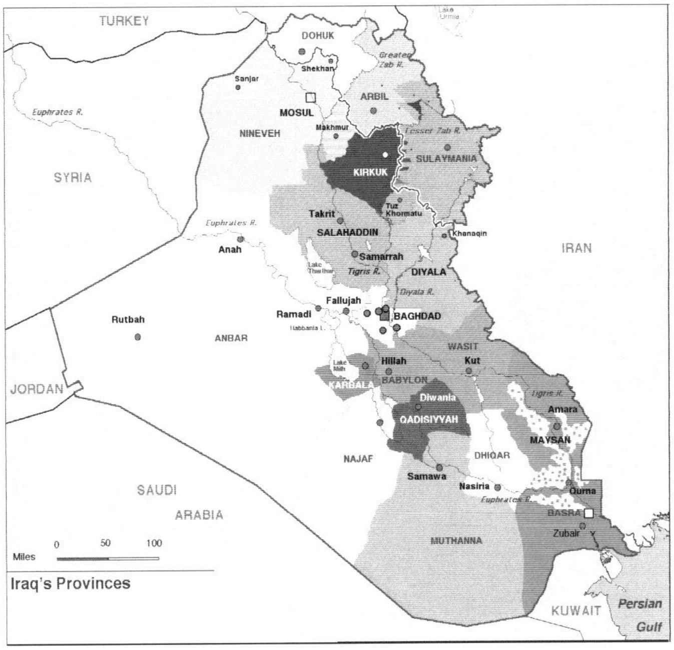

Figure 1. Iraqi map showing Qadisiyah governorate (in dark green).

2.2. Data

Water footprint studies require a wide range of data input. The accuracy of the analysis when investigating the WF of any region is directly related to the quality of data used. Within the scope of this study, many data obtained from [59, 61, 62, 63] for Qadisiyah governorate were used. However, it has been preferred to use actual data measured on-site instead of estimated data in terms of the study area-specific issues such as climate characteristics and crop characteristics. For ten years (2010-2020), meteorological data including monthly maximum, minimum, and average temperatures, relative humidity, wind speed, sundial time, solar radiation, and monthly average rainfall were obtained from the Iraqi Meteorological Organization and Seismology (IMOS) [62] and CLIMWAT 2.0 software [64], which were used to estimate the reference evapotranspiration (ET_0), to model the effective rainfall (Peff) and thus the blue and green evapotranspiration values. Crops coefficients (Kc), sowing and harvesting dates, beginning, development, medium, and last period lengths of the crops are obtained from the relevant national reports [61]. Finally, the total production amounts (tons), cultivated area information (ha), as well as the number of cattle, sheep, and poultry reared throughout the governorate, and the total population has been supplied from [59, 63].

The present study included 27 types of crops; Cereals (wheat, rice, millet, and sesame. Fodder crops, (barley, alfalfa, clover, maize, and sorghum). Vegetables, (eggplant, green peas, green onion, crusty onion, carrots, green pepper, cowpea, okra, cucumber, tomatoes, turnip, lettuce, radish, spinach, and zucchini). Fruits, (watermelons, yellow-melons, and date palm).

Statistics on animals showing the complete annual number of cattle, buffalo, sheep, goats, camels, and poultry had been obtained from the Iraqi Central Statistical Organization [59] and the Iraqi Ministry of Agriculture / Planning and Follow-up Directorate / Statistics Department [63]. Where regional data are unavailable, the worldwide average data was chosen and the national statistics were obtained for the average annual water footprint of each category of animals [65].

2.3. Methods

According to Hoekstra *et al.* water footprint method [16], area-based WF calculations have three main components, they are agricultural WF, WF of livestock, and finally domestic and industrial WF (Eq. 1). The agricultural WF is the largest component of the total WF on any spatial scale and is estimated by considering the total blue and green water requirements of the cultivated crops [66]. On the other hand, when calculating the WF of animal husbandry, the total volume of water processed directly and indirectly by livestock raised in the region is used [67]. Direct water consumption, which consists of drinking and potable water, is generally around 2% and 98% of the total water consumption is used in the production of animal feed. In this context, the vast majority of the water used in the livestock sector is similar to the agricultural WF, again water used in the agricultural activities necessary for the production of barley, corn, and alfalfa, or polluted water. Finally, domestic and industrial WF is the smallest component of the total WF [68]. Contrary to what is known by traditional water use accounts, water allocated by municipalities to meet industrial and urban water needs corresponds to only 10% of total water operated by human means [69].

$$TWF = WF_A + WF_L \tag{1}$$

Here, TWF is the total water footprint; WF_A, agricultural WF and WF_L is the WF of livestock.

2.3.1. Crops water footprint

To estimate the agricultural WF, firstly green and blue ET values should be found. Blue ET can be summarized as evapotranspiration from surface and groundwater, and green ET as evapotranspiration from precipitation. Within the scope of this study, long term meteorological data, mean rainfall average, and characteristics of cultivated crops have been defined as the input to CROPWAT 8.0 program to find evapotranspiration value of each crop, a program is software developed by the Food and Agriculture Organization of the United Nations (FAO) to model the irrigation water requirements of crops [70]. To calculate the reference evapotranspiration (ET_0), the Penman-Monteith method (Eq. 2) was applied to long-term climate data by using the mentioned program. Crop evapotranspiration (ETc) values were calculated by using the CROPWAT 8.0 program by using crop coefficients (Kc), growth period lengths, and ET_0 values (Eq. 3). Blue and green ET values were estimated and the crop water requirement (CWR) option of the related software was preferred. The program used for the calculation of crop water use (CWU) (Eq. 4 and 5) and the USDA-SC method were used for modeling the effective precipitation (P eff) [71].

$$ET_0 = \frac{0.408\Delta(Rn-G)+\gamma\frac{900}{T+273}u^2(es-ea)}{\Delta+\gamma(1+0.34\,u2)} \tag{2}$$

$$ET_c = K_c \times ET_0 \tag{3}$$

$$ET_{blue} = \max\,(0, ET_c - P_{eff}) \tag{4}$$

$$ET_{green} = \min\,(ET_c, P_{eff}) \tag{5}$$

Where, ET_0 is the reference evapotranspiration (mm/day), Rn is the net radiation, G is the soil heat flux (MJ/m^{2}day), Δ is the slope of the vapor pressure curve (kPa/°C), γ is the psychometric constant (kPa/°C). T is the mean daily air temperature (°C) at 2 m height, u^2 is the wind speed at 2 m height (ms^{-1}), es, and ea is the saturation and actual vapor pressure (kPa), respectively. Kc; plant coefficients, ETc; annual evapotranspiration value of the crop (mm/year), P. eff; effective rainfall (mm), and finally, ET_{blue} and ET_{green} are defined as blue and green evapotranspiration values (mm/year), respectively.

After calculating the crop water use (CWU) values (Eq. 6-7), the virtual water contents (VW_C) of the crops were estimated (Eq. 8-9). Virtual water content refers to the total water consumed or operated in the background of a product or process and is generally defined as the volume of water spent per unit product (m^3/ton). In this context, the blue and green VW_C values of each crop in the study area were calculated separately (Eq. 10) and water production of each crop was estimated by using crop cultivated areas and production quantities, and finally the total agricultural WF of Qadisiyah governorate (Eq. 11) calculated. The WF of livestock and poultry production depends on the account of cattle, goats, sheep, camels, and poultry [65, 72].

$$CWU_{blue}\left(\frac{m3}{ha}\right) = 10\sum ET_{blue} \tag{6}$$

$$CWU_{green}\left(\frac{m3}{ha}\right) = 10\sum ET_{green} \tag{7}$$

$$VWC_{blue}\left(\frac{m3}{ha}\right) = \frac{CWU_{blue}}{Y} \tag{8}$$

$$VWC_{green}\left(\frac{m3}{ha}\right) = \frac{CWU_{green}}{Y} \tag{9}$$

$$VWC = VWC_{blue} + VWC_{green} + VWC_{grey} \tag{10}$$

$$WF\,(m^3) = \sum VWC_i\left(\frac{m3}{ton}\right) xc_i(ton) \tag{11}$$

2.3.2. *WF of date's production*

The WF was estimated on the basis that the total annual net use of the date-palm water is around 70 m^3/tree, irrigation 85% (59.5m^3/tree), and 15% rain (10.5m^3/tree) [73, 74, 75].

2.3.3. *Water footprint of livestock and poultry production.*

The overall WF was calculated in the governorate by collecting the number of animals in a given category (or per ton of weight for poultry) multiplied by the annual water footprint of the category live animal (m^3/yr) [65].

The majority of cows are raised for dual purposes in rural Iraq; thus, the total number was evenly divided into dairy and beef cattle [63]. Buffaloes were classified as dairy cattle, camels classified as horses [76], Table 6.

The method for measuring the livestock WF has been presented by [17] as follows using equations (12, 13, and, 14):

$$WF_a = WF_{direct} + WF_{indirect} \tag{12}$$

Where WF_a is the WF of the live animal (m^3/ton), WF_{direct} is the volume of water for drinking, cleaning, or other services during the animal's lifetime. $WF_{indirect}$ is the amount of water used in animal feed production.
WF_{direct} accounts for only 1% of the total WF of livestock [77]. Following the processes of [78], live animal direct and indirect WF are estimated as follows:

$$WF_{direct} = \frac{\int_{birth}^{slaughter} q_a \, dt}{W_a} \tag{13}$$

$$WF_{indirect} = \frac{\int_{birth}^{slaughter} \{\sum_{c=1}^{n} WF_c \cdot Feed_{a\,c}\} dt}{W_a}, \tag{14}$$

Where qa, WF_c, $Feed_{a,c}$, and W_a are the daily water consumption volume (m^3/day), the water footprint of feed $_c$ (m^3/ton), the daily animal consumption of a crop $_c$ (ton/day), and the average weight before the slaughter of live animal a (ton), respectively. The basic data required for animal production was obtained from [63] and the daily consumption of water was obtained from [65].

3. Results and Discussion

3.1. Comparison of WFs of Different Sectors

The total annual average WF of the agricultural production and animal husbandry has been calculated as 995,020,883 Mm^3/yr for 10 years average from 2010 to 2020. The amounts of water in each sector are visualized in Table 1 and Figure 2. Accordingly, the water used within the scope of agricultural (crop production) in Qadisiyah is 383,668,376 m^3 (crops and dates without fodder) and this amount corresponds to 31.3 % of the total agricultural WF.
The agricultural WF is the largest water volume used in the area. The reason for this is that the amount of evapotranspiration (ET) required by vegetation is high. More than 70% of water is extracted as a result of human processes on a global scale [79]. In addition, the agricultural sector is responsible for 90% of total clean water use worldwide [80]. It has been confirmed by similar studies that the WF of agricultural activities is relatively high, especially in arid regions with continental climates [17, 36].

Table 1. The agricultural WF components of Qadisiyah Governorate (m^3/yr) (2010-2020).

Agricultural				
Crops without fodder	Date Palms	Livestock	Poultry	Total
826,541,975	70,595,980	233,878,241	184,185,425	1,500,348,813

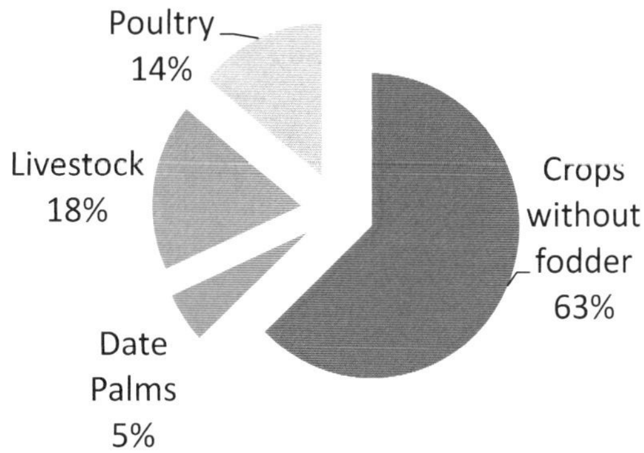

Figure 2. The percentages of agricultural WF components of Qadisiyah Governorate (2010-2020).

The WF of animal husbandry aims to produce meat, milk, or other products from cattle, sheep, goats, camels, and poultry [81, 82] and the total volume of water is used directly and indirectly [83]. The total water used for the production of fodders (forage) crops corresponds to approximately one-third of all agricultural activities on a global scale [84], it is calculated within the livestock WF in the current study. Here, the total WF of animal husbandry (livestock and poultry) was calculated as 311,027,797 m^3 and corresponds to 31.2 % of the total WF, (Table 1 and Figure 2).

Although the WF method takes into account the total volume of water used in all processes, traditional water usage calculations only analyze water drawn directly from surface and underground deposition systems. One of the advantages of WF studies over traditional water usage calculations is the separation of the water used into green, blue and gray components. Green water is mainly evapotranspiration from rainwater; blue water is the amount of evaporation from surface and groundwater. In the present study, the amount of irrigation to be realized within the scope of agricultural activities was included in blue water with the CWR option mentioned in the CROPWAT 8.0 program considering the irrigation water requirement (IWR) of the crops. According to the average results obtained for the 2010-2020 time period, it was calculated that the blue WF represents 83 % of the total crops WF, while the green WF is 17 %, (Table 2). The high values of the blue component of the WF are due to the important role that irrigated agriculture plays, irrigation, and other agricultural activities account for almost 75% of the water use in Iraq [14]. Since it is the most water-consuming activity compared to other sectors, the water use issues of agricultural production in the study area are discussed in this section.

For the average rainfall season in Qadisiyah, the WF blue and green components of the production of the crops were found to be 446,683,192 (83%) and 90,235,105 (17%) million m^3, respectively. It is known that the green water ratio is 78 % in agricultural activities on a global scale [32]. Although the blue WF is expected to be high in the continental climate, this results in excessive use of surface water, limiting agricultural activities, especially in areas with insufficient water resources. Therefore, more effective irrigation methods should be put in place to reduce the amount of blue ET.

Considering the 27 different crops studied here, the annual average WF and rate of each crop for the 2010-2020 period are given in (Table 2). Accordingly, the majority of water use throughout the governorate is based on the production of cereals such as wheat, barley, and rice.

Wheat production has special importance in the study area [85]. It was calculated that the WF of the wheat crop with a total water processing capacity of 137,286,724 million m^3 in the study area corresponds to 25.6 % of the crops in the region and 13.8 % of the total WF. The ratio of green water used for wheat production in the study area was calculated as 23.84 %. When the world average is considered, this rate is reported to be around 70% [30].

Crops with a relatively high footprint of blue water have been identified as barley, wheat, and alfalfa, these three crops account for about 59.2% of the surface water consumed for crops irrigation and 26.6% of all water consumes in the region with 264,644,191 million m^3 of freshwater. Therefore, increasing the irrigation efficiency of these products should be taken into account by the relevant authorities in particular. However, the virtual water content of these products should also be discussed concerning future planning. The virtual water content (VWC) refers to the total water used in the background of the products and is defined as the volume of clean water (m^3) per unit product (ton). The high VWC of any product is directly related to the yield in the region where the product is grown [86]. In terms of product-based strategies to be developed in the plans for the sustainability of water resources, the VWCs of these products in different regions of Iraq, their yields, economic returns, dependence on foreign countries, etc. all cases should be examined together and decided.

3.2. Agricultural water footprint

3.2.1. Crops water footprint

Taking into account 27 different crop types specified in section 2.2, the governorate crop production WF was projected for 2010–2020. Table 2 displays the corresponding annual WFs for the crops being examined. The annual average WF of Qadisiyah for crops is calculated at 536,918,297 Mm3/yr including crops for fodder. Fodder crops have the largest WF, which correlates to 42 percent of the total WF crop with about 225,679,481 Mm^3/yr. Barley holds the highest WF of 139,575,569 Mm^3/yr followed by wheat with 137,286,724 Mm^3/yr, date palms 72,429,560 Mm^3/yr, and alfalfa 56,603,268 Mm3/yr. Fodder crops are calculated within the livestock WF in this study. Other crops with relatively insignificant water footprints were included, (Table 2 and Figure 3).

The blue WF accounts for 83.2 % of the overall WF crops, while the green WF accounts for 16.8 %. The high values of the WF's blue section are due to the important role played by irrigated agriculture, irrigation and other agricultural activities account for nearly 75% of Iraq's water use.

Table 2. The WF annual average values of crop production in Qadisiyah Governorate (2010-2020).

	Production		Area	Water Footprint (m^3/ton)			WF of Production (Mm^3/yr)		
	ton/yr	ton/ha		blue	green	total	blue	green	total
Cereals (60% of the TWF)									
Wheat	270,62	2.544	106,34	796	297	1,093	215,417,50	80,375,625	295,793,125
Rice	100,10	4.372	22,89	3,318	9	3,327	332,155,02	900,963	333,055,989
Sesame	1,639	1.074	1,526	9,115	414	9,529	14,939,485	678,546	15,618,031
Millet	755	0.527	1,431	7,795	840	8,635	5,885,225	634,200	6,519,425
Fodder Crops (24% of the TWF)									
Barley	120,11	1.762	68,16	1,083	401	1,484	130,080,21	48,164,511	178,244,724
Maize	3,886	2.406	1,615	2,540	191	2,731	9,870,440	742,226	10,612,666
Sorghum	2,157	1.06	2,034	633	358	991	1,365,381	772,206	2,137,587
Alfalfa	40,49	14.33	2,824	1,197	79	1,276	48,467,727	3,198,789	51,666,516
Clover	9,094	9.532	954	1,801	118	1,919	16,378,294	1,073,092	17,451,386
Vegetables (10% of the TWF)									
Eggplant	5,015	6.005	835	843	63	906	4,227,645	315,945	4,543,590
Green peas	1,754	3.077	570	1,390	58	1,448	2,438,060	101,732	2,539,792
Green onion	1,182	8.889	133	481	20	501	568,542	23,640	592,182
Crusty onion	2,285	4.445	514	962	40	1,002	2,198,170	91,400	2,289,570
Carrots	74	7.789	9.5	549	23	572	40,626	1,702	42,328

Water melons	32,74(	6.017	5,442	1,158	63	1,221	37,919,868	2,062,998	39,982,866
Yellow melons	25,48(	8.353	3,050	834	46	880	21,250,320	1,172,080	22,422,400
Green pepper	174	2.23	78	3,109	200	3,309	540,966	3,4800	575,766
Cowpea	2,462	4.104	600	1,233	92	1,325	3,035,646	226,504	3,262,150
Okra	5,008	6.713	746	754	57	811	3,776,032	285456	4,061,488
Cucumber	11,10(	9.784	1,135	518	39	557	5,752,908	433,134	6,186,042
Tomatoes	6,367	12.784	498	848	27	875	5,399,216	171,909	5,571,125
Turnip	604	1.967	307	1,348	174	1,522	814,192	105,096	919,288
Lettuce	6,050	2.318	2,610	1,144	147	1,291	6,921,200	889,350	7,810,550
Radish	65	1.625	40	3,113	233	3,346	202,345	15,145	217,490
Spinach	276	1.586	174	1,672	216	1,888	461,472	59,616	521,088
Zucchini	1,102	1.751	629	2,889	216	3,105	3,183,678	238,032	3,421,710
Date palms (m^3/tree) (6% of the TWF)									
	ton/y(	ton/tre(	Dates						
	32,28(	0.032	,008,5	59	11	70	59,502,326	11,093,654	70,595,980
Total			233,12				932,792,50:	153,862,35:	**1,086,654,854**
WF without fodder crops							726,630,44(	99,911,527	**826,541,975**

Water footprints overall crops rely on green water in most countries of the world [84]. Nevertheless, the blue water content of crops was obtained as 83.2 % while the green WF is 16.8 % in this governorate. The hot climate, the lack of rain, and the high evapotranspiration levels increase the demand for water from crop production [88]. Irrigation is needed in the arid regions of Iraq due to the lack of adequate rain, especially during the summer, which is likely to result in the rise of the blue water portion of crops. All the crops rely primarily on the blue water of the study area, except for wheat and barley which are winter crops and partially lower the blue water share.

Ewaid *et al.*, [89] study's results indicate that the WF for Iraqi rice in 2017 was 3091 m^3/ton, higher than the world average (1325 m^3/ton). Approximately 821,737,881 Mm3/yr of water have been used to irrigate rice-growing areas to generate 265,852 tons, green WF is nearly non-existent as rice grows in the dry and hot season, and the blue WF is dominant.

Wheat is one of the world's most harvested cereals and has been domesticated in Iraq [85]. Overall, 70% of global wheat production's WF is based on green water [45]. In this study, the wheat production's share of green water is significantly below the global average. The green water provides 27% of the total water requirement for wheat production, which slightly reduces the reliance on blue water in the area. The total WF of wheat produced in all Iraq is estimated at 4,120,160,334 Mm^3/yr [85]. Also, the governorate of Qadisiyah is an important area for wheat production. The total wheat production has an annual WF of 137,286,724 Mm^3/yr.

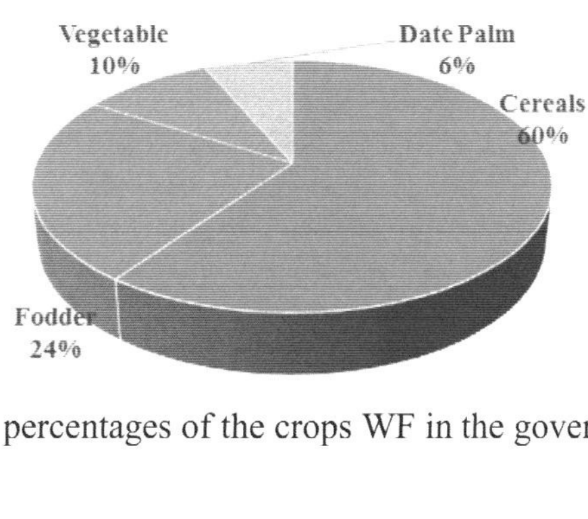

Figure 3. The primary percentages of the crops WF in the governorate (2010-2020).

IOP Conf. Series: Earth and Environmental Science 1029 (2022) 012025 doi:10.1088/1755-1315/1029/1/012025

Policymakers should carefully consider the blue, green, and total WF of each crop for better practices in water management. As a practical approach, crops with lower WFs like vegetables could be encouraged to preserve limited freshwater resources, the economic and nutritional value of crops, food safety issues, and national virtual water content should also be considered to make a better decision on agricultural management of the study area [88].
It is therefore proposed that available water sources should be used and that the main objectives should be to minimize WFs, decrease the amount of evaporation, collect rainwater and provide more sophisticated irrigation systems to reduce WF [14]. Authorities must promote food, goods, and services produced and used at lower WF rates [85]. Increasing awareness, product labeling, and other similar methods may lead to the solution to the problem. Ignoring production methods that have lower productivity levels may increase the efficiency of water use [88].
Blue WF contributed to the governorate's largest share due to the climatic conditions. Cereal production has absorbed half the water withdrawal from the agricultural sector. It is proposed that the production of rice, wheat, and barley be decreased since the productivity of these crops depends comparatively on limited supplies of blue water. In addition, the agricultural sector decision-makers should minimize crop production with high WF, to reduce the demand for water resources. WF crop production depends on productivity and climate conditions. It is therefore recommended that cereal farming should be reduced because the output of these crops is heavily dependent on the limited resources of blue water, vegetable production requires little water, so it should be grown rather than crops consuming much like rice.

3.2.2. Water footprint of date's production
Dates are one of central and southern Iraq's main fruit crops. Date palm plantations have deteriorated in the last 50 years due to negligence and widespread degradation resulting from an increase in human and domestic animal populations [90].
The date palm tree can be used as a naturally renewable resource and can provide a wide range of products [91, 92]. Date palms can grow in a very dry, warm climate and are fairly resistant to salty and alkaline soils, require a long, hot summer with little rain and very low humidity, but with plenty of groundwater close to the surface or irrigation [93]. Table 4 shows the WF of the dates studied here.

Table 3. The date palms average WF annually in Qadisiyah Governorate (2009-2018).

	Production		Trees	Water (m^3/tree)		Footprint	WF of Production (Mm^3/yr)		
	ton/yr	ton/tree		blue	green	total	blue	green	total
Qadisiyah	32,289	0.032	1,008,514	59	11	70	59,502,326	11,093,654	70,595,980

3.2.3. Water footprint of livestock and poultry production.
The volume of cattle and poultry WF reflecting what is consumed for direct and indirect purposes by the animals studied. In this analysis, the average intake of water per animal amount or ton of animal weight for poultry was used to calculate the livestock and poultry WF [65]. Table 5 shows some statistics of poultry production [63] and Table 6 displays the provincial WF for the production of livestock and poultry. The average WF of livestock production is around 271,630,281 Mm3/yr, which corresponds to 28 percent of total WF, (Figure 2).

Table 4. Statistics on poultry production in Qadisiyah Governorate (MOA, 2021).

	projects	sets	Chickens Numbers 1000	Weight per chicken	Quantity of sold chickens (ton) poultry	breeding
Qadisiyah	198	4	7000	1963 g	13741	130

Table 5. The WF of livestock and poultry production in Qadisiyah Governorate (2010-2020).

Animal category	m^3/yr/animal	Number of animals	WF
Beef cattle	630	75,547	47,594,610
Dairy cattle	2,056	75,547	155,324,632
Broiler chickens	26	7,000,000	182,000,000
Layer chickens	33	66,225	2,185,425
Camels	1,599	5,109	8,169,291
Sheep	68	294,283	20,011,244
Goats	32	86,827	2,778,464
Total			418,063,666

Figure 4 indicates the percentages of the governorate's WF for animal production. The largest proportion of water used for animal production was found to be linked to dairy cattle (57 percent), (Table 6).

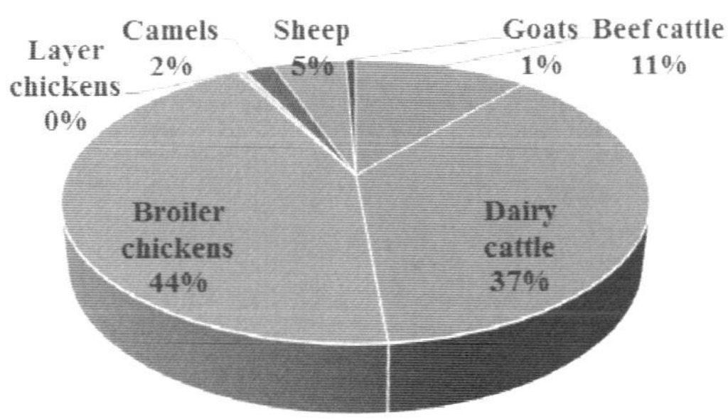

Figure 4. The percentages of the animal production WF (2009-2018).

Animal feeding indirectly uses the highest demand for freshwater for animal production corresponding to 98 percent of the WF [72]. About one-third of the total agronomic freshwater is used to produce feeds and nearly half of WF's global livestock production is tied to beef (33%) and dairy cattle (19%); 20% of WF is linked to the raising of chickens (broiler and layer). On the other hand, the majority of the worldwide consumption of livestock freshwater is green water, which corresponds to 87% of the overall animal production WF [65]. In this study, the blue-water dependency on livestock activity is higher than the worldwide average. Calculating the water productivity of animal production provides a clear picture of the situation and the prospects for development for more efficient use of water resources and higher water productivity in animal production.

4. Conclusions

According to the results of the study, the average water footprint of the governorate is estimated to be 995,020,883 million m^3 per year. It has been determined that the majority of the water used is processed by crop cultivation including fodders and date palms (536,918,297 m^3) with a rate of 54 %. Livestock in the study area; industrial and domestic water footprints were also calculated. Total water usage and water footprint of 27 different plant species cultivated in the region were calculated. The most important agricultural products in terms of water use; barley, wheat, and alfalfa. These products are responsible for 59.2% of the crop's water footprint.

The most important product grown in the study area was determined as wheat which has 13.8 % of the total water footprint of the governorate. When similar studies in the literature are considered, it is concluded that the use of blue water in the region is high. This situation is explained especially by the increase in evapotranspiration amounts since the study area is located in the hot climate zone. The reasons for the high water footprint of any region are mainly due to the climatic characteristics, virtual water contents of produced/used products and lifestyle of people in the region, food habits, and so on [18]. However, the development of strategies and actions to reduce the water footprint and to reduce the amount of blue water used in agricultural activities is of particular importance for the sustainability of water resources, especially in arid or semi-arid regions. For more efficient and rational management of water, the virtual water content and the use of blue water of the major products in Iraq should be reviewed. In this context, increasing the water footprint studies will contribute to the development of detailed water and agricultural product policies.

References

[1] Hoekstra, A. Y., Mekonnen, M. M., Chapagain, A. K., Mathews, R. E., & Richter, B. D. (2012). Global monthly water scarcity: blue water footprints versus blue water availability. PloS one, 7(2), e32688.

[2] Rahi, K. A., Al-Madhhachi, A. S. T., & Al-Hussaini, S. N. (2019). Assessment of Surface Water Resources of Eastern Iraq. *Hydrology*, *6*(3), 57.

[3] Ewaid, S. H., Abed, S. A., & Kadhum, S. A. (2018). Predicting the Tigris River water quality within Baghdad, Iraq by using water quality index and regression analysis. *Environmental technology & innovation*, *11*, 390-398.

[4] Al-Ansari, N., Knutsson, S. (2011). Toward prudent management of water resources in Iraq. J. Adv. Sci. Eng. Res. 2011 (1), 53–67.

[5] Al-Ansari, N. (2013). Management of water resources in Iraq. *Engineering*, *5*(6), 667-684.

[6] Distefano, T., & Kelly, S. (2017). Are we in deep water? Water scarcity and its limits to economic growth. *Ecological Economics*, *142*, 130-147

[7] Salman, D. A., Amer, S. A., & Ward, F. A. (2014). Water appropriation systems for adapting to water shortages in Iraq. *JAWRA J. of the American Water Resources Association*, *50*(5), 1208-1225.

[8] Mekonnen, M.M., Hoekstra, A.Y. (2016). Sustainability: Four billion people facing water scarcity. Sci. Adv. 2.

[9] Almamalachy, Y. S., Al-Quraishi, A. M. F., & Moradkhani, H. (2020). Agricultural drought monitoring over Iraq utilizing MODIS products. In *Environmental Remote Sensing and GIS in Iraq* (pp. 253-278). Springer, Cham.Abed, S. A., Ewaid, S. H., & Al-Ansari, N. (2019). Evaluation of Water Quality in the Tigris River within Baghdad, Iraq using Statistical Techniques. *Journal of Physics: Conference Series* (Vol. 1294, No. 7, p. 072025). IOP Publishing.

[10] Ewaid, S.H.; Kadhum, S.A.; Abed, S.A (2019a). Development and evaluation of irrigation water quality guide using IWQG V.1 software: A case study of Al-Gharraf Canal, Southern Iraq. Environ. Technol. Innov. 2019, 13, 224–232.

[11] Joint, F. A. O. (2018). Challenges and Opportunities for Crop Production in Dry and Saline Environments in the ARABIA Member States (No. IAEA-TECDOC--1841). Joint FAO/IAEA Division of Nuclear Tech. in Food and Agri.
[12] Clark, G. (2007). Evolution of the global sustainable consumption and production policy and the United Nations Environment Program's (UNEP) supporting activities. *Journal of cleaner production*, *15*(6), 492-498.
[13] Ewaid, S. H., Abed, S. A., and Al-Ansari, N. (2019b). Crop Water Requirements and Irrigation Schedules for Some Major Crops in Southern Iraq. Water, 11(4), 756.
[14] Veettil, A. V., & Mishra, A. K. (2018). The potential influence of climate and anthropogenic variables on water security using blue and green water scarcity, Falkenmark index, and freshwater provision indicator. Journal of environmental management, 228, 346-362.
[15] Hoekstra, A. Y., Chapagain, A. K., Mekonnen, M. M., & Aldaya, M. M. (2011). The water footprint assessment manual: Setting the global standard. Routledge.
[16] Chapagain, A.K., Hoekstra, A.Y. (2004). Water footprint of nations. Volume 1: Main report. Value Water Res. Rep. Ser. 1, 1–80.
[17] Hoekstra, A. Y., & Chapagain, A. K. (2006). Water footprints of nations: water use by people as a function of their consumption. Integrated assessment of water resources and global change (pp. 35-48). Springer, Dordrecht.
[18] Hoekstra A.Y., Hung P.Q. (2002). A quantification of virtual water flows between nations in relation to international crop trade, Water Research, 49 (11), 203–209.
[19] Mekonnen, M. M., & Hoekstra, A. Y. (2010a). A global and high-resolution assessment of the green, blue, and grey water footprint of wheat. Hydrology and earth system sciences, 14(7), 1259-1276.
[20] Oki, T., & Kanae, S. (2006). Global hydrological cycles and world water resources. *science*, *313*(5790), 1068-1072.
[21] Vanham, D. (2012). A holistic water balance of Austria–how does the quantitative proportion of urban water requirements relate to other users?. Water Science and Technology, 66(3), 549-555.
[22] Ercin, A. E., & Hoekstra, A. Y. (2014). Water footprint scenarios for 2050: A global analysis. Environment international, 64, 71-82.
[23] Quinteiro, P., Rafael, S., Villanueva-Rey, P., Ridoutt, B., Lopes, M., Arroja, L., & Dias, A. C. (2018). A characterization model to address the environmental impact of green water flows for water scarcity footprints. Science of the Total Environment, 626, 1210-1218.
[24] Novoa, V., Ahumada-Rudolph, R., Rojas, O., Munizaga, J., Sáez, K., & Arumí, J. L. (2019). Sustainability assessment of the agricultural water footprint in the Cachapoal River basin, Chile. Ecological indicators, 98, 19-28.
[25] Johnson, M. B., & Mehrvar, M. (2019). An assessment of the grey water footprint of winery wastewater in the Niagara Region of Ontario, Canada. Journal of cleaner production, 214, 623-632.
[26] Muratoglu, A. (2019). Water footprint assessment within a catchment: A case study for Upper Tigris River Basin. Ecological Indicators, 106, 105467.
[27] Qasemipour, E., & Abbasi, A. (2019). Virtual Water Flow and Water Footprint Assessment of an Arid Region: A Case Study of South Khorasan Governorate, Iran. *Water*, *11*(9), 1755.
[28] Mekonnen M.M., Hoekstra A.Y., (2011a) National water footprint accounts: The green, blue and grey water footprint of production and consumption, Volume 1: Main report 50, Education, UNESCO-IHE Institute for Water Education, Delfth, The Netherlands.
[29] Mekonnen, M.M., and Hoekstra, A.Y. (2010a). The green, blue and grey water footprint of crops and derived crop products, Value of Water Research Report Series No. 47, UNESCO-IHE, Delft, the Netherlands.

[30] Mekonnen, M. M., & Hoekstra, A. Y. (2011b). The green, blue, and grey water footprint of crops and derived crop products. Hydrology and Earth System Sciences, 15(5), 1577-1600.
[31] Mekonnen, M. M., & Hoekstra, A. Y. (2014). Water footprint benchmarks for crop production: A first global assessment. Ecological indicators, 46, 214-223.
[32] Zhang, Y., Huang, K., Yu, Y., Hu, T., & Wei, J. (2015). Impact of climate change and drought regime on the water footprint of crop production: the case of Lake Dianchi Basin, China. Natural Hazards, 79(1), 549-566.
[33] Zhuo, L., Mekonnen, M. M., & Hoekstra, A. Y. (2014). Sensitivity and uncertainty in crop water footprint accounting: a case study for the Yellow River basin. Hydrology and earth system sciences, 18(6), 2219-2234
[34] Dumont, A., Salmoral, G., & Llamas, M. R. (2013). The water footprint of a river basin with a special focus on groundwater: The case of Guadalquivir basin (Spain). Water Resources and Industry, 1, 60-76.
[35] Aldaya, M. M., & Llamas, M. R. (2008). *Water footprint analysis for the Guadiana river basin* (Vol. 3). Delft, The Netherlands: UNESCO-IHE
[36] Hu, T., Huang, K., Yu, Y., Zhang, X., Xu, Y., & Wang, X. (2016). Measuring water footprint on a lake basin scale: a case study of lake dianchi, China. CLEAN–Soil, Air, Water, 44(10), 1296-1305
[37] Hoekstra, A. Y. (2019). *The water footprint of modern consumer society*. Routledge.
[38] Pfister, S., Bayer, P., Koehler, A., & Hellweg, S. (2011). Environmental impacts of water use in global crop production: hotspots and trade-offs with land use. *Environmental science & technology*, *45*(13), 5761-5768.
[39] Chen, J., Shi, H., Sivakumar, B., & Peart, M. R. (2016). Population, water, food, energy, and dams. Renewable and Sustainable Energy Reviews, 56, 18-28.
[40] Kinouchi, T., Nakajima, T., Mendoza, J., Fuchs, P., & Asaoka, Y. (2019). Water security in high mountain cities of the Andes under a growing population and climate change: A case study of La Paz and El Alto, Bolivia. Water Security, 6, 100025.
[41] Hoff H., Döll P., Fader M., Gerten D., Hauser S., Siebert S. (2014). Water footprints of cities indicators for sustainable consumption and production, Hydrology and Earth System Sciences, 18, 213-226.
[42] Abdullah, M., Al-Ansari, N., & Laue, J. (2019). Water Resources Projects in Iraq, Reservoirs in The Natural Depressions: Water Resources Projects in Iraq, Reservoirs in The Natural Depressions. *Earth Sciences and Geotechnical Engineering*, *9*(4), 137-152.
[43] Iraqi CSO (2019). Iraqi Central Statistical Organization yearly statistics, Ministry of Planning Baghdad, at http://www.cosit.gov.iq/en/
[44] Mekonnen, M.M. and Hoekstra, A.Y. (2011 b) National water footprint accounts: the green, blue and grey water footprint of production and consumption, Value 2 of Water Research Report Series No. 50, UNESCO-IHE, Delft, the Netherlands.
[45] Vanham, D. A., & Bidoglio, G. (2013). The water footprint of Milan. Water Science and Technology, 69(4), 789-795.
[46] Xu, M., Li, C., Wang, X., Cai, Y., & Yue, W. (2018). Optimal water utilization and allocation in industrial sectors based on water footprint accounting in Dalian City, China. Journal of cleaner production, 176, 1283-1291.
[47] Liao, X., Zhao, X., Jiang, Y., Liu, Y., Yi, Y., & Tillotson, M. R. (2019). Water footprint of the energy sector in China's two megalopolises. Ecological modelling, 391, 9-15.
[48] Feng, L., Chen, B., Hayat, T., Alsaedi, A., & Ahmad, B. (2017). The driving force of water footprint under the rapid urbanization process: a structural decomposition analysis for Zhangye city in China. Journal of Cleaner Production, 163, S322-S328.

[49] Zhang, F., Zhan, J., Li, Z., Jia, S., & Chen, S. (2018). Impacts of urban transformation on water footprint and sustainable energy in Shanghai, China. Journal of cleaner production, 190, 847-853.
[50] Fang, K., Zhang, Q., Yu, H., Wang, Y., Dong, L., & Shi, L. (2018). Sustainability of the use of natural capital in a city: Measuring the size and depth of urban ecological and water footprints. Science of the Total Env., 631, 476-484.
[51] Qian, Y., Dong, H., Geng, Y., Zhong, S., Tian, X., Yu, Y., ... & Moss, D. A. (2018). Water footprint characteristic of less developed water-rich regions: the case of Yunnan, China. Water Research, 141, 208-216.
[52] Luo, P., Yang, Y., Wang, H., Gu, Y., Xu, J., & Li, Y. (2018). Water footprint and scenario analysis in the transformation of Chongming into eco-island. Resources, Conservation and Recycling, 132, 376-385.
[53] Zhao, X., Tillotson, M. R., Liu, Y. W., Guo, W., Yang, A. H., & Li, Y. F. (2017). Index decomposition analysis of urban crop water footprint. Ecological Modelling, 348, 25-32.
[54] Cai, B., Liu, B., & Zhang, B. (2019). Evolution of Chinese urban household's water footprint. Journal of cleaner production, 208, 1-10.
[55] Pérez, A. J., Hurtado-Patiño, J., Herrera, H. M., Carvajal, A. F., Pérez, M. L., Gonzalez-Rojas, E., & Pérez-García, J. (2019). Assessing sub-regional water scarcity using the groundwater footprint. Ecological indicators, 96, 32-39.
[56] Ding, X., Wang, S., & Chen, B. (2019). The Blue, Green, and Grey Water Consumption for Crop Production in Heilongjiang. Energy Procedia, 158, 3908-3914.
[57] Manzardo, A., Loss, A., Fialkiewicz, W., Rauch, W., & Scipioni, A. (2016). Methodological proposal to assess the water footprint accounting of direct water use at an urban level: A case study of the Municipality of Vicenza. Ecological indicators, 69, 165-175.
[58] Iraqi CSO (2019). Iraqi Central Statistical Organization yearly statistics, Ministry of Planning Baghdad, at http://www.cosit.gov.iq/en/
[59] Bishay, F. K. (2003). Towards sustainable agricultural development in Iraq. The transition from relief, rehabilitation, and reconstruction to development.
[60] DEAT (2019). Guide to Agricultural Operations in Iraq; Department of Extension and Agricultural Training, Ministry of Agriculture: Baghdad, Iraq, (In Arabic)
[61] IMOS (2021). Iraqi Meteorological Organization and seismology (IMOS) unpublished data.
[62] MOA (2021). The Iraqi M. of Agriculture, Dept. of Statistics. Annual statistics of agricultural crops.
[63] FAO, CLIMWAT 2.0, Food and Agricultural Organization of the United Nations, http://www.fao.org/land-water/databases-andsoftware/ climwat-for-cropwat/en/.
[64] Mekonnen, M. M., & Hoekstra, A. Y. (2012). A global assessment of the water footprint of farm animal products. Ecosystems, 15(3), 401-415.
[65] Lovarelli, D., Bacenetti, J., & Fiala, M. (2016). Water Footprint of crop productions: A review. Science of the Total Environment, 548, 236-251.
[66] Ran, Y., Lannerstad, M., Herrero, M., Van Middelaar, C. E., & De Boer, I. J. M. (2016). Assessing water resource use in livestock production: A review of methods. Livestock Science, 187, 68-79
[67] Hoekstra, A. Y. (2015). The water footprint of the industry. In Assessing and measuring environmental impact and sustainability (pp. 221-254). Butterworth-Heinemann.
[68] Postel, S. L., Daily, G. C., & Ehrlich, P. R. (1996). Human appropriation of renewable freshwater. Science, 271(5250), 785-788.
[69] FAO (2019). CropWat 8.0 Model, Food and Agricultural Organization of the United Nations, http://www.fao.org/land-water/databases-and-software/cropwat/en/.
[70] USDA–SCS (1993). Irrigation Water Requirements, Chapter 2, part 623 of the National Engineering Handbook, Soil Conservation Service.

[71] Personal Contact (2019). Personal contact with the engineers of the industrial establishments of the governorate.

[72] Zaid, A. and P. F. de Wet (2002). Botanical and systematic description of the date palm. In A. Zaid and E. J. Arias-Jiménez (eds.), *Date Palm Cultivation. FAO Plant Production and Protection Paper 156 Rev. 1.* Rome, Italy

[73] Alamoud, A. I., Mohammad, F. S., Al-Hamed, S. A., & Alabdulkader, A. M. (2012). Reference evapotranspiration and date palm water use in the Kingdom of Saudi Arabia. *International Research Journal of Agricultural Science and Soil Science*, *2*(4), 155-169.

[74] Al-Omran, A. O., Eid, S., & Alshammari, F. (2019). Crop water requirements of date palm based on actually applied water and Penman-Monteith calculations in Saudi Arabia. *Applied Water Sci.*, *9*(4), 69.

[75] Moioli, B., & Borghese, A. (2005). Buffalo breeds and management systems. *Buffalo*
a. *production and research. Rome: Food and Agriculture Organization of the UN*, 51-76.

[76] Dang, Q.; Lin, X.; Konar, M. (2015). Agricultural virtual water flows within the United States. Water Resour. Res. 2015, 51, 973–986.

[77] Mubako, S.T.; Lant, C.L. (2013). Agricultural Virtual Water Trade and Water Footprint of U.S. States. Ann. Assoc. Am. Geogr. 2013, 103, 385-396.

[78] Tigrek, S., & Kibaroglu, A. (2011). The strategic role of water resources for Turkey. In Turkey's Water Policy (pp. 27-42). Springer, Berlin, Heidelberg.

[79] Ercin A.E. (2012) Governance of globalized water resources: The application of water footprint to inform corporate strategy and government policy, PhD Thesis, University of Twente.

[80] Bosire, C. K., Lannerstad, M., de Leeuw, J., Krol, M. S., Ogutu, J. O., Ochungo, P. A., & Hoekstra, A. Y. (2017). Urban consumption of meat and milk and its green and blue water footprints—Patterns in the 1980s and 2000s for Nairobi, Kenya. Science of the total environment, 579, 786-796.

[81] Ibidhi, R., Hoekstra, A. Y., Gerbens-Leenes, P. W., & Chouchane, H. (2017). Water, land, and carbon footprints of sheep and chicken meat produced in Tunisia under different farming systems. Ecological indicators, 77, 304-313.

[82] Hoekstra, A. Y. (2012). The hidden water resource use behind meat and dairy. Animal Frontiers, 2(2), 3-8.

[83] Mekonnen M.M., Hoekstra A.Y. (2011c) The green, blue and grey water footprint of farm animals and animal products. Volume 1: Main Report, UNESCO-IHE Institute for Water Education. UNESCO-IHE Institute for Water Education, The Netherlands.

[84] Ewaid, S. H., Abed, S. A., & Al-Ansari, N. (2019c). Water Footprint of Wheat in Iraq. Water, 11(3), 535.

[85] Ye, Q., Li, Y., Zhuo, L., Zhang, W., Xiong, W., Wang, C., & Wang, P. (2018). Optimal allocation of physical water resources integrated with virtual water trade in water-scarce regions: A case study for Beijing, China. Water Research, 129, 264-276.

[86] Alwash, A. (2018). Towards Sustainable Water Resources Management in Iraq. The report, 45 p. Iraq Energy Institute. www.iraqenergy.org.

[87] Ewaid, S. H., Abed, S. A., & Al-Ansari, N. (2020a). Assessment of Main Cereal Crop Trade Impacts on Water and Land Security in Iraq. *Agronomy*, *10*(1), 98

[88] Ewaid, S. H., Abed, S. A., Chabuk, A., & Al-Ansari, N. (2021). Water Footprint of Rice in Iraq. In IOP Conference Series: Earth and Environmental Science (Vol. 722, No. 1, p. 012008). IOP Publishing.Zabar, A., & Borowy, A. (2012). Cultivation of date palm in Iraq. *Annales Universitatis Mariae Curie-Skłodowska. Sectio EEE, Horticultura*, *22*(1), 39-54.

[89] El-Juhany, L. I. (2010). Degradation of date palm trees and date production in Arab countries: causes and potential rehabilitation. *Australian Journal of Basic and Applied Sciences*, *4*(8), 3998-4010.

[90] Al-Omran, A. O., Eid, S., & Alshammari, F. (2019). Crop water requirements of date palm based on actually applied water and Penman-Monteith calculations in Saudi Arabia. *Applied Water Science*, *9*(4), 69.
[91] Ewaid, S.H., Abed, S.A., 2017. Water quality index for Al-Gharraf river, southern Iraq. Egypt. J. Aquatic Res. 43 (2), 117–122. http://dx.doi.org/10.1016/j. ejar.201703001.
[92] Ewaid, S.H.; Abed, S.A.; Al-Ansari, N. Crop Water Requirements and Irrigation Schedules for Some Major Crops in Southern Iraq. Water 2019, 11, 756.
[93] Ewaid, S.H.; Abed, S.A.; Al-Ansari, N. Water Footprint of Wheat in Iraq. Water 2019, 11, 535.
[94] Ewaid, S.H.; Abed, S.A.; Al-Ansari, N. Assessment of Main Cereal Crop Trade Impacts on Water and Land Security in Iraq. Agronomy 2020, 10, 98.
[95] Salim, M. A. and Abed, S. A. (2019). The first oriental honey buzzard *pernis ptilorhynchus* (Temminck, 1821) in Iraq. Eco. Env. & Cons. 25 (1) ; pp. (1926-1929).
[96] Ewaid, S.H.; Abed, S.A.; Al-Ansari, N.; Salih, R.M. Development and Evaluation of a Water Quality Index for the Iraqi Rivers. Hydrology 2020, 7, 67.
[97] Ewaid, S.H., Al-Farhani, B.F., Abed, S.A. & Al-Ansari, N. (2020). Modeling of trihalomethane compounds formation in Baghdad water supply network. Scientific Review Engineering and Environmental Sciences, 29 (2), 136-144. doi: 10.22630/PNIKS.2020.29.2.12.
[98] Mudhafar A. Salim et al 2020 J. Phys.: Conf. Ser. 1664 012105.
[99] Salam Hussein Ewaid et al 2020 J. Phys.: Conf. Ser. 1664 012143.
[100] Ahmed Sabah Al-Jasimee et al 2020 J. Phys.: Conf. Ser. 1664 012141.
[101] Mudhafar A. Salim and Salwan Ali Abed 2021 IOP Conf. Ser.: Earth Environ. Sci. 735 012021
[102] Salam Hussein Ewaid et al 2021 IOP Conf. Ser.: Earth Environ. Sci. 722 012008
[103] Salam Hussein Ewaid et al 2021 IOP Conf. Ser.: Earth Environ. Sci. 790 012075
[104] Ahmed Alaa Kandoh and Salwan Ali Abed 2021 IOP Conf. Ser.: Earth Environ. Sci. 790 012034
[105] Ahmed Alaa Kandoh et al 2021 IOP Conf. Ser.: Earth Environ. Sci. 790 012073.
[106] Alamoud, A. I., Mohammad, F. S., Al-Hamed, S. A., & Alabdulkader, A. M. (2012). Reference evapotranspiration and date palm water use in the Kingdom of Saudi Arabia. *International Research Journal of Agricultural Science and Soil Science*, *2*(4), 155-169.
[107] Ali, A. S. A., and Hama, N. N. (2016). Integrated management for major date palm pests in Iraq. *Emirates Journal of Food and Agriculture*, 24-33.
[108] Xu, M., & Li, C. (2020). Application of the Water Footprint: Water Stress Analysis and Allocation. Springer.

Synthesis and Characterization of TiO2 – Ag- Cellulose Nanocomposite and evaluation of photo catalytic degradation efficiency for Congo Red

[1]Nidal Hassoun Thamer and [2]Oraas Adnan Hatem

[1,2] Department of Chemistry, College of Science, University of AL- Qadisiyah, Iraq .

Abstract. Photocatalytic degradation of toxic organic contaminants in aquatic environments is a cost-effective, efficient, and long-term solution for environmental restoration. The capacity to fine-tune the structures and compositions of photocatalysts is vital for optimizing photocatalytic decontamination processes, as is a deep understanding of the mechanisms that govern photocatalytic interfacial molecular changes. We created a ternary TiO2/Ag/cellulose nanocomposite particle by hydrothermal process that acts as a photocatalyst for the photodegradation of congo red CR dye into mineralized small molecules

The morphology , size and crystalline structure of the TiO2 – Ag – Cellulose nanocomposite were investigated by scaning electron microscopy SEM , transmission electron microscopy (TEM) and X-ray diffraction XRD respectively . Surface functional groups were analyzed using FTIR analysis. Specific surface area of the composite was characterized by BET-BJH analysis and the elemental composition of the composite was study using EDX analysis . Interestingly , the prepared TiO2 – Ag – Cellulose nanocomposites showed high catalytic degradation activity for CR dye . The photocatalytic degradation of the azo dye was monitored spectrophotometrically , a high degradation efficiency 81,76 % under visible radiation were obtained for composite after 120 min of reaction compared to 30% for pure TiO2 .

Keywords : Titanium dioxide , cellulose , Photo Degradation , nanocomposite , congo red.

1. Introduction

TiO2 has been extensively investigated and is widely used in industry due to its photoactivity efficiency, stability, and low cost [1, 2]. TiO2 nanostructures such as nanotubes, nanocables, and nanobelts have been found by several researchers as effective catalysts[3-5] . specially TiO2 nanotubes have photocatalytic activity, ion-exchange capacity, a large surface area, and a multilayer wall structure, to name a few physicochemical properties [6]. Increase electron transfer and, as a result, their efficiency in many applications with a broad surface area[7] . High purity, crystallinity, and temperature are all required for these applications [8]. TiO2 is a semiconductor with a 3.2 eV indirect band g and a 3.2 eV direct band gap [9, 10]. Particle size has a significant impact on the structure and properties of TiO2 [11-13]. In dye-sensitized solar cells, small particle sizes are preferable because a higher surface area improves the contact area. By mixing the nanoparticles with the dye, charge transmission is improved and charge recombination is decreased [14-16].

TiO2 with a smaller size has a better redox potential for photocatalytic degradation of organic contaminants when exposed to UV light [17]. TiO2 nanoparticles produced in a neutral pH environment can be used as photocatalysts to clean the environment on an industrial scale [18]. Photocatalytic elimination of organic and inorganic impurities on various types of semiconductors has been studied by a number of researchers [19] .
The oxidative degradation of adsorbed colored organic or inorganic compounds is caused by an indirect reaction at the positive holes [20]. which originates in the semiconductor's valence band and then migrates to the catalyst's surface, where it participates in interfacial oxidation-reduction events, but the band gap was reduced in the presence of nanoparticles, resulting in a reduction in the recombination process. When exposed to visible light, TiO2 has been changed to improve its photocatalytic efficiency [21] .
Non-metallic modification necessitates irradiation. It has been attempted to dope metal and In TiO2 and noble metals, TiO2transition metal is employed as a metal doping agent (Ag,Au,Pd,) Because of its added value as an antibacterial agent, silver has gained a lot of attention among the noble metals, and it has the potential to be highly successful [22-34] . The importance of metallic silvers in medical applications and TiO2's antibacterial activity Scientists should think about creating silver-doped TiO2 [35]. Current research studies concerning the use of Ag in medical devices are now underway. dental implants are a type of dental implant that is used to replace missing Packaging for food [36-40] . Filters for air conditioners are available. Some research has focused on bacterial inactivation. Moreover, dye treatment using visible light irradiation [31, 41].

Cellulose is a type of biopolymer that can be found in plants like wood and cotton cells [42]. It's the most common polymer on the planet. The cellulose polymer is a linear homopolysaccharide made up of 1,4-glycosidic linkages that connect D-anhydroglucopyranose units (AGU) [43]. It's a biodegradable and renewable polymer that's been around for a long time [44]. Nanocellulose has received a lot of interest, particularly as a filler for biocomposites[45]. The advantages of nanocellulose include its long-term survivability, abundance, and mechanical properties such as a high surface-to-volume ratio, high tensile strength and stiffness, and high flexibility, as well as its thermal and electrical properties [46].

2. Experimental part

2.1. Preparation of TiO2 nanoparticles

Tio2 prepared using the hydrothermal method [47]. 2ml of Ticl4 was put in a conical flask in an ice bath, 8ml of isopropanol was added slowly with continuous stirring for 10 min keeping in the ice bath until the end of the reaction and vanishing of the mist to obtain the oiled yellow solution of Titanium tetra isopropoxide TTIP [Ti (OC3H7)4]. Following that, 25ml of distilled water with 25ml of hydrochloric acid (HCL) were mixed with continuous stirring for 10min. Then 2ml of the mixture was added to Titanium tetra isopropoxide TTIP with stirring for 30min. After that transferred 60ml of the solution to the autoclave, the autoclave was set in an oven at 160 c̊ for 4hr. After the reaction was complete, the autoclave was cooled to room temperature. A white precipitate of Tio2 was produced. The product was washed with distilled water and ethanol three-time using a centrifuge 500 rpm and dried at 80c̊ for 30min. To enhance the crystallinity, the product is calcinated at 500 °C for 2hr.

2.2. Preparation of TiO2/ Cellulose

0.06g of cellulose was dissolved in 25ml of distilled water with a stirring well for 15min. The mixture was added to TTIP and the reaction was complete according to the above procedure but without calcinations to avoid damage to cellulose

2.3. Preparation of TiO2 / Ag / Cellulose Nanocomposite

AgNO3 solution was prepared by dissolving 0.102g of AgNO3 in 100ml of distilled water with stirring for 20min. Then, prepared a solution of NaBH4 by dissolving 0.151g of NaBH4 in 500ml of distilled water with stirring for 20 min. PVP solution was prepared by dissolving 0.08 g of PVP in 60ml of distilled water. Following that, 8ml of AgNO3 solution was added to 60ml of NaBH4 solution slowly with stirring for 2h until yellowish-brown Ag solution was obtained.
0.1g of TiO2 / Cellulose was added to 15ml of Ag solution under nitrogen protect for 5min and stirring for 30min, the mixture was transferred to ultrasonicator for 20 min. Then the mixture was heated in a water bath for 30min at 60c̊. Finally, The product was filtered, washed in water and ethanol, then dried and weighed.

2.4. Characterization techniques

The XRD pattern of the TiO2 –Ag-Cellulose nanocomposite was acquired using an X-ray diffractometer Shimadzu model: XRD 6000 with CuK (=0.154 nm) radiation and a diffraction angle between 10° and 70°. MIRA3, Tescan FE-SEM was used to study the morphology of the composite. A Leo A12AB TEM was used to get nanoparticles' size and morphology. UV-vis diffuse reflectance spectra were obtained using a Shimadzu UV2550 recording spectrophotometer with an integrating sphere ranging from 200 to 800 nm (DRS). BaSO4 was used as a reference.
At 423 K, nitrogen adsorption-desorption measurements were performed using a QuantachromeInstruments NOVA 2200e analyzer. The adsorption data were used to determine the multipoint BET surface area.

3. Results and discussion

The XRD pattern of the TiO2-Ag-Cellulose nanocomposite is shown in Figure 1. All of the diffraction peaks can be indexed to the facets of wurtzite phase Ag and cellulose (JCPDS Card no. 80-0006). The intensity of the diffraction peaks is substantially higher in the pattern of TiO2-Ag-Cellulose composites. It's in keeping with the fact that Ag and cellulose can operate as an excellent nucleation substrate while also improving TiO2 crystallinity. The X-ray diffraction spectrum revealed a peak at 22.64°, which is the cellulose peak [48]. In addition to the appearance of the characteristic peaks of titanium oxide at 57.48°, 62.52°, 24.23°, 33.29°, 49.51°, corresponding to the diffraction angles of titanium oxide with the anatase phase [49, 50] it also showed peaks at 46.33°, 39.36° due to silver nanoparticles deposited on the surface [51]

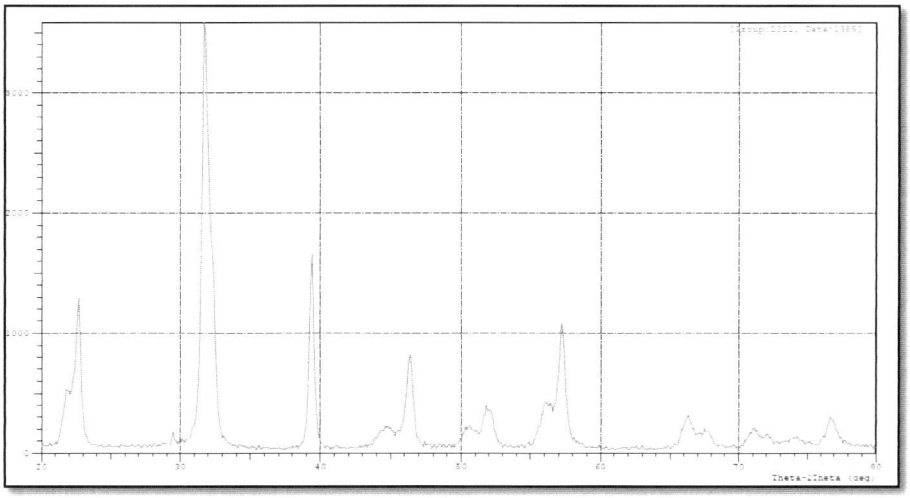

Figure 1. XRD pattern of TiO2/Ag/Cellulose

The surface morphological characteristics of the produced TiO_2-Ag-Cellulose nano-composite were investigated using SEM analysis Figure 2 (low magnification) .

The Ag and TiO2 was discovered to contain spherical nanoparticles and homogenous dispersion over the cellulose surface .The Ag and cellulose are firmly and routinely deposited on the nanoparticles of TiO_2. This shows a robust interaction between Ag , cellulose and TiO_2nanoparticles. The elemental makeup of Ag and TiO2 is shown by the EDS data in Fig. 3 table.1

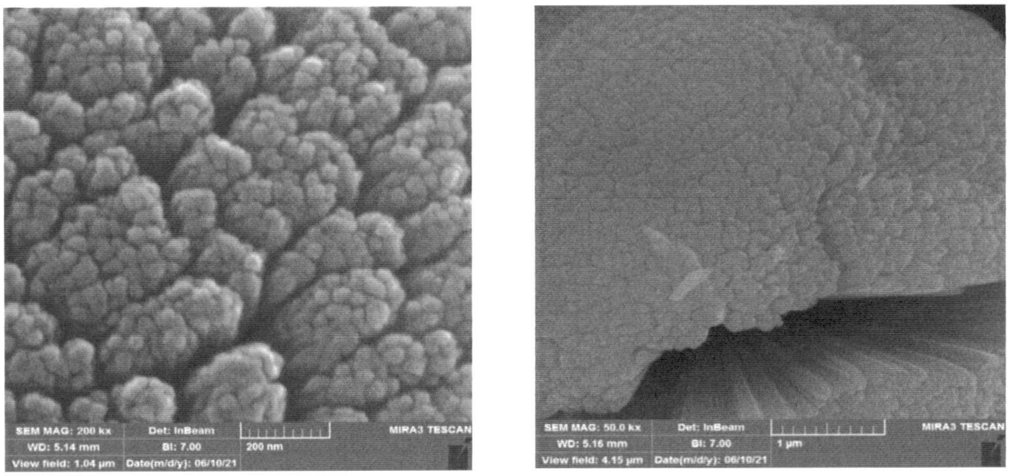

Figure 2. SEM of TiO_2-Ag-Cellulosenanocomposit

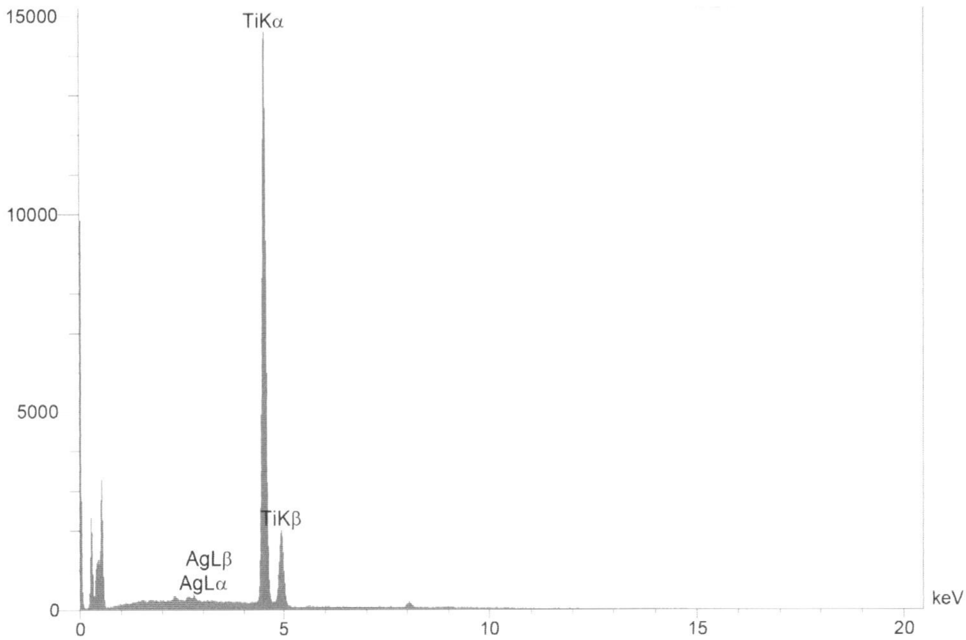

Figure 3. EDS of TiO_2-Ag-Cellulose Nanocomposite

Table 1. Element analysis of the ternary Composite

Material	Element	Weight %	Atomic %
	C	**18.89**	**32.48**
TiO2/Ag/Celle louse	**O**	**38.21**	**49.34**
	Ti	**41.55**	**17.92**
	Ag	**1.35**	**0.26**

The DRS spectra of TiO2-Ag-Cellulose nanocomposites are shown in Figure 4. The absorption edge of the TiO2-Ag-Cellulose nanocomposites is (490 nm), suggesting that Ag has a considerable impact on the samples' absorbance in visible light. As a result, the band gap of the TiO2-Ag-Cellulose composite is 2.85 eV, which is smaller than that of pure TiO2 which equal to 3.2 eV. A narrow band gap photo catalyst has a larger photo response range, which is an excellent feature. The decrease of the band gap can be attributed to TiO2 absorption as well as the electrostatic binding or charge transfer interaction between TiO2 and the specific Ag sites.
When an electron of anatase TiO2 (3.2 eV band gap) is promoted from the valence band to the conduction band by UV stimulation, a hole in the valence band is left behind. The flat band potential of anatase TiO2 was found to be lower than that of metallic Ag .Photo generated electrons moved toward metallic Ag and collected at Ag, forming a Schottky barrier between Ag nanoparticles and TiO2 nanotubes when metallic Ag was coupled with TiO2 nanotubes [52-54].

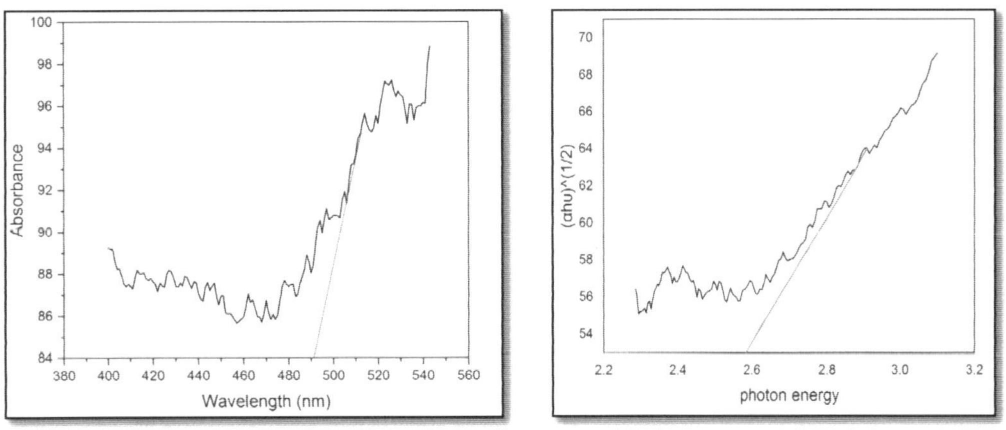

Figure 4. DRS of TiO_2-Ag-Cellulose Nanocomposite

The specific surface area and pore structure of TiO_2-Ag-Cellulosecomposites were
investigated using the nitrogen adsorption–desorption isotherm (Fig. 5 and 6) and the (BET) technique .
Surface area and Pore volume as well as Pore diameter were found to be 26.265 $m^2 g^{-1}$, 0.05628 $g^{-1}cm^3$ and 19.908 nm respectively, allowing for more surface active sites and easier charge carrier transfer .

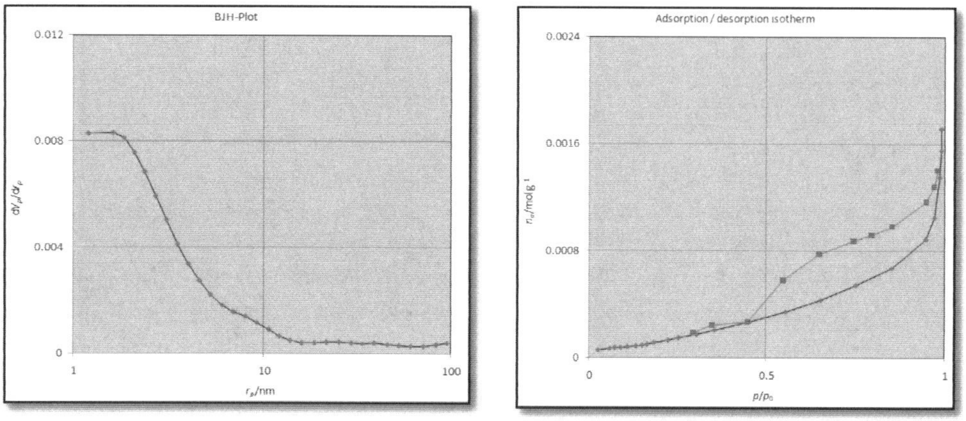

Figure 5. BET-BJH of TiO_2-Ag-Cellulose

Figure 6 shows the photocatalytic degradation efficiency of congo red in TiO_2-Ag-Cellulose nanocomposites when exposed to visible light. In the absence of catalysts, irradiation causes no significant reduction in CR concentration, demonstrating that CR is largely stable under visible light ,also TiO_2-Ag-Cellulose Nanocomposite been discovered to have better photocatalytic performance than pureTiO_2, showing that Ag and cellulose can increase TiO_2photodegradation.In comparison to pureTiO_2, which has a photodegradation efficiency of 30%, the ideal conditions of pH5, dye concentration $3x10^{-5}$, and catalyst mass 0.1 g have a photodegradation efficiency of 81.76% When compared to pure TiO_2 30%.

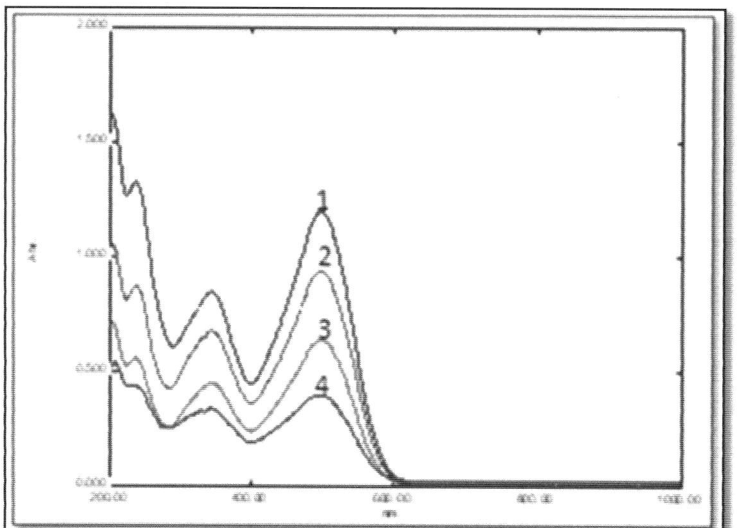

Figure 6. The photo degradation profile of CR in the presence of TiO_2-Ag-Cellulose nanocomposite under visible light (1.at30min 2.at60min 3.at 90min 4.at 120 min)

References

[1] Liu, N., et al., A review on TiO2-based nanotubes synthesized via hydrothermal method: Formation mechanism, structure modification, and photocatalytic applications. 2014. 225: p. 34-51.
[2] Hashimoto, K., H. Irie, and A.J.J.j.o.a.p. Fujishima, TiO2 photocatalysis: a historical overview and future prospects. 2005. 44(12R): p. 8269.
[3] Mor, G.K., et al., A room-temperature TiO2-nanotube hydrogen sensor able to self-clean photoactively from environmental contamination. 2004. 19(2): p. 628-634.
[4] Varshney ,4- G., et al., Nanoscale TiO2 films and their application in remediation of organic pollutants. 2016. 306: p. 43-64.
[5] Camposeco, R., et al., Synthesis, characterization and photocatalytic activity of TiO2 nanostructures: nanotubes, nanofibers, nanowires and nanoparticles. 2016. 266: p. 90-101.
[6] Linsebigler, A.L., G. Lu, and J.T.J.C.r. Yates Jr, Photocatalysis on TiO2 surfaces: principles, mechanisms, and selected results. 1995. 95(3): p. 735-758.
[7] Lei, Y., et al., Preparation and photoluminescence of7- highly ordered TiO 2 nanowire arrays. 2001. 78(8): p. 1125-1127.
[8] Varghese, O.K., et al., Crystallization and high-temperature structural stability of titanium oxide nanotube arrays. 2003. 18(1): p. 156-165.
[9] Chen, D., et al., Mesoporous anatase TiO29- beads with high surface areas and controllable pore sizes: a superior candidate for high-performance dye-sensitized solar cells. 2009. 21(21): p. 2206-2210.
[10] Ismail, A.A. and D.W.J.J.o.M.C. Bahnemann, Mesoporous titania photocatalysts: preparation, characterization and reaction mechanisms. 2011. 21(32): p. 11686-11707.
[11] Tang, H., et al., Urbach tail of anatase TiO 2. 1995. 52(11): p. 7771.11-
[12] Dar, M.I., et al., Controlled synthesis of TiO_2 nanoparticles and nanospheres using a microwave assisted approach for their application in dye-sensitized solar cells. 2014.
[13] Mbonyiryivuze, A., et al., Titanium dioxide nanoparticles biosynthesis for dye sensitized solar cells application. 2015.
[14] Chen, X. and S.S.J.C.r. Mao, Titanium dioxide nanomaterials :14- synthesis, properties, modifications, and applications. 2007. 107(7): p. 2891-2959.
[15] Lan, Y., Y. Lu, and Z.J.N.e. Ren, Mini review on photocatalysis of titanium dioxide nanoparticles and their solar applications. 2013. 2(5): p. 1031-1045.
[16] Gupta, S.16- M. and M.J.c.s.b. Tripathi, A review of TiO 2 nanoparticles. 2011. 56(16): p. 1639-1657.
[17] Taga, Y.J.T.S.F., Titanium oxide based visible light photocatalysts: Materials design and applications. 2009. 517(10): p. 3167-3172.
[18] Fujishima, A., T. Rao, and D.J.J.P.P. Tryk, C, P-Doped titania xerogels as efficient UV-visible photocatalysts. 2000. 1: p. 1-21.
[19] Haider, A.J., L.Z.N. Jameel, and S.Y.J.E.T.J. Taha, Synthesis of TiO2 nanoparticles by using sol-gel method and its applications as antibacterialagents. 2014. 32: p. 418-426.
[20] Verdier, T., et al., Antibacterial activity of TiO2 photocatalyst alone or in coatings on E. coli: the influence of methodological aspects. 2014. 4(3): p. 670-686.

[21] Shaheed, M.A. and F.H.J.J.o.B.U. Hussein, Synthesis21- photocatalytic activity of TiO2 Nanoparticles. 2012. 22(1): p. 423-30.
[22] Razali, M.H., et al., Morphological structural and optical properties study of transition metal ions doped TiO2 nanotubes prepared by hydrothermal method. 2013. 1(4): p. 314-31.8
[23] Ribao, P., et al., TiO 2 structures doped with noble metals and/or graphene oxide to improve the photocatalytic degradation of dichloroacetic acid. 2017. 24(14): p. 12628-12637.
[24] Adán, C., et al., Photocatalytic activity of bismuth vanadates under UV-A and visible light irradiation: Inactivation of Escherichia coli vs oxidation of methanol. 2015. 240: p. 93-99.
[25] Batalović, K., et al., Modification of N-doped TiO 2 photocatalysts using noble metals (Pt, Pd)–a combined XPS and DFT study. 2017 :(10)19 .p. 7062-7071.
[26] Wahyuni, E.T. and R.J.I.T.D.-M.f.a.S.E. Roto, Silver Nanoparticle Incorporated Titanium Oxide for Bacterial Inactivation and Dye Degradation. 2018: p. 331-348.
[27] Grabowska, E., H. Remita, and A.J.P.P.o.M.P. Zaleska, Photocatalytic activity of TiO2 loaded with metal clusters. 2010. 45: p. 29-38.
[28] Maicu, M., et al., Comparative study of the photodeposition of Pt, Au and Pd on pre-sulphated TiO2 for the photocatalytic decomposition of phenol. 2011. 217(2-3): p. 275-283.
[29] Sontakke, S., J. Modak, and G.J.C.e.j. Madras, Photocatalytic inactivation of Escherischia coli and Pichia pastoris with combustion synthesized titanium dioxide. 2010. 165(1): p. 225-233.
[30] Sontakke, S., J. Modak, and G.J.A.C.B.E. Madras, Effect of inorganic ions, H2O2 and pH on the photocatalytic inactivation of Escherichia coli with silver impregnated combustion synthesized TiO2 catalyst. 2011. 106(3-4): p. 453-459.
[31] Li, Y., et al., Preparation of Ag-doped TiO2 nanoparticles by a miniemulsion method and their photoactivity in visible light illuminations. 2011. 129(1-2): p. 501-505.
[32] Li, H., et al., Antibacterial activity of TiO2 nanotubes: influence of crystal phase, morphology and Ag deposition. 2013. 284: p. 179-183.
[33] Anandan, S., et al., Effect of loaded silver nanoparticles on TiO2 for photocatalytic degradation of Acid Red 88. 2008. 92(8): p. 929-937.
[34] Suwarnkar, M., et al., Enhanced photocatalytic activity of Ag doped TiO2 nanoparticles s ynthesized by a microwave assisted method. 2014 :(4)40 .p. 5489-5496.
[35] X., X. Xue, and H.J.A.S.S. Yang, Preparation and characterization of Ag-doped TiO2 nanomaterials and their photocatalytic reduction of Cr (VI) under visible light. 2014. 321: p. 396-403.
[36] Caballero, L., et al., Inactivation of Escherichia coli on immobilized TiO2 using fluorescent light. 2009. 202(2-3): p. 92-98.
[37] Cai, R., et al., Photokilling of malignant cells with ultrafine TiO2 powder. 1991. 64(4): p. 1268-1273.
[38] McCullagh, C., et al., The application of TiO 2 photocatalysis for disinfection of water contaminated with pathogenic micro-organisms: a review. 2007. 33(3): p. 359-375.
[39] Kiwi, J. and V.J.L. Nadtochenko, Evidence for the mechanism of photocatalytic degradation of the bacterial wall membrane at the TiO 2interface by ATR-FTIR and laser kinetic spectroscopy. 2005. 21(10): p. 4631-4641.

[40] Foster, H.A., et al., Photocatalytic disinfection using titanium dioxide: spectrum and mechanism of antimicrobial activity. 2011. 90(6): p. 1847-1868.
[41] Colmenares ,41- J., et al., Synthesis, characterization and photocatalytic activity of different metal-doped titania systems. 2006. 306: p. 120-127.
[42] Li, J., Q.H. Xu, and L.Q. Jin. Research development on hydrophobic modification of cellulose nanofibrils. in AdvancedMaterials Research. 2013. Trans Tech Publ.
[43] Klemm, D., et al., Cellulose: fascinating biopolymer and sustainable raw material. 2005. 44(22): p. 3358-3393.
[44] Moon, R.J., et al., Cellulose nanomaterials review: structure, properties and nanocomposites :(7)40 .2011 .p. 3941-3994.
[45] Varshney, V. and S. Naithani, Chemical functionalization of cellulose derived from nonconventional sources, in Cellulose fibers: bio-and nano-polymer composites. 2011, Springer. p. 43-60.
[46] Li, W., et al., Characteristics46- of self-alkalization in high-rate denitrifying automatic circulation (DAC) reactor fed with methanol and sodium acetate. 2014. 154: p. 44-50.
[47] Bao, X.-w., et al., Preparation of TiO2 photocatalyst by hydrothermal method from aqueous peroxotitanium acid gel. 2005. 59(4): p. 412-415.
[48] Zhang, C., et al., Surface modification on cellulose nanofibers by TiO2 coating for achieving high capture efficiency of nanoparticles. 2019. 9(2): p. 139.
[49] Scarpelli, F., et al., Mesoporous TiO2 thin films: State of49- The art. 2018. 508(1): p. 135-142.
[50] Al-Taweel, S.S. and H.R.J.J.C.P.R. Saud, New route for synthesis of pure anatase TiO2 nanoparticles via utrasound-assisted sol-gel method. 2016. 8(2): p. 620-626.
[51] Evcin, A., et al., Characterization of Ag-TiO2 powders prepared by sol-gel process. 2017. 132(3): p. 608-611.
[52] Zhang, H., et al., Tuning photoelectrochemical performances of Ag− TiO2 nanocomposites via reduction/oxidation of Ag. 2008. 20(20): p. 6543-6549.
[53] Sun, L., et al., Ultrasound aided photochemical synthesis of Ag loaded TiO2 nanotube arrays to enhance photocatalytic activity. 2009. 171(1-3): p. 1045-1050.
[54] Li, H., et al., State-of-the-art progress in diverse heterostructured photocatalysts toward promoting photocatalytic performance. 20 :.p. 998-1013.

Synthesis and characterization of a novel organic reagent and its complexes with Cu (||), Cr(|||), Co(|||) and Fe(|||) metal

Rawaa Ch. Hasan[1], Faiq F. Karam[1], and Sadiq J. Baqir[2]

[1]Departments of chemistry, collage of science University of Al-Qadisiyah, Iraq

[2]Al-mistaqbal university college, medical laboratories techniques department, Iraq

Abstract. This work included the preparation of a series of metal complexes with the new organic azo reagent dye. The donor the prepared ligand is It a derivative of 2-amino-6-methoxy benzothiazolyl mediated by a diazonium solution with a 4-methoxy phenol solution. The preparation of the organic reagent 2-[2¯-(6-Methoxy benzothiazolyl) azo]-4-methoxy phenol (6-MBTAMP), where this reagent characterized by several spectroscopic techniques, which included the H^1-NMR proton spectrum, the C^{13}-NMR carbon spectrum, and also a series of metallic complexes for this ligand were prepared. It included four complexes with its Cu (||), Fe (|||), Cr (|||), Co (|||), ions. Where these lectured complexes were diagnosed by the UV-visible spectrum and the TF-IR spectrum. It has an octahydral geometric shape.

Key words: Azo, H^1-NMR, FT-IR, C^{13}-NMR, UV-VIS,extraction

1. Introduction

Azo compounds, there are organic compounds consisting of two organic groups that are linked by coupling reactions with the azo radical to offer colored compounds and should contain one or more azo groups. Thanks to its name due to its possession on the chromophore group (-N = N-), which may be related to the alkyl or Ariel and therefore the pigment is more stable when linked to the Ariel group [1]. They constitute quite half the dyes used at this time and differ in their degree of complexity consistent with the amount of azo groups, and therefore the number and nature of the exochromic groups present in them. thanks to the likelihood of prepering differing types of those dyes due to the pliability in their preparation, and in their use and their possession of varied properties thanks to the supply of amine, substitutes that are utilized in their preparation and are considered one among the foremost important sorts of dyes at all [2]. Azo dyes are the most important group of industrially prepared organic materials. Where the dyes are linked to the materials to be colored directly or with the assistance of drugs called stabilizers, because the process of linking occurs with different mechanisms, including physical adsorption or mechanical retention of the pigment, or they'll be linked with them through covalent bonds or through the formation of complexes with metals or salts or by the formation of solutions with them [3] [4].

Azo dyes are classified either consistent with the amount of azo groups, as they're mono-azo dyes, di-azo, tri-azo, or poly-azo dyes, consistent with the character of the prevailing oxychrome groups. In it to acid dyes, if it contains groups like (-COOH-SO_3H -OH) or basic if it contains groups like (NR_2 -NRH -NH_2-). And if the pigment contains both sorts of groups, then its classification depends on the amount and strength of those groups. These dyes also can be classified consistent with the tactic of use into direct dyes, stabilizer dyes [4] [5] [6]. The phenotypic dyes that they're composed of, and include (homogeneous ring azo dyes). This sort includes azo compounds are divided consistent with the sort of the ring compounds whose aromatic rings don't contain different atoms (all of them are carbon atoms) and should not contain substitutes as within the azo benzene compound and this sort of dyes the effectiveness of the limiter thanks to the shortage of linking sites in it. (Heterocyclic azo dyes) are relatively recent reagents. They have been widely utilized in the sectore of chemical analyzes because they contain heterozygous aromatic ring located on one or each side of the azo radical with the presence of an atom or donor atom like (S,N,O) or the aromatic rings are replaced by a special group (acidic and basic), or its going to contain the previous types together and within the same ring [7]. The target of this work is to review synthesis a replacement reagent to make complexes with ionic pollutant [8] [9] [10].

2. Material and experimental work

2.1. Materials

All organic chemicals, solvent and inorganic salt used were of a high degree of purity and were used directly without need for further purification. Chemicals supplied from Sigma-Aldrich, Merck, BDH, Fluka, Scharlau ,CDH,and MOH Companies, Stock solutions were prepared by dissolving a required amount of chloride salt of every Fe, Cu , Cr , and Co. Solutions were prepared by dissolving a required amount of acetate salt in ammonium acetate buffer solutions, then dilution of the stock solution in double water with a degree of (1000 mg L^{-1}) for all metallic ions, and these solutions were diluted daily to get working solutions.The H1NMR spectra were administered on a Bruker model 500-UltraShield 300 MHZ spectrometer using DMSO-d, as solvent for measurement and using HDO as an indoor standard reference. UV-Visible spectrophotometer Double beam scanning at range of (200 - 1650 -nm) and single beam scanning at range of (200-1200 nm) both using absolute ethanol as solvent for a measurement. Transform Infrared (FT-IR) spectra were recorded on a Shimadzu model 8400s FT-IR spectrometer scanning at range between (4000-400 cm') via a KBr pellets .And thenC13NMR spectrometer were recorded on a Bruker model 500- MHZ spectrometer using DMSO-d as solvent for a measurement

2.2. Synthesis of azo ligand

2-[2⁻-(6-Methoxy benzothiazolyl) azo]-4-methoxy phenol (6-MBTAMP)

According to the method proposed by the researcher Khaled. J. Al-Adilee and his group [11] [12]. the organic reagent 6-MBTAMP was prepared by making some adjustments to it, where the method of work is divided into two parts, one complementary to the other, to obtain the desired reagent under study.2-[2⁻-(6-Methoxy benzothiazolyl) azo]-4-methoxy phenol (6-MBTAMP)Synthesized of the novel azo ligand with some modification via the diazotization coupling reaction dissolving (1.8g) from 2-amino-6-methoxy benzothiazole in mixture contain (5ml of concentrated of acid HCl) and (30ml) of pre-cooled at temperature between (0-3)C°, then to the present solution was added drop wise an answer of sodium nitrite prepared by dissolving (0.9g, 0.013mol of $NaNO_2$) in 30ml of water, and stirred for (30min) at (0-5)°C. The resulting diazonium chloride solution was added also drop wise while maintaining temperature at (1.6g, 0.01mol of 4-methoxyphenol within the mix contains on 40ml of absolute ethanol and 10ml of 10% NaOH) and coolant to (0-5) °C, after completion of the add-on process the ultimate mixture was stirring a minimum of for (2h) then complete the precipitate of the ligand by modifying the acidic function of the answer to pH=12 and allowed to face overnight. The precipitate was filtered off, washed with distilled water and eventually recrystallized. The freezing point of azo ligand point found to be (90-95) The ligand appearance was as a sepia rod crystal, scheme (1) the step of synthesis of ligand:

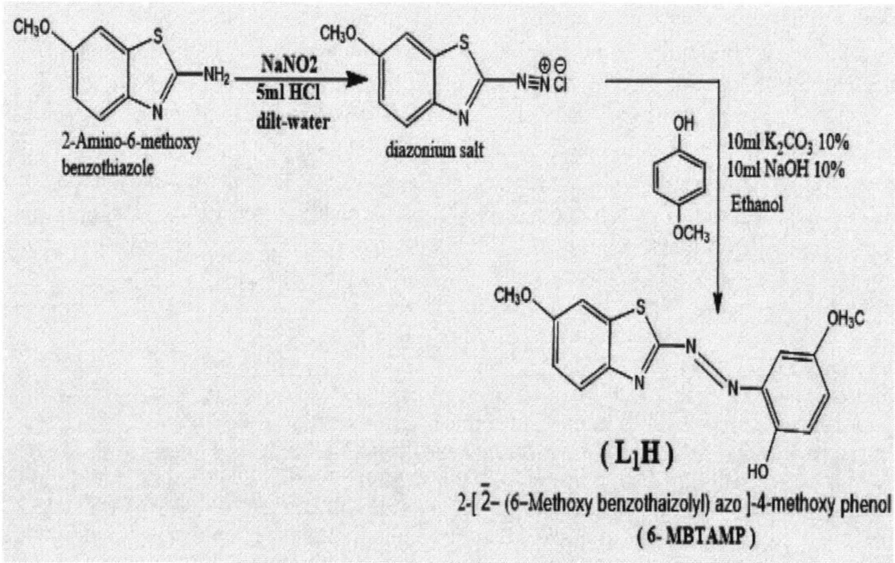

Scheme 1. Synthesis of thiazolyl azo dye ligand (6-MBTAMP).

2.3. Preparing of metallic ions complexes

After knowing the best conditions for the formation of complexes such as optimal concentration, a preferable acidic function, and a molar ratio for both the reagent and the metal complexes. This stage of the study included preparing the selected complexes for the study, and taking a specific weight from the prepared organic reagent, and the amount of 0.63g per 30ml of absolute ethanol where the addition is slow, and continuous stirring to the metallic ion solution as if it was 0.16g for preparing the iron complex, 0.33g for preparing the cobalt complex, 0.18g for preparing the chromium complex, 0.19g for preparing the copper complex for a period of 30min, at a temperature equal to 60 ° C, and after mixing, deposits of these complexes were formed. And finally the orange complex chrome.

3. Result and discussion

3.1. Effect of pH

The pH study included knowing the optimal acidic function for each metallic ion, and a wide range of acidic function was used, ranging from (3-11). Dissolving the metal salts with a solution of a concentration equal to 0.01M of ammonium acetate and the acidity function is changed by using dilute acetic acid and ammonia solution, where the stability of the chromium complex was at pH = 7, the iron complex at pH = 8, the copper complex at pH = 9 and finally the cobalt complex at pH = 10 and these values were relied on as the best because they give a high absorbency and constant concentration. Figures 1-4, illustrate the perfect pH for metal complexes:

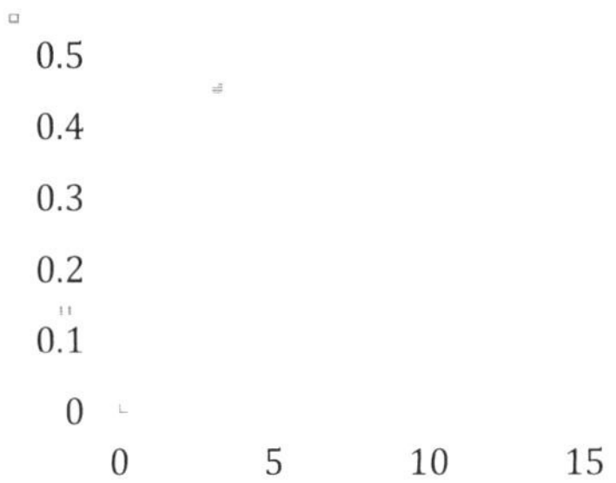

Figure 1. the effect of pH on Copper complex at 353 nm

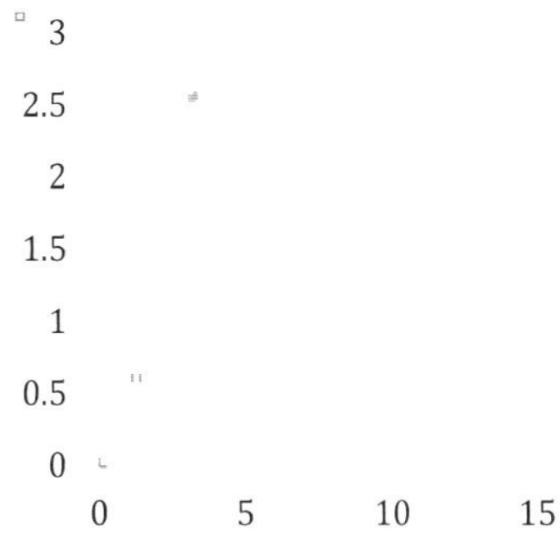

Figure 2. the effect of pH on iron complex at 354 nm

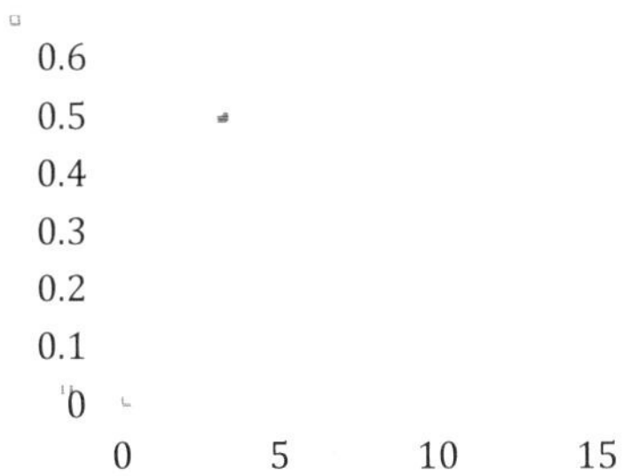

Figure 3. the the effect of pH on cobalt complex at 510 nm

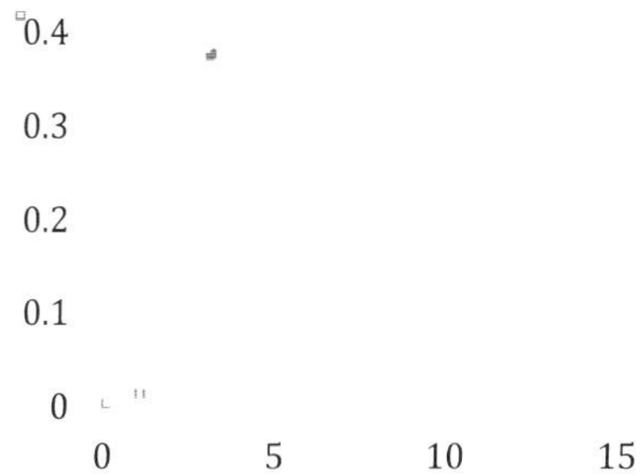

Figure 4. the effect of pH on Chromium complex at 632 nm

3.2. Molar ratio [Metal: Ligand] ratio

Measuring the molar ratio is the best in determining the formulas of complexes in their aqueous solutions by taking the highest values of absorbance to find the molar ratio [reagent: metal] where the absorbance values increase and the color intensity increases gradually and when the color intensity is established at a certain point, it is the point of intersection that indicates the formation of the complex and explains that The molar ratio of all complexes is [2:1] [13] [14] , as shown in the following figures(5-8):

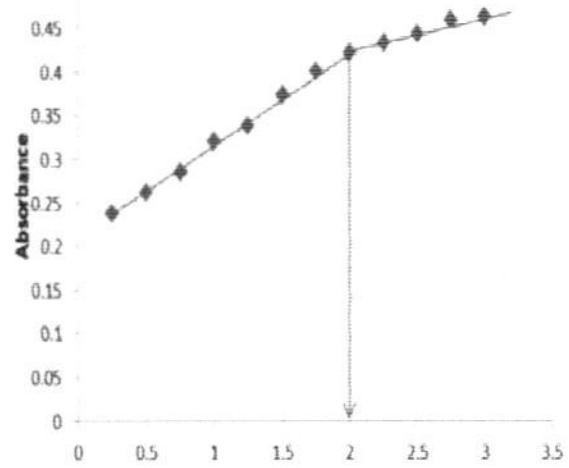

Figure 5. Calibration curve of (Co-6-MBTAMP)

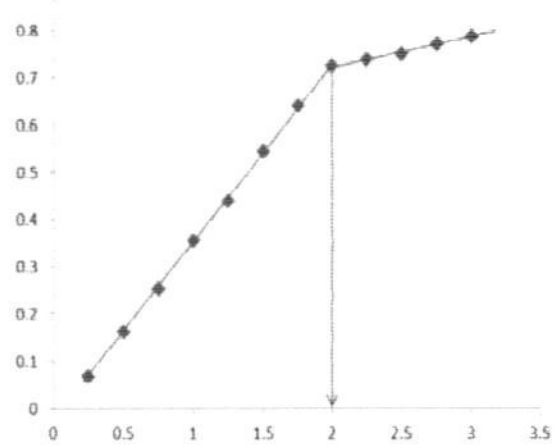

Figure 6. Calibration curve of (Fe-6-MBTAMP)

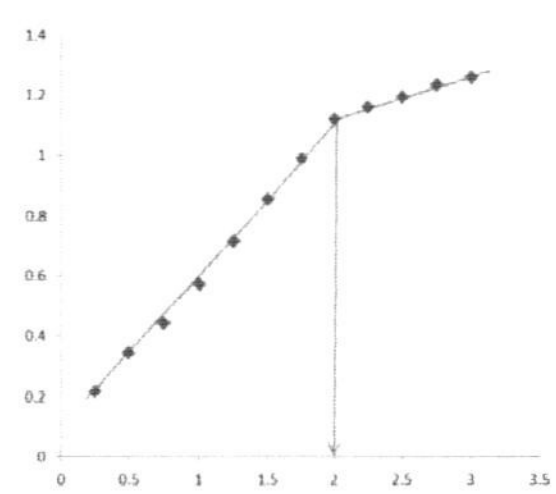

Figure 7. Calibration curve of (Cr-6-MBTAMP)

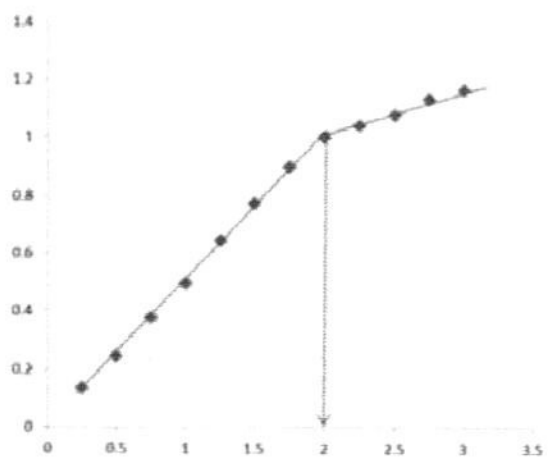

Figure 8. Calibration curve of (Cu-6-MBTAMP)

3.3. Calculation of the metal complexes stability and dissociation onstants

The stability and dissociation constants can be calculated from the impossible values by calculating the molar ratio and knowing the ideal ratio of mixing solutions to the metal ions. The values of the constants of the chelating complexes formed were calculated. It is required to know their stability for the metal ions with the prepared organic reagent [15].

Table 1. Values of stability and dissociation constants

Metal	Am	As	α	ß	Log ß
Fe(III)	**0.787**	**0.726**	**0.077**	**$15*10^{12}$**	**11.17**
Cu(II)	**1.165**	**1.001**	**0.140**	**$4*10^{12}$**	**12.6**
Cr(III)	**1.256**	**1.118**	**0.109**	**$3*10^{13}$**	**13.47**
Co(III)	**0.463**	**0.422**	**0.088**	**$5*10^{10}$**	**10.69**

Through the above table, the results of the stability constants for the metal ion complexes showed us during this study that the metal complexes are highly stable, and that the chromium complex is the most stable, and then come after the copper complex, the iron complex and finally the cobalt complex.

3.4. H^1-NMR

The H^1-NMR assay was performed for the prepared organic reagent 6-MBTAMP using HDO as a standard reference in the assay procedure and obtained in DMSO-d_6 as asolvant which is illustrated in figure (9): a singlet at σ=7.75 ppm corresponding to the hydroxyl group ,a singlet at σ=3.2 ppm corresponding OCH_3 of the phenolic ring, a singlet at σ=3.4 ppm corresponding OCH_3 of benzothazolyl, a singlet at σ=7.38 ppm and σ=7.11 ppm inside the installment of the phenol ring, a dubplate at σ=6.8 ppm of the Composition of benzothiazole, a doblate at σ=8.10 ppm of the near to the thiazole ring and a singlet at σ=7.49 ppm of the Composition of benzothiazole.

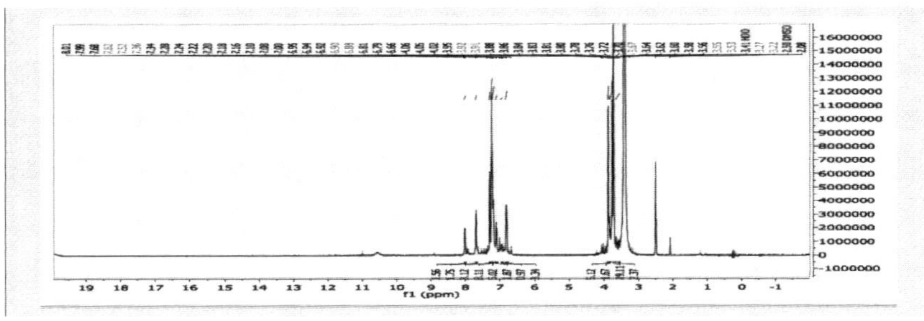

Figure 9. H1-NMR Spectrum of thiazolylazol dye ligand

3.5. Electronic spectra measurement

Absolute ethanol is used to measure the electron spectra of both ligand and its chelates complexes at [10^{-4}M], as the mixing solutions of ions give different distinct colors from one element to another. The free azo reagent towards red wavelengths and the appearance of the absorption peaks located within the boundaries of the visible region of this spectrum, as well as the absorption peaks near the infrared and ultraviolet region, and this is due to the secondary level (d) [11], which is not completely filled with electrons that contain these elements, the spectra electronic is summarized in table (2):

Table 2. electronic spectra for ligand with its complexes

	Absorbance Band Maxima		
Compounds	**Wavenumbercm-1**	**Wavelength (nm)**	**Transition**
MBTAMP-6	**31545.74**	**317**	**π-π***
	27173.91	**368**	**π-π***
	20000.00	**500**	**n-π***
[Cu-6-MBTAMP]	**15576.32**	**642**	**2Eg-2T2g**
	45454.55	**220**	**π-π***
	37878.79	**264**	**π-π***
	32258.06	**310**	**π-π***
[Co-6-MBTAMP]	**45871.56**	**218**	**Center ligand**
	38022.81	**263**	**1A1g-1T1g**
	30959.75	**322**	**1A1g-1T1g**
	15772.87	**634**	**1A1g-1T2g**
[Fe-6-MBTAMP]	**45248.87**	**221**	**Center ligand**
	37735.85	**265**	**Center ligand**
	32258.06	**310**	**2A1g-2T2g**
	18761.73	**533**	**2A1g-2T1g**
[Cr-6-MBTAMP]	**35842.29**	**279**	**π-π***
	33783.78	**296**	**π-π***
	27624.31	**362**	**π-π***
	19157.09	**522**	**4A2g-4T2g**

Figures 10-14, illustrate the electronic spectrum for the reagent and complexes :

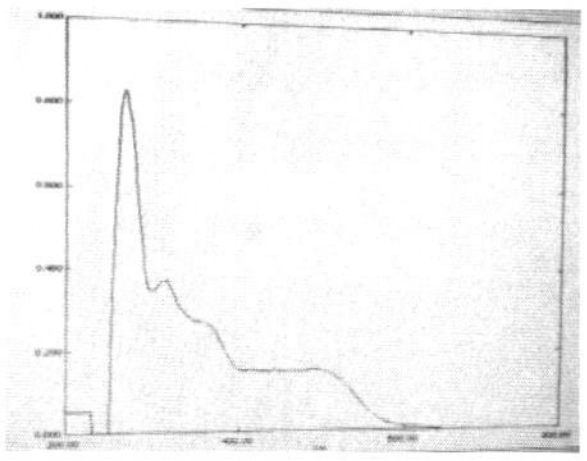

Figure 10. the UV-vis spectrum of the reagent (6-MBTAMP)

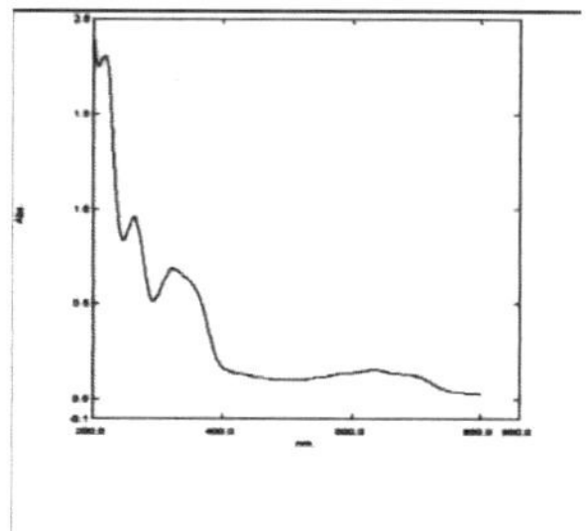

Figure 11. the UV-vis spectrum of the cobalt complex

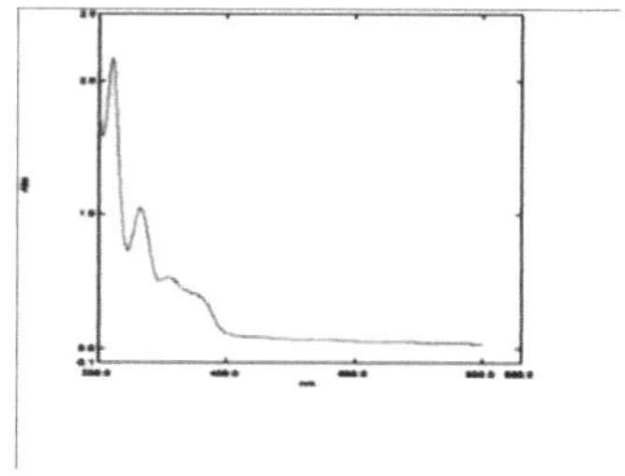

Figure 12. he UV-vis spectrum of the iron complex

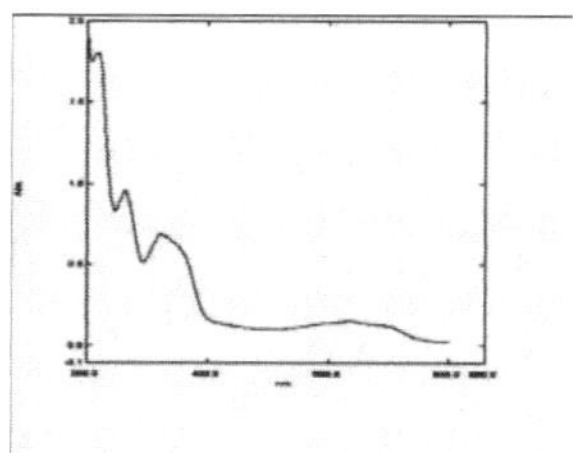

Figure 13. the UV-vis spectrum of the copper complex

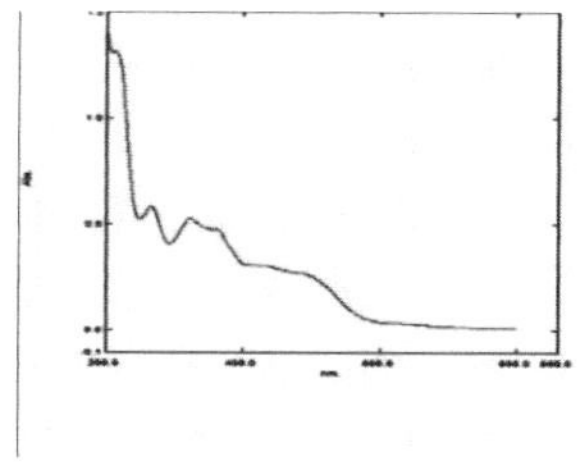

Figure 14. the UV-vis spectrum of the chromium complex

3.6. INFRARED SPECTRA MEASUREMENT

Assay infrared spectra in order to delve into the surface chemistry through information that illustrates the chemical shape and molecular structure of the sample to be diagnosed, the surface of the organic reagent is distinguished by its containment of many different active groups belonging to the reagent itself(6-MBTAMP), the sites of absorption beams in the infrared spectrum of the organic detector and their chelates complexes with Cu(III), Cr(III), Co(III) and Fe(III) at a range of frequencies Ranging between (400-4000cm^{-1}) and included many interactions in the heterocyclic rings and their association with the azo bridge group in addition to the presence of the substituted phenol rings [16] [17]·, As shown in the following table(3): Spectral characterization of reagent 6-MBTAMP and its complexes via Infrared technique:

Table 3. functional groups for ligand and complexes

Group	Ligand	Cr(III)	Co(III)	Cu(III)	Fe(III)
O-H			m-2831.31	m-3309.62	m-3386.77
C-H	S.sh2926.11	s.2923.88	S.2931.60	S.2931.60	S.2931.60
C=C	Sh .1548.89	m.w 1542.95	m.w 1604.65	m.w 1473.51	m.w 1542.95
C-N	S.1330.93	m.w 1319.22	m.w 1265.22	m.w 1110.92	m.1211.21
C=N	S. 1641.48	m.1542.95		w.s 1542.95	m. 1542.95

C-O	**S. 1274.99**	**m.s 1226.64**	**S. 1103.21**	**S. 1026.06**	**m. 1110.92**
C-S	**m.1178.55**	**1118.64**	**Sh 1265.22**	**W 1265.22**	**1226.64**
N=N	**S. 1435.09**	**S.1434.94**	**m.w 1604.66**	**br 1419.51**	**S 1434.94**
M-N	**-**	**W 516.89**	**W416.60**	**m-594.03**	**M 439.74**
M-O	**-**	**M 671.18**	**S.663.47**	**W 671.18**	**M 617.18**
C-H aromatic	**S. 808.20**	**810.05**	**933.48**	**825.48**	**810.05**

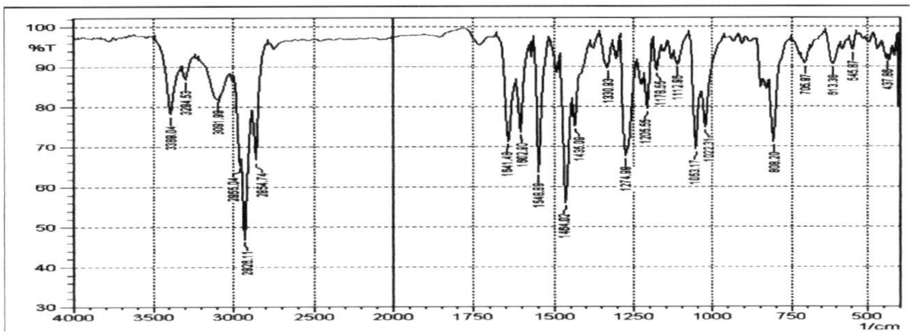

Figure 15. the infrared spectrum of the (6-MBTAMP)

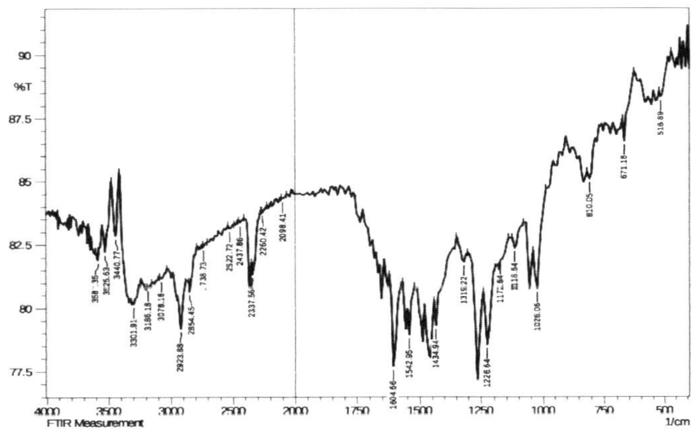

Figure 16. the infrared spectrum of the (Cr-6-MBTAMP)

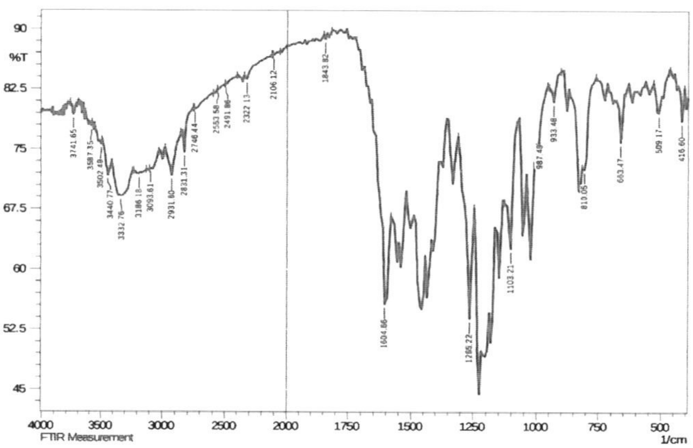

Figure 17. the infrared spectrum of the (Co-6-MBTAMP)

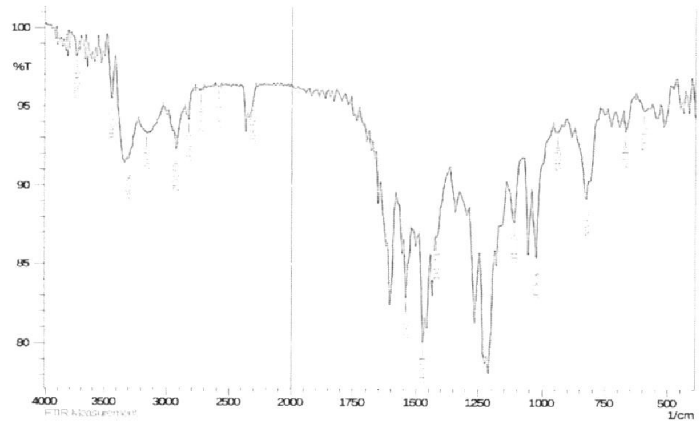

Figure 18. the infrared spectrum of the (Cu-6-MBTAMP)

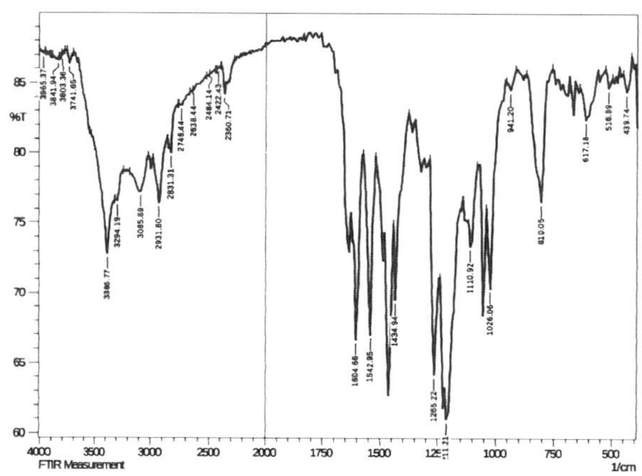

Figure 19. the infrared spectrum of the (Fe-6-MBTAMP)

3.7. C^{13}-NMR

The C^{13} carbon NMR spectrum was performed for the prepared organic reagent under study(6-MBTAMP), which showed two signals at (38.15 ppm and 40.11 ppm) and for two carbon atoms C^{10} and C^{19} methoxide groups -OCh3, and also showed a signal at 55.38 ppm belonging to C^{18} attached to the hydroxide group.

Figure 20, also showed many signs at (113.53 ppm), (113.53 ppm), (ppm106.24), (118.82ppm)) belonging to the C^2-C^9 carbon atoms, the methoxy benzothiazole ring, as for the signals at (125.40ppm), (ppm130.51), (134.62ppm) they refer to the C^{13}- C^{17} carbon atoms of the methoxy phenol ring.

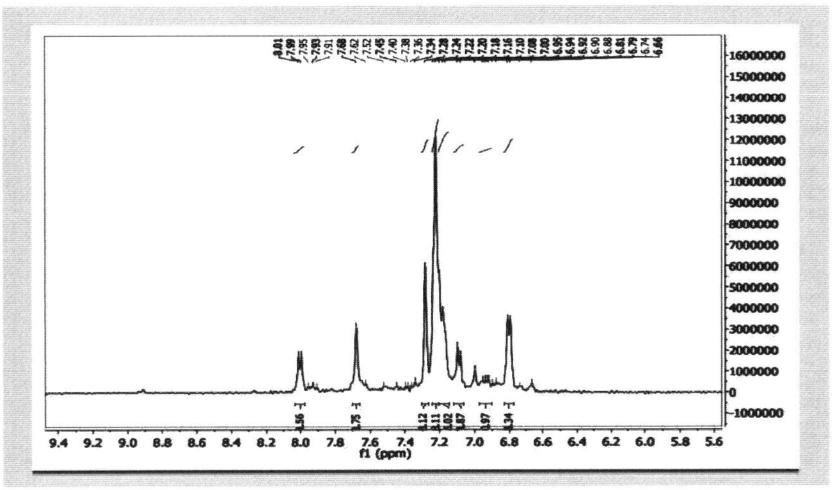

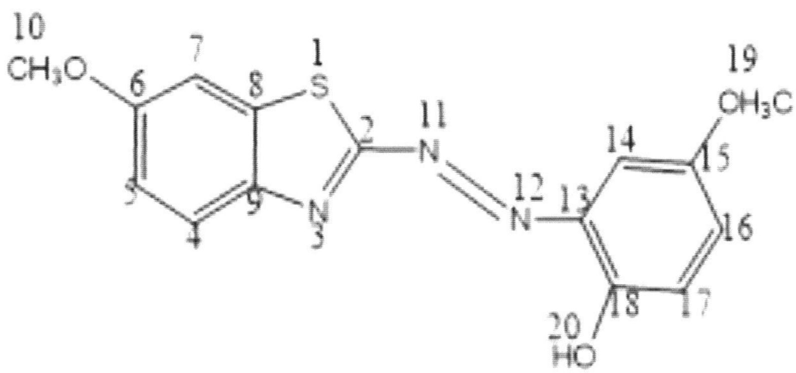

Figure 20. C^{13}- NMR spectrum for the (6-MBTAMP)

4. Conclusion

Based on what we reached during this study, we conclude the preparation of the thiazole azo 6-MBTAMP dye as an analytical chelate organic reagent, and also show us that the tridentate ligand as well as explain to us the proposed synthetic formula of ligand, and these complexes can be prepared easily by knowing the optimal conditions for complexity represented by the optimum concentration and function The best acidity and molar ratio (metal: reagent) were [2: 1] for all complexes, and it also had a high degree of stability.

References

[1] Savić, J. & Vasić, V. Complex formation between Pd (II) and immobilized imidazol-azo-chromotropic acid. Acta Chim. Slov 53, 36–42 (2006)..

[2] Hamzah, M. A. M., Jebur, I. K. & Ahmed, A. K. Synthesis, Characterization and Biological Activity Evaluation of Some New Azo Derivatives from 2-Amino Benzothiazole and Their Derivatives. kirkuk Univ. J. Sci. Stud. 13, (2018)..

[3] Shihad, A. A. A. et al. New Analytical Method for Estimation of ferrous in Ferrous Sulfate drug By Preparation and Using 2- (E- (1H _ benzo (d) imidozol-2-yl) diazenyl -5- (E-4-dimethyl amino benzaliden amino) phenol as a reagent. 10, 2179–2182 (2.

[4] Al-Khateeb, Z. T., Karam, F. F. & Al-Adilee, K. Synthesis and characterization of some metals complexes with new heterocyclic azo dye ligand 2-[2--(5-Nitro thiazolyl) azo]-4-methyl-5-nitro phenol and their biological activities. J. Phys. Conf. Ser. 1294,.

[5] Karam, F. F. & Al-khateeb, Z. T. Journal of Global Pharma Technology Preparation of a Novel Cartridge of SPE Column using a new Surface of Hetrocyclic Azo Ligand with Modified Activated Carbon. 30–44..

[6] Aljeboree, A. M. et al. Photocatalytic of pharmaceutical tetracycline (TCS) by zinc oxide (ZnO). J. Crit. Rev. 7, 960–962 (2020)..

[7] Khataee, A. R. & Kasiri, M. B. Photocatalytic degradation of organic dyes in the presence of nanostructured titanium dioxide: Influence of the chemical structure of dyes. J. Mol. Catal. A Chem. 328, 8–26 (2010)..

[8] Karam, F. F., Saeed, N. H. M., Al-Yasari, A. H., Ahmed, L. M. & Saleh, H. M. Kinetic study for reduced the toxicity of textile dyes (reactive yellow 14 dye and reactive green dye) using UV-A Light/ZnO system. Egypt. J. Chem. 63, 1–12 (2020)..

[9] Jabbar Almukhtar, J. G. & Karam, F. F. Preparation characterization and application of Chitosan nanoparticles as drug carrier. J. Phys. Conf. Ser. 1664, (2020)..

[10] Almukhtar, J. G. J. & Karam, F. F. Kinetic and Thermodynamic Studies for Mebeverine hydrochloride Adsorption from Aqueous Solution using prepared chitosan polymer in delivery drug system. J. Phys. Conf. Ser. 1664, (2020)..

[11] Al-Adilee, K. & Kyhoiesh, H. A. K. Preparation and identification of some metal complexes with new heterocyclic azo dye ligand 2-[2−-(1-Hydroxy-4-Chloro phenyl) azo]-imidazole and their spectral and thermal studies. J. Mol. Struct. 1137, 160–178 (2017)..

[12] Al-Adilee, K. J. & Jaber, S. A. Synthesis, Characterization and Biological Activities of Some Metal Complexes Derived from Azo Dye Ligand 2-[2′-(5-Methyl thiazolyl) azo]-5-dimethylamino Benzoic Acid. Asian J. Chem. 30, 1537–1545 (2018)..

[13] Al-Adilee, K. J. & Fanfon, D. Y. Preparation, Spectral Identification and Analytical Studies of Some Transition Metal Complexes with New Thiazolylazo Ligand and Their Biological Activity Study. J. Chem. Chem. Eng. 6, 1016 (2012)..

[14] Al-Khateeb, Z. T., Karam, F. F. & Al-Adilee, K. Synthesis and characterization of some metals complexes with new heterocyclic azo dye ligand 2-[2--(5-Nitro thiazolyl) azo]-4-methyl-5-nitro phenol and their biological activities. in Journal of Physics: Con.

[15] Ferreira, E. S., Giacomelli, C., Giacomelli, F. C. & Spinelli, A. Evaluation of the inhibitor effect of L-ascorbic acid on the corrosion of mild steel. Mater. Chem. Phys. 83, 129–134 (2004)..

[16] Silverstein, R. M., Webster, F. X. & Kiemle, D. J. Spectrometric Identification of Organic Compounds, 7th editio John Wiley & Sons. (2005)..

[17] Patel, R. D., Patel, M. P. & Patel, R. G. Synthesis, characterization and anti-bacterial activity of some new 2, 3, 6-trisubstituted quinazolin-4 (3H)-ones. (2005)..

Role of Pomegranate peels as a activated carbon for removal of pollutants

Namer A. Hussain[1] · Ahmed Taifi[2] , Ola Kamal A. Alkadir[3] , Nour H. Obaid[4] , Zahraa M. Abboud[5] , Aseel M. Aljeboree[6], Abdul Latif Al Bayaa[7] , Salwan Ali Abed [8] and Ayad F. Alkaim [9]

[1]Ministry of Education, Directorate of Education Babil, AL-Hillah, Babil, Iraq
[2]Al-Manara College For Medical Sciences/Iraq
[3]Al-Nisour University College/Baghdad/Iraq
[4,5,6,9] Department of Chemistry, College of Sciences for Girls, University of Babylon, Hilla, Iraq
[7] The Islamic University, Najaf, Iraq
[9] College of Science, University of Al-Qadisiyah, Iraq

Abstract. Pomegranate peels are considered as solid agricultural residues that were used as adsorbent materials, very effective in get rid of brilliant yellow Dye from aqueous solution after converting it to Acid-Activated Carbon by phosphoric acid. The surface properties were studied through the use of (FTIR) and (FESEM) techniques. Several factors affect the adsorption process have been studied like, effect of dye concentration (10-100 mg/L), where the adsorption efficiency (Qe mg/g) increases with increasing dye concentration (mg/g) and weight effect (0.01-0.1), the adsorption efficiency(Qe mg/g) decreases with increasing surface weight . Desorption of the Brilliant yellow from the ACPP was studied utilizing several type of solvents (acetone 10% (v/v), methanol 10% (v/v) , 0.1N HCl, 0.1 N NaOH,).The adsorption is found through the results that obeys the Friendlich isotherm because it depends on the value of (R2) and also the heterogeneous surfaces.

Keyword: Activated Carbon, Pomegranate peels, Brilliant yellow Dye , Isotherm model.

1. Introduction

Dyes are chemical compounds that have the ability to bind to fabrics or surfaces to impart colour. Usually dyes are complex organic molecules that have the ability to resist, for example the action of detergents [1-3]. Textile dyes are widely used in many industries and advanced technology, such as, various textiles, paper dyeing, leather tanning, cosmetics, food supplies, plastics industry, printing and dyeing industries, and rubber. Recently, the adsorption process has been applied in many agricultural products such as orange peels, lemon peels, banana peels, walnut peels, date kernels, melon peels, pomegranate peels and other solid agricultural waste as an economical and inexpensive way to remove various pollutants [4-9]. These prepared surfaces from agricultural waste have proven their effectiveness in removing Many types of pollutants such as phenol, gases, heavy metals, COD, pharmaceuticals, and dyes. In order to increase the adsorption capacity of the adsorbent materials, many researchers have adopted the use of a different activation process for these surfaces, whether acid or base, to increase the surface area and increase the efficiency of the prepared surface.

Pomegranate shell waste is composed basically of tannins, carboxylic acids, and flavonoids, which are potential sites for biosorption of dye [10-13] . In this research, one of the solid wastes, pomegranate peels, was used and converted into activated carbon as a very effective adsorbent material in remove Brilliant yellow Dye. The effect of concentration, weight, pH and adsorption isotherms were studied.

1.1. Preparation of Activated Carbone

Pomegranate fruits were collected from the local market in Hilla - Iraq. At first, pomegranate peels(PP) are removed manually and washed with hot water to remove dust and color from peels . Then they are washed with distilled water several times and dried under the sun for a week. After that, the peels are ground to obtain a fine powder that is treated with phosphoric acid. concentration (0.1N) with stirring for one hour, then washed with distilled water several times, then dried and burned at a degree (300 °C) for 2hour and kept in a sealed glass container, thus obtaining the powder (ACPP) used in the experiment.

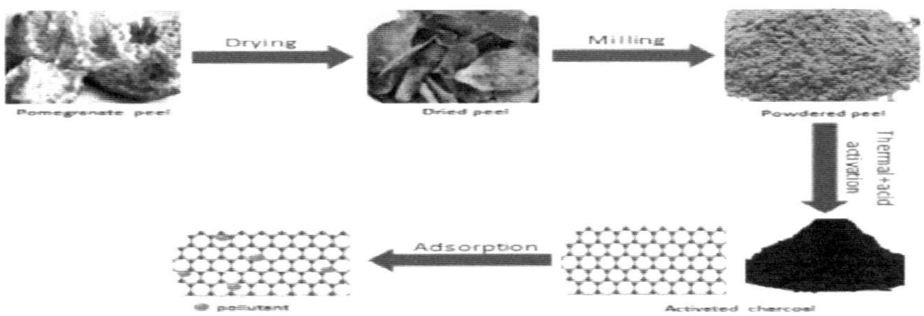

Schema 1. Preparation of Activated carbon from pomegranate peels (ACPP)

1.2. Batch desorption studies

The removal capacity of ACPP for Brilliant yellow Dye was investigated via usual batch adsorption experiments. Stander solutions of Brilliant yellow Dye were prepared via dissolving 0.5 gm of Brilliant yellow Dye in 500 ml DW. Solution of pH adjusted by 0.1 N HCl or 0.1 N NaOH. Taking 100 mL of Solution of dye put in Erlenmeyer flasks 100ml at pH 7.7,study several parameter like equilibrium time, concentration of dye , mass of ACPP , and initial Brilliant yellow Dye concentration. Adsorption experiments were conducted at 20 °C and an agitation rate of 120 rpm on a shaking shaker water bath . After the contact time, the adsorbent was separated from aqueous solutions via centrifugation at 6500 rpm for 15 minute. The residual concentration of dye in each aliquot was estimation by using spectrophotometer UV-visible . The percentage removal and adsorption capacity of Brilliant yellow Dye calculate by equation 1,2:

$$\textbf{Removal Percentage E\%} = \frac{\textbf{Co} - \textbf{Ce}}{\textbf{Co}} \times \textbf{100} \qquad (1)$$

$$\textbf{Adsorption capacity Qe} = \frac{(\text{Co} - \text{Ce})\text{Vml}}{\text{W gm}} \qquad (2)$$

2. Results and Discussion

2.1. FTIR

The IR spectra of adsorbent PPs among 3500 and 500 cm–1 is find to study the characteristics surface of adsorbent. FTIR spectrum of PPs particles taken before adsorption of Brilliant yellow appear the involvement of functional groups present on the of PPs surface[14, 15] . In the spectrum of PPs before adsorption there is a strong peak at 2450 cm−1 representing the −CH stretching as appear in Figure 1. And IR spectra of adsorbent ACPP among 4000 and 400 cm–1 is find to study the characteristics surface of adsorbent. A comparison of the FTIR spectrum of ACPP particles taken before and after adsorption of Brilliant yellow appear the involvement of several functional groups present on the of ACPP surface[14, 15] . The FTIR spectra of Brilliant yellow –treated ACPP appear stretching and bending vibrations of several functional groups denoting their involvement in the adsorption method appear in Figure 1.

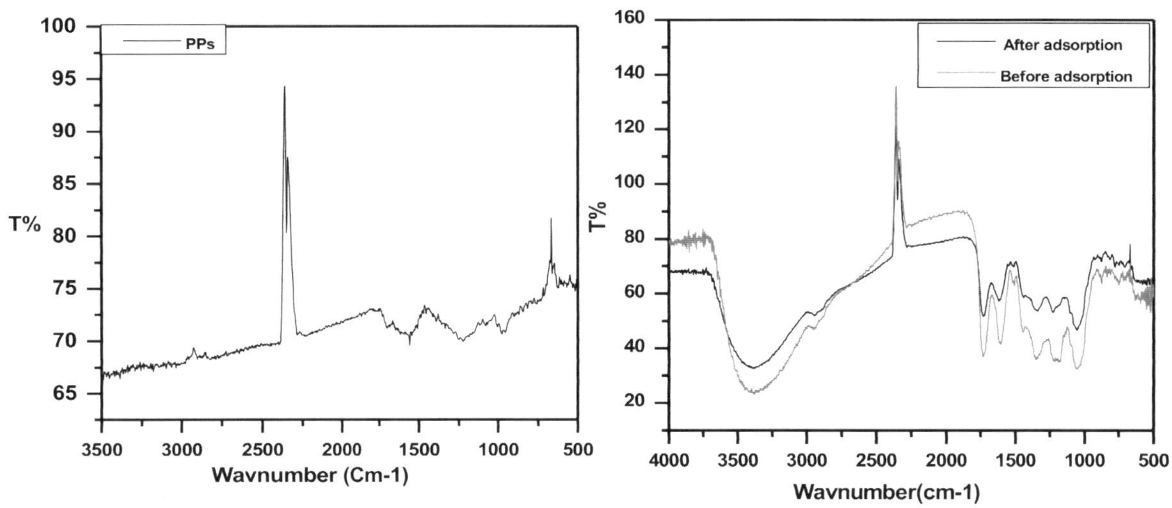

Figure 1. FT-IR image of PPs before adsorption and ACPP before and after adsorption of Brilliant yellow .

2.2. FESEM

Technique SEM is used to determine the morphological properties of the surface of the activated carbon from ACPP before and after the adsorption process, that the surface before the adsorption process is rough containing agglomerations and inactive active sites. surface, as shown in the figure 2

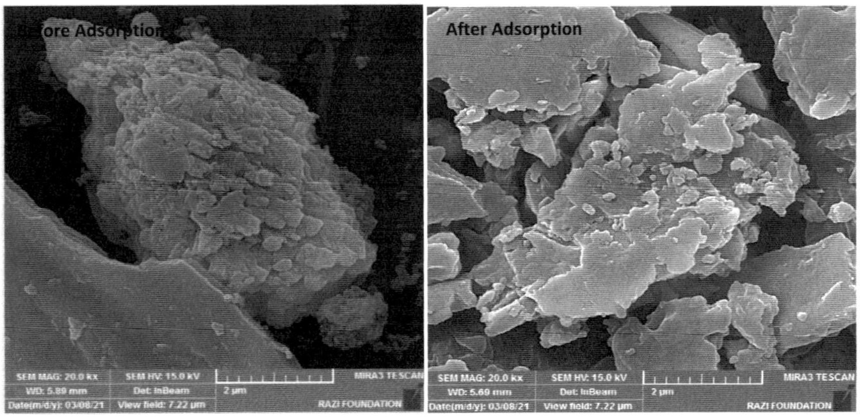

Figure 2. FESEM Image of Activated carbon from pomegranate peels before and after adsorption

2.3. Effect of brilliant yellow concentration

The effect of Initial concentration on the removal of **brilliant yellow** was analyzed at (10-100 mg/L). 100 ml of solution brilliant yellow were put in elementary flask at (20°C) and were agitated at 120 rpm by use 0.05 g weight of ACPP as shown in Figure 3 . it is probable to verify that the quantity adsorbed of brilliant yellow raise by means of the increase of the primary concentration of brilliant yellow . When the primary concentration of brilliant yellow increases from 10 to 100 mg /L , the adsorption capacity increase from 54.44 to 95.55 mg /g due to the increase of driving forces. Also the removal percentage E% decrease from (98.88 to 52.54 %) with increase concentration of dye[14, 16, 17] .

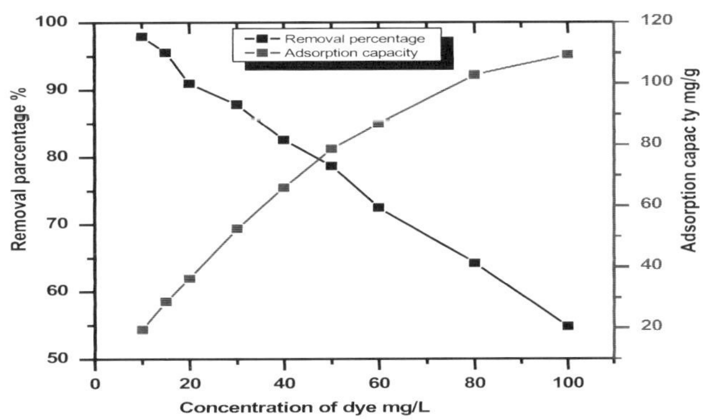

Figure 3. Effect of initial concentration of Brilliant yellow Dye on the adsorption capacity and percentage removal of ACPP (Exp. Condition: equilibrium time 1 hr., Temp. = 25°C)

2.4. Effect of weight activated carbon

The effects of the weight ACPP onto Brilliant yellow Dye at fixed concentration of Brilliant yellow Dye (50 mg L^{-1}) as appear in Figure 4. When the mass of ACPP was increased from 0.01- 0.12 gm an increase in the Brilliant yellow Dye from 50.543 to 249.45 mg/g was observed. This behavior could be explained considering that by weight ACPP increases it occurs an increase on the surface, therefore, more active sites are available to bind Brilliant yellow from aqueous solution[18, 19] . At masses over 0.05 gm, the quantity of Brilliant yellow remained almost constant probably due to the aggregation of particles of bio-sorbent in the solution, consequently, a decrease in the surface for Brilliant yellow uptake occurs. This result is in agreement with result reported via Islam and Ishikawa. For this reason, the weight 0.05 gm was chosen for the subsequent experiments [20, 21].

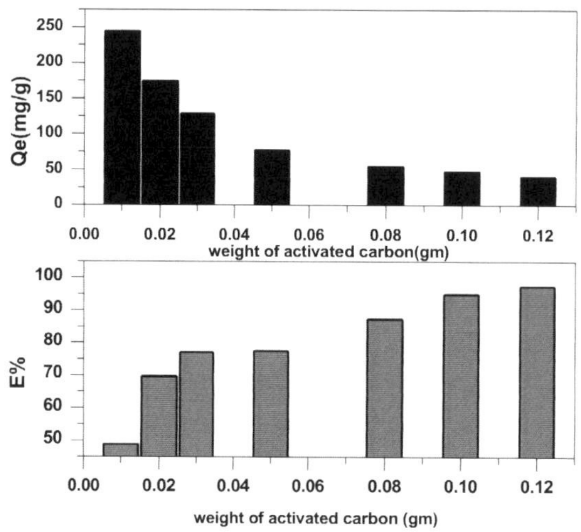

Figure 4. Effects of the weight ACPP on the amount of adsorbed brilliant yellow (Initial dye concentration = 50 mg. L^{-1}, equlibruim time 1hr and Temp . = 20°C)

2.5. Regeneration and Desorption Studies

Desorption of the Brilliant yellow from the ACPP was studied utilizing several type of solvents (acetone 10% (v/v), methanol 10% (v/v) , 0.1N HCl, 0.1 N NaOH,). The find data are appear in table 1. Desorption capability of methanol was find higher than that of the other solvents. Desorption efficiencies higher than 90% were get in a short time (3 min) and in a one-step elution using 5 ml of HCl [22, 23].

Table 1. Comparative of desorption efficiency of various eluents for the Brilliant yellow dye onto surface of ACPP

Regeneration and Desorption	E%
HCl	80.345
acetone	78.765
methanol	66.67
NaOH	60.98

2.6. Adsorption isotherms

The analysis of the model result via fitting them to several models is an significant step to found the suitable isotherm that can be utilized for design purpose. In this research , two model like Langmuir model and Freundlich model , in nonlinear forms were applied to the equilibrium result of Brilliant yellow onto ACPP as show in Figure 5.

The Langmuir model nonlinear equation given in equation :

$$Q_e = \frac{Q_m\,K_L\,C_e}{1 + K_L\,C_e} \qquad (3)$$

The isotherm Freundlich model was applic-able in hetero-geneous surface adsorption and The Freundlich model nonlinear equation given in equation :

$$Q_e = k_f\,C_e^{\frac{1}{n}} \qquad (4)$$

The value R2 and respective constants find for Isotherm Freundlich and Isotherm Langmuiras appear in Table 2. The low correlation coefficient obtained for Freundlich isotherm suggests that this model might not be applicable to the present system. The value of R2 get for sotherm Freundlich and Isotherm Langmuiras were close to one suggesting best fitted to the equilibrium result [2, 12, 24]. The best fit to isotherm Freundlich (R2= 0.9833) .

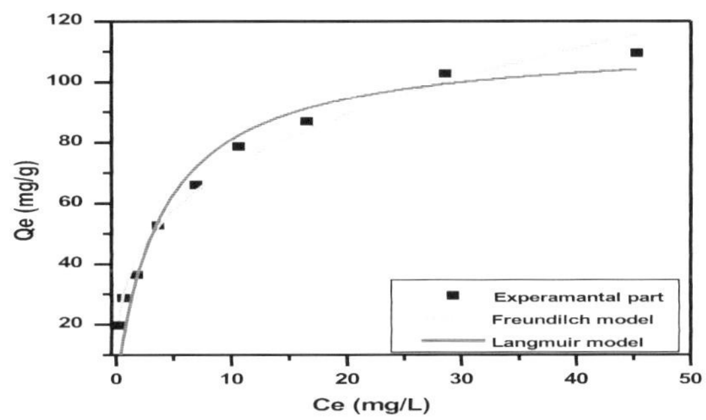

Figure 5. Adsorption models nonlinear fitted of adsorption **dye** on to **ACPP**, conc. = 30 mg. L^{-1}, Temp. = 20°C, and mass 0.05 gm).

Table 2. Freundlich and Langmuir model parameter of **dye** adsorbed on to **ACPP**

Isotherm mode	**Parameters**	**ACPP**
	K_F	34.761 ± 1.841
Freundlich	1/n	0.321 ± 0.018
	R^2	0.9833

	q_m (mg/g)	113.323 ± 7.522
Langmuir	K_L(L/mg)	0.244 ± 0.062
	R^2	0.9377

3. Conclusion

ACPP can be utilized as adsorbent for remove Brilliant yellow dye from water. The ACPP is characterized by high efficiency and effectiveness in remove Brilliant yellow; Cheap and available. Through the results of the best weight for the surface of ACPP at (0.05 gm), removal percentage E% about (80.11%). Mathematical statistics were applied to two types of adsorption isotherms models. It was found that fitted to the Friendlich model depending of the value of (R2) to be close to one and also the Friendlich model is applied to heterogeneous surfaces.

Reference

[1] R. Huang, B.W., B. Yang, D. Zheng, Z. Zhang, *Equilibrium, kinetic and thermodynamic studies of adsorption of Cd (II) from aqueous solution onto HACC-bentonite.* Desalination, 2011. **280**: p. 297-30.

[2] Adejumoke A. Inyinbor, F.A.A.a.G.A.O., *Adsorption of Rhodamine B Dye from Aqueous Solution on Irvingia gabonensis Biomass: Kinetics and Thermodynamics Studies.* S. Afr. J. Chem., 2015. **68**: p. 115-125.

[3] Aseel M. Aljeboree, A.F.A., *Role of plant wastes as an ecofriendly for pollutants (Crystal Violet dye) removal from Aqueous Solutions* Plant Archives 2019 **19**(2): p. 902-905.

[4] I. Ali, M.A., T.A. Khan, *Low cost adsorbents for the removal of organic pollutants from wastewater.* J. Environ. Manag., 2012. **113**: p. 170-183.

[5] M.A. Ahmad, R.A., *Removal of malachite green dye from aqueous solution using rambutan peel-based activated carbon: equilibrium, kinetic and thermodynamic studies.* Chem. Eng. J., 2011. **171**: p. 510-516.

[6] R. Lakshmipathy, N.A.R., N.C. Sarada, *Optimization of brilliant green biosorption by native and acid-activated watermelon rind as low-cost adsorbent.* Desalination and Water Treatment, 2015. **3**: p. 1-10.

[7] Mohammed A. Jawad, A.J.K., Nadher D. Radia, *Role of Sodium Alginate-g-poly (Acrylic acid-fumaric acid) Hydrogel for Removal of Pharmaceutical Paracetamol from Aqueous Solutions by Adsorption* International Journal of Pharmaceutical Quality Assurance, 2021. **12**(3): p. 202-205.

[8] Aseel M Aljeboree, A.N.A., Ayad F. Alkaim, *ACTIVATED CARBON (AS A WASTE PLANT SOURCES)–CLAY MICRO/NANOCOMPOSITE AS EFFECTIVE ADSORBENT: PROCESS OPTIMIZATION FOR ULTRASOUND-ASSISTED ADSORPTION REMOVAL OF AMOXICILLIN DRUG.* Plant Archives 2019 **19**(2): p. 915-919.

[9] Aljeboree A.M. , A.A.N., Alkaim A.F., *Removal of pharmaceutical amoxicillin drug by using (Cnt) decorated clay/ fe2 o3 micro/nanocomposite as effective adsorbent: Process optimization for ultrasound-assisted adsorption.* International Journal of Pharmaceutical Research, 2019. **11**(4): p. 80-86.

[10] Gecgel U, K., H., *Adsorption of Remazol Brilliant Blue R on activated carbon prepared from a pine cone.* Nat Prod Res., 2012. **26**: p. 659-664.
[11] Aseel M. Aljeboree, N.D.R., Layth S. Jasim,Ayad F. Alkaim, *Removal of Pharmaceutical by Adsorption Process; Study Effect of Various Parameters and Determination of Pharmaceutical by Oxidative Coupling Reaction.* International Journal of Drug Delivery Technology, 2021. **11**(3): p. 912-916.
[12] Aljeboree, A.M., Alkaim A F, *Removal of Antibiotic Tetracycline (TCs) from aqueous solutions by using Titanium dioxide (TiO2) nanoparticles as an alternative material.* Journal of Physics: Conf. Series 2019. **1294** p. 052059.
[13] Md. Maksudur Rahman Khan, M.W.R., Huei Ruey Ong, Ainihayati Binti Ismail, Chin Kui Cheng, *Tea dust as a potential low-cost adsorbent for the removal of crystal violet from aqueous solution.* Desalination and Water Treatment, 2015. **2**: p. 1-11.
[14] Yahya A. Faleh , N.D.R., *Removal of Metformin hydrochloride from Aqueous Solutions by using Carboxymethyl cellulose-g-poly(acrylic acid-co-acrylamide) Hydrogel: Adsorption and Thermodynamic Studies.* IOP Conf. Series: Earth and Environmental Science 2021. **790**: p. 012062.
[15] Gupta, V.K. and Suhas, *Application of low-cost adsorbents for dye removal â€" A review.* Journal of Environmental Management, 2009. **90**(8): p. 2313-2342.
[16] M. Ghaedi, A.S., H. Tavallali, F. Shojaiepoor, B. Keshavarz, H. Hossainian, *Activated carbon and multiwalled carbon nanotubes as efficient adsorbents for removal of arsenazo(III) and methyl red from waste water.* Toxicol Environ Chem, , 2011. **93**: p. 438-449.
[17] Aljeboree A.M., A.A.N., Alkaim A.F., *Activated carbon (as a waste plant sources)-clay micro/nanocomposite as effective adsorbent: Process optimization for ultrasound-assisted adsorption removal of amoxicillin drug.* Plant Archives, 2019. **19**: p. 915 - 919.
[18] Kamil, A.M., Mohammed, H.T., Alkaim, A.F.,Hussein, F.H., *Adsorption of Congo red on multiwall carbon nanotubes: Effect of operational parameters.* Journal of Chemical and Pharmaceutical Sciences, 2016. **9**(3): p. 1128-1133.
[19] Fabryanty, R., Valencia, C., Edi Soetaredjo, F., Shella Permatasari Santoso,J. N. P., Yi-Hsu Ju, Suryadi Ismadji. A. K., *Removal of crystal violet dye by adsorption using bentonite – alginate composite.* Journal of Environmental Chemical Engineering, 2017. **5**: p. 5677–5687.
[20] F. Long, J.L.G., G.M. Zeng, L. Chen, X.Y. Wang, J.H. Deng, Q.Y. Niu, H.Y. Zhang, X.R. Zhang, *Removal of phosphate from aqueous solution by magnetic Fe-Zr binary oxide.* Chem. Eng. J. , 2011. **171**: p. 448-455.
[21] Salam Hussein Ewaid et al 2021 IOP Conf. Ser.: Earth Environ. Sci. 722 012008
[22] Salam Hussein Ewaid et al 2021 IOP Conf. Ser.: Earth Environ. Sci. 790 012075
[23] Salam Hussein Ewaid et al 2020 J. Phys.: Conf. Ser. 1664 012143.
[24] Ewaid, S.H.; Abed, S.A.; Al-Ansari, N. Water Footprint of Wheat in Iraq. Water 2019, 11, 535.
[25] Hadi, Z.A., Aljeboree, A.M., Alkaim, A.F., *Adsorption of a cationic dye from aqueous solutions by using waste glass materials: Isotherm and thermodynamic studies.* International Journal of Chemical Sciences, 2014. **12**(4): p. 1273-1288.
[26] Cardoso NF, L.E., Pinto IS, Amavisca CV, Royer B, Pinto RB, Alencar WS, Pereira SFP *Application of cupuassu shell as biosorbent for the removal of textile dyes from aqueous solution.* J Environ Manage 2011. **92**: p. 1237-1247.
[27] Layth S J and Aljeboree, M.A., *Removal of Heavy Metals by Using Chitosan/ Poly (Acryl Amide-Acrylic Acid)Hydrogels: Characterization and Kinetic Study .* NeuroQuantology, 2021. **19**(2): p. 31-37.
[28] J. Chatterjee, N.R., S. K Sar, *Kinetic isotherm of amoxicillin antibiotic through adsorption and its removal by electrocoagulation.* Orient. J. Chem.,, 2014. **30**(2): p. 775-784.

Effect of Biofertilizers on Vegetative Growth and Flower Yield of African Marigold *Tagetes erecta* L. c.v Pusa Narangi Gainda

[1]Razzaq Owayez Idan , [2]Ali jabbar Abdulsada and [1]Sabah Abd Fleih

[1]Department of Horticulture, College of Agriculture, Karbala University, Ministry of Higher Education and Scientific Research, Iraq

[2]Ministry of Science and Technology, Agricultural Research Directorate. Iraq – Baghdad

Abstract. This experiment was conducted to identify the effect of three different biofertilizers and their interaction on flower yield and vegetative growth of African marigold *Tagetes erecta* L. c.v Pusa Narangi Gainda. The results showed that biofertilizer treatments have been significantly effect on spread plant, plant height, number of leaves/ plant number of branches/ plant, flower yield/ plot number of flowers/ plant, flower yield/ plant, and flower yield/ hectare. The highest values were obtained (107.17cm), (41.67 flowers/plant), (80.12cm), (21.50 branches/plant), (106.17 leaves/plant), (458.83g flower yield/ plant), (4129.50g flower yield/ plot) and (41.30 flower yield/ hectare) respectively by the treatment (T_5) with (Azotobacter + Azospirillum+ Mycorrhizae). While the lowest values were obtained (67.69cm), (45.86cm), (10.50 branches/ plant), (47.03 leaves/ plant), (19.00 flowers/ plant), (152.00 g flower yield/ plant), (1368.00 g flower yield/ plot), and (13.6 flower yield/ hectare) respectively by the treatment (T_1) with control.

Key words: Marigold, Azotobacter, Azospirillum, Mycorrhizae

1. INTRODUCTION

Around different detached flowers, those involves of marigold highest point mossy cup oak position Furthermore may be Exceedingly esteemed for their stupendous flower, splendid colour, delightful appearance, myriads for sizes, shapes, forms, fragrance, keeping caliber and will be enriched with expansive range for business potentialities done medicinal and modern division. Marigold is basically developed to India, tropical Africa, srilanka and Madagascar. India involves 15 percentage of the territory to accepted bloom in the universe [1].

Marigold is a standout amongst the A large portion usually developed blossoms for The greater part nations what's more utilized extensively ahead religious and social works in distinctive types. Due to their simplicity clinched alongside cultivation, totally versatility will changing dirt What's more climatic conditions, long span from claiming blooming Also engagingly shaded blooms for phenomenal keeping quality, the marigolds have get to be a standout amongst those practically mainstream blossoms for our organizations in the nation. Blooms need aid sold in the business likewise detached alternately concerning illustration garlands. Since its variable tallness and color marigold is particularly utilize to enhancement also included in scene plants [2].

Nowadays, the vitality Furthermore administration of valuable microorganisms need expanded and need prompt the stronghold for business patterns around that reality [3]. Biofertilizers could however assume a critical part on accomplishing maintainable benefit toward low cosset [4].

Biofertilizers are naturally animated item holding specific strains about micro-organisms which might help supplements of the plants through microbial action. These supplements for compost would moderately less expensive Also assistance to decrease concoction fertilizers utilization. If biofertilizers can't totally displace accepted fertilizers Anyhow for The majority yields upto 20-25 percentage of the nitrogen prerequisite might a chance to be met through biofertilizers [5]. Microorganisms for example, such that Rhizobium, Azotobacter, Azospirillum, Pseudomonas, Also bacillus is regularly utilized Likewise biofertilizers will enhance crop yield. It is news person that 10-20 % about crop yield camwood make expanded for the utilization for biofertilizer concerning illustration and supplement [6]. Biofertilizers also control Furthermore smother soil borne ailments Furthermore A percentage from claiming inoculants generate antibiotics [7].

2. MATERIALS AND METHODS

This research has been done at Agricultural Research Directorate, Ministry of Science and Technology located in Baghdad, Iraq through season 2018. Pusa Narangi Gainda cultivar of African marigold has been used in this experiment. Before planting of seedling, experimental plot has been ploughed one month later. To bring the land to a fine tilth, three harrowing has been given. The experimental area has been located in flat beds of (1x1m) size. The single of the plot was $1.0m^2$, length of the experiment was 12.60m, width of experiment was 5.00m, gross area of experiment was 63.00m, width of main irrigation channel was 1.00m, width of sub irrigation channel was 0.05m, the space for plant to plant was 30cm, total number of plants/plot was 9, total number of plants in the field 243 and net cultivated was 23m ^{2}. NPK has been used with rate 100 -100- 100Kg. Before transplanting, half dose of N and total dose of P2O5 and K2O have been applied. Then, the second dose of N applied as top dressing of 30 day after planting. Randomized Complete Block Design (RCBD) has been used in this experiment. The means were compared using LSD analysis at 5% (Al-Sahuki and Wahib, 1990). The SAS program has been used to analysis the data (SAS, 2001).

2.1. Prepare the seedlings

First of all, 3.0 x 1.0 m nursery beds have been prepared for our experiments. On 27/6/2018, the seeds were sown. The nursery beds have been maintained systematically until the seeds were ready for use. On 12/8/2018 seedlings were transplanting. Gap filling was done when ever required during the first two weeks after transplanting. After 30 days of planting, first weeding has been done. After transplanting, a light irrigation was applied immediately depending on soil upon the moisture requirements

2.2. Prepare the biofertilizers

Biofertilizers collected from the Center of Biotechnology/ Agricultural Research Directorate, Ministry of Science and Technology. Baghdad- Iraq. When real leaf appeared, the bio fertilizers have been added to African marigold transplants at a rate of 5ml of bacterial suspension for each one (80-100x 103cfu/ g incoula for each gram of bacteria) with 10 g Mycorrhizae was added to the African marigold transplants at planting. The trays have been placed in the plastic house till transferring to the permanent places.

2.3. Treatments

The biofertilizers

T1 without added biofertilizers (control)

T2 *Azotobacter chroococcum*

T3 *Azospirillum brasilense*

T4 Mycorrhizae (Glomus mossa)

T5 *Azotobacter chroococcum* + *Azospirillum brasilense* + Glomus mossa

Parameters of vegetative growth and yield

For study the parameters, 5 plants have been selected randomly from each treatment, and measured the plant height (cm), spread plant (cm), the branches number/ plant, number of flowers/ plant, number of leaves/ plant, flower yield/ plot, flower yield/ plant, and flower yield/ hectare.

3. RESULT AND DISCUSSION

Plant height (cm)

The results of Table 1 showed effect of Biofertilizers treatments were significantly differences in plant height (cm). The highest plant (107. 17cm) recorded in tratment (T5) (Azotobacter + Azospirillum + Mycorrhizae), compared with control was recorded (67. 69 cm).

Plant spread (cm)

Table 1 showed that there are significant differences between Biofertilizer treatments in plant spread. The greatest plant spread (80.12cm) recorded in tratment (T5) (Azotobacter + Azospirillum + Mycorrhizae), compared with (T1) control was recorded (45. 86cm).

Number of branches/ plant

The different treatments indicate that biofertilizers significantly affect the number of branches/ plant. The results are clear in Table 1 that among the different treatments, the highest number of branches (21.50) in the treatment (T5) (Azotobacter + Azospirillum + Mycorrhizae). The lower number of branches/ plant was recorded in (T1) Control (10. 50).

Number of leaves / plant

It is evident from (ANOVA) table that the number of leaves/ plant has been greatly significantly impact with the different biofertilizer treatments. As well as the evident from table 1 that the large number of leaves/ plant has been recorded (106. 17) in the treatment (T5) (Azotobacter + Azospirillum + Mycorrhizae). Treatment (T1) Control recorded with its minimum (47. 03).

Table 1. Effect of bioflterlizers on parameters of growth of African marigold (*Tagetes erecta* L.) c.v Pusa Narangi Gainda.

Treatments	Plant height	Plant spread	No. of branches	No. of leaves

T1	67.69	45.86	10.50	47.03
T2	87.98	61.10	16.25	69.42
T3	86.88	60.03	15.83	68.83
T4	86.62	58.95	15.33	68.17
T5	107.17	80.12	21.50	106.17
F-test	S	S	S	S
S Ed	0.21	0.05	0.05	0.11
C D	0.43	0.11	0.10	0.23

The role of the biofertilizers in absorption, availebility and concentration of nutrient such as phosphorus and nitrogen increased most of shoot characteristics of plants with biofertilizer inoculation (Azospirillum+ Azotobacter+ Mycorrhizae). In stimulating production of growth regulators, the biofertilizers play an important role that positively reflect in elongation, increased divisions, and cells expansion, which reflected on shoot growth [8] [9] [10] [11]. Increasing vegetative growth depends on availability of elements in leaves that increase activity and management of photosynthesis, which leads to increase the CO_2 in the leaves which is base units for building the amino acids, carbohydrates, and protein structures [12] [13] [14]. So, the results of this manuscript were agreed with related articles [15] [16] [17].

Number of flowers/ plant

The numbers of flowers/ plant were affected significantly with different organic manures treatments. Table (3) indicates that among the different treatments, the large number flowers were (41. 67 flowers/ plant) in the treatment (T5) (Azotobacter + Azospirillum + Mycorrhizae). The lower number of flowers/ plant has been recorded in (T1) and control (19 flowers/ plant).

Flower yield/ plant (g)

The flower yield/ plant of marigold were significantly affected compared with different treatments (Table 2). The highest flower yield/ plant (g) was recorded (458. 83 g) in the treatment (T5)

(Azotobacter + Azospirillum + Mycorrhizae), while, minimum flower yield/ plant was recorded in treatment (T1) control (152. 00 g).

Flower yield/ plot (g)

From the data that presented in table no 2, it is observed that (T5) treatment (Azotobacter + Azospirillum + Mycorrhizae) gave highest flower/ plot (4129. 50 g). However, the minimum flower yield/ plant (1368. 00g) were recorded in treatment (T1) Control.

Flower yield (to/ha)

The data analysis in Table 2 showed that highest flower yield **(**41. 30 to/ha) in the treatment (T5) (Azotobacter + Azospirillum + Mycorrhizae) has been recorded. But, treatment (T1) control has been recorded (13. 68 to/ha).

Table 2. Effect of bioflterlizers on parameters of flower yield of African marigold *Tagetes erecta* L. c.v Pusa Narangi Gainda

Treatments	No. Flowers/plant	Flowers yield/plant	Flowers yield/plot	Flowers yield t/ha
T1	19.00	152.00	1368.00	13.6
T2	30.00	276.26	2486.37	24.86
T3	28.83	259.56	2336.01	23.36
T4	28.50	249.33	2244.0	22.44
T5	41.67	458.83	4129.50	41.30
F-test	S	S	S	S
S Ed	0.15	1.46	13.15	0.15
C D	0.30	3.05	27.42	0.32

The using of biofertilizers lead to increase and encouragement of plant growth indicators. These strategies have been used in the biological system. The advantages of this technique lead to availability of nutrients through nitrogen fixation in soil, increase the resistance of plants to biotic and abiotic stresses , phosphorus solubilization and growth regulators production such as IAA and GA_3. All the factors above have been contributed to increasing number of fruits, weight of fruit, and diameters and lengths of fruits. Therefore, biofertilizers lead to increased total yield and plant growth. Furthermore, growth regulators contribute increased the shoot and root growth. This is due to division and elongation of cells and tissues. [18] [19] [20] [11] [16].

References

[1] Jawaharlal, M. (2004) Traditional Floriculture. In: Emerging Trends in Ornamental Horticulture, ISOH, New Delhi, 5-10..

[2] Anon, (2006) Marigold cultivation for higher income. Indian Agricultural Research Institute, New Delhi..

[3] Alarcon, A. and Ferrera, C.R. (2000). Biofertilizer: Importance and management in Agriculture. Agriculture Tecnica-en-Mexico., 26 (2): 191-203..

[4] Sharma, A.K. (2003). A search of sustainability. Newsletter Agribios. I (XI), p. 31..

[5] Sivakumar, U.; Natarajan, T.; Chandrayan, K.; Llamurugu, K. and Kumar, G. (2002). Demand, marketing, adoption and promotion of biofertilizer in Tamil Nadu, Newsletter Biofertilizer, 11-12..

[6] Brown, M.E. (1972). Plant growth substances produced by micro-organisms of soil and rhizosphere. J. Appl. Bact., 35: 443.

[7] Singh, P.; Vasudev, R.S. and Bajaj, B.S. (1965). Seed bacterization and biological activity of bulbiformin. Ann. Appl. Biol., 55: 89..

[8] Mirzakhani, M., Ardkani,M.R; Aeene Band, A; Rejali, F and Shirani Rad, A.H. (2009). Response of spring safflower to co-inoculation with Azotobacter chroococcum and Glomus intraradices under different levels of nitrogen and phosphorus. American J. Agricult.

[9] Salhia, B. M. (2010).The effect of Azotobacter chrococcum as nitrogen biofertilizer on the growth and yield of Cucumis Sativus, MS.Thesis of Biological Sciences / Botany The Islamic University, Gaza..

[10] Allawi ,M.M. (2013).Impact of bio,organic and chemical fertilization on the roots architectural and growth and yield of pepper plant (Capsicum annuum L.) Ph.D . Thesis. Horticulture and Landscape Department .College of Agriculture., University of Baghdad.

[11] Shashi, K; Kumar, M . kumar, R. and Raghav ,M.(2018). Effect of biofertilizers on growth and yield of tomato (Lycopersicon esculentum Mill). International Journal of Current Microbiology and Applied Sciences 7 (2), 2542- 2545..

[12] Jarak, M. Durić, S and Dorđević, B. (2010). Benefits of inoculation with Azotobacter in the growth and production of tomato and pepers. Matica srpska Journal for Natural Sciences, 119: 71-76..

IOP Conf. Series: Earth and Environmental Science 1029 (2022) 012029 doi:10.1088/1755-1315/1029/1/012029

[13] Lalitha, M .,Karunakar Babu,M .,Ravi sankar,C and Ashoka Rani,Y.(2004) . Effect of bio-regulation on chlorophyll content and keeping quality of betelvine (Piper betle L.) Andhra Agric .J ,51(1-2): 165-71..

[14] Taiz, L, and Zeiger, E. (2010). Plant Physiology , 5Th end. Sinauer, Sunderland , Massachussetts.USA..

[15] El-Hifny, I .M and El-Sayed, M.A. (2011). Response of sweet pepper plant growth and productivity to application of ascorbic acid and biofertilizers under saline conditions. Australian Journal of Basic and Applied Sciences, 5(6): 1273-1283.

[16] Dharmendra, S.(2014). Effect of Organic Manures andBiofertilizers on Growth, Yield and Quality of tomato (Solanum lycopersicum L.) .MSc. Thesis (Agriculture). Department of Horticulture, College of Agriculture. Jawaharlal Nehru Krishi Vishwa Vidyalaya, Ja.

[17] Saeed, F. H.; Hadi, M. A and Kazem, D. H.(2014). The Detection of Auxin and Cytokine Hormons in Leachate Growth of Some Biogenic Fertilizers. 1st National Women's Scientific Conference (Special Issue): 186-192..

[18] Saeed, F. H.(2015). Integrated Management of Chemical, Organic and Biological Fertilizers and Their Impact on the Growth and Productivity of Some Genotypes of Cucumber Plant. Ph.D Thesis. Horticulture and Gardening Engineering Department / College of Agri.

[19] Saharan,B.S and Nehra,V.(2011). Plant growth promoting Rhizobacteria: A critical review. Life Sciences and Medicine Research,J. (21) p:1-29.

[20] Tosic, I; Golic, Z and Radosavac, A. (2016). Effects of the application of biofertilizers on the microflora and yield of lettuce (Lactuca sativa L.). Acta Agriculturae Serbica, ,21 (42): 91-981.

[21] Aldrich, J. and Baker, R. (1970). Biological control of Fusarium roseum f. sp. Dianthi by Bacillus subtilis. Pl. Dis. Reporter., 54: 446..

[22] Alkaff,H.A. And Hassan.A,A.(2003).). Effect of biofertilizers and organic fertilizer on the growth and yield of okra plants. Uni. Of Aden. Of natural and applied Sci., 7 (1): 25-35..

[23] Gupta, A., K. Srinivas and V. Shukla. (1981). Response of okra (Abelmoschus esculentus L. Moench) to plant spacing and nitrogen, phosphorus fertilization. Indian J. Hort. Sci. 38(3-4): 218-222..

[24] Meena, M. L; Gehlot,V,S.; Meena, D.C.; S., Kishor, S; Kumar, K. S. and Meena, J.K.(2017). Impact of biofertilizers on growth, yield and quality of tomato (Lycopersicon esculentum Mill.) cv. Pusa Sheetal. Journal of Pharmacognosy and Phytochemistry . 6(4.

[25] Nuru Z. M., Islam, M. Z., Islam M. R. (2003). Field efficiency of bio fertilizers on the growth of okra (Abelmoschus esculentus L. Moench). J. Pl. Nutrition and Soil Sci., 166(6), 764-770..

[26] Thakar, N. A, Patel. H. R. And Patel, C. C. (1987). Azolla in management of rppt-knot disease in okra. Indian J. Nematology, 17 (1): 136-137..

[27] Vimala, B. And Natarajan, S. (2001). Effect of nitrogen, phosphorus and biofertilizers on pod characters. Yield and quality in pea. South Indian Horti, 48(1-6): 60-63..

Synthesis And Characterization of New Mn(II), Co(II), Cd(II) and Hg(II) Complexes with ligand [N-(pyrimidin-2-ylcarbamothioyl)benzamide] and their Anti-bacterial Study

Bushra Mukhlif Fayyadh1, Noor Alhuda Basim Abd[2] and Basima Muhsin Sarhan[3]

[1]Supervision Department,General Directorate of Education,Baghdad,AL-karkh,Iraq
[2]Medical Laboratory Techniques Department, Collage of Medical Technology, The Islamic University, Najaf, Iraq
[3]Department of Chemistry,College of Education for Pure Science (Ibn Al-Haitham), Univ. of Baghdad,Iraq

Abstract. A new ligand [N-(pyrimidin-2-ylcarbamothioyl)benzamide](L) has been synthesized through the reaction of the benzoyl iso-thiocyanate with a 2-Amino pyrimidine, The ligand had been identified by the elemental micro-analysis CHNS, UV-Vis, FT-IR, and ^{1}H, ^{13}CNMR spectra, some of the transition metal complexes of that ligand have been produced and characterized with the use of the UVvis spectra, FT-IR, magnetic susceptibilities, conductivity measurements, and atomic absorptions. Ligand's behavior is similar to bidentate coordinating via an S and O donor, From the results that have been obtained, molecular formula of all of the complexes that have been prepared have been $[M(L)_2Cl_2]$ (M^{+2} =Mn, Cd, Hg and Co),the suggested geometrical structure for all of the complexes have been octahedral. The metal and ligand complexes exhibit good activities towards bacteria; *Escherichia coli* and *Staphylococcus aureus*.

KeyWords: Spectral studies, 2-Amino pyrimidine, Ammonium thiocyanate.

1. Introduction

In the past, we have worked with 3- , 5- and 2-amino-pyridines as organic bases and have discovered various packing patterns. Families of the related complexes have been observed, in which packing motifs have been maintained, producing iso structural complexes, whereas in the others, seemingly small changes in organic base have led to massive changes in crystal packing. The pyrimidine as well as its derivations in a set of the nitrogen that contain aromatic ligands have attracted much interests. In spite of simple structure, they have a variety of the characteristics and exhibit themselves as mono dentate as well as the bridging ligands. Their effectiveness in bonding of the ions of the metal and capability for the formation of the discrete and poly nuclear compounds have been quite common. The presence of the coordination and the hydrogen bonds in complexes that have derivations of the pyrimidine promotes assembly to stable coordination as well as supra molecular ensembles [1] [2] [3].

Lately, much interest was directed towards complexing capability of the 2-amino-pyrimidine with the transition metal ions, fundamentally as a result of the many different applications in the medicinal, biological and industrial areas. Numerous therapeutically important compounds include the system of pyrimidine ring. For instance, the sulfadiazine, an antibiotic, and barbiturate, a phenobarbital, include system of the pyrimidine ring and have significant impacts on some of the processes like drug interaction catalysis with the bio-molecules. Pyrimidine provides possible sites of binding for the metal ions as well as their coordination characteristics providing information about interactions of the metal ions with the nucleic acids in the biological system. Researches on metal compounds of the nucleic acid bases were mainly performed on purine complexes and a lower degree of interest was directed towards the pyrimidine derivatives' metal complexes [4] [5] [6]. The complexe Schiff base (6-[2Hydroxy-benzylidene)-amino]-pyrimidine2,4diol) (L) with Zn(II), Cu(II), Hg(II) and Cd(II) had been made. Schiff bases and complexes had been identified by the FTIR, ^{1}HNMR, Mass spectra, UV-Vis, elemental microanalyses (C.H.N.), magnetic moment, and molar conductance [7]. Innovative complexes of the bi-nuclear metals of a general formula of $[M_2(PymL)X_3]$ (M represents Cu(II), Co(II), Ni(II), or Zn(II); PymL $=C_{13}H_{17}N_4O_6$ and X= CH_3CO_2 or Cl) have been synthesized with a use of template condensations of the Schiff base (L) that has been obtained from the glycine with the use of the 2,3-butanedione, 5methyl -2,6pyrimidine-dione and metal chloride/acetate salt in 1:1:2stoichiometric ratios. The synthesized complexes have been identified through the elemental analyses, magnetic measurements, conductance, UVvis, FTIR, ^{1}H, ^{13}C-NMR spectral researches [8]. The objective of the present study is preparation and characterization of new ligand [N-(pyrimidin-2ylcarbamo thioyl) benzamide] (L), in addition to its metal complexes with the ions of Co(II), Cd(II), Mn(II), and Hg(II).

2. Experimental

2.1. Materials and reagents

All of the reagents have been analar or chemical pure grades by the BDH, Fluka and Merck. All of the metal chloride salt and solvent types have been bought from Merk and Fluka com.,and utilized with no purifications.

2.2. Physical measurements

^{1}H and^{13}C–NMR were recorded with the use ofthe Ultra Shield 300MH$_z$ Switzerland at Univ. of Aal el-Bait, Jordan. The melting point has been logged with the use of the apparatus of the Stuart- melting point. FTIR spectra were registered at a range of (4000cm^{-1} -400cm^{-1}) and (4000cm^{-1} -200cm^{-1}) ranges of the ligand and metal complexes on Shimadzu IR470 spectra utilizing the CsI and KBr pellets. Electronic spectra have been obtained with the use of the UV160 Shimadzu spectro-photometer at 25°C for 10^{-3}M solution DMSO with 1.000cm ± 0.001cm matched quartz cell. Molar Conductivity has been evaluated at a temperature of 25°C for 10^{-3} M of the DMSO solution with the use of Philips PW. Digital. microelemental analysis (CHNS) have been conducted with the use of the Acrlo Erba 1106elemental analyzer. The measurements of the magnetic susceptibility have been obtained through the balance magnetic susceptibility model MSBMKI. The complexes' metal contents have been specified with the atomic absorption approach, with the use of the Shimadzu (AA-680G).

2.3. Method for the ligand synthesis

Ligand has been prepared using 2 steps (Scheme-1)

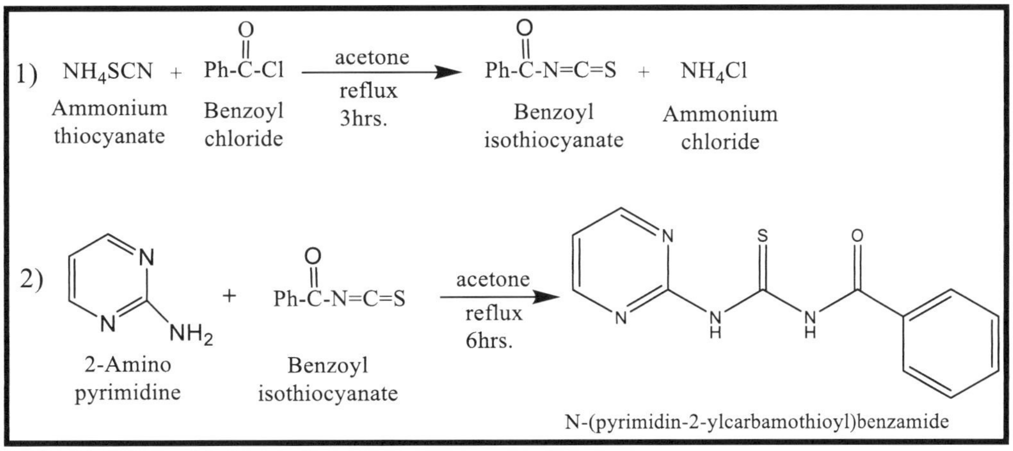

Scheme 1. Synthesis route of the ligand (L)

(A)- Synthesis of the (Benzoyl iso-thiocyanate) [9]

Mix of benzoyl chloride (3.65ml,26mmol) and ammonium thiocyanate (2 g, 1 mmol) in (25.0ml) of acetone has been stirred and refluxed for 3hrs then filtered after that, this filtrate has been utilized for additional reactions.

(B)- Syntheses of [N-(pyrimidin-2-ylcarbamothioyl)benzamide] (L)

(2.472g, 1mmol) of 2-amino pyrimidine in (20.0ml) acetone has been added rapidly into the benzoyl iso-thiocyanate and attaining reflux. Following a 6 hrs refluxing, resulting solid has been gathered, washed by the acetone and then undergone re-crystallization from the ethanol, yield (75%), (mp=160-162)°C, H% found (3.87) calculated (3.99), C% found (55.79) calc.(56.21), S% found (12.4) calc.(12.7) and N% found (21.69) calc.(22.0).

2.4. Method for syntheses of the metal complexes

Syntheses of [Mn(L)$_2$Cl$_2$] complex

A solution of (0.17g, 26mmol) $MnCl_2.4H_2O$ in (10.0ml) ethanol has been added into the solution of (6.70g, 26mmol) (L) in (10 ml) of the ethanol. This mix has been stirred for 6hrs at the temperature of the room, dark green solid has been obtained through the filtration, washed by (1:1) mix of the water: ethanol, which has been re-crystallized from the ethanol and dried in the oven (50°C).

Syntheses of [Co(L)$_2$Cl$_2$], [Cd(L)$_2$Cl$_2$], [Hg(L)$_2$Cl$_2$]complexes.

A similar approach to the one that has been mentioned for preparing [Mn(L)$_2$Cl$_2$] complex has been utilized for the preparation of complexes of [Co^{+2}, Hg^{+2} and Cd^{+2}] ions with (L), Table1 shows a set of the physical characteristics of prepared compounds.

Table1. Some physical characteristics of ligand (L) and its complexes.

Compound	M.wt (gm/mole)	Color	M.P(℃) or dec.	M% Calculation (Found)	Molar Cond. $Ohm^{-1}cm^{2}mol^{-1}$ in DMSO	μ_{eff} (B.M)
$C_{12}H_{10}N_4OS$ (L)	258.3	Yellow	160-162℃	-	2.1	-
$[Mn(L)_2Cl_2]$	642.44	Deep green	270 (dec)	8.55 (8.52)	19	5.92
$[Co(L)_2Cl_2]$	646.43	Deep green	283 (dec)	9.12 (9.01)	18	4.79
$[Cd(L)_2Cl_2]$	699.91	Yellow	198-200℃	16.06 (16.98)	11	0
$[Hg(L)_2Cl_2]$	788.09	Yellow	266 (dec)	25.45 (25.82)	13	0

Dec. =decomposition

3. Results and Discussions

3.1. Ligand (L)

^{1}H-NMR spectrum of free ligand (L),Figure (1) that has been found in the DMSO-d_6 solvent has shown some signals, such as: singlet peak at (δ2.5ppm) for the (DMSO), multiplet peaks at (δ6.50- δ9.03ppm) for the aromatic protons, singlet peak at (δ10.39ppm) for (^{1}H,NH sec. amide), singlet peak at (δ10.45ppm) for (1 H, NH -CS).

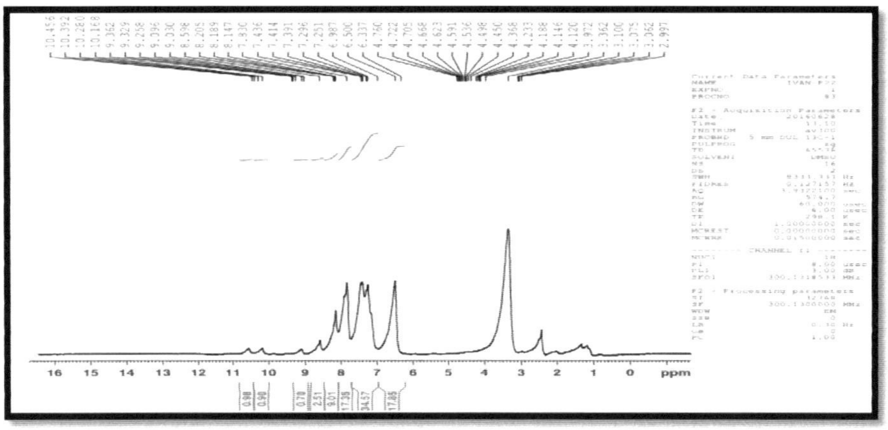

Figure 1. ^{1}H-NMR spectrum of ligand (L)

^{13}CNMR spectra of free ligand (L),Figure(2) that has been recorded in the DMSOd_6 solvent have shown some signals, such as: signals at (δ38.65- δ40.32ppm) for the (DMSO), signals at δ(127.00 and 128.67ppm) for the aromatic carbons, signal at (δ178.11ppm) for (C=O sec. amide), signal at (δ184.34ppm) for (C=S) [10] [11].

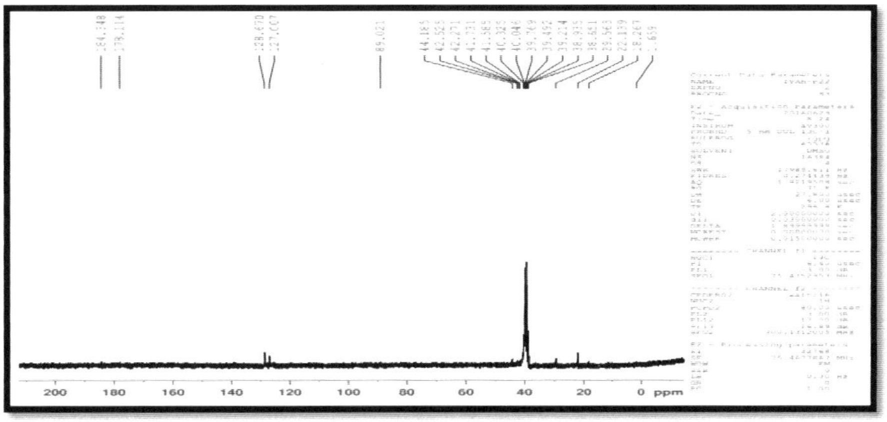

Figure 2.^{13}C-NMR spectrum of ligand (L)

FTIR spectra of free ligand (L), Figure3 showed the bands at (1666cm^{-1}) and (1180cm^{-1}) due to υ C=O(amide) and υ C=S, whereas a different band of absorption at (3363cm^{-1}) might be explained as υ N-H [12] [13] [14]. FTIR spectra data of free ligand have been given by table2.

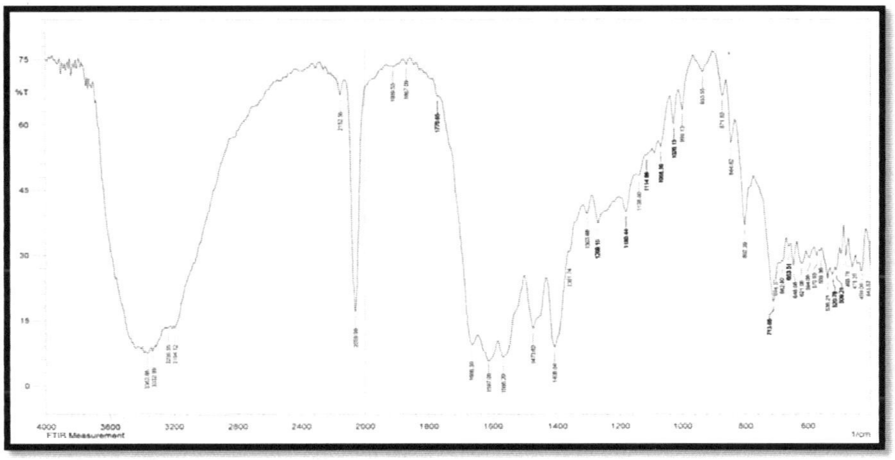

Figure 3. Infrared spectrum of ligand (L)

Table 2. The characteristic infrared band for free ligand (L) and its metal complexes.

Compound	**υ(N-H)**	**υ(C=O) Amide**	**υ(C=S)**	**υ(M-O)**	**υ(M-S)**	**υ(M-Cl)**
Ligand (L)	3363(s)	1666(s)	1180(s)	—	—	—
$[Mn(L)_2Cl_2]$	3360(s)	1616(s)	1203(m)	455(w)	424(m)	255(w)
$[Co(L)_2Cl_2]$	3367(m)	1620(w)	1230(s)	497(m)	416(w)	279(w)
$[Cd(L)_2Cl_2]$	3365(s)	1612(s)	1238(m)	470(s)	424(m)	265(m)
$[Hg(L)_2Cl_2]$	3360(s)	1624(w)	1215(m)	447(s)	424(m)	262(w)

b=broad s=strong m=medium w=week

UVvis

spectrum of free ligand (L). Figure 4 depicts a high intense peak of absorption at ($33557 cm^{-1}$) that could be
a result of electronic transition of the $\pi \longrightarrow \pi^*$ type [15]. Data of free ligand (L) electronic spectrum have been given in table3.

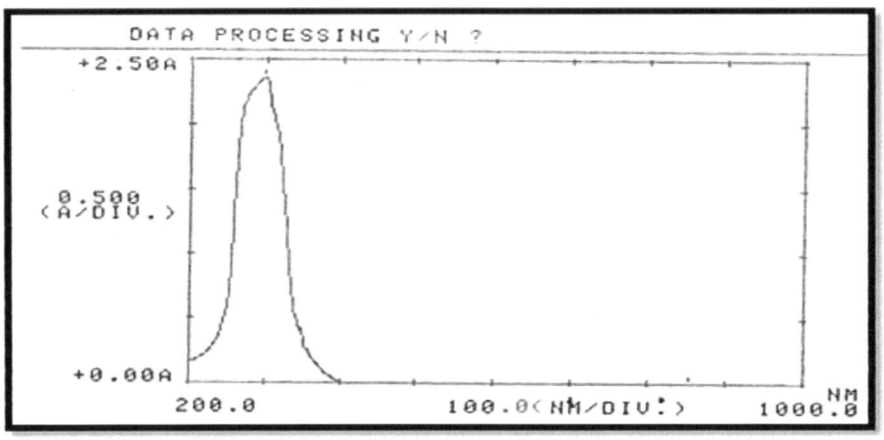

Figure 4. U.V. spectrum of ligand (L)

Table 3. The peak electronic transitions and structure geometries of ligand (L) as well as its complexes

Compounds	λ_{max} nm	Wave number cm^{-1}	ABS	ε_{max} $molar^{-1}cm^{-1}$	Transitions
Ligand(L)	298	33557	2.357	2357	$\pi \rightarrow \pi^*$
$[Mn(L)_2Cl_2]$	297	33670	2.213	2213	Intra ligand
	720	13888	0.020	20	$^6A_1g \rightarrow {}^4T_2g_{(G)}$
	939	10649	0.013	13	$^6A_1g \rightarrow {}^4T_1g_{(G)}$
$[Co(L)_2Cl_2]$	300	33333	2.369	2369	Intra ligand
	364	27472	0.435	435	C.T mixed with $^4T_1g_{(F)} \rightarrow$
	678	14749	0.021	21	$^4T_1g_{(P)}$
	835	11976	0.012	12	$^4T_1g_{(F)} \rightarrow {}^4A_2g$
					$^4T_1g_{(F)} \rightarrow {}^4T_2g_{(F)}$
$[Cd(L)_2Cl_2]$	297	33670	2.273	2273	Intra ligand
$[Hg(L)_2Cl_2]$	273	36630	1.667	1667	Intra ligand
	367	27247	0.094	94	Intra ligand

L.F= Ligand field **C.T = Charge transfer**

3.2. Ligand (L) Complexes

Solid complexes soluble in some of the common solvents like DMSO, DMF, and rather thermally stable. The values of the molar conductivity of all of the complexes in the solvent of the DMSO in 10^{-3}M at 25°C (Table1) indicates non-electrolyte nature [16]. The measurements of the atomic absorption for all of the complexes have given approximate values in the case of comparing it to the theoretical values, Table1 includes physical characteristics for ligand as well as its complexes.

FTIR Spectra of complexes

The spectra of the FT-IR had shown noticeable differences between the bands Figure 5 illustrates stretching vibration of the υ(C=O amide) ranging from (1624 to 1612cm^{-1}) shifted lower frequency values by (54-42cm^{-1}), which suggests a potential to coordinate the ligand via oxygen atom at the group of the carbonyl [17] whereas the band that has been caused by the υ(C=S) emerged between (1238 and 1203cm^{-1}) shifted to higher values of the frequency by (58-23cm^{-1}) indicating coordination of the ligand with sulfur atom at thiol group to central ion [18]. The band of stretching vibration υ(N-H) showed either no changes or quite limited in their frequency values (3365-3360cm$^{-1)}$ there for the indication don't coordinate to metal ion. Metal-sulfur, Metal-oxygen, and Metal-Chlorine bonds have been established through a presence of a stretching tremor of υ(M-S), υ(M-O), & υ(M-Cl) about (497-447cm^{-1}) ,(424-416cm^{-1}) and (279-255cm^{-1}) [19]. Table2 lists the important bands and assignments for all of the compounds that have been prepared.

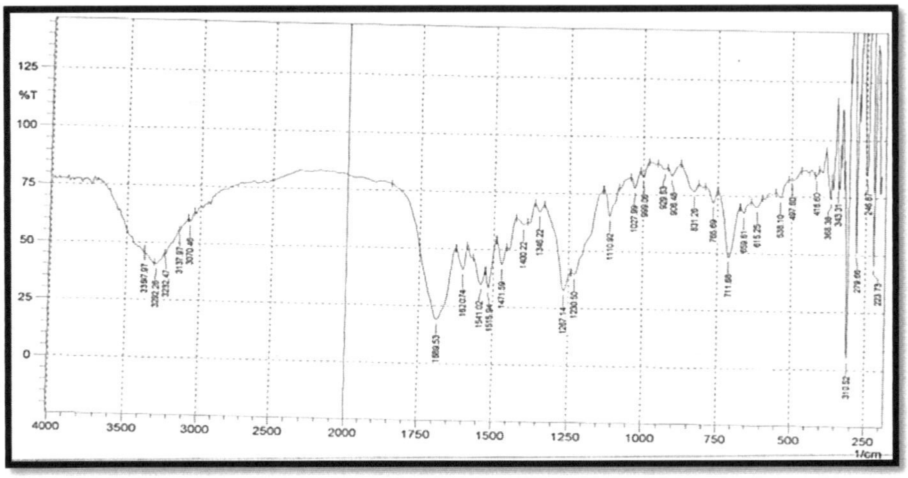

Figure 5. Infrared spectrum of complex $[Co(L)_2Cl_2]$

Electronic spectra of complexes

-$[Mn(L)_2Cl_2]$ d^5

The Deep green Mn(II) complex exhibits the band at ($33670 cm^{-1}$) that results from the intra ligand as well as the other bands at ($13888 cm^{-1}$) and ($10649 cm^{-1}$), resulting from electronic transfers $^6A_1g \longrightarrow {}^4T_2g_{(G)}$ and $^6A_1g \longrightarrow {}^4T_1g_{(G)}$ respectively, which suggest an octahedral geometry around ion of Mn(II) [20].

-$[Co(L)_2Cl_2]$ d^7

The Deep green complex spectrum had given 4 bands at $33333 cm^{-1}$, $27472 cm^{-1}$, $14749 cm^{-1}$ and $11976 cm^{-1}$ that result from the intra ligand, C.T mixed with $^4T_1g_{(F)} \longrightarrow {}^4T_1g_{(P)}$, $^4T_1g_{(F)} \longrightarrow {}^4A_2g$ and $^4T_1g_{(F)} \longrightarrow {}^4T_2g_{(F)}$ respectively and rachinter parameter of electronic repulsion (B^-) has been discovered to be equal to $419.53 cm^{-1}$, from equation $\beta = B^- / B^0$, has been found equal (0.43),those parameter have been accepted to Co(II) octa-hedral complex [21].

-The complexes of $[Cd(L)_2Cl_2]$ and **$[Hg(L)_2Cl_2]$**, Figure (6) exhibit only the charge transfer of the ($M \rightarrow L$) in a range of (36630-$27247 cm^{-1}$) [22] [23]. All the transitions as well as their assignments have been summarized in Table3.

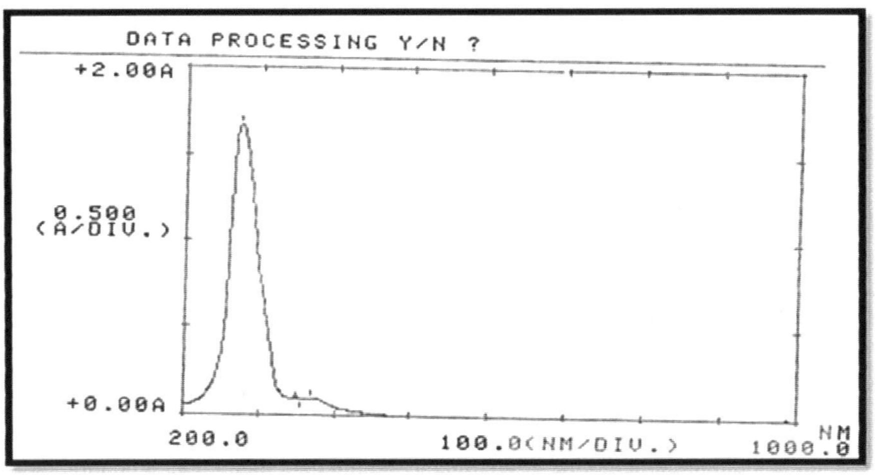

Figure 6. UV spectrum of complex $[Hg(L)_2Cl_2]$

Complexes' magnetic moment

The assessed effective magnetic moment (μeff) and magnetic susceptibility values for Co(II), Mn(II) complexes have been listed in Table1. Mn(II), Co(II) complexes exhibit μeff (5.92, 4.79) B.M, that may be normal values for the octahedral complexes with a high spin. [24]

Biological Activities of ligand (L] and related Complexes

Ligand (L] and its complexes' biological activities have been researched with the use of the approach of inhibition [25] [26], 2 gram bacteria types, which are negative (*E. coli*) and positive (*S. aureu*). Ligand (L) and metal complexes had shown good activities for the inhibition of the growth of the bacteria which have been researched at the concentration that has been prepared. The following points have been concluded and observed in results that have been listed in Table4, Figure (7)

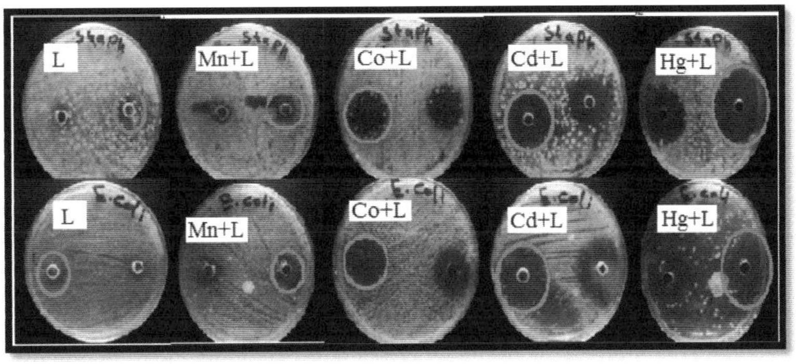

Figure 7. Effects of S. Aureus (gram positive) E. Coli (gram negative) on the ligand and its complexes.

Table 4. Diameter of the inhibition circle for bacteria after 24hrs in the cubation paid and 37°C.

Compounds	**Inhibition Zone Diameter (mm)**	
	***S. Aureus* (gram positive)**	***E. Coli* (gram negative)**
L	6	5
$[Mn(L)_2Cl_2]$	8	12
$[Co(L)_2Cl_2]$	14	15
$[Cd(L)_2Cl_2]$	24	25
$[Hg(L)_2Cl_2]$	31	32

4. Conclusion

The proposed structures for the complexes based on a magnetic moment, molar conductivity, spectroscopic researches (FTIR, atomic absorption and UV-Vis) and (^1H – ^{13}CNMR for the ligand (L) only) for ligand and all of the complexes which have been prepared, it has been suggested that ligand (L's) behavior is like a bidentate on the coordination with the ions of Co(II), Mn(II), Hg(II) and Cd(II) through the (C=O amide) group's oxygen atom and (C=S) group's sulfur atom, which suggests the octahedral form around ions of the metal for all of the complexes that have been prepared, Figure (8).

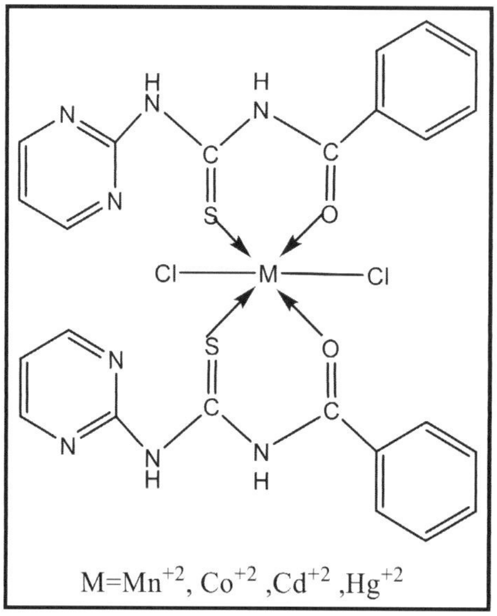

Figure 8.: Proposed chemical structure formula of complexes

References

[1] E. M. Miyuki, J. P. Barry, and M. T. Mark, J. Coord. Chem. 55(11) (2002) 1337–1351..

[2] Yu. V. Kokunov, Yu. E. Gorbunova, and V. V. Kovalev, Russian Journal of Inorganic Chemistry 58(10) (2013) 1187–1192 ..

[3] J. J. Samson, and P T. M. ackianathan, Jenniefer and Muthiah Chemistry Central Journal 7(35) (2013) 1-15..

[4] Y. Sindhu,; C. J. Athira, and M. S. Sujamol, Synthesis and Reactivity in Inorganic, Metal-Organic, and Nano-Metal Chemistry, 43 (2013) 226–236..

[5] G. M. Rajkumar, G. V. Suresh, and K. K. Gopal, ejpmr. 5(03) (2018) 225-235..

[6] M. O.Abd El-Fatah, S. A. Mayada, and S. S. Mamdouh, Journal of the Korean Chemical Society, 54(4) (2010) 402-420..

[7] N. R. Al-Khafaji, RJPBCS.9(2) (2018) 189-195..

[8] N. S. Abhay, P. S. Netra, and K. S.Chandra, Arabian Journal of Chemistry 9(1) (2016) 48-61..

[9] A. Kabbani,; H. Ramadan, and A. Hanuom, Journal of the University of Chemical Technology and Metallurgy 40(4) (2005) 339-344..

[10] D. Seenaiah, B.C. Venkatesh, A. Padmaja. and V. Padmavathi, Indian Journal of Chemistry 52B,(2013) 942-948..

[11] S. Ahmad, and M. Amr, The Open Bioactive Compounds Journal, 6 (2018) 1-10..

[12] B.M. Sarhan,; S.M. Lateef and E.J. Waheed, Ibn Al-Haitham J. for Pure & Appl. Sci, 28 (2) (2015) 102-115..

[13] R.M. Silverstein, G.C. Bassler, and T.C Morrill Spectroscopic identification of organic compounds, 4th Ed., John Wiley and Sons, New York. ,(1981).

[14] B.M. Sarhan,; S.M. Lateef and E.J. Waheed, DJPS. 12(1) (2016) 28-42..

[15] R.J Dyes, Prentice –Hall, Inc., Englewood cliffs, N, J. London (1996).

[16] Y. Yu-Ye, and L. Jian-Feng, Molecules, 14 (2009)1747-1754..

[17] Nakamoto, K; Infrared spectra of inorganic and coordination compounds, 4th Ed., John Wily and Sons, New York. (1996).

[18] B.M. Sarhan,; E.J. Waheed, and N.J. Kadhim, Journal University of Karbala, 14(3) (2016)91-101..

[19] N. Nawar,; I.I. El-swwah, , N.M. Hosny, and M.M. Mostafa, Arabian Journal of chemistry,17(1) (2011) 434–438..

[20] B .M. Sarhan, A.H. Abed, and R.M. Rumez, , AL-Mustansiriyah Journal of science, 24(4) (2013) 65-74..

[21] A.B.P. Lever, Inorganic electronic spectroscopy, Elsevier publishing company Amsterdam, London, New York (1968)..

[22] F. Arjmamd, S. Parveen, and D.K. Mohapatra, Inorg.Chem.Acta, 388(15) (2012) 1-10..

[23] A.D. Naik, J. Beck M.N. Dirtu, and C. Bebrone, Inorg.Chem.Acta, 368(2011) 21-28..

[24] L. Mulay, Magnetic Susceptibilit , John wiley and sons; New York; part1,vol(4) (1977)..

[25] S. Tauber, and R. Nau, Immunomodulatory properties of antibiotic, Current molecular pharmacology, 1(2008) 68-75..

[26] M. Neha, A. Imran, and K. Ashok, International Archive of Applied Sciences and Technology, 5 (2014) 37-42..

Toxicity Remediation of Petroleum refinery Wastewater by Electrooxidation Techniques Using Graphite Electrode

Ehab Dheyaa radhy[1] and Sarmad T Najim[2]

[1] Middle Refinery Company,najaf, Iraq
[2] Dept. of Chemical Engineering, Univ. of Al-Nahrain, baghdad,Iraq.

Abstract. The present experimental study used a successful advanced electrochemical oxidation process (AEOP) and a parallel-plate electrochemical reactor to investigate the potential of electrochemical oxidation (EO) treatment method for real produced wastewater produced by the Al-Samawah petroleum refinery. In electrochemical oxidation, several electrodes type were employed, with carbon felt and graphite plate serving as the anode. Tests of total organic compounds and chemical oxygen demand have confirmed this. The influence of numerous operational factors on performance, such as current density, supporting electrolyte, the distance between electrodes, and temperature, has been investigated, as well as the energy consumption. After 1 hour of treatment, results were obtained at 40°C, and 15 mA/cm^2 using graphite plate anode, that the COD removal was 77.509%. And the TOC removal was 68.13%, by using the direct Electrooxidation technique and COD removal 86.802%, TOC removal 70.662% by using indirect Electrooxidation.

Keywords: Electrochemical oxidation processes. Wastewater treatment. Direct oxidation. Total organic compounds

1. Introduction

Petroleum pollution of waterbodies and groundwater may be occur from a variety of sources, including truck and train spills, pipeline and storage tank leaks and bursts, and the Untreated wastewater from petroleum refineries that is dumped into the environment *[1] [2]*. Total organic compounds and salinity are common characteristics of the effluent from petroleum refineries, making it a substantial cause of environmental degradation. Because there are numerous harmful components in this water, it must be treated carefully before being released into receiving bodies *[3]*. As the global demand for petroleum products rises, our environment becomes increasingly vulnerable to harmful effects. One of these concerns stems from the massive amounts of wastewater created during crude oil processing, which contains significant levels of pollutants in the range of 0.6–1.4 gallons per ton of oil produced *[4]*. Different hydrocarbon components, oil, detergents, solvents, sand, organic acids, salt, suspended mineral solids, and sulfur compounds, catalyst particles, chemical transformation products, and equipment corrosion products are all found in petroleum refinery effluents *[5]*.

Since the end of the 1980s, the oxidation of different organic dissolved in water pollutants has been studied on a variety of anodic materials, with a special focus on electrochemical destruction *[6] [7]*. Electrochemistry has recently attracted a lot of interest in the treatment of various industrial effluent, such as refinery wastewater, due to its special capability to oxidize or decrease contaminants within the water close to the electrode *[3]*. Electrochemical oxidation processes powerful method for recovering and recycling these fluids. Because of its appealing characteristics such as flexibility, energy economy, automation capability, and environmental compatibility, Electrooxidation of refractory effluents has gotten a lot of attention *[8] [9]*. Freshwater consumption and capital costs are reduced by using recycled water from petroleum waste in cooling towers and boilers. Electrochemical technology may be used to clean and reuse effluent from petroleum refineries. It uses fewer chemicals and produces less sludge as a result. In addition, as compared to other traditional approaches, the time necessary to attain the standard limit for reuse/recycle water is less *[10] [11]*.

Hydroxyl radicals can be created either directly (anodic oxidation) or indirectly in EAOPs. Water when in contact directly with the anode produces OH in a heterogeneous manner. Processes are commonly used the OH to treat a variety of wastewaters, with a mineralization efficiency of about 100% in most situations. They may be used in treating a wide range of effluent, and they can also remove contaminants in gas, liquid, and solid form *[12] [13]*. These contaminants are treated by reactions that take place on the anode. The following are the reactions: (a) Water breakdown and hydroxyl radical generation (b) Active oxygen generation via OH oxidation (b) manufacture of anode oxides MO (d) oxygen release from the breakdown of MO (e) Organic contaminants are burned by hydroxyl radicals (f) Organic substances are broken down, as in Figure 1-1 *[14] [15]*.

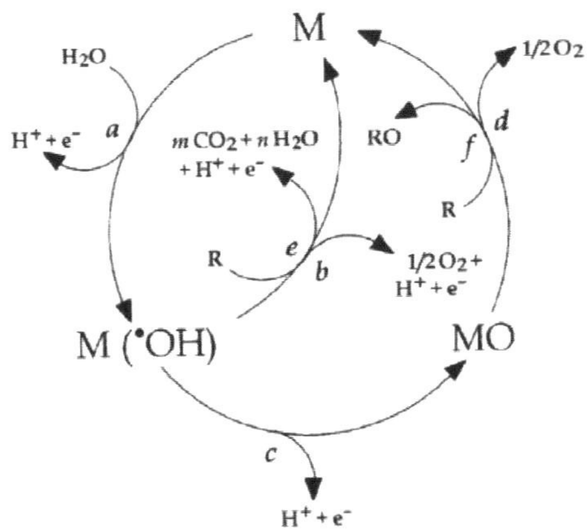

Figure 1. Direct oxidation

M + H2O→ M(HO) + H+ + e- (1)

The M electrode interacts with hydroxyl radical to form the MO electrode oxide

M(HO•) → MO + H+ + e- (2)

If the electrode is active, the reactions are as follows

(a) Metal oxide oxidizes the organic pollutants

MO + R →M + RO (3)

(b) Addition to the decomposition of metal oxide to release oxygen

MO→M + 1/2 O_2 (4)

If non-active electrode

M (HO•) + R → M + mCO_2 + nH_2O + H+ + e- (5)

(c) Rate of formation of anode oxides

M (HO•) → M + 1/2 O_2 + H+ + e- (6)

Factors affecting direct oxidation: (a) The rate of formation of HO• (b) Activity of the anode electrode (c) The rate of formation of anode oxides [14].

The primary objective of the research was to determine how efficient EO is in breaking down suspended particle material in real petroleum refinery effluent. The researchers investigated how effectively EO worked with a graphite anode to remove pollutants and the toxicity of petroleum refinery effluent. Also appraises the operating conditions and performance of oxidation rate for the removal of contaminants indicated as total organic compounds (TOC) chemical oxygen demand (COD). Graphite electrodes were used to conduct the direct and indirect oxidation and also graphite as the cathode.

2. Experimental

2.1. wastewater

The national oil refinery at al Samawah, Iraq, ejects large amounts of petroleum refinery effluent, totaling more than 100 m3 each day (PRW). The waste water used in the research was prepared from the current treatment unit for physical separation. The water was taken from the last horizontal sedimentation tank, where it is used to separate the oily upper layer from the sewage, to serious environmental problems of groundwater and soil. Effluents contains dangerous chemicals, greases, and oils.

Table 1. Shows the characteristics of the polluted water.

Parameter	value (mg/L)
$SO4^-$	178
CL^-	98.6
Oil and Grease	136.7
TDS	1497
TSS	713
COD	538
TOC	317
BOD	28
Heave matel	
Pb	0.04
Ni	0.12
Co	0.03
Fe	9.6
Zn	0.8
TPH	119.5

2.2. Electrochemical oxidation procedure

At atmospheric pressure, the oxidation process was carried out in a Pyrex glass batchstirred reactor. Graphite, commercially obtained for laboratory, used as anode and cathode, each having an effective area of 60 cm^2. The electrodes are placed in the form of parallel to each other, with varying distances (30, 25, 15 mm), connected with a DC power supply QJ3005XII (I = 0–3 A) where the current densities (5, 10, 15 mA/cm^2). In each experiment, 450 ml of real effluent water is placed in the reactor. And samples were taken at certain intervals on a regular basis.

IOP Conf. Series: Earth and Environmental Science 1029 (2022) 012031 doi:10.1088/1755-1315/1029/1/012031

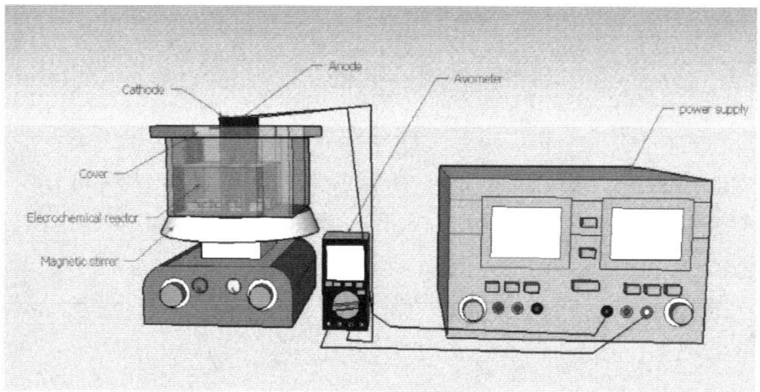

Figure 2. The electrochemical system

2.3. Analytical techniques

Samples were taken every 10 minutes for testing (COD) using the spectrophotometer (MD200, Lovibond). COD was measured in petroleum refinery wastewater by adding (2 ml) of wastewater to K2Cr2O7 as an oxidizing agent, which is placed in a reactor at 150 °C for a period of up to 120 minutes COD (RD125, Lovibond). Then the sample is left to cool to room temperature, the sample is read using a spectrophotometer (MD200, Lovibond).

Total Organic Compounds using the DR500 spectrophotometer. The DRB200 switch is turned on and the TOC system is selected. 10 mL of sample is added to an Erlenmeyer flask, then 0.4 mL of Buffer Solution is added to it. The pH should be 2 and mixed for ten minutes. TOC Persulfate Powder is added to the detection vial. The blank is prepared, 1 ml of pure water is added to the detection flask, and then the sample is prepared in the same way. The detection flask is closed and placed in the reactor. The reactor is closed as the process in the reactor requires two hours at a temperature of 103-105 °C. After two hours, the samples are taken out of the reactor and left to cool for one hour, then they are examined in the DR500 spectrophotometer

3. Results & Discussion

3.1. Effect of the current density

One of the most important factors to consider in the operating factor. In electrochemical oxidation is the current density because it indicates the main factor behind charge migration. Its influence was investigated in the range of 5 and 15 mA/cm^2 for TOC and COD removal as shown in Figure.3 the applied charge of real wastewater at a distance of 1.5 cm with 40 °C temperature. The findings reveal that when the current density rises, the rate of TOC and COD removal rises as well. In terms of COD analysis, from 53.71% to 77.509% removal, In terms of TOC analysis (Figure.3) from 51.73% to 68.13% removal was accomplished after 60 minutes of oxidation in the experiment with 15 mA cm^2, by using graphite electrode.

It is well knowledge that as current density increases, the production of oxidizing radicals on the anode's surface rises, resulting in maximal deterioration. As a result, the greater removal efficiency was achieved at a higher current density. As increases the current density, the operating voltage will rise, and therefore oxygen evolution will increase.

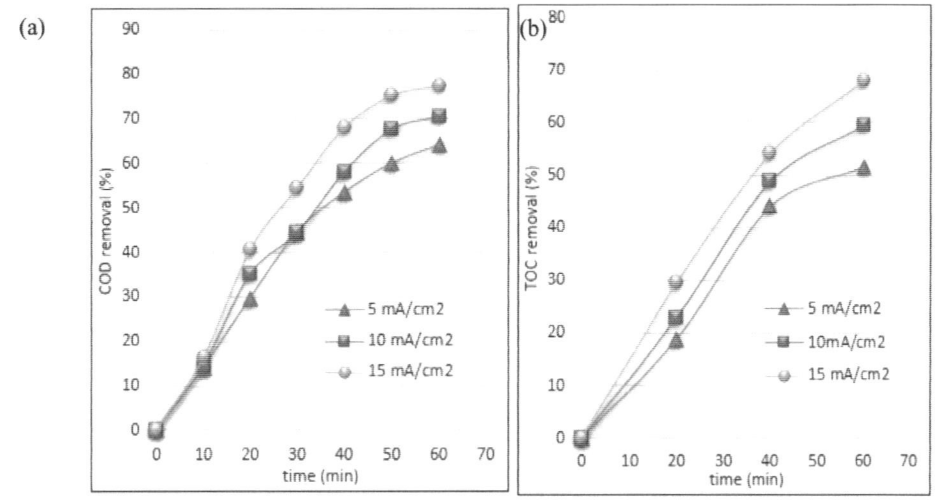

Figure 3. Effect current density operating condition (1.5 cm distance; 40°C; and at agitation speed 250 rpm) (a) COD removal % (b) TOC removal %.

3.2. Effect of Electrolyte Temperature

The impact of temperature in TOC and COD removal, two temperatures level were studied, 25 and 40 °C. The findings are shown in Figure 4. With graphite electrodes at a distance of 1.5 cm and 15 mA/cm^2 current density, the final COD content of wastewater was reduced to 151, 121 ppm, respectively, resulting in COD removal of 71.93, and 77.50 percent, and TOC removal of 62.46, and 68.13 percent at 25, and 40°C, respectively, during of 60 minutes.

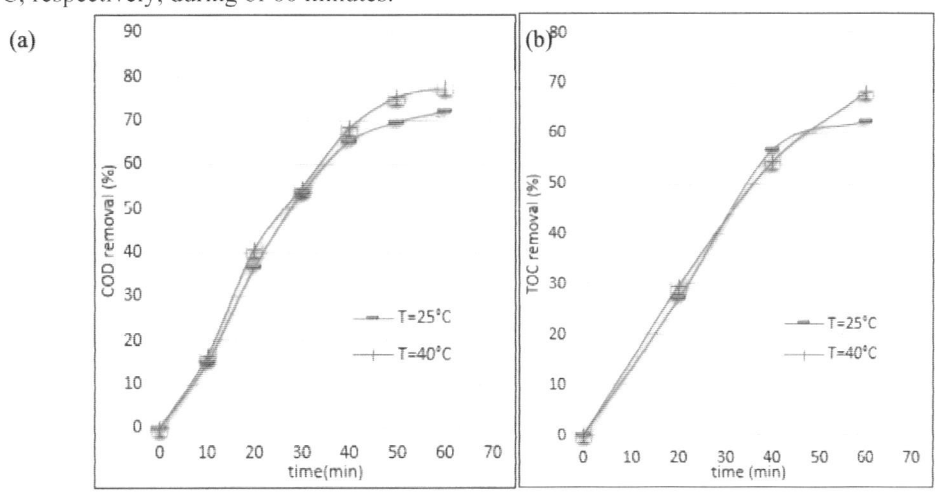

Figure 4. Effect electrolyte temperature operating condition (1.5 cm distance; 15 mA/cm^2; and at agitation speed 250 rpm) (a) COD removal % (b) TOC removal %.

3.3. *Effect of Distance between Electrodes*

As seen in Figure 5, decreasing the electrode spacing from 30 mm to 15 mm demonstrating that the closer the electrodes, the greater the effectiveness of pollution removal. If the distance is 1.5 cm, using graphite electrode the COD removal percentage will rise from 71.93% to 77.50%, at a current density 15 mA/cm^2, 40°C, and the higher the distance, the decrease in the removal occurs. The optimal electrode distance was about 15 mm, where the applied electric field intensity was greater. Short circuits might be produced by either direct electrode connection or conducting contaminants in the water if the electrode distance is too short.

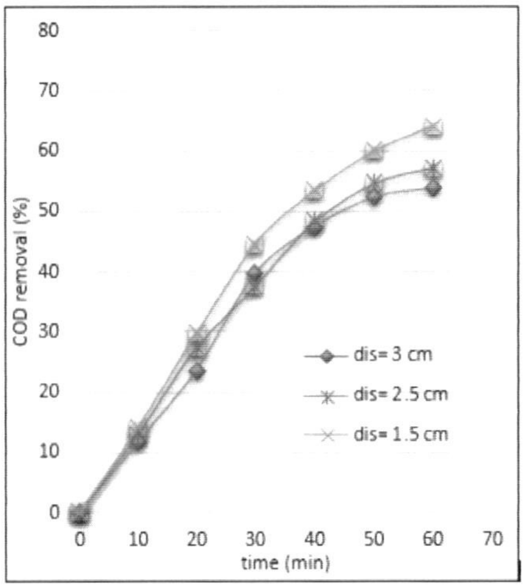

Figure5. Effect distance between electrodes on COD removal, operation conditions (current density 15 mA/cm^2; 40 °C).

3.4. Effect of Supporting Electrolyte (Indirect oxidation)

For indirect oxidation, active chlorine is one of the most prevalent electrochemically generated oxidants. The effluent of rising NaCl conc. on electro-oxidation was investigated in order to evaluate the exact effects of indirect oxidation mediated by active chlorine. Figure (6 (a)) shows the effluent of NaCl conc. on the percentage of COD removal and figure (6 (b)) shows the effluent of NaCl conc. on the percentage of TOC removal. It was found that as the conc. of NaCl was increased from 0 to 1.5 g, the COD removal% enhanced from 77.509 to 86.802%. The COD removal rate is clearly increasing with rising NaCl conc. up to 1.5 g, as shown in the figure 6.

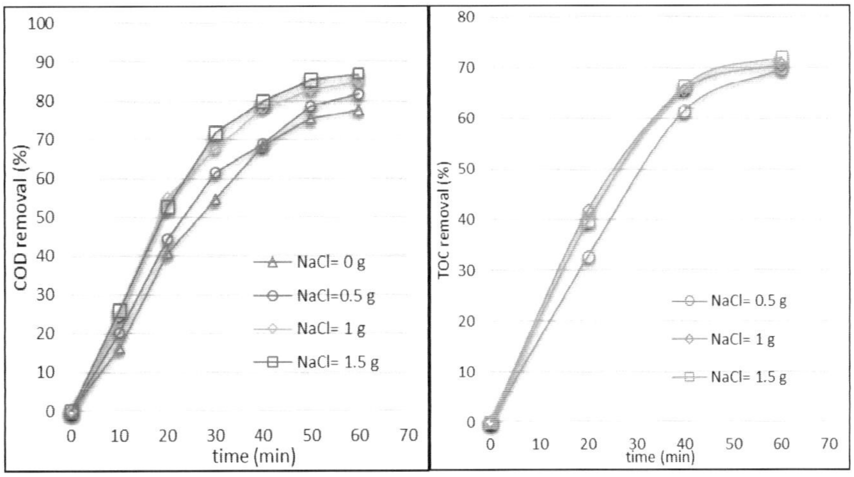

Figure 6. Effect of NaCl concentration operating condition (current density 15 mA/cm^2, temperature 40 °C, distance 1.5 cm; and at agitation speed 250 rpm) (a) COD removal % (b) TOC removal %.

4. Conclusions

The deterioration of refinery effluents was assessed using the TOC and COD removal efficiency. Several operating conditions were used to investigate the pollutants deterioration in PRW such as current density, the distance between electrodes, temperature, and various electrode material types. the results proved that the direct and indirect oxidation process was accepted graphite electrode was used to remove contaminants from PRW with COD and TOC removal rates of 77.50% and 68.138% respectively, using graphite plate, at distance between electrodes 15 mm, 40°C, and 15mA/cm2 current density. The removal efficiency of COD and TOC is improved by adding NaCl because of the generated active chlorine in the solution by electro. It was increased in COD 10.705% and TOC 3.57% compared to direct treatment. The estimated energy consumption of PRW treatment by Electrooxidation process of maximum 11.5821 Wh/l when applying a current density of 15 mA/cm2 and 10.48 Wh/l for indirect oxidation.

Acknowledgements

I thank Almighty Allah for providing me with the capacity to perform this job. My family, supervisor Asst. Pro. Dr. Sarmad Talib deserves particular appreciation for their support, assistance, and guidance during this project.

References

[1] B. Rajasekhar, I.M. Nambi, S.K. Govindarajan, Investigating the degradation of nC12 to nC23 alkanes and PAHs in petroleum- contaminated water by electrochemical advanced oxidation process using an inexpensive Ti/Sb-SnO2/PbO2 anode, Chem. Eng. J. 404 (2021.
[2] S. Jafarinejad, Petroleum Waste Treatment and Pollution Control, 2016.
[3] L. Yan, H. Ma, B. Wang, Y. Wang, Y. Chen, Electrochemical treatment of petroleum refinery wastewater with three-dimensional multi-phase electrode, Desalination. 276 (2011) 397–402.

https://doi.org/10.1016/j.desal.2011.03.083..
[4] S. Chakawa, M. Aziz, Investigating the result of current density, temperature, and electrolyte concentration on cod: Subtraction of petroleum refinery wastewater using response surface methodology, Water (Switzerland). 13 (2021). https://doi.org/10.3390/w.
[5] S. V. Meshcheryakov, E.A. Mazlova, V.L. Zavorotnyi, D.A. Miroshnichenko, Treatment of wastewaters in oil and gas plants, Chem. Technol. Fuels Oils. 36 (2000) 143–145. https://doi.org/10.1007/BF02725265..
[6] C.A. Martínez-Huitle, S. Ferro, Electrochemical oxidation of organic pollutants for the wastewater treatment: Direct and indirect processes, Chem. Soc. Rev. 35 (2006) 1324–1340. https://doi.org/10.1039/b517632h..
[7] G. Chen, Electrochemical technologies in wastewater treatment, Sep. Purif. Technol. 38 (2004) 11–41. https://doi.org/10.1016/j.seppur.2003.10.006..
[8] A.N. Ghanim, A.S. Hamza, Evaluation of Direct Anodic Oxidation Process for the Treatment of Petroleum Refinery Wastewater, J. Environ. Eng. 144 (2018) 04018047. https://doi.org/10.1061/(asce)ee.1943-7870.0001389..
[9] C. Martínez-Huitle, S. Ferro, A. De Battisti, Electrochemical incineration of oxalic acid, Electrochim. Acta. 49 (2004) 4027–4034. https://doi.org/10.1016/j.electacta.2004.01.083..
[10] D. Skban Ibrahim, P. Seethala Devi, C. Veerababhu, N. Balasubramanian, Treatment of petroleum effluent using a tubular electrochemical reactor, Pet. Sci. Technol. 32 (2014) 1932–1939. https://doi.org/10.1080/10916466.2012.701693..
[11] N. Mohan, N. Balasubramanian, C.A. Basha, Electrochemical oxidation of textile wastewater and its reuse, J. Hazard. Mater. 147 (2007) 644–651. https://doi.org/10.1016/j.jhazmat.2007.01.063..
[12] L. Feng, E.D. van Hullebusch, M.A. Rodrigo, G. Esposito, M.A. Oturan, Removal of residual anti-inflammatory and analgesic pharmaceuticals from aqueous systems by electrochemical advanced oxidation processes. A review, Chem. Eng. J. 228 (2013) 944–964.
[13] Y. He, H. Lin, Z. Guo, W. Zhang, H. Li, W. Huang, Recent developments and advances in boron-doped diamond electrodes for electrochemical oxidation of organic pollutants, Sep. Purif. Technol. 212 (2019) 802–821. https://doi.org/10.1016/j.seppur.2018.11.056.
[14] Bassam Faron Alfarhani *et al* 2019 *J. Phys.: Conf. Ser.* 1294 052045
[15] Bassam Alfarhani, Maha Al-Tameemi, Hector C. Goicoechea, Fernando Barbosa, Andres D. Campiglia, Direct analysis of benzo[a]pyrene metabolites with strong overlapping in both the spectral and lifetime domains, Microchemical Journal, 137 2018, 51-61,
[16] B. Alfarhani, M. Al-tameemi, A.V. Schenone, H.C. Goicoechea, F. Barbosa, A.D. Campiglia. Microchem. J., 2016 129 83–89. https://doi.org/10.1016/j.microc.2016.06.010.
[17] Zeina. M. Kadam (2021). Assessment of Electrochemical Activity of some Amino acids on the Modified Carbon Electrode Surface. Research Journal of Pharmacy and Technology, 14(10):5325-9. doi: 10.52711/0974-360X.2021.00928
[18] Z. M. Kadhim and A. H. Ali, (2018), Studying of Optical and Structural Properties of a Novel Ligand (5-ClCPAI) Thin Films by Spray Pyrolysis Method. Journal of Engineering and Applied Sciences, 13(14), 11032-11040.
[19] Z. M. Kadam, C. D. Gwenin, (2017), Polymer Membranes based on Ionophore-impregnated for Nutrients Detection by Electrochemical Methods, Der Pharma Chemica, 9(20):29-33
[20] Ali A Alsudani and Ghaidaa Raheem Lateef Al-Awsi, 2020. Biocontrol of Rhizoctonia solani (Kühn) and Fusarium solani (Marti) causing damping-off disease in tomato with Azotobacter chroococcum and Pseudomonas fluorescens. Pakistan Journal of Biological Sciences, 23: 1456-1461.
[21] Ghaidaa Raheem Lateef Al-Awsi *et al* 2021 *IOP Conf. Ser.: Earth Environ. Sci.* 790 012013.

[22] Bassam F. ALfarhani *et al* 2021 *IOP Conf. Ser.: Earth Environ. Sci.* 790 012037
[23] C.A. Martinez-Huitle, S. Ferro, Electrochemical Oxidation of Organic Pollutants for the Wastewater Treatment: Direct and Indirect Processes, ChemInform. 38 (2007). https://doi.org/10.1002/chin.200714276..
[24] M. Shestakova, M. Sillanpää, Electrode materials used for electrochemical oxidation of organic compounds in wastewater, Rev. Environ. Sci. Biotechnol. 16 (2017) 223–238. https://doi.org/10.1007/s11157-017-9426-1..

Synthesis, characterization and evaluation of photo degradation efficiency of Reduced Graphene Oxide –Cadmium sulfide Nanocomposite

Duhak Ahmed Alwan[1] , Oraas Adnan Hatem[2]

[1,2]Department of Chemistry, College of Science, University of AL-Qadisiyah, Iraq.

Abstract. A simple hydrothermal process has been followed to prepare rGO/CdS Nanocomposite. The structural, chemical and surface properties of the prepared nanocomposites were diagnosed and studied using X-ray diffraction spectroscopy. Surface functional groups were analyzed using FTIR and Raman analysis. FE-SEM was employed to illustrate the particle size and surface morphology of the (rGO/CdS) nanocomposite. Surface area and porosity analysis (BET, BJH)and the elemental composition of the composite was study using EDX analysis .

rGO sheets effectively improve the photocatalytic degradation efficiency of Crystal violet dye under visible light irradiation , rGO–CdS nanocomposite possessed high photodegradation efficiency of dye about 85% which was 2.8 time higher than the control CdS

rGO sheets accepting photo induced electrons from cadmium sulfide which reduces the recombination of (electron-hole) in the CdS. These well separated photocatalytic charges also demonstrated the photocatalytic activity of the rGO/CdS composite.

Keywords:, Cadmium Sulfide, Reduced Graphene Oxide, Degradation efficiency, Nanocomposite , Crystal violet

1. Introduction

Graphene has attracted technological and strong scientific interest, because of its excellent conductive, electronic, optical and thermal Properties, high specific surface area and superior chemical stability,[1-3] as well as potential applications in energy storage,[4-6], photocatalyst,[7-10]and solar cells[11].
Graphene is the name given to the flat monolayer of carbon atoms tightly packed in a two-dimensional (2D) lattice[12]. it is a basic material for graphite. It can be rolled into 0D fullerenes, it can be rolled into 1D nanotubes or stacked in 3D graphite[13].

Graphene Oxide It is a chemical compound consisting of (carbon, oxygen, and hydrogen) in different proportions. It consists of an oxidized graphene layer. The graphene oxide is a single-layered material of the carbon atom and is two-dimensional due to the oxidation of graphite. Graphene oxide is widely used in many fields, and it is due to The presence of effective oxygen groups resulting from (epoxy, carbonyl and hydroxyl) groups linked in the middle of the structure and the carbonyl group at the ends. Therefore, graphene oxide has a hydrophilic property in nature, ease of dispersal in water and the ability to introduce water molecules between GO sheets, this depends on the internal space between GO sheets that reaches from (0.6 to 1.2) nm[14, 15]

Advantages have entice the attentiveness of much research . Graphene can be reduced to reduced graphene oxide or can the incorporation of rGO with other nanoparticles .The combination of semiconductor nanostructure rGO ,can supply ideal, arbiter for active transportation from carrier where by carrier traps or scattering will be reduced. nevertheless .The simple in situation expansion of 1-D nanostructures having control able structure is for the time being, profit significance by the research society[16-20].

CdS is one of the committed photocatalysts due to its high photocatalytic execution and photo response to visible light. It has a narrow energy gap (2.42 eV), which explains its interesting photocatalytic activity[20]. However, photogenerated charge carriers assemble easily due to their small energy gap, and thus they respond to visible light[21, 22] .modern, many methods were reported to synthesizer GO–CdS nanocomposite[23, 24]. Explicitly, composites of graphene or its derivatives with CdS driven more attention to combine

CdS/rGO nanocomposite were synthesized using a simple hydrothermal method. The effect of integration oreGO nian-sheet on the properties of the synthesized composites was extensively thoughtful in expression of FT-IR, X-ray, BET, FE-SEM, EDX and UV–visible spectroscopy.

2. Experimental Part

2.1. Prepare of (Graphene Oxide)

GO was prepared by oxidizing(graphite powder) through (modified Hammer method)[25], where 2 g of graphite was mixed with 1 g of sodium nitrate(NaNO3) and 46 mL of sulfuric acid (H2SO4) was added in a 250 ml flask with continuous stirring in temperature of less than 5 °C, after To finish mixing, add 6 g of potassium permanganate (KMnO4) gradually with continuous stirring. Then the ice bath was raised and the process of stirring the solution continued for 12h at 35 °C, then added 250 mL from deionized water with continuous stirring, the temperature of the solution was raised to 90 °C with continuous stirring for 1.5h. 10 mL of hydrogen peroxide H2O2 at a concentration of %30 was added. The precipitate formed was washed several times for HCl at a concentration of %30 and then with water until a neutral solution (pH=7), and dried in oven at 60 °C for 24 hours.

2.2. Preparation of(rGO)

Reducing graphene oxide was prepared by reacting 0.2 g of graphene oxide with 2.5 g of ascorbic acid and adding 75 mL from deionized water. Indicates the reduction of graphene oxide. Separate the solution using a centrifuge at 4000 rpm for 10 minutes, and then wash of (deionized water and absolute ethanol) and dry it at 70 °C[26].

2.3. Preparation of CdS nanoparticles:

Cadmium sulfide was prepared by reacting aqueous cadmium sulfate (3CdSO4.8H2O) with thiourea (NH2)2CS) at a concentration of 0.02M and 0.05M, respectively, in a volume of 100ml.The solution remains under continuous stirring until it becomes homogenous and begins to appear yellow. The temperature is raised to 70 °C with continuous stirring for 1.5 h. Then the stirring is stopped and the heating continues at 70 °C for 3 hours, after which the formed precipitate is washed and dried[27].

2.4. Preparation of Binary Composite (rGO/CdS nanocomposite)

The composite was prepared by adding 0.2 g of rGO to 100 mL of deionized water and then placed in an ultrasonic device for an hour. Cadmium sulfate 3CdSO4.8H2O solution and thiourea (NH2)2CS solution were prepared as in the previous paragraph, and gradually added to the rGO solution. Ammonium hydroxide was added in the form of drops until a pH=10 was obtained and a precipitate appeared in the solution .The solution remains under continuous stirring until it becomes homogeneous and begins to appear yellow, the temperature is raised to 70 °C while stirring for 2 h, then the stirring is stopped and the heating continues at 70 °C for 3 hours, the precipitate formed is washed and dried.

3. Results and discussion

3.1. The XRD patterns of RGO-CdS nanocomposite.

X-ray diffraction technique was used to find out the crystalline properties of the prepared solids (crystal size, crystalline nature, as well as the interlayer distances).
The X-ray diffraction spectrum of reduced graphene oxide was studied. Where the results showed the disappearance of the characteristic graphene oxide band at $2\theta=10.5°$ and the appearance of a prominent and wide peak at $2\theta=25.7°$, which indicates the success of the peeling process and the formation of graphene oxide sheets in addition to the removal of functional groups containing oxygen.in(fig1.a)
As for the XRD results of cadmium sulfide, it showed the presence of four prominent peaks attributed to the planar hexagonal crystals of CdS at each of (52.0°)., (43.8°), (27.4), (25.0°) in (fig1.b).In rGO-CdS binary composite, similar peaks of the X-ray diffraction spectrum of CdS appeared at (53.2°), (46.2°), (27.3°). Compound molecules It has also been noted that the beams of the composite are wide, indicating that the crystal size of the nanocomposites molecules is relatively small. The absence of the rGO peak in the binary complex may be due to the relatively low diffraction intensity in addition to the overlap of the peaks and their obscuration by the diffraction angles located within the range $2\theta=25°-28°$ belonging to

cadmium sulfide.The high intensity of the deflection angles in the case of the binary complex relative to cadmium sulfide gives a clear indication of the fact that reduced graphene oxide provides an ideal support surface for the formation of cadmium sulfide crystal cores .(Fig 1.c)

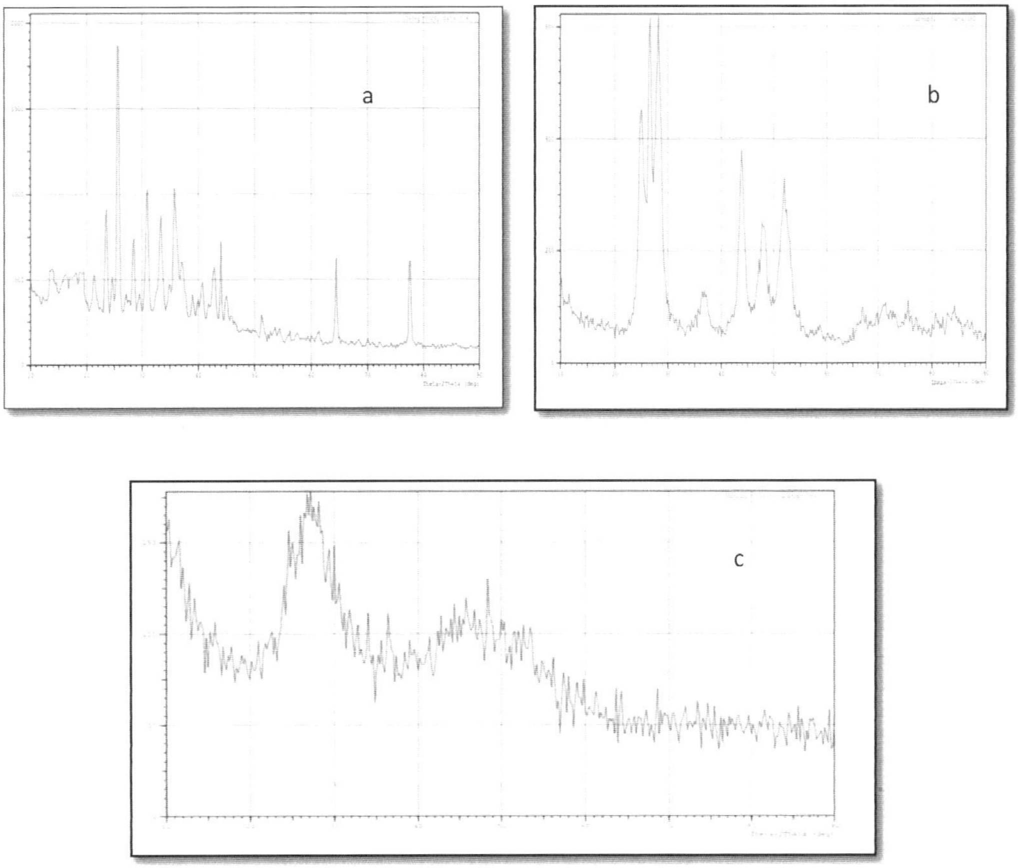

Figure 1. XRD patterns of a)rGO, b)CdS c) RGO-CdS Nanocomposite

3.2. Morphological characterization of rGO -CdS nanocomposites

This technique was used to diagnose the morphology of the prepared samples (particle size and particle distribution, as well as crystal defects and surface structure). FE-SEM has many advantages such as high accuracy and great control over the degree of amplification. The measurement was made by using a voltage acceleration (kV200) and a secondary electron detector, and a magnification power of (2.5-40 kx) SEM of the binary complex (rGO/CdS), showing the effect of rGO on CdS morphology. Where the binary complex appeared in the form of spherical cadmium sulfide particles distributed on the reduced graphene oxide sheets with an average particle size ranging from (10.25-29.20) nm.

Reduced graphene oxide has a paper-like shape with a wrinkled surface that may be due to its multiple functional groups in addition to the extremely small thickness of the layer. With many folded edges with a high degree of curvature, the spherical CdS particles are distributed on the rGO sheets.(Fig 2).

Figure 2. SEM of RGO-CdS Nanocomposite

3.3. Analysis of element (EDX) studies of rGO/CdS nanocomposite

Elemental analysis EDX is used to analyze the proportion of the elements in the composite,fig.3. The results showed the EDX analysis of the binary composite the ratio of cadmium, sulfur, oxygen and carbon.

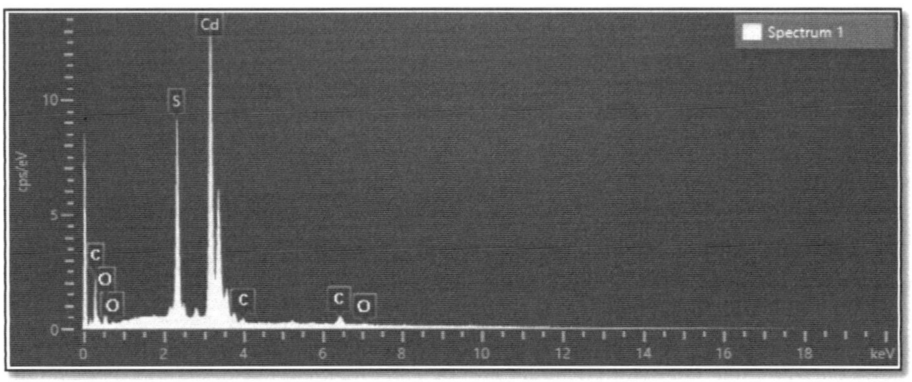

Figure 3. EDX of RGO-CdS Nanocomposite

3.4. DRS spectra for the RGO–CdS nanocomposites

The UV and visible diffraction spectra give information about the optical properties as well as knowledge of the measurement of the energy gap (Eg)of the prepared composites. Figure4. shows the DRS spectrum of the rGO-CdS binary composite. It is clear that rGO has a significant effect on enhancing the visible light absorption in the λ<500 nm region. Compared with pure CdS, where the rGO-CdS binary composite ,shows a clear path ochromic shift towards higher wavelengths, as a result, the energy gap of the rGO-CdS composite becomes narrower than that of pure CdS, so the energy gap of the pure CdS and the rGO-CdS binary composite is approximately 2.42, 1.984eV respectively .

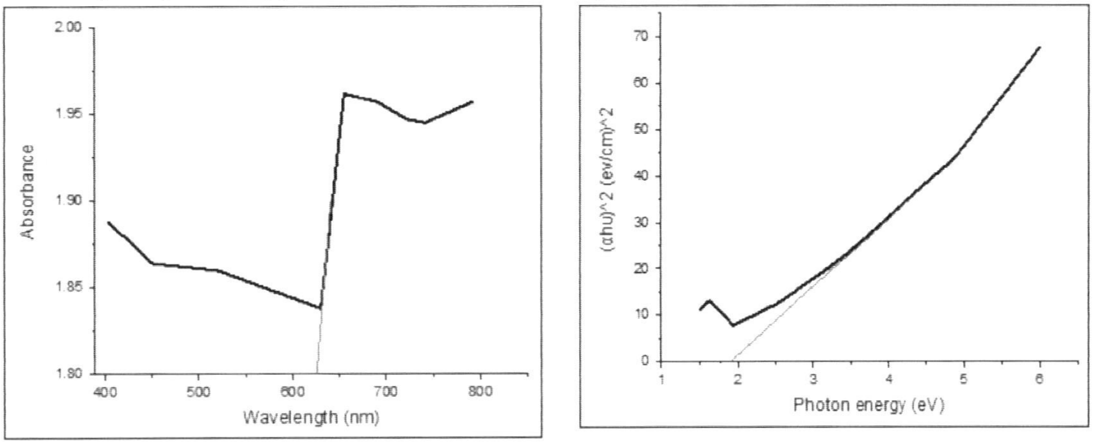

Figure 4. DRS of RGO-CdS Nanocomposite

3.5. Surface area analysis studies of rGO–CdS nanocomposite

As shown in Fig. 5. Demonstrates the BET isotherms for nitrogen adsorption and adsorption on the surface of the binary composite. Describes the pore size distribution on the surface of the binary composite. It is noted that the isotherms (adsorption - adsorption) of the binary composite are of the fourth class (IV). Which indicates multi-layer adsorption, as well as the hysteresis loops of type (H3) of (0.9>P/P°>0.3) and this indicates that the surface pores are in the form of non-solid plate-like assemblies Through the BJH method, it was found that the surface of the binary composite ,with a pore size of 0.3137 cm3/g, a surface area of m2/g55,956 and an average pore diameter of 22.422 nm.

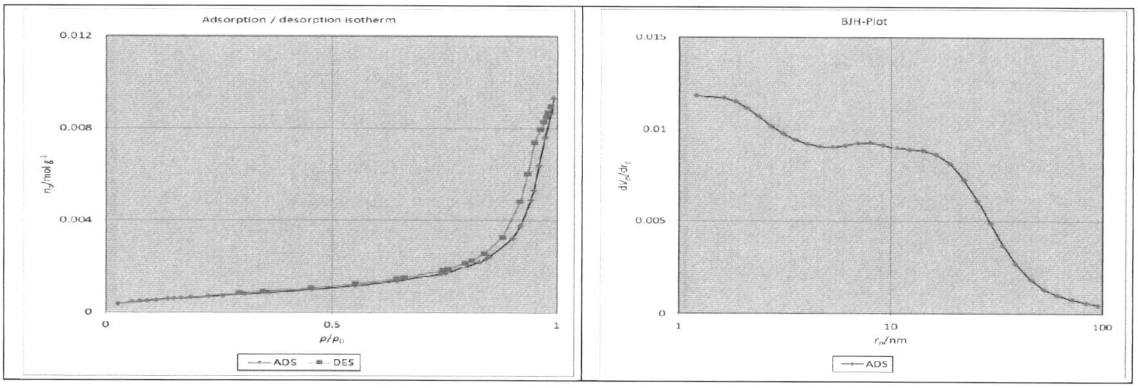

Figure 5. BET –BJH of RGO-CdS Nanocomposite

3.6. Raman spectra of rGO/CdS composite.

As shown in Fig. 6,Represents the Raman spectrum of the binary composite and observes the appearance of two prominent peaks at 294 cm^{-1} and 595 cm^{-1}, corresponding to the vertical photon mode longitudinal optical transition of the fundamental frequency (1-LO) and the upper frequencies (2-LO) respectively for the CdS nanoparticles formed on the oxide sheets graphene. The Raman spectra of the binary composite prove the presence of rGO within the composite matrix, as two peaks are observed in range(1372 cm^{-1} and 1594 cm^{-1}), representing the (D and G) bands of reduced graphene oxide, respectively .

The Raman spectrum of reduced graphene oxide is usually analyzed by these two peaks which represent the G pattern of the sp2 carbon and the D pattern of the sp3 carbon. The ratio between D/G gives a clear picture of the amount of distortion in the graphene oxide network, where the high value of D/G corresponds to more defects and disturbances in the graphite structure.

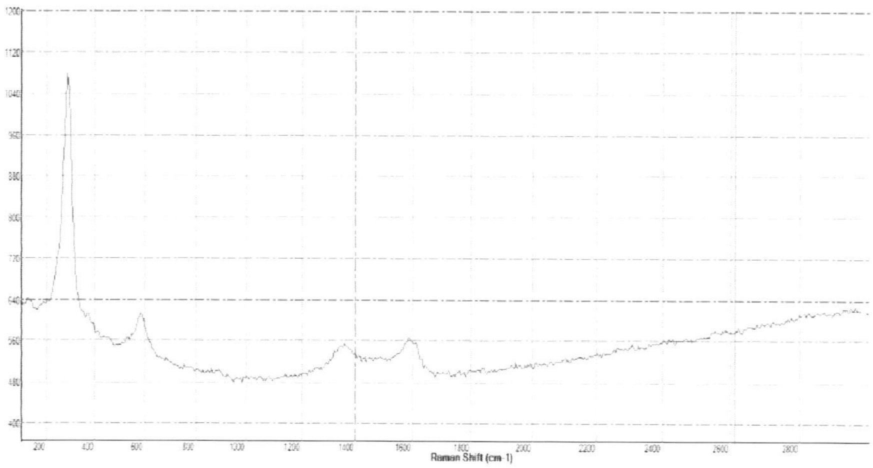

Figure 6. Raman spectrum of RGO-CdS Nanocomposite

3.7. Photocatalytic degradation efficiency of crystal violet by RGO–CdS nanocomposites under visible-light irradiation

A little decrease in the CV concentration after irradiation in the absence of catalysts, which propose which CV is fully stable under visible light. It is as wells potted that the rGO–CdS nanocomposites exhibit better photocatalytic execution than the pure CdS, whose mean that RGO can perfect the photo degradation ability of CdS. The optimum condition were pH9 ,dye concentration $4x10^{-5}$, 10° C and catalyst mass 0.3 g which shows the highest photodegradation efficiency of 85% in comparison with pure CdS 30% it is clearly that RGO sheet prevent the recombination (of electron–hole) pairs so increase the efficiency of photo catalytic degradation (Fig 7).

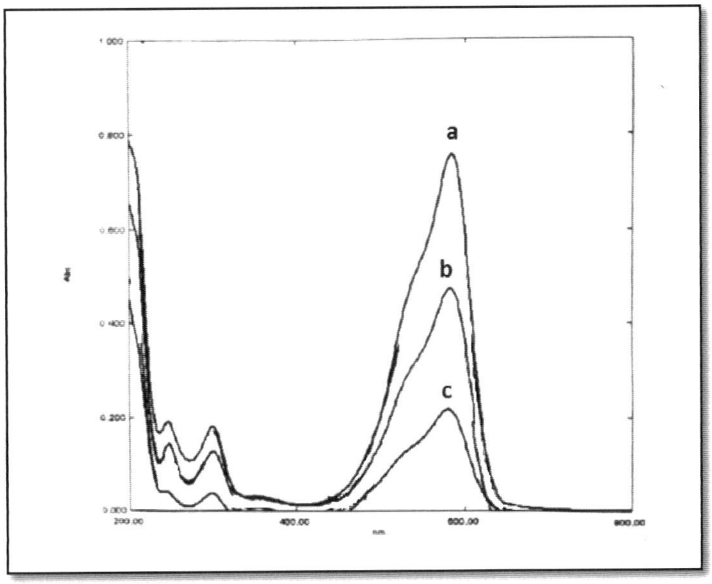

Figure 7. Photocatalytic degradation of crystal violet in a. apsence of catalyst b. Cds c. RGO–CdS nanocomposites under visible-light irradiation

References:

[1] Shen, J., et al., *Synthesis of amphiphilic graphene nanoplatelets.* 2009. **5**(1): p. 82-85.

[2] Liu, X., et al., *Microwave-assisted synthesis of CdS–reduced graphene oxide composites for photocatalytic reduction of Cr (VI).* 2011. **47**(43): p. 11984-11986.

[3] Shen, J., et al., *One step hydrothermal synthesis of TiO 2-reduced graphene oxide sheets.* 2011. **21**(10): p. 3415-3421.

[4] Zhang, K., et al., *Surfactant-intercalated, chemically reduced graphene oxide for high performance supercapacitor electrodes.* 2011. **2** :(20)1p. 7302-7307.

[5] Lee, J.K., et al., *Silicon nanoparticles–graphene paper composites for Li ion battery anodes.* 2010. **46**(12): p. 2025-2027.

[6] Zhang, K., et al., *Graphene/polyaniline nanofiber composites as supercapacitor electrodes.* 2010. **22**(4): p.1401-1392 .

[7] Xiong, Z., et al., *Photocatalytic degradation of dyes over graphene–gold nanocomposites under visible light irradiation.* 2010. **46**(33): p. 6099-6101.

[8] Jiang, G., et al., *TiO2 nanoparticles assembled on graphene oxide nanosheets with high photocatalytic activity for removal of pollutants.* 2011. **49**(8): p. 2693-2701.

[9] Xiong, Z., L.L. Zhang, and X.S.J.C.A.E.J. Zhao, *Visible-Light-Induced Dye Degradation over Copper-Modified Reduced Graphene Oxide.* 2011. **17**(8): p. 2428-2434.

[10] Xiang, Q., J .Yu, and M.J.N. Jaroniec, *Enhanced photocatalytic H 2-production activity of graphene-modified titania nanosheets.* 2011. **3**(9): p. 3670-3678.

[11] Zhu, G., et al., *Graphene-incorporated nanocrystalline TiO2 films for CdS quantum dot-sensitized solar cells* :(2)650 .2011 .p. 248-251.

[12] Wehling, T., et al., *Molecular doping of graphene.* 2008. **8**(1): p. 173-177.

[13] Singh, V., et al., *Graphene based materials: past, present and future.* 2011. **56**(8): p. 1178-1271.

[14] Ambrosi, A., et al., *Chemically reduced graphene contains inherent metallic impurities present in parent natural and synthetic graphite.* 2012. **109**(32): p. 12899-12904.

[15] Khan, Z.U., et al., *A review of graphene oxide, graphene buckypaper, and polymer/graphene composites: Properties and fabrication techniques.* 2016. **32**(4): p. 336-379.

[16] Geim, A.K.J.s., *Graphene: status and prospects.* 2009. **324**(5934): p. 1530-1534.

[17] Rout, C.S., et al., *Synthesis and characterization of patronite form of vanadium sulfide on graphitic layer.* 2013. **135**(23): p. 87.8725-20

[18] Xiang, Q. and J.J.T.J.o.P.C.L. Yu, *Graphene-based photocatalysts for hydrogen generation.* 2013. **4**(5): p. 753-759.

[19] Kamat, P.V.J.T.J.o.P.C.L., *Graphene-based nanoassemblies for energy conversion.* 2011. **2**(3): p. 242-251.

[20] Liu, F., et al ,*.Rapid microwave assisted synthesis of graphene sheets supporting CdS nanocrystals for highly efficient visible light photocatalyst.* 2014. **18**(6): p. 443-450.

[21] Al-Baradi, A.M., et al., *Structural and optical characteristic features of RF sputtered CdS/ZnO thin films.* 2020. **29**(8): p. 080702.

[22] Zeng, P., et al., *One-pot synthesis of reduced graphene oxide–cadmium sulfide nanocomposite and its photocatalytic hydrogen production.* 2011. **13**(48): p. 21496-21502.

[23] Yan, S., et al., *Hydrothermal synthesis ofCdS/functionalized graphene sheets nanocomposites.* 2013. **570**: p. 65-69.

[24] Afreen, G., et al., *Effectiveness of reactive oxygen species generated from rGO/CdS QD heterostructure for photodegradation and disinfection of pollutants in waste water.* 2020. **1** :08p. 110372.

[25] Chen, W., L. Yan, and P.R.J.C. Bangal, *Preparation of graphene by the rapid and mild thermal reduction of graphene oxide induced by microwaves.* 2010. **48**(4): p. 1146-1152.
[26] Altaa, S.H.A., H.A.H. Alshamsi, and L.S.J.J.D.W.T. Al-Hayder ,*Synthesis and Characterization of rGO/Co3O4 composite as nanoadsorbent for Rhodamine 6G dye removal.* 2018. **114**: p. 320-331.
[27] Pawar, R.C., C.S.J.M.C. Lee, and Physics, *Sensitization of CdS nanoparticles onto reduced graphene oxide (RGO) fabricated by chemical bath deposition method for effective removal of Cr (VI).* 2013. **141**(2-3): p. 686-693.

A Numerical Study for Solving Mathematical Model in Chemistry and Environment

Mohammed S. Mechee[1], Ahlam A. Mahdi[2] and Ali N. Al-Mousawi[3]

[1]Information Technology Research and Development Centre (ITRDC) - University of Kufa.
[2]Dep. of Math., Faculty of Education, University of Kufa, Najaf, Iraq.
[3]Dep. of Math., Faculty of Education, University of Kufa, Najaf, Iraq.

Abstract. The purpose of this work is to demonstrate how the first order ordinary differential equation may be use as a mathematical model to explain chemical processes and mixture situations. The application of the ordinary differential equation of the first degree was analyze in the modeling of some chemical phenomena, such as clarifying the application in the problems of mixing solutions in both single and multiple tank systems. These problems were develope with the presentation of the first form of the model to include n- tank.

Keywords: Mathematical Model; Ordinary Differential Equation.

1. Introduction

The history of differential equations (DEs) goes back to the 17th century when the calculus was developed by Isaac Newton (c.1665). Differential equations arise naturally as mathematical models in the fields of engineering. science, economics and other subjects of biological systems, physical systems and economic systems, and equations have become the main tool for application in all fields of mechanics, thermodynamics and electrical [1]. A differential equation is an unknown function equation that includes not only function but its derivatives as well, for example, there are Newton's equations, Maxwell's rates for magnetism, Lagrange's rates for classical mechanics, Einstein's equation for the ratio of general gravity, and the Schrödinger equation for quantum mechanics [2]. Such equations are called ordinary differentin equations shortly ODEs. Differential equations (DEs) have significant rule the history of mathematics and sciences due to some models of nature are expressed through DEs, so we can say that DEs are the means by which scientists describe and understand the world. Some experts in these fields have used DEs to make some of the problems under research more understandable in today's world. Nowadays, there are many statistical, physical and dynamic processes that use mathematical modeling for the problems of initial value and dietary limits. [3] Mathematical courses such as numerical techniques, differential equations, and statistics are commonly utilized in chemical careers to solve difficulties with modeling real-world chemistry and/or biochemistry problems. Many scientists have offered a set of conjectures, provided mathematical modeling, and provided many instances connected to chemical mathematics and their understanding, The use of chemical problems allows for a better and more accurate understanding in less time and at a lower rate. These systems are utilized in a variety of mathematical modeling models and applications, including differential equations, probability, and statistics [4]

2. Basic Concepts

2.1. *Mathematical Model* : Mathematical Model is Process in which real-life situations and relations in these situations are expressed by using mathematics [5].

2.2. Linear Systems of ODEs [6]

For linear systems, we consider the effects of two unknown functions on each other. A two by two system of linear ordinary differential equations has the form

$$\begin{cases} \frac{dx(t)}{dt} = ax(t) + by(t) \\ \frac{dy(t)}{dt} = cx(t) + dy(t) \end{cases} \quad or \quad \frac{d}{dt}\begin{bmatrix} x(t) \\ y(t) \end{bmatrix} = \begin{bmatrix} a & b \\ c & d \end{bmatrix}\begin{bmatrix} x(t) \\ y(t) \end{bmatrix},$$

We set Y to be the unknown vector and A be the coefficient matrix, then we have a matrix equation

$$\frac{dy(t)}{dt} = Ay(t).$$

2.3. Remark [2]:

i. The derivative of a a vector valued function is defined as

$$x'(t) = \begin{bmatrix} x_1'(t) \\ \vdots \\ x_n'(t) \end{bmatrix},$$

ii. By the definition of the matrix-vector product, Equation (1.6) can be written as

$$\begin{aligned} x_1'(t) &= a_{11}(t)x_1(t)+a_{12}(t)x_2(t) + \cdots a_{1n}(t)x_n(t) + b_1(t), \\ x_2'(t) &= a_{21}(t)x_1(t) + a_{22}(t)x_2(t) + \cdots a_{2n}(t)x_n(t) + b_2(t), \\ &\vdots \\ x_n'(t) &= a_{n1}(t)x_1(t)+a_{n1}(t)x_2(t) + \cdots a_{nn}(t)x_n(t) + b_n(t). \end{aligned}$$

3. Application of Ordinary Differential Equation

We can describe the differential equations applications in real life such that it was used in many fields, including chemistry, and the following proverbs explain one of the applications of these equations:
Example 1: The Problem of a Single-Tank Mixture [7]
In general, suppose that:
$r_1(\tau)$ = rate of enter the solution,
c_1 = rate of enter concentration,
r_2 = rate of out solution,
c_2 = rate of out concentration,
n = number of gallons

rate of input in tank = rate of output,
Amount of the salt in the tank = amount of the input-amount of the output,
For example
Initially, a 20-pound sugar solution dissolved in 165 gallons of water was placed in a 200-gallon tank. At a rate of 8 gal/min, a sugar solution with a concentration of (2 pounds/min) enters the tank, and the resultant mixture leaves at a rate of (3 gal/min). Count and plot the quantity of sugar in the tank as a function of time to determine how much sugar is in the tank when it starts to overflow.

Solution
Assume $r_{in}(\tau)$ = rate of entry, and $\alpha(\tau)$ is the amount of pounds of sugar in the tank at time τ. The tank holds 20 pounds of sugar at the start, thus $\alpha(0) = 20$ lbs. The rate at which the solution enters the tank is $r_{in}(\tau) = 8\frac{gal}{min}$, whereas the solution concentration is $c_{in}(\tau) = 2\frac{gal}{min}$. $r_{out}(\tau) = 3\frac{gal}{min}$ is the rate at which the solution exits the tank. As a result, the tank's rate of filling is increasing:.

$$r(\tau) = r_{in}(\tau) - r_{out}(\tau) = 5\frac{gal}{min}$$

Let $n(\tau)$ at time t, be the number of gallon of solution in the tank. Because the tank initially held 165 gallons and the rate of growth equals
$5\frac{gal}{min}$, $n(\tau) = 165 + 5\tau$ As a result, the sugar solution concentration in the tank and flowing out of the tank at time τ equals

$$c_{out}(\tau) = \frac{\alpha(\tau)}{n(\tau)} = \frac{\alpha(\tau)}{165 + 5\tau},$$

As a result, the first condition issue that must be addressed is

$$\frac{d\alpha}{d\tau} = r_{in}(\tau)c_{in}(\tau) - r_{out}(\tau)r_{out}(\tau), \tag{1}$$

$$= (8\frac{gal}{min}\left(2\frac{lbs}{gal}\right) - (3\frac{gal}{min})\left(\frac{\alpha(\tau)}{165+\tau}\right)\frac{lbs}{min},$$

$$= 16\frac{lbs}{min} - \frac{3\alpha(\tau)}{165+5\tau}, \tag{2}$$

Since $\alpha(0)=20$
The tank is full when $n(t) = 165 + 5\tau = 200$. Solving for t, we find

$$\tau = \frac{35}{5} = 7 \text{ min}$$

Using solve IVP to solve the initial value issue (2) on the interval [0, 7] by associating $\alpha(\tau)$ with $y(\tau)$,we were able to solve the initial value problem (2) on the interval [0, 7]./

$$f(x, y) = \frac{16 - 3y}{165+5x}, \tag{3}$$

by making the integration interval [0,7] and the starting values y (0) = 20 Figure 1 shows a graph of the sugar level in the tank. For the precise answers, n(7) = 123.7934 pounds is discovered.
So, when we solve Equation (2)

$$\alpha(\tau) = 10\tau + 330 - \frac{c}{\sqrt[5]{(\tau+33)^3}}, \tag{4}$$

Therefore, we conclude

The Numerical Results :

$$\alpha(\tau) = 10\tau + 330 - \frac{2526.214}{(\tau + 33)^{\frac{3}{5}}},$$

1- $\lim_{\tau\to 10}\left(10\tau + 330 - \frac{2526.214}{(\tau+33)^{\frac{3}{5}}}\right) = 165.5223457$

2- $\lim_{\tau\to\infty}\left(10\tau + 330 - \frac{2526.214}{(\tau+33)^{\frac{3}{5}}}\right) = \infty$

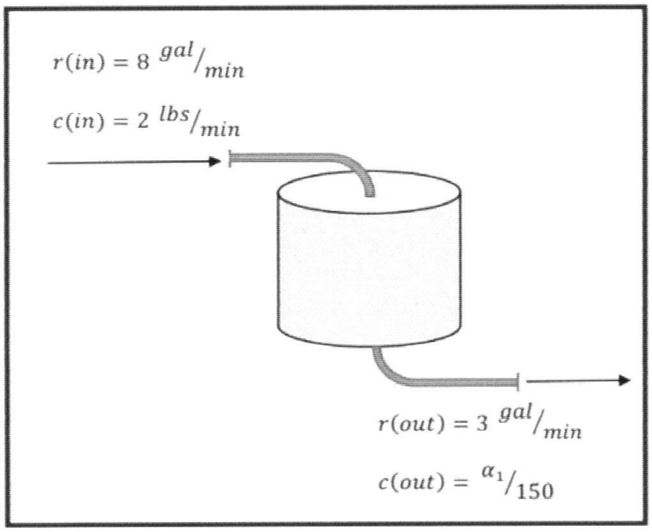

Figure 1. Diagram for a One Tank Mixture Problem

Example 2: [7]*Two-Tank Mixture Problem*

In a 150-gallon tank, a 60 percent sugar solution is first introduced. A 10% sugar solution is fed into the tank at a rate of 5 gal/min. The combination flows out of the tank at the same pace and into a 75-gallon tank previously filled with clean water. The second tank's mixture is pumped at a rate of 5 gallons per minute. Calculate and graph the quantity of sugar in the second tank in the interval [0, 125]. When does the sugar level in the second tank reach its maximum? What is the maximum quantity of sugar in the second tank.

Solution

In general, suppose the following:

$\alpha_1(\tau)$ = in the first tank, the amount of gallons of salt,

$\alpha_2(\tau)$ = in the first tank, the amount of gallons of salt,

$r_{1_{in}}(\tau)$ = rate of mixture entered first tank,

$r_{2_{in}}(\tau)$ = rate of mixture entered second tank,

$r_{1_{out\tau}}(\tau)$ = rate of solution out from first tank,

$r_{2_{out\tau}}(\tau)$ = rate of solution out from second tank,

$c_{1_{in}}(\tau)$= concentration inside first tank,

$c_{2_{in}}(\tau)$= concentration inside second tank,

$c_{1_{out\tau}}(\tau)$ = rate of concentration leave first tank,

$c_{2_{out\tau}}(\tau)$ = concentration leave second tank,

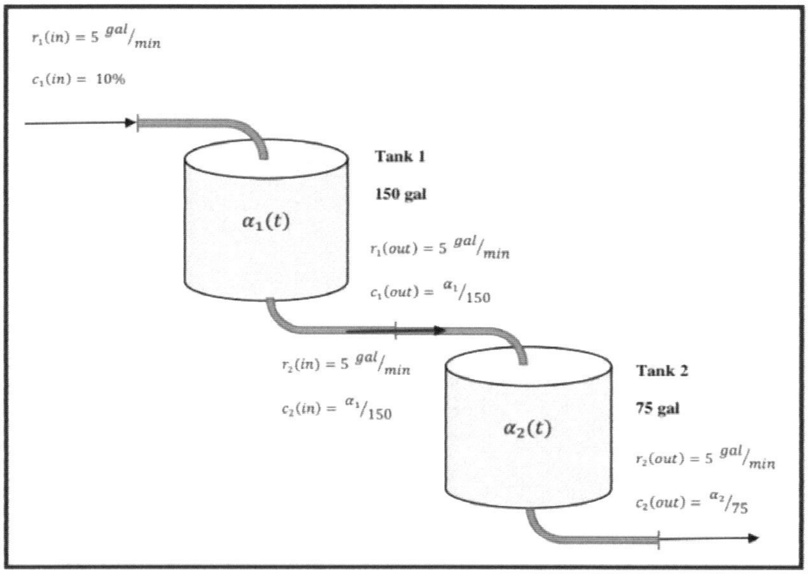

Figure 2. Mixture Situation in Two Tanks

$\frac{d\alpha_1}{d\tau}$ = rate amount of salt in the first tank,

$\frac{d\alpha_2}{d\tau\tau}$ = rate amount of salt in the second tank,

Amount of the salt in the tank = amount of the input - amount of the output,

$r_{in} = r_{out\tau}$,

$\frac{d\alpha_1}{d\tau} = r_{1_{in}}(\tau)c_{1_{in}}(\tau) - r_{1_{out\tau}}(\tau)c_{1_{out\tau}}(\tau)$,

$\frac{d\alpha_2}{d\tau} = r_{2_{in}}(\tau)c_{2_{in}}(\tau) - r_{2_{out\tau}}(\tau)c_{2_{out\tau}}(\tau)$,

For Example:

Let $\alpha_1(\tau)$ and $\alpha_2(\tau)$represent the number of gallons of sugar in tanks 1 and 2, respectively, at time t. At time t=0, the original amount of liters of sugar in tank 1 is$\alpha_1(0) = 60\% \times 150 \text{ gal} = 90 \text{ gal}$

$$\alpha_1(0) = 60\% \times 150 \text{ gal} = 90 \text{ gal}$$

Due to the fact that the second tank is originally filled with pure water $\alpha_2(0) = 0$ gal. for tank 1, we've got:

$$r_{1_{in}}(\tau) = r_{1_{out}}(\tau) = 5\frac{\text{gal}}{\text{min}},\ c_{1_{in}}(\tau) = 10\% = 0.1,$$

As a result, the total amount of sugar in tank 1 must fulfill the differential equation:

$$\frac{d\alpha_1}{d\tau} = r_{1_{in}}(\tau)c_{1_{in}}(\tau) - r_{1_{out}}(\tau)c_{1_{out}}(\tau), \tag{5}$$

$$= 5\ \frac{gal}{min} \times 0.1 - 5\ \frac{gal}{min}\left(\frac{\alpha_1}{150}\right), \tag{6}$$

$$= \left(5 - \frac{\alpha_1(\tau)}{30}\right)\frac{gal}{min},$$

Concerning of tank 2, we have:

$$r_{2_{in}}(\tau) = r_{2_{out}}(\tau) = 5\frac{\text{gal}}{\text{min}},$$

$$c_{2_{in}}(\tau) = \alpha_1(\tau),\ c_{2_{out}}(\tau) = \left(\frac{\alpha_2(\tau)}{75}\right),$$

As a result, the quantity of gallons of sugar in tank 2 meets the requirements of the differential equation:

$$\frac{d\alpha_2(\tau}{d\tau} = r_{2_{in}}(\tau)c_{2_{in}}(\tau) - r_{2_{out}}(\tau)c_{2_{out}}(\tau), \tag{7}$$

$$= 5\ \frac{gal}{min}\left(\frac{\alpha_1(\tau)}{150}\right) - 5\ \frac{gal}{min}\left(\frac{\alpha_2(\tau)}{75}\right),$$

$$= \left(\frac{\alpha_1(\tau)}{30} - \frac{\alpha_2(\tau)}{15}\right)(gal/min),$$

Hence, number of sugar gallons in the tank 1 and 2 must simulation fulfill the two issues of starting value

$$\frac{d\alpha_1(\tau)}{d\tau} = 5 - \frac{\alpha_1(\tau)}{d\tau(\tau)}\quad,\ \alpha_1(0) = 90$$
(8)

$$\frac{d\alpha_2(\tau)}{d\tau} = \frac{\alpha_1}{30} - \frac{\alpha_2}{15}\quad,\ \alpha_2(0) = 0 \tag{9}$$

By solve Equation (8) we find :

$$\alpha_1(\tau) = 75\, e^{\frac{-\tau}{30}} + 15,$$

$\alpha_2(\tau) = 75e^{\frac{-\tau}{15}} + 7.5 - 82.5e^{\frac{-\tau}{15}}$

Numerical Results

1-$\lim_{\tau\to\infty} \alpha_1(\tau) = \lim_{\tau\to\infty}\left(75\, e^{\frac{-\tau}{30}} + 15\right) = 15$

$2 - \lim_{\tau\to\infty} \alpha_2(\tau) = \lim_{\tau\to\infty}(75e^{\frac{-\tau}{15}} + 7.5 - 82.5e^{\frac{-\tau}{15}}) = 7.5$

And

1- $\lim_{\tau\to 10} \alpha_1(\tau) = \lim_{\tau\to 10}\left(75\, e^{\frac{-10}{30}} + 15\right) = 68.73984829$

$2 - \lim_{\tau\to 10} \alpha_2(\tau) = \lim_{\tau\to 10}(75e^{\frac{-10}{15}} + 7.5 - 82.5e^{\frac{-10}{15}}) = 3.64937161$

Example 3: Three Tanks Mixture Problem

Three tanks connected to each other have a mixture of water and sugar with the capacity of $(150, 75, 100)$ $gallons$ respectively, , The first tank (A) will be filled with a 60 percent sugar solution. A 10% sugar solution enters tank (A) at a rate of 5 gal/min, with the solution exiting at the same rate into tank (B), which was originally filled with clean water. The solution comes out from the second tank and enters the third tank (C) which was start filling with pure water at the rate $5\frac{gal}{min}$. What is the amount of sugar in the third tank?

Solution

Assume a volume of tanks $s_1, s_2, s_3$with correspondingly, and let$\alpha_1(\tau), \alpha_2(\tau), \alpha_3(\tau)$ represent the number of gallons of sugar in tanks at time and with the same assumptions as the previous example two tank.

$$\frac{d\alpha_1}{d\tau} = r_{1_{in}}(\tau)c_{1_{in}}(\tau) - r_{1_{out}}(\tau)c_{1_{out}}(\tau), \quad (10)$$

$$= 5\,\frac{gal}{min} \times 0.1 - 5\,\frac{gal}{min}\left(\frac{\alpha_1}{150}\right),$$

$$= \left(5 - \frac{\alpha_1(\tau)}{30}\right)\frac{gal}{min},$$

Concerning of tank 2, we get :

$$r_{2_{in}}(\tau) = r_{2_{out}}(\tau) = 5\frac{\text{gal}}{\text{min}},$$

$$c_{2_{in}}(\tau) = \alpha_1(\tau),\ c_{2_{out}}(\tau) = \left(\frac{\alpha_2(\tau)}{75}\right),$$

As a result, the amount of gallons of sugar in tank 2 meets the differential equation:

$$\frac{d\alpha_2(\tau)}{d\tau} = r_{2_{in}}(\tau)c_{2_{in}}(\tau) - r_{2_{out}}(\tau)c_{2_{out}}(\tau), \quad (11)$$

$$= 5\ \frac{gal}{min}\left(\frac{\alpha_1(\tau)}{150}\right) - 5\ \frac{gal}{min}\left(\frac{\alpha_2(\tau)}{75}\right),$$

$$= \left(\frac{\alpha_1(\tau)}{30} - \frac{\alpha_2(\tau)}{15}\right)(gal/min)\,, \qquad (12)$$

For tank (C):

$$\mathrm{r}_{3_{\mathrm{in}}}(\mathrm{t}) = \mathrm{r}_{2_{\mathrm{out}}}(\mathrm{t}) = \mathrm{r}_{3_{\mathrm{out}}}(\mathrm{t}) = 5\,\frac{\mathrm{gal}}{\mathrm{min}}\,,$$

$$c_{3_{in}}(\tau) = c_{2_{out\tau}}(\tau) = \frac{\alpha_2(\tau)}{s_2} = \frac{\alpha_2(\tau)}{75}\,,$$

Then,

$$\frac{d\alpha_3(\tau)}{d\tau} = r_{3_{in}}(\tau)c_{3_{in}}(\tau) - r_{3_{out\tau}}(\tau)c_{3_{out\tau}}(\tau), \qquad (13)$$

And $c_{3_{in}}(\tau) = c_{2_{out\tau}}(\tau) = \frac{\alpha_2(\tau)}{s_2} = \left(\frac{\alpha_2(\tau)}{75}\right),$

Hence, we have the next order:

$$\frac{d\alpha(\tau)_1}{d\tau} = 5 - \frac{\alpha_1(\tau)}{30}\,,$$

$$\frac{d\alpha_2(\tau)}{d\tau} = \frac{\alpha_1(\tau)}{30} - \frac{\alpha_2(\tau)}{15}\,,$$

$$\frac{d\alpha_3(\tau)}{d\tau} = \frac{\alpha_2(\tau)}{15} - \frac{\alpha_3(\tau)}{20}\,,$$

With the initial condition: $\alpha_1(0) = 90,\ \alpha_2(0) = 0,\ \alpha_3(0) = 0.$

Therefore, by solve this system we get

By solving this system, we get the following results:

$$\alpha_1(\tau) - 150 + 60\,e^{\frac{-\tau}{30}}\,,$$

$$\alpha_2(\tau) = 60\,e^{\frac{-\tau}{30}} - 60\,,$$

$$\alpha_3(\tau) = 240\,e^{\frac{-\tau}{30}} - 160\,e^{\frac{-\tau}{20}} - 80\,,$$

Numerical Discussion

1- $\lim_{\tau\to 10}(\alpha_1(\tau)) = \lim_{\tau\to 10}(150 + 60\,e^{\frac{-\tau}{30}}) = 192.9918786$

2- $\lim_{\tau\to 10}(\alpha_2(\tau)) = \lim_{\tau\to 10}\left(60\,e^{\frac{-\tau}{30}} - 60\right) = -\,17.00812137$

$\lim_{\tau\to 10}(\alpha_3(\tau)) = \lim_{\tau\to 10}\left(240\,e^{\frac{-\tau}{30}} - 160\,e^{\frac{-\tau}{20}} - 80\right) = -5.077391016$

And when τ approaches infinity we get :

1- $\lim_{\tau\to\infty}(\alpha_1(\tau)) = \lim_{\tau\to\infty}(150 + 60\, e^{\frac{-\tau}{30}}) = 150$

2- $\lim_{\tau\to\infty}(\alpha_2(\tau)) = \lim_{\tau\to\infty}\left(60\, e^{\frac{-\tau}{30}} - 60\right) = -60$

4. $\lim_{\tau\to\infty}(\alpha_3(\tau)) = \lim_{\tau\to\infty}\left(240\, e^{\frac{-\tau}{30}} - 160\, e^{\frac{-\tau}{20}} - 80\right) = -80$

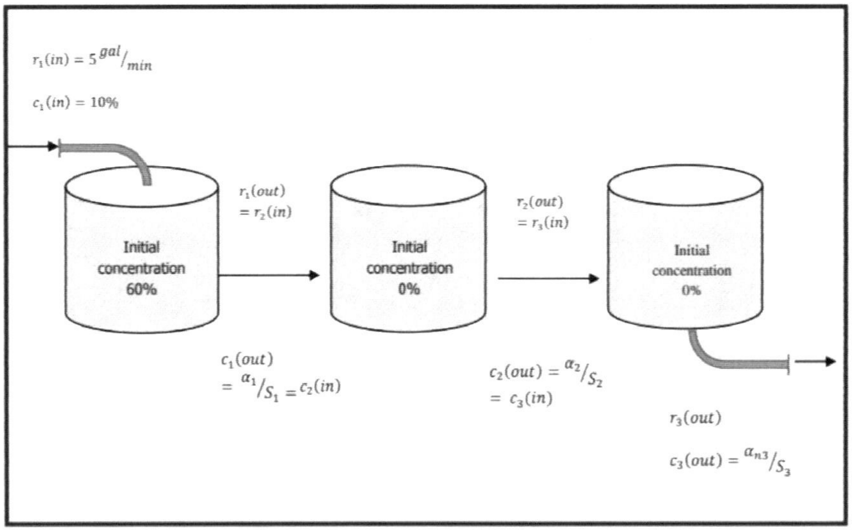

Figure 3. Three Tanks Mixture Problem

Example 4 : n-Tanks Mixture Problem

So the general formula will be as follows

$\mathrm{r}_{\mathrm{n}_{\mathrm{in}}}(\mathrm{t}) = \mathrm{r}_{\mathrm{n}_{\mathrm{out}}}(\mathrm{t}) = r,$

$$c_{n_{in}}(\tau) = c_{n-1_{out\tau}}(\tau) = \frac{\alpha_{n-1}(\tau)}{s_{n-1}},$$

$$c_{n_{out\tau}}(\tau) = \frac{\alpha_n(\tau)}{s_n},$$

Therefore,

$$\frac{d\alpha_1}{d\tau} = \frac{rc_1}{s_1} - \frac{r\alpha_1}{s_1} = r\left(\frac{c_1}{s_1} - \frac{\alpha_1}{s_1}\right) = \frac{r}{s}(c_{1-}\alpha_1),$$

$$\frac{d\alpha_2}{d\tau} = \frac{r\alpha_1}{s_1} - \frac{r\alpha_2}{s_2} = r\left(\frac{\alpha_1}{s_1} - \frac{\alpha_2}{s_2}\right) = \frac{r}{s}(\alpha_1 - \alpha_2), \quad (14)$$

$$\frac{d\alpha_3}{d\tau} = \frac{r\alpha_2}{s_2} - \frac{r\alpha_3}{s_3} = r\left(\frac{\alpha_2}{s_2} - \frac{\alpha_3}{s_3}\right) = \frac{r}{s}(\alpha_2 - \alpha_3),$$

$$\vdots \qquad\qquad\qquad\qquad \vdots$$

$$\frac{d\alpha_n}{d\tau} = \frac{r\alpha_{n-1}}{s_{n-1}} - \frac{r\alpha_n}{s_n} = r\left(\frac{\alpha_{n-1}}{s_{n-1}} - \frac{\alpha_n}{s_n}\right) = \frac{r}{s}(\alpha_{n-1} - \alpha_n), \tag{15}$$

The previous system can be written as follows:

$$\begin{pmatrix} \acute{\alpha}_1 \\ \acute{\alpha}_2 \\ \acute{\alpha}_3 \\ \vdots \\ \acute{\alpha}_n \end{pmatrix} = r \begin{pmatrix} \frac{-\alpha_1}{s_1} & 0 & 0 & 0 \\ \frac{\alpha_1}{s_1} & \frac{-\alpha_2}{s_2} & 0 & 0 \\ 0 & \frac{\alpha_2}{s_2} & \frac{-\alpha_3}{s_3} & 0 \\ \vdots & \ddots & \ddots & 0 \\ 0 & 0 & \frac{\alpha_{n-1}}{s_{n-1}} & \frac{-\alpha_n}{s_n} \end{pmatrix} \begin{pmatrix} \alpha_1 \\ \alpha_2 \\ \alpha_3 \\ \vdots \\ \alpha_n \end{pmatrix} + \begin{pmatrix} rc_1 \\ 0 \\ 0 \\ \vdots \\ 0 \end{pmatrix}, \tag{16}$$

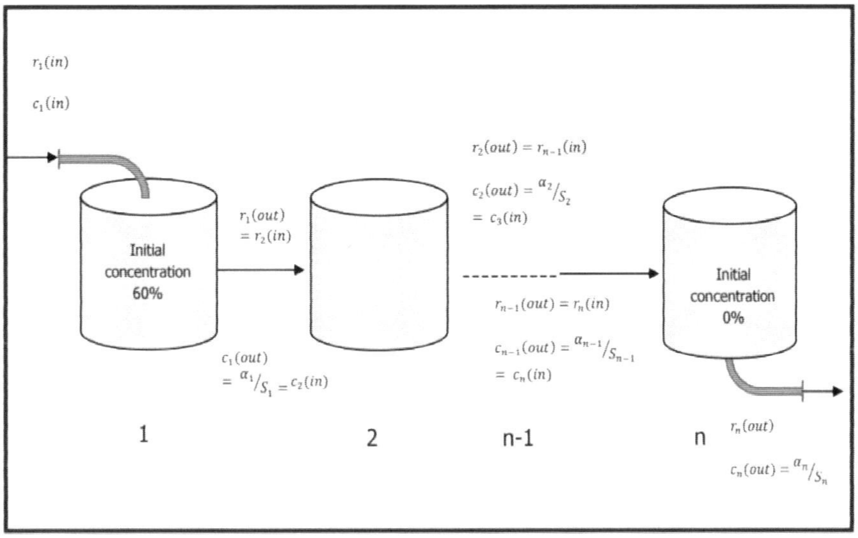

Figure: 4. Diagram for n-tanks mixture problem

4. Conclusions and Discussion

We presented mathematical modeling to provide a better understanding and predict the solution to find the concentration of some substances in the problems and ordinary differential equations of the first degree were used to find the rates of these substances present in the chemical mixture. It's also clear that determining the concentration of the mixed solution after a certain amount of time results in a well-mixed solution. Finally, this work believe that ordinary differential equations will be use to solve many future technological issues and predicting outcomes.

Acknowledgements

First and foremost, I want to thank Allah Almighty for his unending blessings and assistance in completing this project. Prof. Dr. Mohammad Sahib Mechee, my supervisor and mentor, deserves special gratitude and recognition, for his effective contribution, continuous cooperation and support. I also extend my sincere thanks to my family, my mother, my brothers, my sisters, my children and my friends. I also express my sincere gratitude to every person who gave me instructions during my study and supported me even if with one word.

References

[1] J.David Logan, A First Course in Differential Equations, Edition, 2015..

[2] G. Nagy, ordinary differential equation, 2021..

[3] Bajpai, A.C., Mustoe, L.R., & Walker, D. (1975), Mathematical education of engineers,: Part I. A critical appraisal. International Journal of Mathematical Education in Science and Technology, 6(3), 361–380.

[4] Martinez-Luaces, V. (2006). Analytic solving of real P.D.E., problems applied to Mass Transfer,: Experiences in project-work with Chemical Engineering and Food Technology Engineering students. In Hughes-Hallett, Vakalis, & Arikan (Eds.), Proceedings of th.

[5] Haines, Christopher, and Rosalind Crouch. Mathematical modelling and applications: Ability and competence frameworks. Modelling and applications in mathematics education. Springer, Boston, MA, 2007. 417-424..

[6] Michael J. Coleman, Population Modeling with Ordinary Differential Equations, November 6, (2006)..

[7] Roberts, Charles. Ordinary differential equations: applications, models, and computing. CRC Press, 2011..

Investigation of Heavy Metals Pollution in Euphrates River (Iraq) By Using Heavy Metal Pollution Index Model

Laith Hemed Kamel[1] , Sadiq Kadhum Lafta Al-Zurfi[2] and Mahmood Basil Mahmood[1]

[1]Department of Biology, Collage of Science, University of Baghdad
[2] Department of Ecology and pollution, Faculty of Science, University of Kufa, Iraq.

Abstract. This study aims to investigate heavy metals (HMs)pollution in the 'Euphrates' river which is located from Al-Kifl city to Kufa city, in Iraq. We analyzed six heavy metal parameters (Cd, Pb, Zn, Cr, Fe, and Mn) in water, every month several samples were taken from six sites on the Euphrates river and covered the period of time from July 2020 to January 2021. Heavy metal analysis results were assessment according to the values of the maximum given in Water Quality Regulation. The obtained outcome of the analysis revealed that cadmium values surpassed the limit values for sites 1 and 5. The overall pollution level of the HMs pollution index model appeared to be helpful to Evaluate with respect to HMs and the values were found higher than the critical pollution index is100 in site 1.

Keywords: Euphrates river; Heavy metal; Iraq; Index Model

1. Introduction

Most pollutants are heavy metals that cause problems in the environment because of their permanency, biological toxicity, inability to degradability and capacity to penetrate the food chain [1]. A few poisonous compounds collected in water bodies of soil and silt when discharged into the environment [2] [3]. Beneath certain conditions, overwhelming metal poisons respond with a few natural substances to change over them into indeed more poisonous metal-organic complex poisons [4] [5].

Since water bodies are in near contact with each other, overwhelming metals can enter sea-going life forms effortlessly primarily in three distinctive ways.

1. Oceanic creatures take overwhelming metal particles broken down within the water into their bodies by gills and other organs. At that point sea-going living beings collect overwhelming metals on surface skin or transmit them through the blood into various organs or tissues within the body.
2. Through their stomach related tracts and so on, oceanic creatures retain overwhelming metals in their bodies by nourishment admissions.

3. Heavy metals (HMs) are absorbed into the subcutaneous layers of the animals and then enter the organisms through osmotic exchanges between HMs and the surfaces of aquatic organisms affected by them [6].

In a study conducted by [7] on the Euphrates, the results of the heavy pollution index showed that the water was truly contaminated with cadmium, chromium and lead.

Al-Asadi [8] study showed that the quality of water revers the combined impacts of natural and human factors. As the fast evolution of petroleum production and plants of electric power and increase number of cars in the region produce an increase HMs levels in river sediments. Whereas [9] study results were mentioned of some HMs (Cu, Cd, Ni, and Pb) in the river and drinking water in the Nasiriyah city south of Baghdad high risk of Pb concentration in all water samples and a slight hazard for nickel and cadmium. Allafta & Opp [10] study data appeared on Shatt Al-Arab River, Southern 'Iraq' no noteworthy relationship between metals concentrations and seasonal. Either consider of [11] study on the 'Euphrates' waterway was found that the rates of concentrations of HMs dissolvable in water in from the examined Al-Hindia stations, Al-Kifl and Kufa were as take after:(Fe< Pb < Ni< Co< Cd). Whereas in the study [12] they come about were uncovered that the concentrations of the examined HMs within the water of Al- Mussaib Channel was minus than the contamination levels and taken after they arrange: Lead< nickel < Copper< Chromium < Mangnase < Zinc. The study by Omran et al [13] considers in the 'Euphrates' river that the toxic heavy Metal values that determined have been being affected by domestic and industrial activities in study areas. Water classifications have also been determined as fourth-grade water contamination of Chromium, Selenium and nickel metals, third-grade Cobalt metal water contamination and other second and first-grade water contamination according to the Intracontinental Water Contamination Control Criteria. The current study aimed to evaluate heavy metals contamination for the Euphrates stream by utilized of contamination indices.

2. Material and Methods

The Euphrates Waterway is one of the foremost noticeable streams in Southwest Asia and the longest, with a length of approximately 2,800 km expands interior Iraq represents about 35% of the whole length of the waterway and irrigates huge regions about 765831 km^2, with a seepage rate of up to 818 m^3 / s [14]. The Euphrates Stream is provided with water from its exterior source of Iraqi borders and included to it from the water put away within the Tharthar store and other lakes, making it with a huge substance of sediment and tiny fish floating with water [15]. Water samples were collected from the 'Euphrates' River from 6 sites were recognized by GPS gadgets (Garmin, USA) (Table 1) each month from July 2020 to January 2021. These samples have been collected in 1 L plastic bottles and add 1 ml of 1:1 nitric acid diluted (From 68 per cent HNO3, BDH) to keep samples [16]. The Atomic Absorption Spectrophotometer, (AAS) SHIMADZU Japan were utilized for HMs analyses.

2.1. Pollution assessment indices:

Contamination indexes were utilized to assess the quality of water. Markers utilized are the heavy metals Pollution Index (HPI) and the HMs evaluation Index (HEI).

2.2. Heavy metals index (HPI)

The HPI assessment the compound effects of HMs on the water quality, as a ranking technique or weightage for each chosen parameter (Wi). The ranking system is randomly between 0 and 1 and it depends on the significance of the ranking framework to each quality criteria [17] [18]. In deciding the HPI data of quality of water, the concentration limits i.e. the permissible (Si) and the elevated desired value (Ii) were derived from WHO benchmarks. Each parameter HPI is set up with the taking after expression[18]:

The calculate of the heavy metal index (HPI) is as follows.

$$HPI = \sum_{i=1}^{n} Q_i W_i \;/\; \sum_{i=1}^{n} W_i$$

Where Qi is the ith parameter's sub-index. Wi is the unit weight and n is the number of parameters taken under consideration. The subscription index (Qi) is calculated according to the formula:

$$Q_i = \sum_{i=1}^{n} \frac{\{M_i(-)I_i\}}{S_i - I_i} * 100$$

Where Mi is the monitoring of HMs and Si (standard value) of the ith parameter. Index of evaluation of heavy metal (HEI). The HEI gives an overall water quality according to HMs concentration and the calculated according to the following formula [19]:

$$HEI = \sum_{i=1}^{n} \quad Ci/(MAC)i$$

Where Ci is the monitoring value of the ith parameter. And the MAC is maximum admissible concentration.

3. Results and Discussion

3.1. Concentration of HMs:

The most elevated esteem of Cd 3.98 µg/l was recorded within the winter season and the least esteem within the Harvest season was 1.76 µg/l (Fig.1). Either spatially variations for Cd metal, the elevated esteem 8.9 µg/l in site1 were recorded and the reduced esteem was 1.73 µg/l in site 5 (Fig.3). Seasonal values were recorded at a higher level in winter 3.98 µg/l compare with other seasons (Fig:2) The statistical analysis (ANOVA at $P< 0.05$) appeared that there was a critical contrast among locales for Cd metal during the study period.

The Cd affects plant metabolization and can bioaccumulate and enter the food chain in aquatic organisms [20]. The concentration achieved during the work is above the permissible limits in site 1. High levels pose a significant threat to valuable flora and fauna. This study observes similar remarks; Cd concentrations are greater than those recommended by the WHO especially in winter on site 1. The results of the comparison suggest that Cd levels are similar to Hassan et al study [21] that was reported in the Euphrates River from Al-Hindiya to Al-Kufa region, Al-Khafaji et al study [22] in Al-Nassiriya City and Habeeb et al study [23] in Abo-Garak to the south of the Kifil city and lower than Salah et al study [7] in Amiriyah Fallujah. The study region can be exposed to river dumps, rainfall and local sources of pollution. The Euphrates River showed less pollution, suggesting maybe found various hydrological, geological and human activities compared with other rivers around the world. High exposure to Cd can lead to liver and kidney damage and acute effects on health can occur [24].

The higher value of lead (Pb) metal was recorded 2.5 µg/l in site 5 and the lower value was recorded (0.58 µg/l in site 1 (fig:1). Either seasonal variations were recorded highest value was 2.1 µg/l in Autumn and the lowest value was 1.03 µg/l in summer (Fig.2). The statistical analysis ($P< 0.05$) showed that there was a significant difference among seasons and sites. Lead causes a lot of diseases and is highly toxic. With the application of new technology, it must be remedied to phase out Pb from different industries. These differences are due to the monthly variations in water discharge, rain and/or waste disposal. The current study recorded Pb levels within the concentration recommended by WHO [25].

The higher value of Chromium metal was recorded 0.98 µg/l in site 5 and the lower value was recorded 0.28 µg/l in site 3 (fig:1). Either seasonal variations were recorded the highest value of 0.78 µg/l in summer and the lowest value was 0.32 µg/l in Autumn (Fig.2). The statistical analysis ($P< 0.05$) showed that there was a significant difference among seasons and sites. Chromium is present in dissolved form as either anionic trivalent $Cr(OH)_3$ or as hexavalent CrO_{42} [26]. Trivalent chromium is most abundant in natural waters [27]. chromium compounds can be found in waters only in trace amounts. Chromium is a dietary requirement for a number of organisms [26]. This however only applies to trivalent chromium. Hexavalent chromium is very toxic to flora and fauna [28].

The higher value of Manganese metal was recorded 6.52 µg/l in site 6 and the lower value was recorded 5.02 µg/l in site 2 (fig:1). While seasonal variations, the highest value were recorded 7.7 µg/l in Autumn and the lowest value was 3.9 µg/l in summer (Fig.2). The statistical analysis ($P< 0.05$) showed that there was a significant difference among seasons and sites. For normal development, immune function of the immune system, digestion, and bone strength, manganese is needed [29]. Too much manganese exposure could have neurological effects. Some recent epidemiological studies have shown a link between exposure to manganese and intellectual disabilities in the drinking water and poorer neurobehavioral function in young children [30].

The higher value of Fe metal was recorded 3.7 µg/l in site 5 and the lower value was recorded 1.5 µg/l in site 1 (fig:1). While seasonal variations, the highest value were recorded 2.95 µg/l in Autumn and the lowest value 1.02 µg/l in summer (Fig.2). The statistical analysis ($P< 0.05$) showed that there was a significant difference among seasons and sites. Iron (Fe) is considerd a fundamental element of our body's health, the most well-known of them is the production of the hemoglobin, which trans the oxygen from our lungs to carry it throughout the body. Insufficient or excess iron can affect the functions of the body [31].

The higher value of Zn metal was recorded 3.4 µg/l in site 1 and the lower value was recorded 0.09 µg/l in site 4 (fig:1). While seasonal variations, the highest value were recorded 1.78 µg/l in winter and the lowest value 0.11 µg/l in summer (Fig.2). The statistical analysis ($P< 0.05$) showed that there was a significant difference among seasons and sites. Zinc is a trace element which is essential for the maintenance of human health in small quantities. Zinc helps in hormonal production, growth, immune and digestive system improvement. A higher or lower zinc concentration than the required quantity of drinking water leads to unwanted health effects. Increased zinc can lead to differentiated health problems like stomach cramps, inflammation of the skin, vomiting, nausea, anaemia, etc. [32]. Zinc oxide is used in the pharmaceutical, paint, rubber, cosmetics, plastic, tin, soap, battery and textiles industries. These industries are responsible for the high concentration of zinc metal unless effective treatment is applied to remove zinc from industrial waste [33].

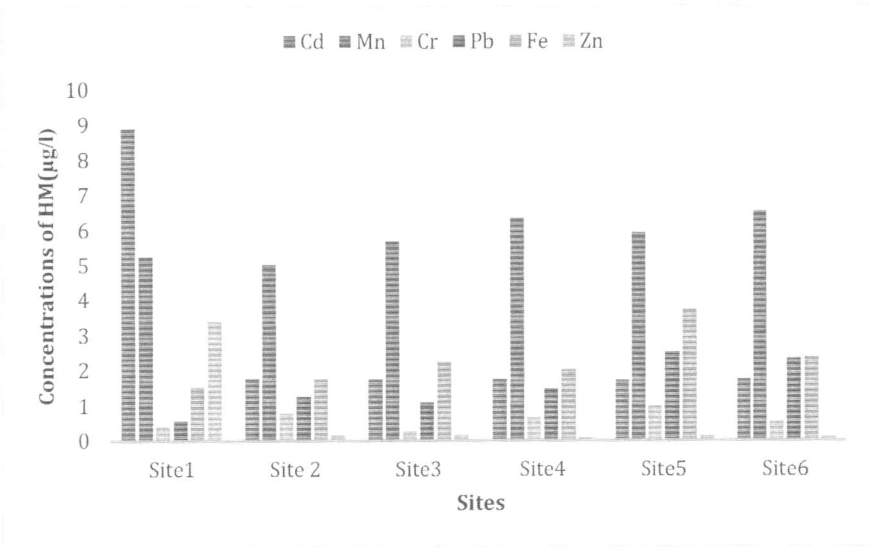

Figure 1. Mean the heavy metals concentrations in study area

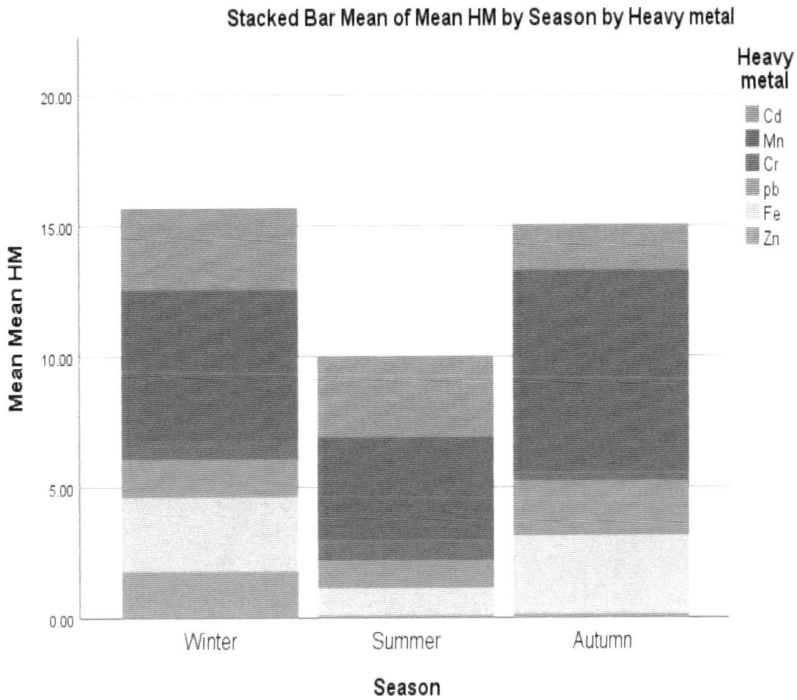

Figure 2. Seasonally variations of the heavy metals concentrations in study area

3.2. Pollution indices:

The spatial distribution values of HPI index were presented in Figure 3. The highest value of HPI index was recorded 209 in site 1 and lowest value 43.8 in site 3. While Temporal variations of HPI were recorded the highest value 76.7 in winter, it was classified as Medium, and the lowest value was 46.3 in Autumn, it was classified as low (Fig: 4). HMs pollution assessment by HPI indicated more serious in sites 1 and it was classified as less serious in Autumn season compared to other seasons. The highest values recorded in the site 1 due to this site is contaminated with sewage water as well as agriculture wastes and effluents of industrial, this site is considered the sources of pollution of the Euphrates River in the study area. [34]. indicated that some heavy metal concentrations (Mn, Fe, Cr, Cd, Cu, Pb, Ni,and Zn) were was most likely caused by agricultural wastewater and industrial effluents. The development of industrialization and Urbanization has a clear adverse environmental influence. After 2003, in Iraq, the number of electricity generators and vehicles increased significantly and more pollutants including heavy metals were emitted from fuel combustion [35].

The spatially variations of HEI, the highest value were recorded 2.96 in site 1 of Cd and the lowest value 1.82×10^{-5} of Zn in site 4 and it was classification of pollution as low of all heavy metals in all sites (Table 1). While Temporally variations of HEI, the highest value were recorded 1.04 of cadmium (Cd) in winter season while the lowest value 2.27×10^{-5} of Zinc (Zn) in summer and it was classification of pollution according HEI as low (Table 2).

The highest hazard of induced HMs contamination is likely to be in food web complexes in the Euphrates River. This alarming concentration could also lead to uncomfortable conditions for people living near this river. Therefore, strict measures should be taken into account to protect the sustainability of the environment of this Euphrates River.

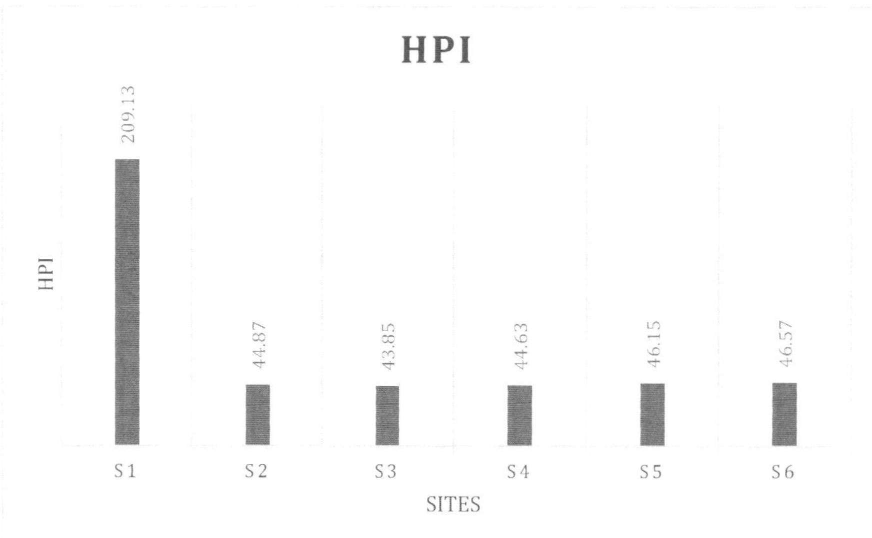

Figure 3. Mean of HPI in sites of study area.

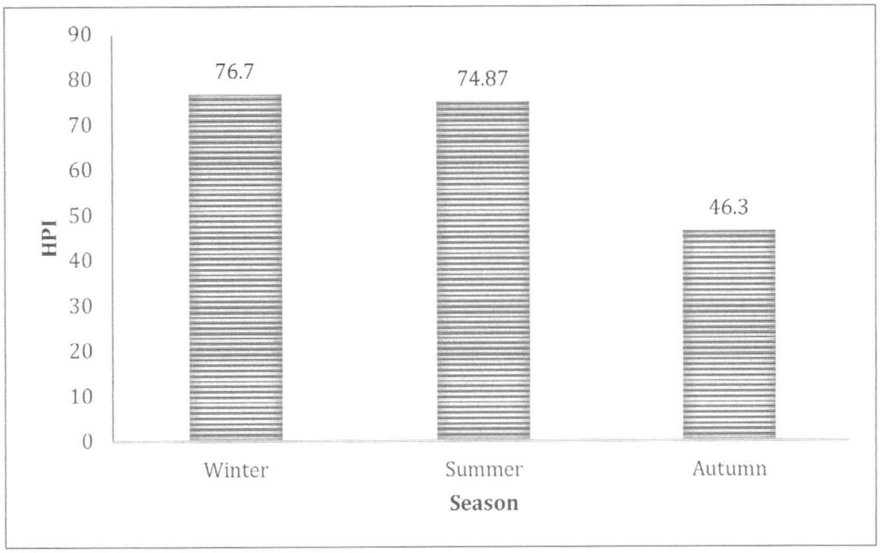

Figure 4. Mean of temporally variations of HPI values in study area

Table 1. Spatially variations of HEI values in Euphrates River

Sites	Heavy metal(μg/l)	Mean Con.	The highest permitted value (MAC)	HEI	HEI classification
Site1	Cd	8.9	3	2.966667	Low
	Pb	0.5846	50	0.011692	Low
	Cr	0.423333	50	0.008467	Low
	Fe	0.045983	300	0.000153	Low
	Zn	3.4133	5000	0.000683	Low
	Mn	5.262778	100	0.052628	Low
Site2	Cd	1.79	3	0.596667	Low
	Pb	1.2737	50	0.025474	Low
	Cr	0.793704	50	0.015874	Low
	Fe	0.052318	300	0.000174	Low
	Zn	0.1644	5000	3.29E-05	Low
	Mn	5.022222	100	0.050222	Low
Site3	Cd	1.761667	3	0.587222	Low
	Pb	1.0913	50	0.021826	Low
	Cr	0.282778	50	0.005656	Low
	Fe	0.0667	300	0.000222	Low
	Zn	0.1722	5000	3.44E-05	Low
	Mn	5.678889	100	0.056789	Low
Site4	Cd	1.756667	3	0.585556	Low
	Pb	1.4741	50	0.029482	Low

	Cr	0.657037	50	0.013141	Low
	Fe	0.06095	300	0.000203	Low
	Zn	0.0911	5000	1.82E-05	Low
	Mn	6.343333	100	0.063433	Low
Site5	Cd	1.730556	3	0.576852	Low
	Pb	2.5007	50	0.050014	Low
	Cr	0.988333	50	0.019767	Low
	Fe	0.111917	300	0.000373	Low
	Zn	0.1389	5000	2.78E-05	Low
	Mn	5.923333	100	0.059233	Low
Site6	Cd	1.763889	3	0.587963	Low
	Pb	2.3257	50	0.046514	Low
	Cr	0.534444	50	0.010689	Low
	Fe	0.070783	300	0.000236	Low
	Zn	0.1156	5000	2.31E-05	Low
	Mn	6.527222	100	0.065272	Low

Table 2. Mean of temporally variations of HEI values in study area

Season	Heavy metal(μg/l)	Mean Con.	The highest permitted value (MAC)	HEI	HEI classification
Winter	Cd	3.135	3	1.045	Low
	Pb	1.4625	50	0.02925	Low
	Cr	0.7286	50	0.014572	Low
	Fe	2.8361	300	0.009454	Low
	Zn	1.7806	5000	0.000356	Low
	Mn	5.729444	100	0.057294	Low
Summer	Cd	3.103056	3	1.034352	Low
	Pb	1.030556	50	0.020611	Low
	Cr	0.7869	50	0.015738	Low
	Fe	1.0241	300	0.003414	Low
	Zn	0.1133	5000	2.27E-05	Low
	Mn	3.949167	100	0.039492	Low
Autumn	Cd	1.768056	3	0.589352	Low
	Pb	2.132037	50	0.042641	Low
	Cr	0.3243	50	0.006486	Low
	Fe	2.9581	300	0.00986	Low
	Zn	0.1539	5000	3.08E-05	Low
	Mn	7.700278	100	0.077003	Low

References

[1] Bazrafshan, E., Sobhanikia, M., Mostafapour, F. K., Kamani, H., & Balarak, D. (2017). Chromium biosorption from aqueous environments by mucilaginous seeds of Cydonia oblonga: kinetic and thermodynamic studies. Global Nest Journal, 19(2), 269-277..

[2] Mahipal S S, Mayuri K, Manisha N, Rajeev K, Prashant A (2016) Heavy Metals Contamination in Water and their Hazardous Effect on Human Health-A Review. Int.J.Curr.Microbiol.App.Sci. 5(10): 759-766..

[3] Hosseinpour, J., & Mahdavy-Moghaddam, H. (2020). Computational study of magnetic field effects on the nozzle of hydrogen micro flame. Combustion and Flame, 220, 247-256..

[4] Maceda-Veiga, A., Monroy, M., Navarro, E., Viscor, G., & de Sostoa, A. (2013). Metal concentrations and pathological responses of wild native fish exposed to sewage discharge in a Mediterranean river. Science of the Total Environment, 449, 9-19.

[5] Chowdhury, S., Mazumder, M. J., Al-Attas, O., & Husain, T. (2016). Heavy metals in drinking water: occurrences, implications, and future needs in developing countries. Science of the total Environment, 569, 476-488..

[6] Long, Y., Luo, Y., Xiao, J., Guo, Z., Chen, L., & Xiao, Y. (2016). Research advances in effects of heavy metals stress on fish. Journal of Southern Agriculture, 47(9), 1608-1614.

[7] Salah, E. A. M., Al-Hiti, I. K., & Alessawi, K. A. (2015). Assessment of Heavy Metals Pollution In Euphrates River Water, Amiriyah Fallujah, Iraq. Assessment, 5(15).

[8] Al-Asadi, S. A., Al-Qurnawi, W. S., Al Hawash, A. B., Ghalib, H. B., & Alkhlifa, N. H. A. (2020). Water quality and impacting factors on heavy metals levels in Shatt Al-Arab River, Basra, Iraq. Applied Water Science, 10(5), 1-15.

[9] Abd Al-ameer, A. S., & Ali, S. A. (2020). Heavy metals determination and pre-concentration in river and drinking water samples at Nassiriyah City.EurAsian Journal of BioSciences. 14:6635-6641..

[10] Allafta, H., & Opp, C. (2020). Spatio-temporal variability and pollution sources identification of the surface sediments of Shatt Al-Arab River, Southern Iraq. Scientific reports, 10(1), 1-16.

[11] Al-Zughaiby, H. H. S., Jawad, H. J., & Al-Awadi, J. H. H. (2020). The relationship between concentrations of some trace elements in the Euphrates River of Iraq. In AIP Conference Proceedings (Vol. 2290, No. 1, p. 020044). AIP Publishing LLC..

[12] Omran, I. I., Al-Saati, N., Hashim, K., Al-Saati, Z., Kot, P., Al Khaddar, R. M., ... & Aljefery, M. (2019). Assessment of heavy metal pollution in the Great Al-Mussaib irrigation channel. Desalination and Water Treatment..

[13] Dinç H, & DÖRTBUDAK, M. Y. (2020) Investigation of Toxic Effects of Heavy Metal Level in Atatürk Dam. Experimental and Applied Medical Science, 1(1), 23-28..

[14] UNESCO. (2002). The challenge of achieving gender parity in basic education: a statistical review, 1990-1998. Unesco..

[15] Daoudy, Marwa (2005), Le Partage des Eaux entre la Syrie, l'Irak et la Turquie. Négociation, Sécurité et Asymétrie des Pouvoirs, Moyen-Orient (in French), Paris: CNRS, ISBN 2-271-06290-X.

[16] Eaton, A. D., Clesceri, L. S., Rice, E. W., Greenberg, A. E., & Franson, M. A. H. (2005). Standard methods for the examination of water and wastewater. American public health association, 1015..

[17] Reddy, S.J. (1995): Encyclopedia of environmental pollution and control. - Environmental media, Karela, India..

[18] Mohan, S.V., Nithila, P., Reddy, S.J. (1996): Estimation of heavy metal in drinking water and development of heavy metal pollution index. - Journal of Environmental Science and Health 31: 283-289..

[19] Edet, A.E., Offiong, O.E. (2002): Evaluation of water quality pollution indices for heavy metal contamination monitoring. A study case from Akpabuyo-Odukpani area, Lower Cross River Basin (Southeastern Nigeria). - GeoJournal 57: 295-304..

[20] Oliveira Ribeiro C.A., Vollaire Y., Sanchez-Chardi A., Roche H. (2005) Bioaccumulation and the effects of organochlorine pesticides PAH and heavy metals in the eel (Anguilla anguilla) at the Camargue Nature Reserve, France. Aqua Toxicol;74:53–69..

[21] Hassan, F., Saleh, M. and Salman, J. (2010), A study of physicochemical parameters and nine heavy metals in the Euphrates River, Iraq, E-Journal of Chemistry 7, 685-692..

[22] Al-Khafaji, B., Mohammed, A.& Maqtoof, A.(2011), Distribution of some heavy metals in water ,sediment & fish Cyprinus carpio in Euphrates river near Al- Nassiriya city center south Iraq, Baghdad Science Journal 8, 552-560..

[23] Habeeb, M., Al-Bermani, A. & Salman, J. (2015), Environmental study of water quality and some heavy metals in water, sediment and aquatic macrophytas in lotic ecosystem, Iraq, Mesopotamia Environmental Journal 1, 66-84..

[24] Boateng, T.K.; Opoku, F.; Acquaah, S.O.; Akoto, O.(2015) Pollution evaluation, sources and risk assessment of heavy metals in hand-dug wells from Ejisu-Juaben Municipality, Ghana. Environ. Syst. Res., 4, 1–12..

[25] World Health Organization(WHO) (2011). Guidelines for drinking-water quality. WHO chronicle, 38(4), 104-108.

[26] World Health Organization. (2020). Chromium in Drinking-water (No. WHO/HEP/ECH/WSH/2020.3). World Health Organization.

[27] Sharma, A., Kapoor, D., Wang, J., Shahzad, B., Kumar, V., Bali, A. S., ... & Yan, D. (2020). Chromium bioaccumulation and its impacts on plants: an overview. Plants, 9(1), 100.

[28] Ray, R. R. (2016). Adverse hematological effects of hexavalent chromium: an overview. Interdisciplinary toxicology, 9(2), 55.

[29] Pereira, C. G., Rabello, C. B. V., Barros, M. R., Manso, H. E. C., Santos, M. J. B. D., Faria, A. G., ... & Fireman, A. K. (2020). Zinc, manganese and copper amino acid complexed in laying hens' diets affect performance, blood parameters and reproductive.

[30] Rahman, S. M., Kippler, M., Tofail, F., Bölte, S., Derakhshani Hamadani, J., & Vahter, M. (2017). Manganese in drinking water and cognitive abilities and behavior at 10 years of age: a prospective cohort study. Environmental health perspectives, 125(5), 0.

[31] Anonymous. (2008). Assessment of surface water for drinking quality. Mughalpura, Lahore: Directorate of Land Reclamation Punjab, Irrigation and Power Department, Canal Bank..

[32] Jamshaid, M., Khan, A. A., Ahmed, K., & Saleem, M. (2018). Heavy metal in drinking water its effect on human health and its treatment techniques–a review. International Journal of Biosciences, 12(4), 223-240.

[33] Briffa, J., Sinagra, E., & Blundell, R. (2020). Heavy metal pollution in the environment and their toxicological effects on humans. Heliyon, 6(9), e04691.

[34] Yan, X., Liu, M., Zhong, J., Guo, J., & Wu, W. (2018). How human activities affect heavy metal contamination of soil and sediment in a long-term reclaimed area of the Liaohe River Delta, North China. Sustainability, 10(2), 338.

[35] Qazar, I. A., Salem, F. M., & Aied, I. A(2018). Spatiotemporal variations in the heavy metal pollution index (HPI) values southern Iraq. 8th National Conference on Environment and Natural Resources / 14-15/11/1028.

Depositional Environments of the Mishrif Formation in Faihaa Oil Field, South of Iraq

Hiba A.Kareem Mohammed[1] , Manal Sh. Al-kubaisi[1] and Ghazi Hasan Alshar'a[2]

[1]Department of Geology, College of Science, University of Baghdad, Baghdad, Iraq
[2]Ministry of Oil, O.E.C, Baghdad, Iraq

Abstract. The Mishrif Formation is one of the Iraqi major oil reserves in southern. This formation consists of limestone deposited during a Lower Cretaceous period as part of the main series of transgression depositions. Faihaa oil field is located close to Iraqi-Iranian borders, about 50km to the north of Basra city center. The area of this field is about 900 km^2. It is bordered by the north Majnoon oil field, the south Sindibad field, the west Nahr Umr field, and the east husynia Iranian field. The microfacies and depositional environment of the Mishrif Formation in the Faihaa oil field were analyzed in this study. In Mishrif Formation, this study clarified several microscopic facies; more than 273 thin sections of core samples from one well (Faihaa-1) were examined, represented by Mudstone Microfacies Wackestone Microfacies and Packstone-Wackestone Microfacies. They were analyzing diagenesis processes and identifying Mishrif Formation. The two identified diagenetic environments that affected Mishrif reservoir in an open platform and restricted platform environments. The Mishrif Formation recognized five diagenetic processes, each with positive and negative implications on the quality of the reservoir. By creating and improving porosity and permeability, dissolution and dolomitization significantly impacted reservoir quality. Compaction and cementation have negative consequences by decreasing porosity and permeability, leading to reduced reservoir quality.

Keywords: Depositional Environments; Mishrif Formation; South of Iraq.

1. Introduction

The Mishrif Formation (Cenomanian-Early Turonian) is the most important carbonate reservoir in south east Iraq and contain oil at 32 structures. The largest accumulation is in the Rumaila North, Rumaila South, West Qurna, Zubair, Majnooon, and Halfaiya fields, located on large -scale north-south trending anticline. At least 15 other commercial oil accumulation in the Mishrif Formation have been discovered in south east Iraq: Abu Ghirab, Ahdab, Amara, Buzurgan, Dujaila, Jabel Fauqi, Gharraf, Hawaiza, etc. [1] In this research there is an attempt study the 15 other commercial oil accumulation in the Mishrif Formation have been discovered in south east Iraq: Abu Ghirab, Ahdab, Amara, Buzurgan, Dujaila, Jabel Fauqi, Gharraf, Hawaiza, etc [1]. In this research there is an attempt study the microfacies analysis, diagenesis processes and depositional environment of Mishrif formation of one well in Faihaa oil field by studying the petrographic and stratigraphic characteristics of more than 273 thin sections from Faihaa- 1.

2. Geology of the study area

Faihaa oil field is located close to Iraqi-Iranian borders, about 50km to the north of Basra city center. The area of this field is about 900 km2. It is bordered by the north Majnoon oil field, the south Sindibad field, the west Nahr Umr field, and the east Husynia Iranian field. It covers an area of (2866 Km) which is a convex fold, see Figure 1.

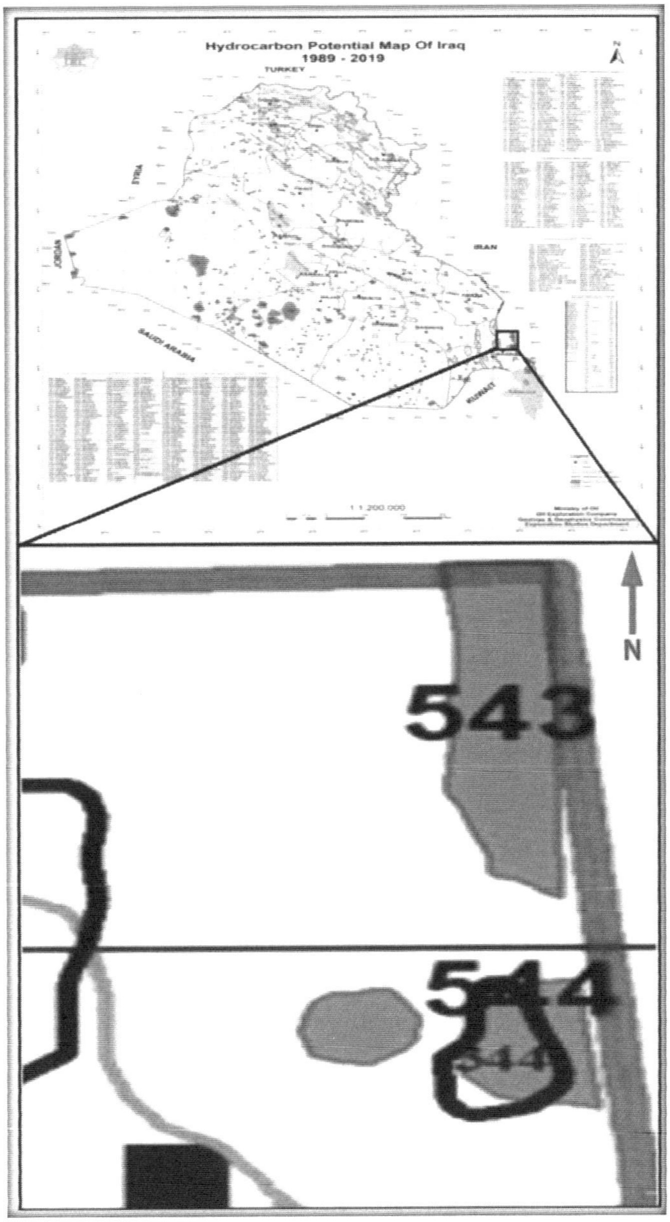

Figure 1. Location map of Study area (Faihaa oil field).

3. Petrography

Petrography is the branch of geology, which collaboration with the description and systematic classification of whole types of rocks, essentially microscope description for thin section [2].
The petrography of Mishrif formation components was analyzed depending on an examination more than (273) thin sections of cores samples for one well (Faihaa-1), were supplied by (Oil Exploration Company) and organized by Author. Microfacies analysis for Mishrif formation depending on the thin section consideration for one well from Faihaa oil field (Faihaa-1), classify the Microfacies, Microfacies distribution and depositional environments, classify the pores types and petrographic diagenetic processes and affected on Petrophysical properties of Mishrif formation.

4. Microfacies of the Mishrif Formation

The rocks of the Mishrif formation were studied in the Faihaa-1 exploration well to provide the lithology slides for this well only as well as the fossils and minerals contained in these rocks, in addition to determining their sedimentary facies and their region facies as it was based on **[3]**as well as on **[4]**to classify these facies whilst diagnosing the Diagenesis Processes affecting these rocks.
The thickness of Mishrif formation in Faihaa-1 (Figure 2), an exploration well is (322 m), where (273 lithology slides of the core) and (10 slides of lithology fragments) by examining the available lithology slides the sedimentary facies were identified, and they are as follows:

- Mudstone Microfacies: These facies were observed relatively less in the lithology of Mishrif formation. This fact consisted of (Mudstone or micrite clay) as it was represented by the presence of benthic Foraminifera such as Miliolid and Nezzazata sp. as shown in (Plate1 A)
  they contain some little or few bioclast or none.
- Wackestone Microfacies: These facies in this formation were observed for a more extended period than Mudstone Microfacies with lithology formations at different depths. The ground of these facies consists of micrite or sparite. It was represented by the presence of different genesis of bottom Foraminifera, such as Nezzazata sp, Chrysalidina sp Dicyclina sp, Cisalveolina sp, Quinqueloculina sp. (Plate1B and C). Some bioclast from echinoderms and mollusks (Plate1 D), and a small percentage of algae and Rudists remains, and these facies contained oil show and some organic materials.
- Packstone -Wackestone Microfacies: The wacky facies overlapped with the compacted limestone facies in terms of the percentage of grains present, and these facies contained a geneses bottom Foraminifera such as miliolid and the geneses Praealveolina sp., Nezzazata sp., Cisalveolina sp., Iraqia sp., Mollusks, Rudists, Echinoderms, and Peloids (Plate1 E, F, G, and H).

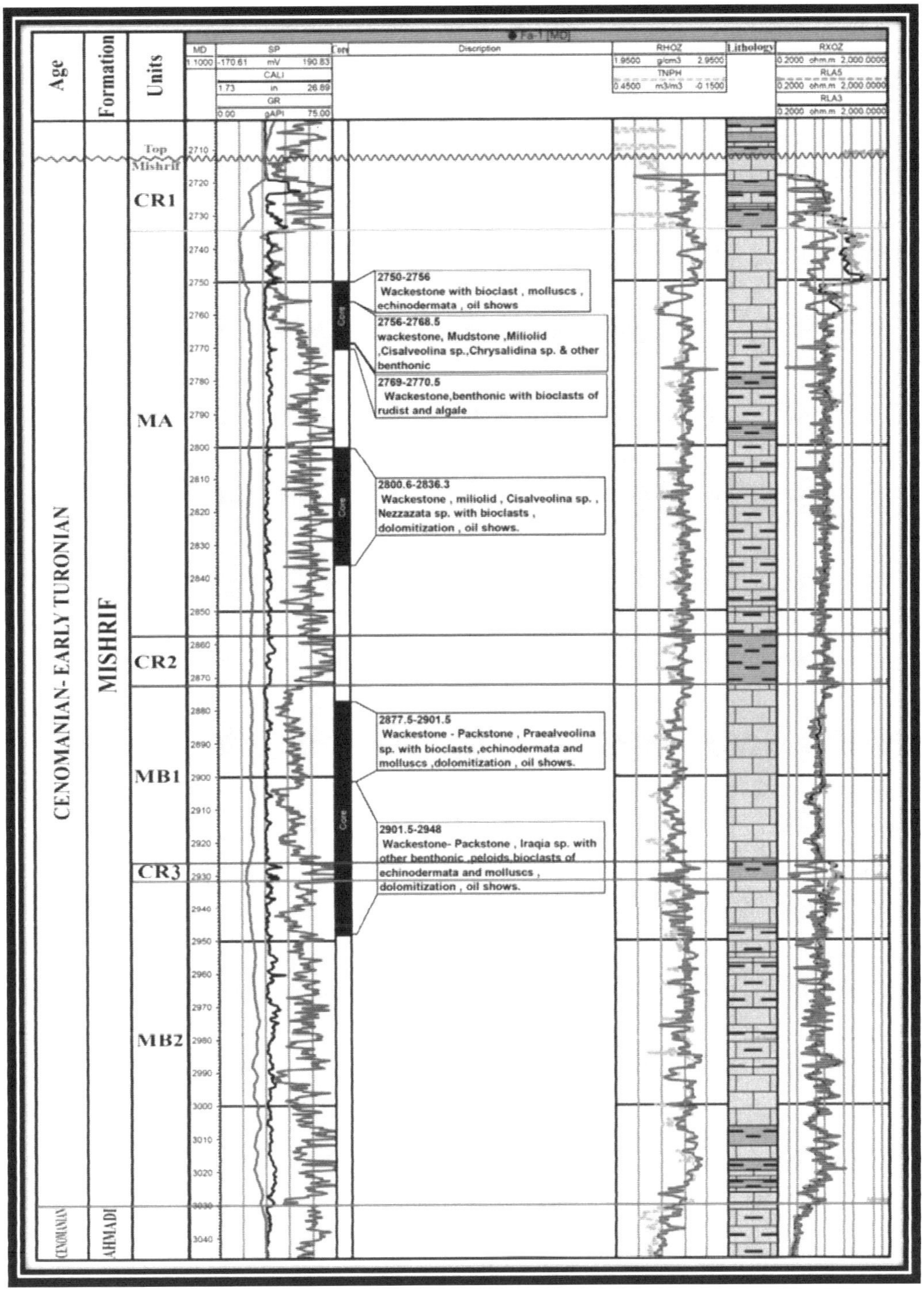

Figure 2. Well section shows the Microfacies of the Mishrif Formation in Faihaa-1.

5. Diagenesis processe affecting the rocks of Mishrif Formation

Diagenesis processes (Figure 3) represent the chemical, physical and biological changes that affect sediments during and after sedimentation and before the metamorphosim **[4]**
Mishrif Formation presented several adaptive processes that affected the quality of the reservoir formation properties, whether negatively or positively, according to the type of diagenesis process. The essential diagenesis processes are:

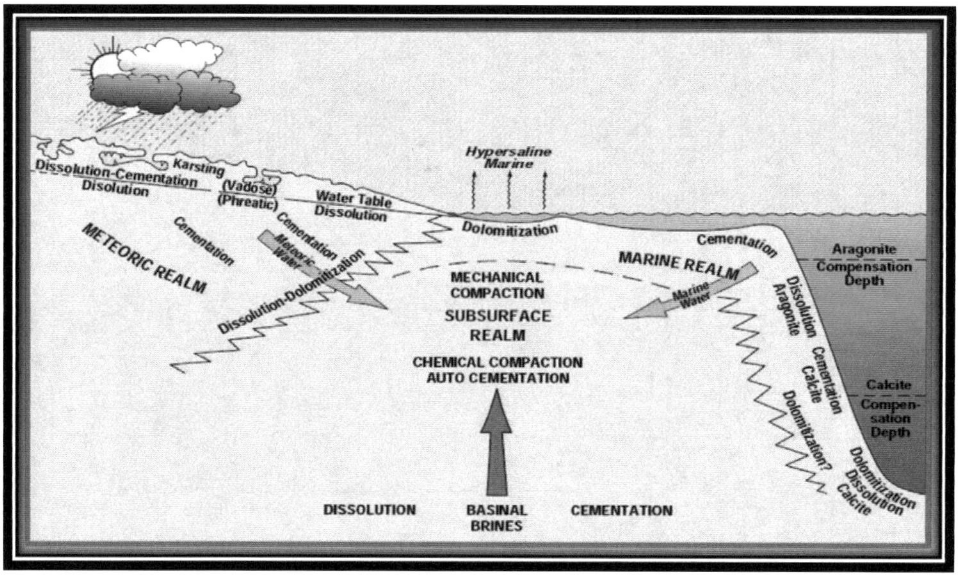

Figure 3. Schematic diagram illustrating the common diagenetic environments within which post-depositional porosity modification and evolution occurs [5].

5.1. Dissolution:

The dissolution process improves the reservoir characteristics by dissolving the rocks and their components by forming different porosity types, as limestone is one of the rocks most affected by this process. The solutions work to dissolve minerals and fossil shells (Figure 6, H).

5.2. Porosity:

It represents the percentage between the total voids of the rock to its full volume, and according to **[6]**the porosity is divided into two types (Table 1 and Figure 4):

Table 1. Porosity types **[7]**.

Time of Formation	Type	Origin
During deposition (primary)	Intergranular Intragranular	Sedimentation
After deposition (secondary)	Intercrystalline Fenestra	Cementation
	Vuggy Moldic	Solution
	Fracture	Tectonics

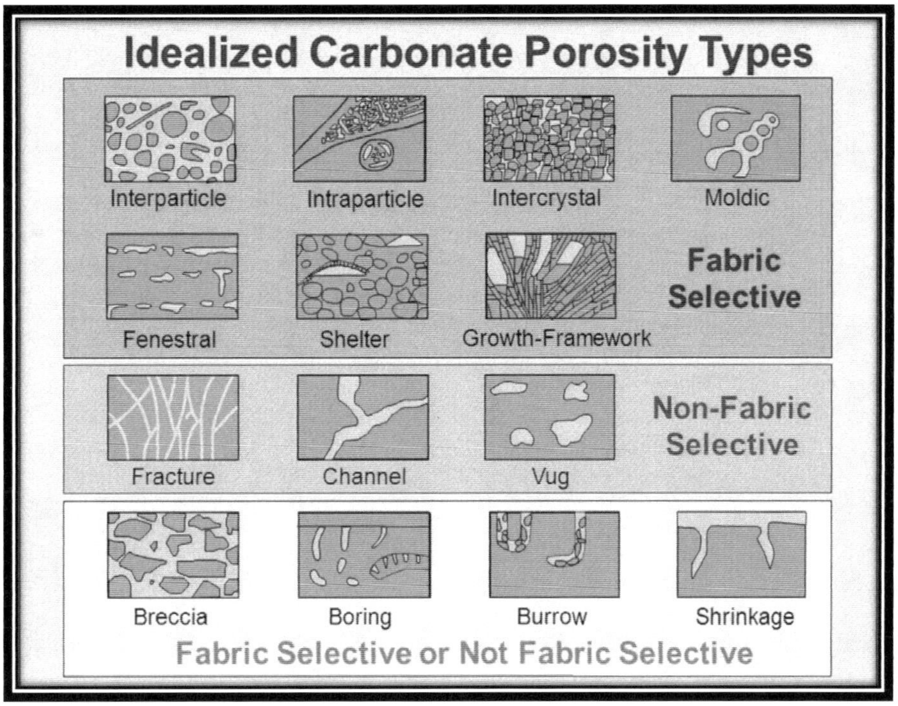

Figure 4. Porosity classification [8].

Primary porosity arises during the sediments localization process while secondary porosity is formed after the sediments localization due to dissolution; in addition, tectonic activities and the resulting fractures are the main types of porosity observed in the rocks formation:

A- Intraparticle Porosity: This porosity is formed by drilling or dissolving. The structural grains are subjected to a partial dissolution process leading to the dissolution of some of the internal parts of the fossil (Figure 7, A).
B- Fractures Porosity: It is secondary porosity formed by dissolution, compression, and tectonic force (Figure 7, B). These fractions may be filled with calcite cement forming veins (Figure 7, C).
C- Vug Porosity: It is a secondary porosity that is relatively large, and most of these pores represent the expansion or enlargement of the size of other types of pores (Figure 7, D).
D - Intercrystal porosity: It is secondary porosity observed between dolomite single crystals (Figure 7, E).
E- Intraparticle porosity: it is secondary porosity as it is observed between non-structural grains such as impurities (Figure 7, H).

5.3. Cementation:

This process is the process of filling the pores formed by the partial or complete dissolution of cement materials deposited chemically directly from solutions saturated with the substances of that cement, where the cement is deposited either between the structural components or within them **[4]** and this process naturally leads to Reducing the percentage of porosity in the rocks, but the effect of the cementation process on the rocks of Mishrif formation in the well of Faihaa-1 was small, as we only observed two types of cement, namely
A- Blocky cement: This cement resembles a mosaic of crystals consisting of amorphous crystalline facets of large sizes and deposited during the later stages of the Diagenesis processes (Figure 7, F)

B- Syntaxial Overgrowth Cement: This type of cement refers to the overgrowth of a crystal so that this original crystal and the growth over it is a large crystal that shares the same crystalline axes and is often observed around the echinoderms pieces (Figure 7, G) this type of cement consists in the early Diagenesis process stages.

5.4. Compression:

The compression process leads to distorting and reducing the total volume of sediments. The compression or tightening occurs due to the weight of the sedimentary cover as this causes a decrease in the size of the pores and expulsion of fluids trapped within the sediments. It is one of the processes that contributed to reducing the reservoir specifications through the compacting and sealing of rock particles with each other. There are two types of distinguished compression:

5.4.1. Physical Compaction: This type of compression c
an be observed from the tightness of the granules accumulated with each other and the distortion with breakage of the granules (Figure. 7, H).

5.4.2. Chemical Compaction:

This compression is represented by the formation of melting surfaces or stylolites that is the result of the chemical compaction process because it includes the dissolution of contact points between mineral grains as the pressure resulting from the weight of the sedimentary cover and surfaces with suture and zigzagged seams appear. These surfaces were observed by accumulating insoluble clay materials such as dolomite crystals, around the skin and bottom of Foraminifera (Figure 8, A).

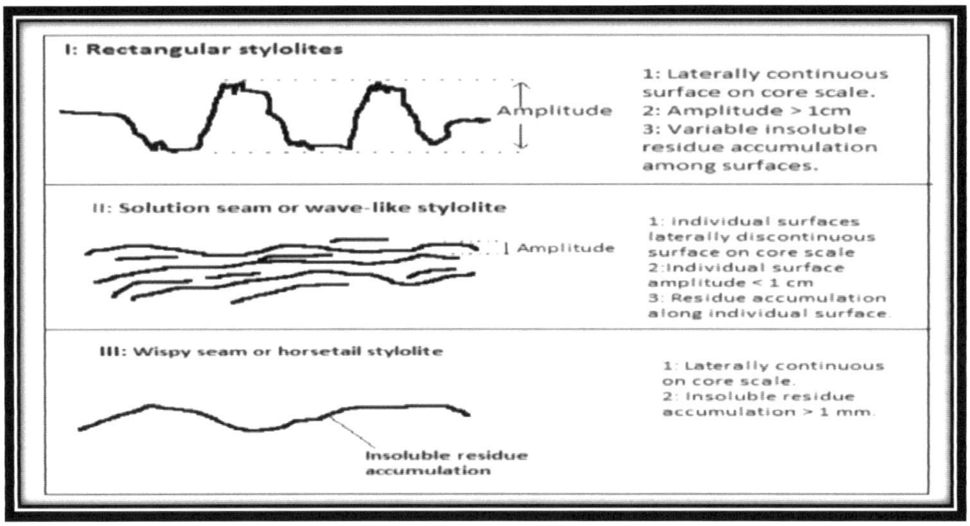

Figure 5. Three types of stylolites with various in the lower Cretaceous is reproduced from [9].

5.5. Dolomitization:

Dolomitization is considered a replacement process that includes dissolving existing minerals whilst depositing new minerals at the same time. Dolomitization occurs when the magnesium ion (Mg + 2) is replaced by calcium ion (Ca + 2) in calcareous deposits while simultaneously with the sediments or during Early Diagenesis processes, examples as follows:

- Shortly after sedimentation, before rock hardening, or within late Diagenesis processes.
- Long after sedimentation, the importance of the Dolomitization process comes from its increase in the porosity of the rocks affected by this process as long as there is no compacting or cementing process afterwards.

This study relied on the classification of researchers **[10]** in the classification of dolomite tissues and the description of the different types of dolomite tissues are as follows:

A. Floating-Rhomb fabric: It consists of single dolomite lozenges floating in a micrite floor (Figure 8, B).
B. Contact - Rhomb Fabric: The diamonds of the dolomite crystals are in contact with each other and floating within the ground, therefore the effect of dolomite is observed on the ground and fossils of the formation rocks (Figure 8, C).
C. Sieve Mosaic Fabric: This tissue consists of dolomite lozenges that were completely faceted, had little compactness and high porosity between the dolomite crystals (Figure 7, E).
D. Sutured Mosaic Fabric: This tissue consists of tightly packed dolomite crystals with little or no pores between the crystals (Figure 8, D).

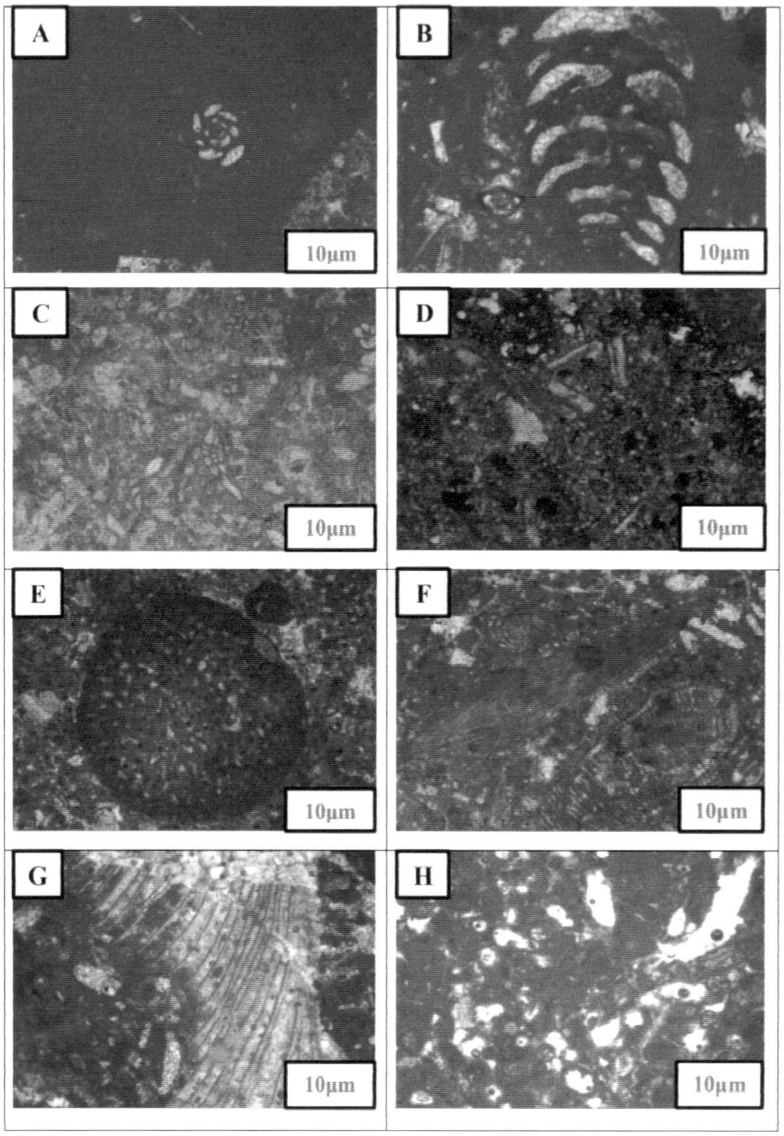

Figure 6. A –Mudstone Microfacies, Miliolid, Benthonic Foraminifera (Mishrif Formation,Core ,Faihaa-1,2767m,4XPL). B– Wackestone Microfacies with benthic Foraminifera Chrysalidina sp., Miliolid (Mishrif Formation,Core ,Faihaa-1,2961.5m,4XPL). C –Wackestone Microfacies, benthic Foraminifera Nezzazzata sp. (Mishrif Formation,Core ,Faihaa-1,2814m,4XPL). D – Wackestone Microfacies, with bioclast of Molluscs,oil shows,organic matter (Mishrif Formation,Core ,Faihaa-1,2754.5m,4XPL). E – Bioclastic Wackstone- Packstone Microfacies, Iraqia sp. (Mishrif Formation,Core ,Faihaa-1,2901.5m,4XPL). F – Bioclastic Wackstone- Packstone Microfacies, Praealveolina sp. (Mishrif Formation,Core ,Faihaa-1,2877.5m,4XPL). G – Bioclastic Wackstone- Packstone Microfacies, with Rudist (Mishrif Formation,Core ,Faihaa-1,2895.5m,4XPL). H – Wackstone- Packstone Microfacies, pelloid with intraparticle porosity created by dissolution (Mishrif Formation,Core ,Faihaa-1,2946m,4XPL).

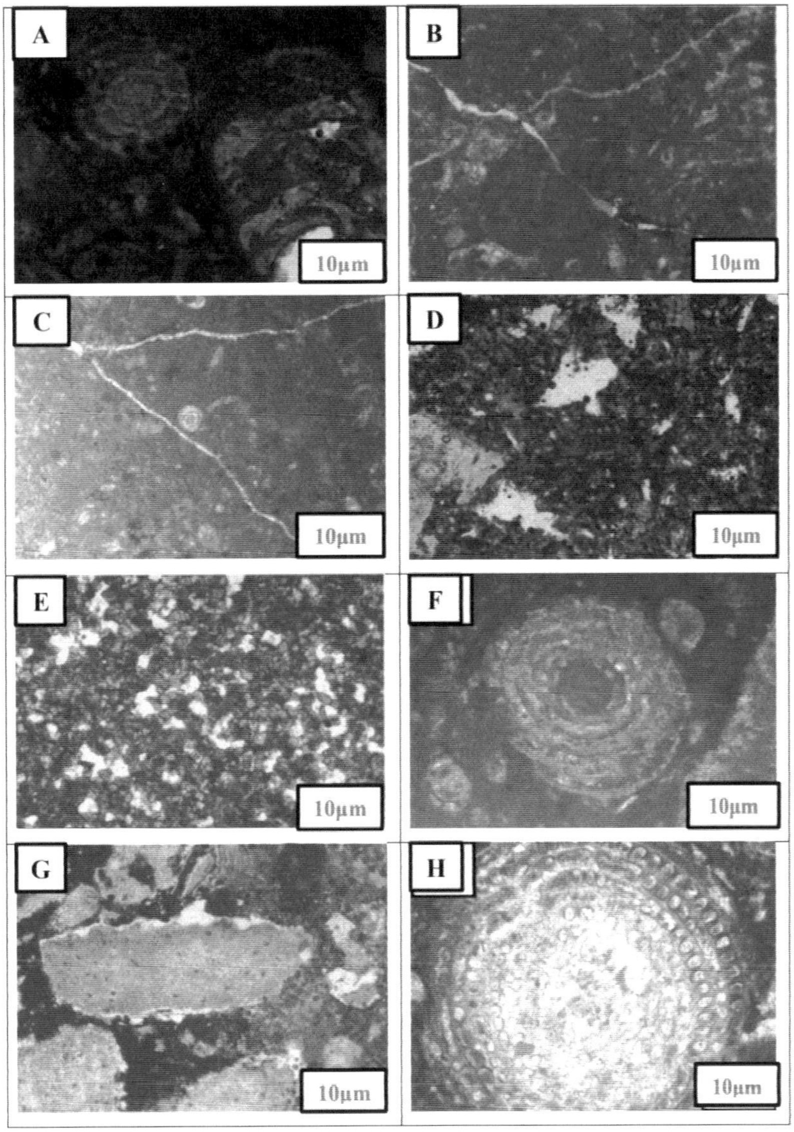

Figure 7. A – Wackestone Microfacies and there is intraparticle porosity (Mishrif Formation,Core ,Faihaa-1,2924m,4XPL). B– Wackestone Microfacies with fractures porosity (Mishrif Formation,Core , Faihaa-1,2761.5m,4XPL). C –Wackestone Microfacies with veins and Miliolid (Mishrif Formation,Core , Faihaa-1,2810.5m,4XPL). D – Wackestone Microfacies, with vugy porosity (Mishrif Formation, Core, Faihaa-1,2755m,4XPL). E – Wackestone Microfacies, Sieve Mosaic Fabric with intraparticle porosity between dolomite crystals(Mishrif Formation,Core ,Faihaa-1,2943m,4XPL). F– Wackestone Microfacies with benthic foraminifera Cisalveolina sp., rudist and miliolid (Mishrif Formation, Core, Faihaa-1,2812m,4XPL). G– Bioclastic Wackstone- Packstone Microfacies, with syntaxial overgrowth cement (Mishrif Formation,Core ,Faihaa-1,2936m,4XPL). H-Wackstone Microfacies, with benthic foraminifera Cisalveolina sp. (Mishrif Formation,Core ,Faihaa-1,2813 m,4XPL).

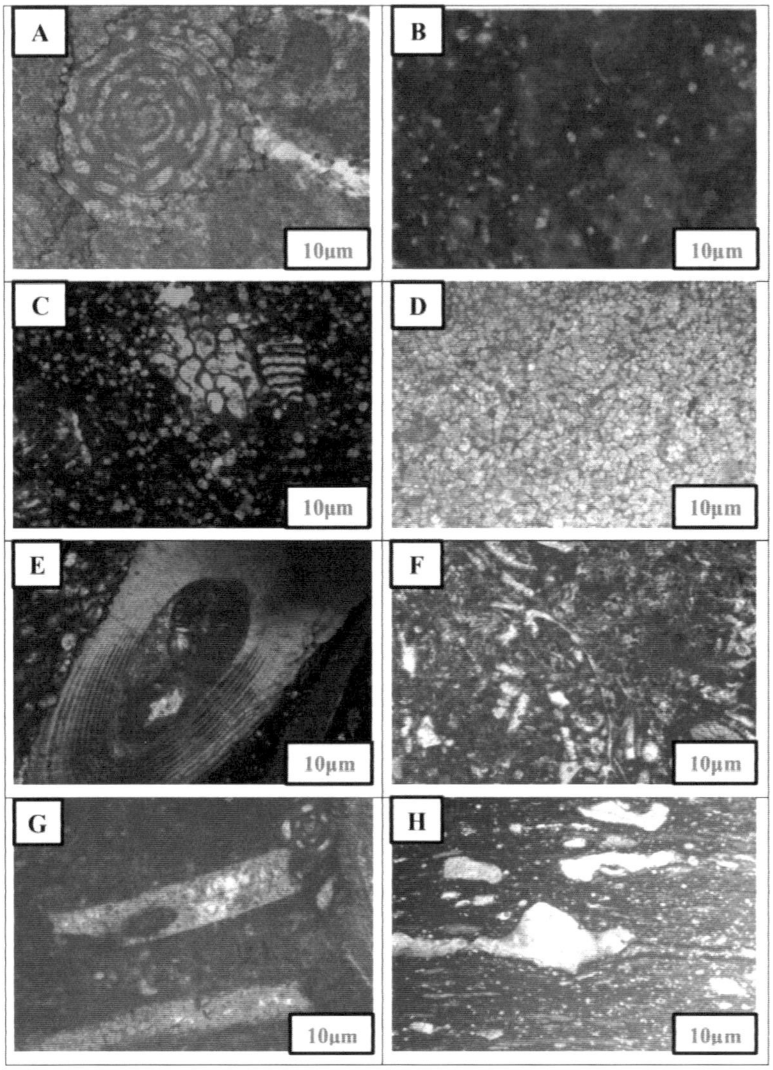

Figure 8. A- Wackstone Microfacies, with dissolve surrounds Cisalveolina sp., (Mishrif Formation, Core, Faihaa-1,2928.5m,4XPL). B- Wackstone Microfacies, with floating-rhomb fabric texture (Mishrif Formation, Core, Faihaa-1,2803m,10XPL). C- Wackstone Microfacies, with contact - rhomb fabric texture and oil shows (Mishrif Formation, Core, Faihaa-1,2815m,4XPL). D- Wackstone Microfacies, here we see sutured mosaic fabric texture (Mishrif Formation, Core, Faihaa-1,2941m,4XPL). E- Wackstone Microfacies, with benthic Foraminifera and bioclast of Molluscs, oil shows (Mishrif Formation, Core, Faihaa-1,2828.5m,4XPL). F- Bioclastic Wackstone Microfacies, with micritization for the floor and spceulas sponge, shell fragments, echinoderms (Mishrif Formation, Core, Faihaa-1 ,2880.5m,4XPL). G- Wackstone Microfacies, with Rudist and Miliolid (Mishrif Formation, Core, Faihaa-1 ,2835m,4XPL). H- Bioclastic Wackstone Microfacies, with spceulas sponge, shell fragments, echinoderms, bitumen, organic matter, rudist (Mishrif Formation, Core, Faihaa-1, 2811m,4XPL).

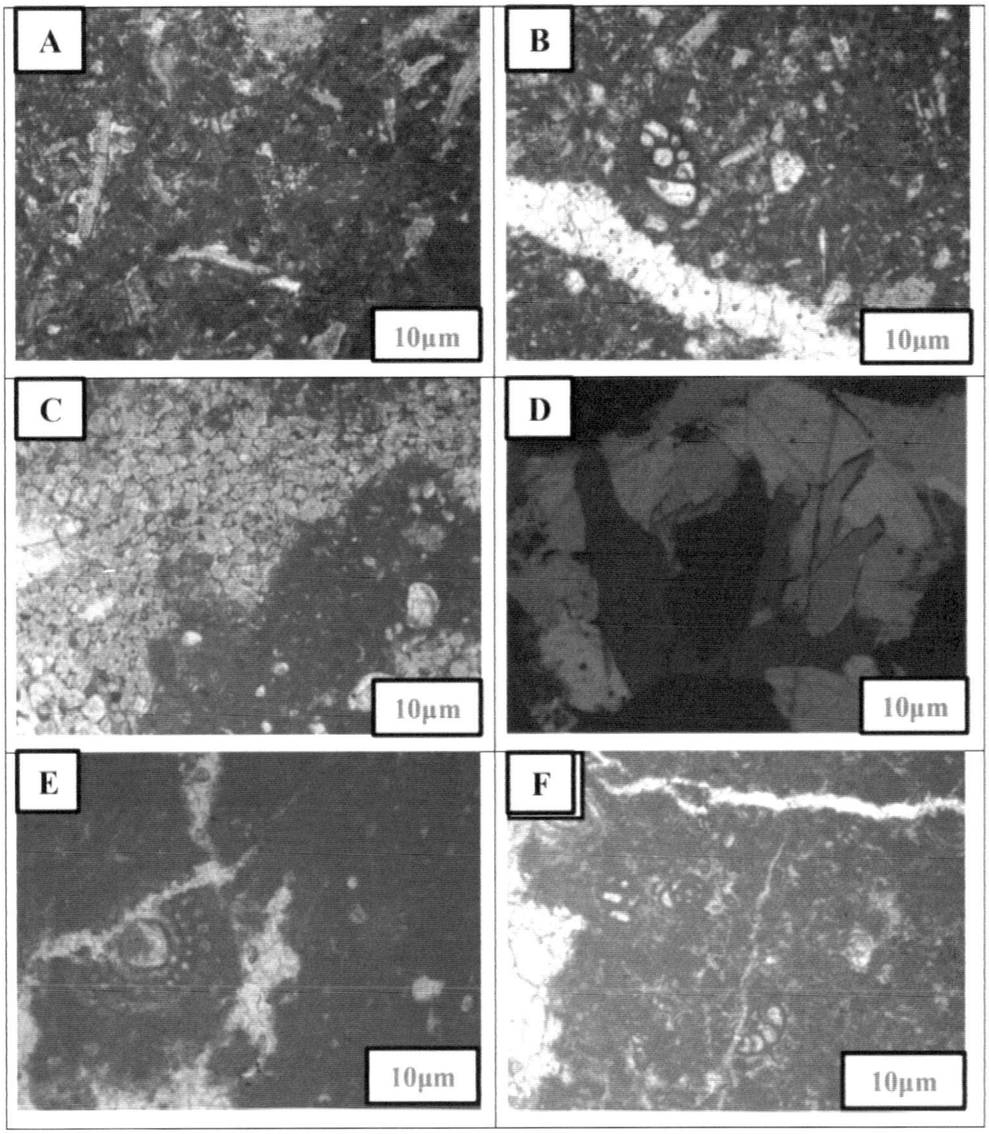

Figure 9. A- Syntaxial overgrowth cement (Faihaa -1, 2916m, XPL). B- Druzy mozaic cement (Faihaa-1, 2913m, XPL). C- Contact - rhomb fabric dolomite (Faihaa-1, 2833.5m, XPL). D- Blocky cement (Faihaa-1 ,2768m, XPL). E- Physical compaction (Faihaa-1 ,2760.5m, XPL). F- Stylolites (Chemical compaction) (Faihaa-1,2945m, XPL).

6. Depositional Environments of Mishrif Formation

The diagnosis of sedimentary microscopic facies that contain biological residual is very important in determining the type of sedimentary environment through these factors the sedimentary environments and the Lagoons zone for Mishrif formation in the study area were determined by relying on the available rock slides, where two types of sedimentary environments were identified. The processes helped Diagenesis (dissolving, porosity and Dolomitization) in increasing the porosity and permeability of the formation rocks deposited in these environments.

I. Open Platform Environment: This environment is characterized by the fact that it contains limestone sediments containing an abundance of bottomed Foraminifera (Miliolid) fossils indicating marshy environment, in addition to mollusks, pieces of algae, echinoderms, shell nuts, plaids, and a small proportion of Rudist that are represented by the facets of clay, wacky and grains.

II. Restricted Platform Environment: The sediments of this environment are composed of clay limestone to wacky and agglomerated limestones, with the presence of benthic Foraminifera rarely.

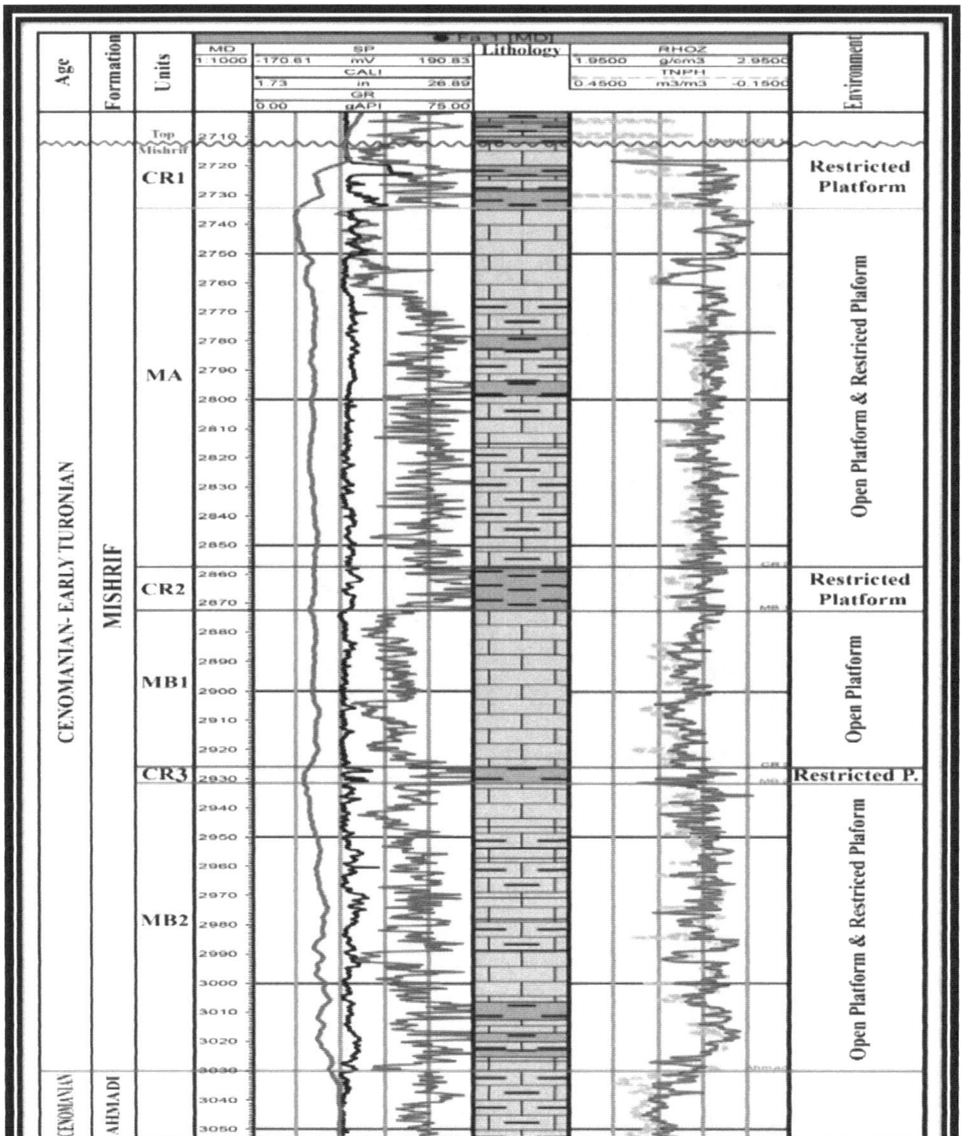

Figure 10. Three types of stylolites with various in the lower Cretaceous is reproduced from [9].

7. Conclusions

The Microfacies analysis of Mishrif Formation have been studied from thin section, the existence two main paleoenvironments which are (Open platform and Restricted platform). Mishrif Formation including many porosity types, primary porosity such as (Interpartical and Interapartical) and secondary porosity such as (Vugy and Fracture).The Mishrif Formation recognized five diagenetic processes, each with positive and negative implications on the quality of the reservoir. By creating and improving porosity and permeability, dissolution and dolomitization significantly impacted reservoir quality. Compaction and cementation have negative consequences by decreasing porosity and permeability, leading to reduced reservoir quality.

References

[1] A.A. Aqrawi, Goff, J.C., Horbury, D. & Sadooni, F. , The Petroleum Geology of Iraq, Scientific Press Ltd, (2010)..
[2] S. Boggs Jr, S. Boggs, Petrology of sedimentary rocks, Cambridge university press2009..
[3] R.J. Dunham, Classification of carbonate rocks according to depositional textures, (1962)..
[4] E. Flügel, Microfacies data: fabrics, Microfacies of carbonate rocks, Springer2004, pp. 976..
[5] C.H. Moore, Carbonate reservoirs, porosity evaluation and diagenesis in a sequence stratigrphic framework, Developments in sedimentology, (2004)..
[6] M.E. Tucker, Sedimentary petrology, an Introduction, Black well Scientific Publications, (1981) 252..
[7] V. Storvoll, K. Bjørlykke, D. Karlsen, G. Saigal, Porosity preservation in reservoir sandstones due to grain-coating illite: a study of the Jurassic Garn Formation from the Kristin and Lavrans fields, offshore Mid-Norway, Marine and Petroleum Geology, 19.
[8] P.W. Choquette, L.C. Pray, Geologic nomenclature and classification of porosity in sedimentary carbonates, AAPG bulletin, 54 (1970) 207-250..
[9] A.S. Alsharhan, J.L. Sadd, Stylolites in Lower Cretaceous carbonate reservoirs, UAE, Society for Sedimentary Geology Special Publication, 69 (2000) 185-207..
[10] A.F. Randazzo, L.G. Zachos, Classification and description of dolomitic fabrics of rocks from the Floridan aquifer, USA, Sedimentary Geology, 37 (1984) 151-162..

Environmental ID and Phylogenetic Studies of *Lactobacillus* spp. Najaf Insulates By means of the 16s rRNA Gene

Huda Saad Azzawi[1] and Jenan Mohammed Hussein[2]

[1,2]University of Kufa

Abstract. Lactobacillus is one of the most important genera in the human and animal digestive tract. They can be found in dairy and non-dairy probiotic foods, and they aid in the restoration of intestinal microflora, which is good for one's health. Lactobacillus strains were phenotypically identified and defined using biochemical characteristics. For the current study, 120 fecal samples were collected from healthy breast-feeding infants aged two years and under. We looked at a couple of primers on complex communities to see if they could service our investigation of complicated micro-ecosystems. The two primers provided positive results with L.acidophilus and L. helveticus. Their strains came from the feces of breast-fed litters. The remains obtained from DNA samples were sequenced and uploaded to the GenBank "NCBI" database after PCR augmentation.

Keywords: GenBank; Intestinal Microflora; *Lactobacillus*; Molecular Identification; Phylogenetic; Probiotics; 16S rRNA

1. Overview

Lactobacillus spp. are bacteria with a long history of use in dairy technology and as a natural preservative in dairy products due to their ability to inhibit the growth of dangerous bacteria and hence extend the shelf life of the products [1]. Additionally, established probiotic properties that give customers with a health advantage have sparked customers' interest.

advantage Lactobacillus is a gram-positive, catalase-negative lactic acid bacteria genus that produces lactic acid as the main end product after carbohydrate fermentation [2] [3]Species of this genus have been isolated from milk, dairy products, sewage, plants, food, and animal intestines. This genus now contains roughly 180 species [4].

Conversely, there is a great deal of variety among the species in this genus, which complicates things [5]

Lactobacillus is categorized based on morphological and physiological appearances, as is usual repetition. Traditional criteria are time-consuming and may fail to correctly identify germs. Biochemical features have recently been employed to provide a new way for bacterial identification [6] in addition to Kits like the API 50 CHL system and molecular approaches like polymerase chain reaction (PCR).

As a result, the purpose of the study was to isolate and categorize different Lactobacillus species using both classical (morphology and biochemical parameters) and molecular (PCR technique) methods.

2. Supplies and Approaches

2.1. Samples assembly

A total of 120 samples were occupied from breast-fed neonates' excrement, raw milk, and specific dairy products. The infants' ages ranged from days to two years. The samples were stored in sterile glass containers and sent to the lab in an icebox within two hours, where they were examined the same day.

2.2. Lactobacilli culture isolation

Using the GENBox anaerobickit, eachsample was s.erially dil-uted in citrate buffer and plated on "MRS" agar m-edium [7]before being incubated at37°Cfor48hours under anaerobic conditions. Colonies were chosen at random, streaked twice on MRS agar plates, and pre-incubated under the same conditions as before to ensure complete purification.

2.3. Isolated culture identification

About "90" insulates were confirmed for Gram staining [8], catalase activity [9], and cell shape. In order to gain a full physiological representation of the insulates, the making of acid and CO2 from glucose [10], growing at 15, 35, and 45°C, glucose fermentation, and biochemical analyses were all calculated.

2.4. Lactobacillus characterization at the molecular level.

2.4.1. Isolation and quantitation of genomic DNA: Total genomic DNA was isolated from ten isolates (promega kit,USA). The manufacturer's instructions for DNA extraction were strictly followed. A 1% agarose gel was used to determine the value of total DNA. Because they included fragmented DNA, samples with a smeared appearance were eliminated. A "shimadzu" UV visible spectrophotometer was used to regulate the concentration of DNA.

2.4.2. Lactobacillus Identification Primer Selection:

16S rRNA sequencing was used to select primers for Lactobacillus species., PCR amplification of 16S rRNA: A total of 20 l of amplification reactions were produced.

In sterile ddH2O, the primer pair (Macrogen/Korea) used in this study (Table 3-9) was dissolved. The stock solution (100 pmol/l) was created by adding ddH2O to the vial containing the lyophilized primer, and the working stock (ten pmol/l) was created by combining 10 l of the stock primer with 90 l of ddH2O.

2.5. PCR Reaction Mixture

The polymerase chain reaction was employed to enlarge DNA fragments using the Maxime PCR PreMix Kit (i-Taq) (Intronbio/Korea) (PCR). The template DNA, forward and reverse primers, and RNase-free water were added to a PCR tube encompassing i-Taq DNA polymerase, dNTP mixture, and reaction buffer mixed together as in the table.

Table 1. Contents of the Reaction Mixture

Contents of reaction mixture	Volume (µl)
Master Mix	
Template DNA	4
Forward primer (10 pmol/µl)	2
Reverse primer (10 pmol/µl)	2
Nuclease free water	Up to 20 ul

2.6. Polymerase Chain Reaction (PCR) Condition

Conventional PCR was used to amplify a target DNA using specific primer pairs. PCR generally consists of three sequential steps (denaturation, annealing, and elongation) of repeated cycles to yield PCR product (amplicon(

The PCR thermal cycling conditions are listed in the table (3-11). Electrophoresis using 1TBE buffer and simplysafe dye on a 1.5 percent (w/v) agarose gel was used to determine the size of PCR products (5L). The sizer100 bp DNA ladder (Intronbio/Korea) was used to determine the product's size.

Table 2. PCR conditioned for amplification of 16S rRNA gene

Step	Temperature (°C)	Time (Second)	No. of cycles
1- Initial Denaturation	95	120	1
2-Denaturation	95	30	
3-Annealing	53	30	30
4-Extension	72	150	
5-Final extension	72	300	1

2.7. Agarose Gel Electrophoresis

The most effective way is to separate DNA fragments on an agarose gel electrophoresis. The amount of agarose in a gel varies from 0.5 percent to 2 percent, depending on the size of the DNA fragments to be separated. A 0.7 percent gel was used to separate genomic DNA (5-10 kb) after extraction, while 1.5 percent -2 percent gels were employed to acquire excellent resolution for tiny PCR product fragments (0.2-1 kb). However, to the indicated weight of agarose, 100 mL of 1TBE buffer was added, which was then heated in the microwave until the solution turned clear. The agarose was cooled to 50-55°C and 5l of simply safe dye (10 mg/ml) was added to 100 ml of melting agarose gel to achieve a final concentration of 0.5g/ml (Sambrook & Rusell, 2001).

Before drying, the agarose was placed into the gel tray, sealed at the ends, and the comb was inserted as needed. The samples went into their separate wells on the gel, whereas the marker went into a single well. (The agarose gel electrophoresis time for genomic DNA is 45 minutes, and the PCR DNA sequencing time for 16sRNA of tested isolates is 1 hour and 30 minutes.) The electrodes were properly connected, and the run was carried out in accordance with the gel % and size.

Sequence submissions and accession numbers: The" NCBI" received the __sequences fr*om this studyvia the Bankitprogram (http://www.ncbi.nlm.nih.gov/BankIt/).

Phylogenetic tree construction: A phylogenetic tree based on 16S ribosomal RNA sequence comparisons of the PCR-amplified length polymorphism and Database sequences was built using the blast tree build in https://www.ncbi.nih.gov/blast/treeview based on Fast Minimum Evolution.

Table 3. the primer which used in the study.

Primer name	Sequence 5′-3′	Reference
27F	AGAGTTTGATCCTGGCTCA	(Loy *et al*., 2002)
1492R	GGTTACCTTGTTACGACTT	

3. Results and Discussion

3.1. Cultural morphological and biochemical identification of bacterial isolates

Using Gram staining and other phenotypic testing, 90 strains were identified as Lactobacillus species. Gram-positive and catalase-negative isolates were found in all of the samples.

Many other bacterial species, as well as Lactobacillus strains, have been phenotypically identified using biochemical features. Through the fermentation of glucose, all iso- lates from faecal samples were also able to produce acid and gas.

Table 4. The morphological and biochemical tests used for documentation of insulates.

Samples	**No. of isolates**	**Morphology**		**Gram staining**		**Catalase production**		**Gelatin Liquefied**	**Indole Produced**
		Rods	**Cocci**	+	-	+	-		
Feces (F)	90	+	-	+	-	+	-	-	-

+: Positive reaction; -: Negative Reaction.

3.2. Molecular revisions

PCR findings for a fragment of 16S rRNA 1450bp of 10 isolates from feces using 16S rRNA specific primers. Ventura et al. found that six isolates were L. acidophilus and four were L. helveticus (Figure 4). (2001). Because bacteria differ significantly in size and sequence, 16S rRNA PCR amplification has proven to be a useful approach for bacterial species-specific typing.

Primer sensitivity and specificity: Before proceeding to the wet lab, in silico data should be validated. PCR Primer Stats takes a list of PCR primers and creates a report with information about each one, including melting temperature, GC content percent, and PCR appropriateness. It helped us assess the potential of the PCR primers chosen by Walter, et al. (2001), McOrist, et al. (2002), and Tilsala and Alatossava (1997) **[11]**from a list of primers selected by Walter, et al. (2001), McOrist, et al. (2002), and Tilsala and Alatossava (1997) **[11]**. (See Table 3).

After being evaluated for specificity and efficiency, a primer is confirmed in the wet lab. We used pairs of species-specific primers in this study, and the results showed that the primers had the same sensitivity and specificity as the strain-specific primers. Lactobacillus acidophilus strains and Lactobacillus helveticus strains Gene Bank submissions and accession numbers: The finding of L.acidophilus and L. helveticus from diverse locations in Iraq utilizing molecular identification based on feces analysis, followed by sequencing and submission to the NCBI Genbank Database with three new accession numbers is described in this paper. We detected this strain in the feces of a breastfed infant under accession numbers.

According to the data, L.acidophilus, L. helveticus has a high capacity to survive passage through the gastrointestinal tract, showing that this strain has a higher ecological fitness for adaptation to a range of hosts and likely ecological niches **[12]**. We discovered that 16S rRNA sequences might be used to determine the diversity of Lactobacillus communities found in human intestine samples. The 16S rRNA sequence reveals information regarding Lactobacillus isolates' accurate identification and distribution, as well as inter-subject variability and differences between two intestinal regions of the same person in different human subjects.

The 16S rRNA sequences of various Lactobacillus spp. in the Gene Bank database showed a high degree of similarity, according to our findings. A non-redundant Gen Bank search yielded 100 matches for sequences identical to the strains, showing a high similarity percentage of 100% with L.acidophilus, L. helveticus strain LRB, and L. helveticus.

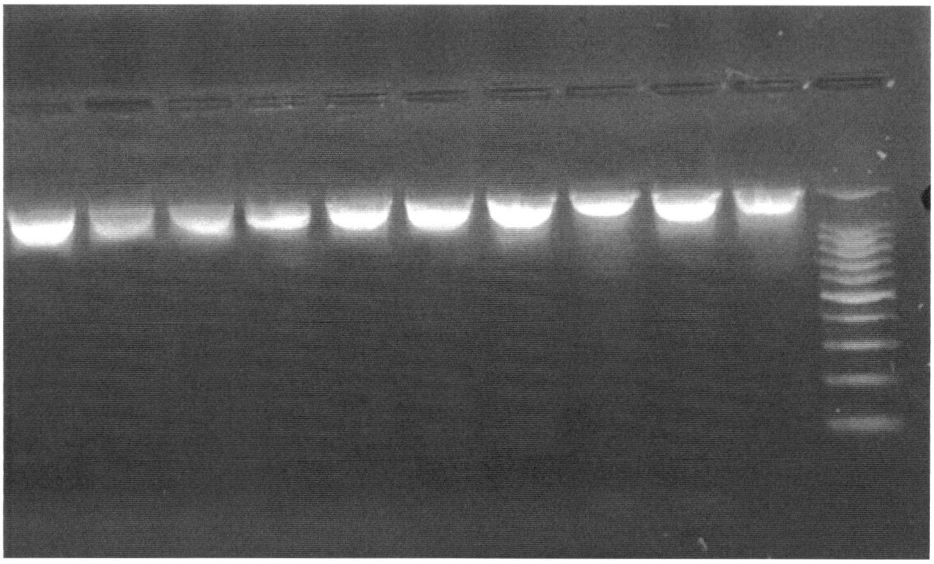

Figure 1. PCR detection of 16srRNA genes in Lactobacillus ,PCR product was 1450 pb

3.3. Phylogeneticrelation-ship of the Genus Lactobacillus spp.

Lactobacillus' diversity may be shown in the genus' wide phenotypic and genotypic diversity, as well as its adaptation to different environments. The 16S/23S rRNA gene sequences of lactobacilli were compared to indicate evolutionary links.

Collins et al. were the ones who originally defined the phylogenetic groups Lactobacillus delbrueckii, Lactobacillus casei–Pediococcus, and Leuconostoc (1991). Lactobacillus buchneri, Lactobacillus reuteri, Lactobacillus casei, Lactobacillus Plantarum, Lactobacillus salivarius, and Lactobacillus delbrueckii were all renamed Lactobacillus acidophilus [13]. Lactobacillus vitulinus–catenaformis, Lactobacillus perolens, and Pediococcus are the three phylogenetic groupings of the Lactobacillus genus [14]

Scientists could employ 16S-rRNA sequences to determine the relationships between closely related species, according to our findings. In most cases, phylogenetic evidence shows that strains belonging to the same species are more closely related than strains belonging to other species. 16S-rRNA sequence divergence rates have been proven to differentiate across subspecies of the same species.

Lactobacillus casei is a genus of Lactobacillus bacteria that consists of three species. This group is homogeneous since it includes both facultative and heterofermentative members, with GC percentages ranging from 45 to 47 mol percent and Lys-d-Asp as the most common peptidoglycan type. Additionally, they are all capable of producing acetoin and L (+)-lactic acid.

Lactobacillus casei includes Lactobacillus rhamnosus, Lactobacillus paracasei, Lactobacillus casei, and Lactobacillus zeae. A thorough taxonomic analysis of this subgroup previously resulted in the species name L. parachuting being temporarily rejected. These taxa are still classified as separate species within their genus [15] [16] [17] [18]. The most common food-borne lactobacilli isolates from the L. casei group, especially from cheese, are L. paracasei and L. rhamnosus. L. paracasei and L. casei are also discovered in silage and are detected in the gastrointestinal systems of both animals and humans.

The L. acidophilus category contains nearly all homofermentative lactobacilli. Lactobacillus delbrueckii, Lactobacillus crispatus (subspecies bulgaricus, lactis, delbrueckii, and indicus), Lactobacillus helveticus, Lactobacillus johnsonii, and Lactobacillus gasseri are some of the other industrially important species described. Although some strains of the species L. Acidophilus have long been known to play an important role in human health and nutrition because to their beneficial effect on gut flora, L. delbrueckii and L. helveticus are well-known starters for vegetable and dairy fermentation, respectively.

Table 5. genetic distance of matrix based on 16S ribosomal RNA sequences, shown the relationships of members of Lactobacillus isolates

	1	2	3	4	5	6	7	8	9	10
1. Lactobacillus acidophilus 1										
2. Lactobacillus acidophilus 2	2.2333									
3. Lactobacillus acidophilus 3	2.4767	2.0633								
4. Lactobacillus helveticus 4	3.3541	2.9865	3.9949							
5. Lactobacillus acidophilus 5	3.942	2.7881	3.013	0.2259						
6. Lactobacillus acidophilus 6	4.0602	3.0184	2.7033	4.3033	4.4648					
7. Lactobacillus 7	2.3437	3.9837	3.8465	2.0646	2.907	3.5964				
8. Lactobacillus helveticus 8	3.3956	2.0867	3.0738	1.3889	1.3829	3.8517	3..7465			
9. Lactobacillus helveticus 9	4.234	1.6156	3.3298	2.9541	2.6961	3.8723	3.8014	1.0181		
10. Lactobacillus helveticus 10	4.2513	3.1449	3.7538	3.6261	3.351	3.0822	3.3658	4.1245	3.1429	

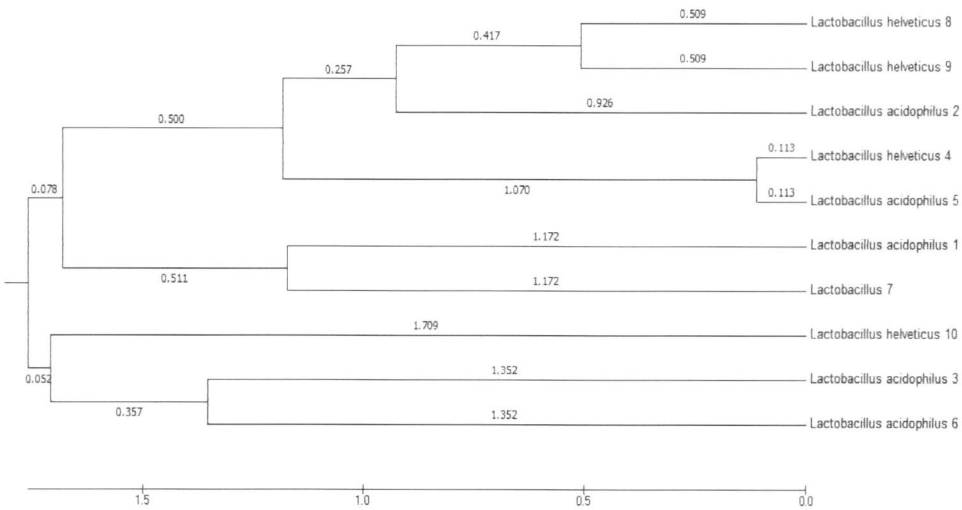

Figure 2. Phylogenetic based on 16S ribosomal RNA sequences, shown the relationships of members of Lactobacillus isolates

References

[1] Anas M., et al. Antimicrobial activity of Lactobacillus species isolated from Algerian raw goat's milk against Staphylococcus aureus. World Journal of Dairy and Food Sciences 3.2 (2008): 39-49..

[2] Bernardeau M., et al. Safety assessment of dairy microorgan- isms: The Lactobacillus genus. International Journal of Food Microbiology 126.3 (2008): 278-285..

[3] Patel AK, Ahire JJ, Pawar SP, Chaudhari BL, Chincholkar SB. Comparative accounts of probiotic characteristics of BACILLUS spp. isolated from food wastes. Food Res Int 2009;42:505e10..

[4] FAO/WHO evaluation of health and nutritional properties of powder milk and live lactic acid bacteria. In: Food and Agriculture Organization of the United Nations and World Health Organization expert consultation report Rome; 2001..

[5] Miele E, Pascarella F, Giannetti E, Quaglietta L, Baldassano RN, Staiano A. Effect of a probiotic preparation (VSL#3) on induction and maintenance of remission in children with ulcerative colitis. Am J Gastroenterol 2009;104:437e43..

[6] Shintaro M., et al. Genomic characterization of a fructophilic bee symbiont Lactobacillus kunkeei reveals its niche-specific adaptation. Systematic and Applied Microbiology 39.8 (2016): 516-526..

[7] Awaad AM., et al. Identification of Bifidobacterium Animalis Ssp. Lactis from Egyptian Women Breast Milk and Feces of Breast Fed Infant Based On 16S-23S rRNA Gene. Advances in Nutrition and Food Science 1.1 (2016): 1-8..

[8] Brock TD., et al. Cell Biology. In Biology of Microorganisms. Prentice Hall Int. Editions. Toronto, Canada: (1994): 43..

[9] Herrero M., et al. Evaluation of technologically important traits in lactic acid bacteria isolated from spontaneous fer- mentations. Journal of Applied Bacteriology 81.5 (1996): 565- 570..

[10] Bergey's Manual of Determinative Bacteriology. Ninth Edition. 428 East Preston street Baltimore, Maryland 21202, U.S.A.: Williams and Wilkins; (1994)..

[11] Ventura M., et al. Specific identification and targeted char- acterization of Bifidobacterium lactis from different environ- mental isolates by a combined multiplex- PCR approach. Ap- plied and Environmental Microbiology 67 (2001): 2760-2765..

[12] Haschke F., et al. Clinical trials prove the safety and efficacy of the probiotic strain Bifidobacterium Bb12 in follow-up formu- la and growing-up milk. MonatsschrKinderheikd 146(1998): 26-30..

[13] Schleifer KH., et al. Phylogeny of the genus Lactobacillus and related genera. Systematic and Applied Microbiology 18.4 (1995): 461-467..

[14] Pot B and Tsakalidou E. Taxonomy and metabolism of Lac- tobacillus. In: Ljungh, A., Wadstro¨m, T. (Eds.), Lactobacillus Molecular Biology: From Genomics to Probiotics. Caister Aca- demic Press, Norfolk, (2009): 3-58..

[15] Dellaglio F., et al. The status of the species Lactobacillus casei(Orla-Jensen 1916) Hansen and Lessel (1971) and Lactobacillus paracasei Collins et al. (1989): request for an opinion. International Journal of Systematic Bacteriology 52, 285e287..

[16] Desai AR., et al. Discrimination of dairy industry isolates of the Lactobacillus casei group. Journal of Dairy Science 89 (2006):3345-3351..

[17] De Vos P., et al. Judicial commission of the international com- mittee on systematics of prokaryotes, 10th international con- gress of bacteriology and applied microbiology, minutes of meetings 28, 29 and 31 July and 1 August 2002, Paris, France. In.

[18] Pot B., et al. Taxonomy of Lactic Acid Bacteria. In: De Vuyst L., Vandamme E.J. (eds) Bacteriocins of Lactic Acid Bacteria. Springer, Boston, MA (1994)..

[19] Collins CH., et al. Collins and Lyne's microbiological meth- ods. Butterworth and Co Ltd; (2004)..

[20] Walter J., et al. Detection and Identification of Gastrointes- tinal Lactobacillus Species by Using Denaturing Gradient Gel Electrophoresis and Species-Specific PCR Primers. Applied and Environmental Microbiology 66.1 (2000): 297-303..

[21] McOrist AL., et al. A comparison of five methods for extrac- tion of bacterial DNA from human faecal samples. Journal of Microbiological Methods 50.2 (2002): 131-139..

[22] Tilsala-Timisjärvi A., et al. Development of oligonucleotide primers from the 16S-23S rRNA intergenic sequences for identifying different dairy and probiotic lactic acid bacteria by PCR. International journal of Food Microbiology 35.1 (1997): 49-56..

[23] Ogunshe AA. Bioinihibition of diarrhogenic Gram-positive bacteria pathogens indigenous probiotics in industrial infant weaning food. Asian Pacific Journal of Tropical Medicine 1.2 (2008): 7-11..

[24] Ozgun D., et al. Identification of Lactobacillus strains isolated from faecal specimens of babies and human milk colostrum by API 50 CHL system. Journal of Medical Genetics 3.3 (2011): 46-49..

Survey of Khor Al-Zubair and Umm Qasr Mudflat Birds: Effect of Tidal Cycle and Seasonal Variations on Their Assemblages and Abundance, Basrah, Iraq

Malik Hassan Ali[1], Hanaa Hussein Mohammed[2] and Mudhafar A. Salim[3]

[1,2] Marine Science Center, University of Basrah, Basrah, Iraq

[3]Iraqi Organization for Conservation of Nature (IOCN), Iraq.

Abstract. Ornithological surveys were conducted at tidal mudflats that nationally represents one of the marine key biodiversity (KBA) areas at Khor Al-Zubair (KBZ) and Umm Qasr south-east of Basrah City, Iraq. On the basis of hourly counting for 7h monitoring period at day time and covering an area of 7.5 km^2 of the intertidal zone, a 7 surveys were conducted from March 2019 until January 2020. A total of 7658 individual birds belongs to 54 species have been recorded. The highest numbers of birds per a day were recorded in Autumn and Winter 4869, 1999 respectively, which coincided with the migration seasons. The lowest numbers were recorded during the Summer season just where only 61 birds have been observed. The birds assemblages and their abundance during the high tide times and during the low tide times were compared statistically and the results weren't different ($t < 0.05$). Additional observations of different surveys targeted different sites of the KAZ mudflat over the period 2005-2020 have resulted in a final list consists of 102 bird species, a result reflecting the importance of this habitat to various migratory and resident birds.

Key words: Intertidal zone, Birds, Khor Al-Zubair, Um Qasr, mudflat.

1. Introduction

The intertidal mudflat and the large marshland area of Mesopotamia, Southern Iraq provides and supports significant populations of large numbers of resident and intercontinental migratory bird species [1]. [2] in an ornithological survey conducted from 2005-2008 in the marshes of Southern Iraq, they recorded 159 species among them eight species are globally threatened. The freshwater Mesopotamian marshlands represent the majority wetlands area of Iraq, 15000-20000 km^2 [1] and were attentive to many researchers and instituditional's investigations that are related to the diversity and migration of birds [3][4]. However, less attention was given to the role of the bird species at the intertidal mudflat habitat which was nationally designated as Key Biodiversity Area (KBA) [2], and as Important Bird Area IBA [5]. As part of the total mudflat areas of Iraqi coast, Khor Al-Zubair KAZ and Umm Qser's UQ intertidal zones extended along about 40 km navigation channel, and the southern shores of Fao (Fig 1). The area is rich of sediment of fluviatile origin transported by Shatt Al-Arab River [6]. Recently [7] reported a big loss of macrobenthic invertebrates in this ecosystem during the past five decades, simultaneously, the area was found inhabited by high population density of two mudskipper species mixed with at least six crab species and the ecosystem was appeared very productive.

Globally, the tidal flats assigned as a very productive habitat and of great importance to the shorebirds [8[9][10]. This article shed light on the shorebirds assemblages and their status in one of the most important coastal/marine KBA of Iraq.

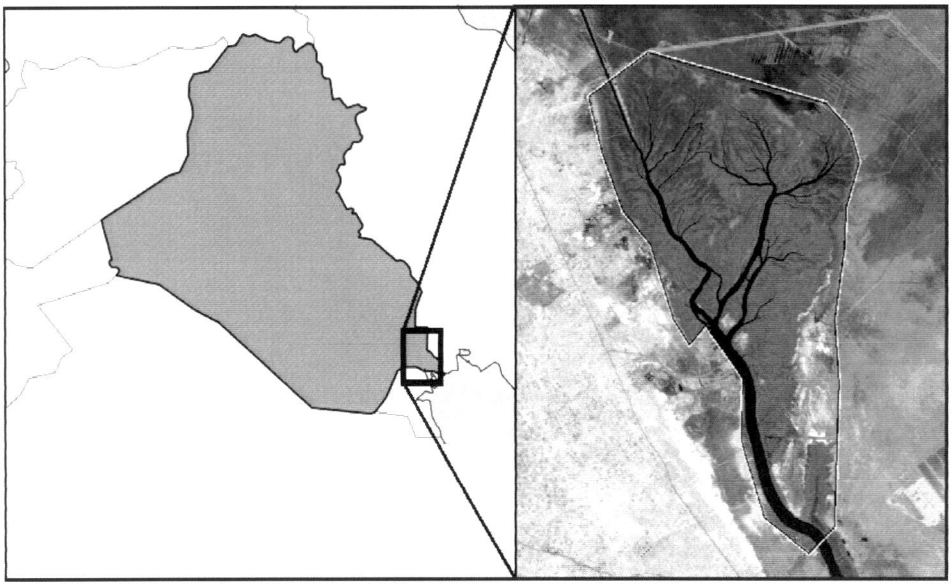

Figure 1. map of the study area, Khor Al-Zubair and Umm Qasr, Basrah, Iraq

2. Materials and Methods

2.1. Study Area

The present study was done in a tidal mudflat area located at Khor Al-Zubair (KAZ) and Umm-Qasr (UQ) which situated in the south-west of Basrah city, Iraq (Fig. 1). Two sites (A_& B) were chose as a vantage point for birds monitoring which they located at the east bank of KAZ water channel (A 30° 13′ 62" N ; 47° 91' 88" E); (B 30° 15' 13" N; 47° 90' 88" E). the site (A) was located just near a dump area, recently established by the General Company for ports of Iraq, and the site (B)was 5 km along the coast northern the site (A). The sediments of KAZ comprise mainly silt and clays of fluviatile origin [11][6].

Figure 2. Photos show the general landscape of the study area. (Photos by Hanaa H. Mohammed).

2.2. Study Period

The data of the study were collected during the period between 21th February 2019 and 29 January 2020. In addition to the pre-survey visit of 21th February 2019 in which a vantage points were identified, seven visits were undertaken in the following dates: In the Spring season at March 7th and, March 14th in Summer at July 15th (2019) , in Autumn at September 18th 2019 and in Winter season at December 8th , December 24th (2019) and January 29th 2020.

The two sites were surveyed in one day for 7 hours, starting from the morning. The tidal cycles were obtained from public record (Fishing forcast https://tides4fishing.com/as/iraq/umm-qasr). The results of different surveys that have targeted Khor Al-Zubair have also been included in this study. These surveyed were conducted in different years over the period 2005-2020, and have covered all the four seasons.

2.3. Birds Counts

Birds were counted on hourly bases for 7 hours per day periods. Birds were identified and counted with aid of high-resolution zoom cameras (Nekon 700D, Nekon 84X optical zoom and Canon Power shot D 10), and by binocular (7 x 50 mm). And birds photographs were taken for further confirmation of bird species. The main taxonomic references used were [12], and [13]. The total birds count, at each site (A & B) were done to cover a study area of nearly 7.5 km length x 1 km width. Since the areas were open intertidal zone with only little plant *Salicornia perennans* there was clear good visibility of birds. A walk-over survey on the intertidal zone were undertaken during the visits for closer search for bird nests, however, the soft mud has restricted the free movement of the survey team in some areas. All bird species were photographed with the cameras for confirming the bird species and their number counts.

3. Results

3.1. Abundance

A total of 7658 individual birds representing 54 species where recorded from the two sites of the mudflat (Table 1). The highest mean number of birds per one day visit was recorded during Autumn season 4869 ind./7h./7.5 km^2. Relatively very low density was observed during Summer 61 ind./7h./7.5 km. (Fig. 2). Birds were observed at all tidal levels but mostly they were found at the area between the low tidal level and the mid tidal level.

3.2. Birds Assemblage

Gulls were the dominant bird group that have been observed during all seasons.

Four *Larus* species were identified, the Slender Billed Gill *Larus genei* was the most abundant 1229 (Table 1). Herons, mainly *Egretta gularis* and *Ardea cinerea* were a common group present in all seasons 172 and 93 total number respectively. Terns particularly Gull-billed *Gelochelidon nilotica* were also observed in large numbers 582 total number during most seasons. However, the following birds were the abundant species: the cormorant *Phalacrocorax carbo,* the Common Curlew *Numenius arquata,* and the Kentish Plover *Charadrius alexandrines*. Some other groups, the genus *Tringa* and the genus *Oenanthe*, although they were relatively less abundant but they were represented by 4 and 3 species in the bird assemblage. However, the list consist very rare species, they even were observed one time during the entire investigation period, for example, the Hoopoe *Upupa epops*.

Table 1. Numbers of the Birds species in the mudflat, Khor Al-Zubair/ Umm Qasr, during 2018-2020, in addition to other bird species observed during different periods at the same area.

	Scientific Name	Common Name	Autumn season abundance birds /7h./7.5 km^2..	Summer season abundance birds /7h./7.5 km^2.	Spring season abundance birds /7h./7.5 km^2.	Winter season abundance birds/7h./7 .5 km^2.	Total no.	IUCN
1	*Tadorna tadorna* *	Common Shelduck	42	-	-	37	79	LC
2	*Anas clypeata*	Shoveler	-	-	-	150	150	LC
3	*Phoenicopterus ruber*	Greater Flamingo	30	-	-	50	80	LC
4	*Ciconia ciconia*	White Stork	-	-	-	1	1	LC
5	*Platalea leucorodia*	Spoonbill	-	-	1	-	1	LC
6	*Ardea purpurea*	Purple Heron	-	-	17	-	17	LC
7	Ardeola ralloides	Squoacco Heron	-	-	4	-	4	LC
8	*Ardea cinerea* **	Gry Heron	42	6	12	33	93	LC
9	*Ardea alba*	Great White Egret	6	-	5	-	11	LC
10	*Egretta garzetta* *	Little Egret	7	-	5	2	14	LC
11	*Egretta gularis* **	Western reef Heron	61	-	45	66	172	LC
12	*Phalacrocorax carbo* **	Cormorant	26	2	8	18	54	LC
13	*Circus aeruginosus*	Western Marsh Harrier	-	-	1	1	2	LC
14	*Bueto rufinus*	Long-legged Buzzard	-	-	1	-	1	LC
15	*Falco tinnunculus*	Common Kestrel	-	-	-	1	1	LC
16	*Dromas ardeola* *	Crab Plover	5	2	5	-	12	LC
17	*Phalaropus lobatus*	Red-necked Phalarope	-	-	12	-	12	LC
18	*Himantopus himantopus* *	Black-winged Stilt	352	-	-	17	369	LC

19	*Ricurvirostra avosetta* *	Avocet	219	-	-	256	475	LC
20	*Pluvialis fulva*	Pacific Golden Plover	-	-	-	7	7	LC
21	*Charadrius hiaticula*	Common Ringed Plover	560	-	-	-	560	LC
22	*Charadrius alexandrines* **	Kentish Plover	105	**5**	11	94	215	LC
23	*Charadrius leschenaultii*	Greater Sand Plover	-	-	-	5	5	LC
24	*Limosa limosa*	Black-tailed Godwit	-	-	-	31	31	NT
25	*Numenius arquata* **	Common Curlew	48	1	32	16	97	NT
26	*Tringa erythropus* *	Spotted Redshank	-	-	11	195	206	LC
27	*Tringa nebularia*	Greenshank	21	-	-	7	28	LC
28	*Tringa totanus*	Redshank	27	-	-	-	27	LC
29	*Tringa stagnatilis*	Marsh Sandpiper	28	-	-	4	32	LC
30	*Xenus ciénereus*	Terek Sandpiper	3	1	-	-	4	LC
31	*Calidris minuta* *	Little Stint	2226	-	6	-	2232	LC
32	*Calidris alpina* *	Dunlin	277	-	1	-	278	LC
33	*Larus genei* **	Slender Billed Gull	410	30	126	663	1229	LC
34	*Larus ridibundus* **	Black-headed Gull	8	-	95	5	108	LC
35	*Larus fuscus* *	Lesser Black-backedGull	2	-	42	35	79	LC
36	*Larus armenicus* **	Armenian Gull	45	-	6	116	167	NT
37	*Gelochelidon nilotica* **	Gull-billed Tern	291	-	137	154	582	LC
38	*Sterna caspia*	Caspian Tern	2	-	1	3	6	LC
39	*Chlidonia leucopterus*	White-winged Tern	2	1	19	-	22	LC
40	*Streptopelia docaocto*	Collared Dove	-	7	3	-	10	LC
41	*Upupa epops*	Hoopoe	-	-	-	1	1	LC
42	*Lanius collurio*	Red-backed Shrike	-	-	4	-	4	LC

43	*Galerida cristata* *	Crested Lark	3	2	3	4	12	LC
44	*Hirundo rustica*	Barn Swallow	2	-	94	-	96	LC
45	*Hippolais pallida*	Olivacious Warbler	-	-	4	-	4	LC
46	*Phylloscopus trochilus*	Willow Warbler	-	-	2	-	2	LC
47	*Muscicapa striata*	Spotted Flycatcher	-	-	1	-	1	LC
48	*Oenanthe oenanthe*	Northern Wheatear	-	-	-	2	2	LC
49	*Oenanthe* deserti *	Desert Wheatear	6	-	5	5	16	LC
50	*Oenanthe finschii*	Finsch's Wheatear	-	-	-	1	1	LC
51	*Passer domesticus* *	House Sparrow	5	4	10	16	35	LC
52	*Motacilla flava*	Yalow Wagtail	8	-	-	-	8	LC
53	*Motacilla alba*	White Wagtail	-	-	-	1	1	LC
54	*Anthus campestris*	Tawny Pipit	-	-	-	2	2	LC
	Total No.		**4869**	**61**	**729**	**1999**		**7658**
	Mean no. of Bird_counts		**4869**	**61**	**364.5**	**666.3**		

* = ... ** = ...

3.3. Tidal Variation

The bird assemblages and their abundance during the high tidal (HT) and the low tidal (LT) periods were compared statistically (Table 2). The calculated means of the number of the bird species observed during HT periods 8.9 no./ 7.5 km^2 and LT periods 9.0 no./7.5 km^2 weren't different $t < t$ 0.05 (2), 18, $P > 0.05$, also the means of the total birds counted during the HT periods and the LT periods 182 and 133.5 respectively, were compared and the results was statistically insignificant, $t < t$ 0.05 (2), 18 , $P > 0.05$.

Table 2. Tidal variation of bird assemblages and the total number of birds (density) during the periods of high tide (HT) and low tide (LT) at the KAZ, Basrah, Iraq mudflat area.

Mean no. of bird species/ 7.5 km^2 X		Mean total no. of Birds/ 7.5 km^2 X	
HT	LT	HT	LT
Mean X= 8.9 no./ 7.5 km^2	X = 9.0	T N = 182	13.5
N = 10	N = 10	N = 10	10
S = 2.28	S = 1-33	S = 149.6	53.6
t = 0.286, P > 0.05 i.s.		t = 0.354, P > 0.05 i.s.	

i.s. insignificant

3.4. Seasonal Abundance

Figure 3. shows the results of the quantitative study conducted on 2018-2020 (7 daily surveys). In Autumn and Winter birds appear most abundant 4869 and 666.3 birds/7h./7.5 km^2 respectively, whereas in Summer they were very few 61 birds./7h./7.5 km^2. .

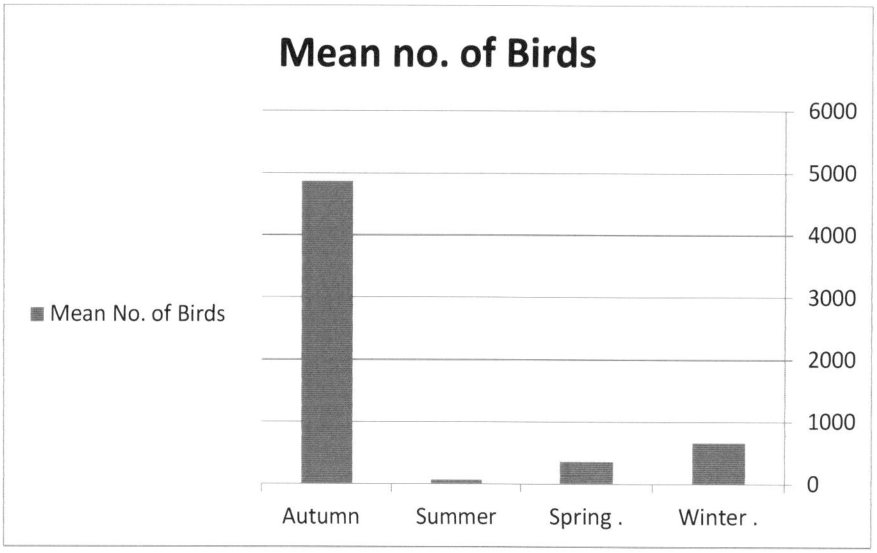

Figure 3. Mean birds density (No./7h./ 7.5 km^2) recorded in the 7 (daily surveys) during four seasons in the periods 2018-2020, KAZ, Basrah, Iraq.

3.5. Additional list of birds species

Through the period 2005-2019 many visits were done by M. Salim in various sites of KAZ's KBA where further birds species have been recorded (Table 3).

Table 3. Birds observed in the study area (By M. Salim) (2005-2019).

	Scientific Name	Common Name	IUCN Status
1	*Anas crecca*	Eurasian Teal	LC
2	*Anas strepera*	Gadwall	VU
3	*Marmaronetta angustirostris*	Marbled Duck	VU
4	*Tachybaptus ruficollis*	Little Grebe	LC
5	*Ciconia ciconia*	White Stork	LC
6	*Plegadis falcinellus*	Glossy Ibis	LC
7	*Nycticorax nycticorax*	Black-crowned Night Heron	LC
8	*Milvus migrans*	Black Kite	LC
9	*Circus pygargus*	Montagu's Harrier	LC
10	*Buteo buteo vulpinus*	Steppe Buzzard	LC
11	*Aquila nipalensis*	Steppe Eagle	EN
12	*Gallinula chloropus*	Common Moorhen	LC
13	*Recurvirostra avosetta*	Pied Avocet	LC
14	*Vanellus spinosus*	Spur-winged Lapwing	LC
15	*Vanellus indicus*	Red-wattled Lapwing	LC
16	*Vanellus leucurus*	White-tailed Lapwing	LC
17	*Pluvialis squatarola*	Grey Plover	LC
18	*Charadrius dubius*	Little Ringed Plover	LC
19	*Gallinago gallinago*	Common Snipe	LC
20	*Tringa totanus*	Common Redshank	LC
21	*Tringa ochropus*	Green Sandpiper	LC
22	*Tringa glareola*	Wood Sandpiper	LC
23	*Actitis hypoleucos*	Common Sandpiper	LC
24	*Arenaria interpres*	Ruddy Turnstone	LC
25	*Calidris alba*	Sanderling	LC
26	*Calidris temminckii*	Temminck's Stint	LC
27	*Calidris ferruginea*	Curlew Sandpiper	NT
28	*Philomachus pugnax*	Ruff	LC
29	*Larus michahellis*	Yellow-legged Gull	LC
30	*Larus cachinnans*	Caspian Gull	LC
31	*Sternula albifrons*	Little Tern	LC
32	*Sterna hirundo*	Common Tern	LC
33	*Chlidonias hybrida*	Whiskered Tern	LC
34	*Columba livia*	Rock Dove	LC
35	*Streptopelia senegalenesis*	Laughing Dove	LC
36	*Ceryle rudis*	Pied Kingfisher	LC
37	*Merops persicus*	Blue-cheeked Bee-eater	LC
38	*Merops apiaster*	European Bee-eater	LC
39	*Lanius isabellinus*	Daurian Isabelline Shrike	LC
40	*Lanius meridionalis*	Southern Grey Shrike	VU
41	*Riparia riparia*	Sand Martin	LC
42	*Phylloscopus collybita*	Common Chiffchaff	LC
43	*Prinia gracilis*	Graceful Prinia	LC
44	*Turdoides huttoni*	Afghan Babbler	LC
45	*Sturnus vulgaris*	Common Starling	LC

46	*Cercotrichas galactotes*	Rufous-tailed Scrub Robin	LC
47	*Saxicola maurus*	Siberian Stonechat	LC
48	*Anthus spinoletta*	Water Pipit	LC

4. Discussion

The lists of birds species of the present study consists 102 different migrated and resident species, this number formed nearly 2/3rd of the total national list (320) of the bird species noted in KBA surveys conducted in 2005 to 2008 [2]. However, in a more recent study [14] reported 168 bird species in the Bahr Al-Najaf area and they mentioning that this number represent more than 40% of the national check list of birds in Iraq [15]. Despite to some species included in the list were rare but such high species diversity heighten the importance in the KAZ mudflat ecosystem as a resting and feeding site of the migratory shorebirds. Also it is remarkable to mention that we expected a higher values of diversity and abundance can be recorded if surveys were conducted more frequently, a longer time in each day of monitoring (> 7h/ day) and covering a wider areas, especially during the migratory season..

The peak of birds density was observed in Autumn season 4869 bird/7h./7.5 km^2. In Winter the density was also relatively high 666.3 bird/7h./7.5 km^2 (Fig. 2). Obviously, Its coincided with the migratory seasons of birds in Iraq which represented one of the key Inter-continental migratory flyways of birds in the world [1][14].

The effects of tidal cycle on the bird's density and their assemblage were examined, and the results reveals no significant differences are there between the HT and LT periods indicating that birds have the same habitat use whether the mudflat area is exposed or flooded. This is also might be due to the vast mudflats areas where the birds can find alternative foraging areas. As far as we studied the area no evidence was found that the recorded birds have used the tidal area for breeding activity i.e. eggs or nests. Hence, it seems that the utilization of mudflat area by these bird species and populations is restricted to feeding and resting. Recently [7] were investigated the benthic community of KAZ/ UQ intertidal mudflats and demonstrating that a high densities of a mixed populations of at least 6 species of crabs: Uca sindensis, Macrophthalmus dentipes, Leptochryseus kuwaitensis, Parasesarma, plicatum, Eurycarcinus orientalis and Nasima dotiliformis, and two species of mudskippers Periophthalmus waltoni and Boleophthalmus dissumieri were actively living in this low diversity but high productive habitat. However, many regional and global reports on mudflat habitats have had noticed the predation of shorebirds on the macrobenthos [16][10]. As a conclusion this habitat is rich with benthic crabs and mudskippers and these preys were available in large number and in different sizes almost the year around as well as during the low and high tide conditions. However, catching such burrowing preys is not easy due to the relatively complex burrows structure they made and their active hidden behavior .

Figure 4. Western reef heron *Egretta gularis* catching mudskipper fish from KAS mudflat area, Basrah (Photo by Hanaa H. Mohammed).

Acknowledgment

The authors are grateful to Basrah Gas company for funding part of this work.

References

[1] UNEP, Partow, H. 2001. The Mesopotamian marsh lands: Demise of an ecosystem. Early Warning and Assessment. Technical Report, UNEP/DEWA/ TR.01-3. Nairobi (Kenya).

[2] Salim, M. A., Perter, R., Rubec, C. 2009. A summary of birds recorded in the marshes of southern Iraq, 2005-2008. BioRisk 3: 205-219, https//doi.org:10.3897/biorisk.3.14

[3] Scott, O. A. and Carp, E. 1982. A midwinter survey of wetlands in Mesopotamia, Iraq: 1979. Sandgrouse No. 4.

[4] Fazaa, N. A., Dunn, J. C. and Wihttingham, M. J. 2017. Distributions and community composition of birds in Iraq's Central Marsh. International Journal of Biodiversity, 4198690, 1-28. https://doi.org/10.1155/2017/4198690

[5] BirdLife International 2019. Important Bird Area factsheet: Khor Az Zubayr. Downloaded from http://www.birdlife.org on 19/06/2019

[6] Alianachi, I. and Albadran, B. N. 2009. Stability of River Branch (Creek) Banks in Khor Al-Zubair Area, NW Arabian Gulf. Electronic Journal of Geotechnical, 9, 1-12. Bundle, G.

[7] Ali, M. H.; Almudafar, N. A.; Mohammed, H. H.; Helmuth, B. and Dwyer, A. M., (2021). Winners and Losers: Post Conflict Biodiversity in the Stressed Ecosystem of Khor AL-Zubair, IRAQ. Unpublished.

[8] Norazlimi, N. A. and Ramli, R. 2014. The relationship between morphological characteristics foraging behavior in four selected species of shorebirds and water birds utilizing tropical mudflats, The Scientific World Journal. Article. Vol. (2015), ID105269, 7pp. https://doi.org/10.1155/2015/105296.

[9] Dissanayake, N. G., Fird, C. L. J., Drylie, T. P. and Casswell, B. A. 2010. Ecological functioning of mudflat: global analysis reveals both regional differences and widespread conservation of functioning, Marine Ecology Progress series.Vol. 604, 1-20.

[10] Botty, G. M., Underwood, J. G. C., Paris, A., Davies, R. G. and Tolhurst T. J. 2020. Shorebirds effect ecosystem functioning on an intertidal mudflat. Front Marine Science, 7, 685, 1-15.

[11] Aqrawi, A. A. 1994. Petrography and mineral content of sea-floor sediments of the sediments of Khor Al-Zubair Estuary, NW Arabian Gulf, Estuarine Coastal Shelf Science, 38(6): 569-582.http://www.aiscience.org/journal/jepsd . American Institute of Science.

[12] Salim, M. A., R Porter, P Schiermacker-Hansen, S Christensen & S Al-Jbour. 2006. Field Guide to the Birds of Iraq. NI & BirdLife International.

[13] Porter, R. F., Christensen, S., Chiermacker-Hansen, P. 1996. Field guide to the birds of Iraq]. NI/Bird Life International, Baghdad. [In Arabic]. Beirut, Lebanon Society for the protection of Nature in Lebanon. Poyser, London, 460 pp.

[14] Salim, M. A. and Abed, S. A. 2017. Avifauna diversity of Bahr Al-Najaf and the surrounding areas. Jordan Journal of Biological Sciences, 10(3): 167-176.

[15] Salim, M. A., OF Al-Sheikhly, KA Majeed & RF Porter. 2012. An Annotated Checklist of the Birds of Iraq. Sandgrouse 34(1): 4-43.

[16] Al-Sayed, H., Naser, H. A. and Al-Wedaei, K. 2008. Observations on macrobenthic invertebrates and wader bird assemblage in a protected marine mudflat in Bahrain. Aquatic Ecosystem Health and Management , 11(4): 450-456.

Pd-Au nanoparticles supported reduced graphene oxide nanoplatelets toward aerobic selective oxidation of benzoyl alcohol

Hadeel Altaee[1] and Hassan A. Alshamsi[2]

[1]Education of Al-Nassiriya, Ministry of Education, Iraq
[2]Department of Chemistry, College of Education, University of Al-Qadisiyah, Diwaniya 1753, Iraq

Abstract. In this paper, mono-metallic catalysts, palladium (Pd) and gold (Au) and bi-metallic, Au-Pd supported on reduced graphene oxide (rGO) synthesized by Sol Immobilization technique. The as fabricated catalysts (Pd/rGO, Au/rGO and Au-Pd/rGO) have been tested by various techniques (Raman Spectroscopy, FE-SEM, XRD and TEM) to examine the structural, compositional and morphological characteristics. The performance of catalysts on the production of benzaldehyde via benzyl alcohol (BnOH) oxidation was investigated adopting H2O2 as an oxidizing agent. In general, Au-Pd/rGO catalyst presented a remarkable superiority over Au/rGO and Pd/rGO. The Au-Pd/rGO catalyst gave the best reaction conversion ratio of 24.3%, while its selectivity for benzaldehyde was 87.5% using 10 mg catalyst mass, reaction temperature 80 oC and 50% H2O2

Keywords: Selectivity, Oxidation, Benzoyl alcohol, Au-Pd/rGO, Hydrogen peroxide, benzaldehyde, Sol immobilization

1. Introduction

Traditional oxidizing agents such as potassium permanganate or chromium oxide have traditionally been used in typical organic synthetic reactions, but due to their high cost and toxic environmental effects, these agents are no longer desirable in conducting such reactions [1] [2] [3].

It is noteworthy that oxidation of primary and secondary alcohols to their corresponding aldehydes and ketones, respectively, is very significant laboratory reactions, because of the importance of carbonyl compounds as intermediate materials for many reactions in addition to their commercial importance, especially in perfumery [4] [5] [6]. Due to the aforementioned restrictions associated with the oxidation of alcohols, there was persistent work to develop a highly effective and selective reaction in moderate reactive conditions, and scientists found their shadows in the noble elements where they used them as heterogeneous catalysts [7] [8] [9] [10] with the possibility of using molecular oxygen or hydrogen peroxide as alternative oxidizers to conventional oxidizers. At the present time, this path is one of the most flexible routs in terms of industrial modification and development in the fabrication of aldehydes resulting from the oxidation of corresponding primary alcohols [11] [12] [13].The previously prevalent conclusion was that Au cannot be acts an effective catalyst for oxidation processes, but subsequent studies in 1980s [14] have proven that the effectiveness of Au NPs has improved the oxidation reactions of alcohols [15]. The reason behind this improvement in catalytic activity is that nano Au is the most stable among noble elements. Moreover, Au possesses a very distinct surface chemistry [16] [17] [18]. Several studies have proven the efficiency of Au NPs for a large number of organic reactions [19] [20], but subsequent studies confirmed that the mixture of Au NPs and Pd NPs has significantly increased the catalytic and selective activity of oxidation and reduction processes [21]. Despite that bare Au alone has a high catalytic and selective activity in many reactions, especially the oxidation of alcohols, however, Au and Pd enhance catalytic activity in an estimated twenty-five times while maintaining the selective advantage high [22] [23]. In the contest, bare Pd is a sufficient catalyst in the direct hydrogenation of molecular oxygen (O_2) for the fabrication of hydrogen peroxide [24].

The researchers found that the type and nature of the supports in addition to the morphology and size of nanoparticles significantly affecting on the catalytic activity of synthesized catalysts [25] [26] [27]. For example, carbon materials have a high catalytic performance because they possess a wonderful crystalline structure [28], which makes them possess exceptional chemo-physical features. In particular, rGO has a special electronic structure and a large surface area [29]. however, there is no great interest in nanoparticles supported on rGO [30] [31] [32]. The current work aims to obtaining rGO as a support for construction of Pd/rGO, Au/rGO and Au-Pd/rGO) nanocomposites as a promise catalysts for selective aerobic oxidation of benzoyl alcohol. Additionally, the research aims to investigate the properties of synthesized catalysts using different advanced analytical tools to relate the catalytic efficacy with these properties.

2. Methodology

2.1. Chemicals

Graphite powder (G, 60 mesh, purity of 98.5 %) was obtained from LOBA Chemical, India. Palladium

(II) chloride ($PdCl_2$, purity of 99%), gold (III) chloride trihydrate ($HAuCl_4.3H_2O$, purity of 99.9%), sodium borohydride ($NaBH_4$, purity >98.0%) and poly vinyl alcohol (PVA, MW ~ 8000-10000 g mol^{-1}) were purchased from Sigma Aldrich. Benzyl alcohol (BnOH, purity of 99%. Mesitylene (Mes, purity of 98%) were provided by Merck, USA. Hydrogen peroxide (30 % and 50%) and (HCl, 37%) were received from Scharlou.

2.2. Characterization of synthesized catalysts

The X-ray powder diffraction of synthesized catalysts was conducted for structure phase investigation using Bruker D6 Advance diffractometer. The diffraction patterns were examined recorded in a range of 2θ angles 5 -75° adopting Cu Kα radiation source with wavelength of 1.541 Å. The surface morphological of as synthesized catalysts were studied using field emission-scanning electron microscopy (FESEM, JEOL JSM-630LV). The particle sizes and morphology of catalysts were evaluated using transmittance electron microscopy (TEM, FEI Tecani G2 F30 S-twin).Raman spectra were recorded in the wavenumbers range from 1000 to 2000 cm^{-1} employing Renishaw Invia Raman Microprobe supplied with 514 nm Ar ion laser beam. Bruker Vector-22 Fourier-transformed infrared spectrophotometer was used to carry out FT-IR spectra of carbonaceous materials throughout the range 4000-500 cm^{-1}.

2.3. Preparation of GO and rGO nanolayers

Firstly, GO nanolayers were prepared using graphite by as a precursor adopting modified Hummer method. Afterwards, GO has been reduced to rGO obtaining sodium borohydride as an efficient reductant material. Both GO and rGO nanolayres were prepared as described in our previously reported work [33]. The main advantage of the synthesis of rGO is to supporting the deposition of metal particles (Au and Pd) on its surface, material for the deposition of Au and Pd metals for the construction of monometallic catalysts (Au/rGO and Pd/rGO) and bimetallic catalyst (Au-Pd/rGO) as explained in the next step.

2.4. Synthesis of catalysts

The Pd(1%)/rGO was synthesized as reported previously [34] and the same method was adopted to fabrication of Au(1%)/rGO catalyst with some modification. .The Au(0.5%)–Pd(0.5%)/rGO catalyst was synthesized adopting sol immobilization process .The gold-palladium bimetallic catalyst (1% wt Au-Pd / rGO). Typically, 1.67 mL of aqueous solution of palladium chloride ($PdCl_2$: Pd 6 mg / mL) and 1 mL of aqueous solution of gold salt ($HAuCl_4$. $6H_2O$: Au 8.9 mg / mL) was dissolved into 800 mL of DDW. After 5 min, a suitable amount of PVA solution (1 wt. %) solution was added while the solution was kept stirring. Then, freshly prepared $NaBH_4$ solution (0.1 M) / (Au + Pd) (mol / mol) = 5) was added followed by addition of 1.98 g of GO. During the addition, a color change was observed, which refers primarily to reduction of both Au and Pd ions and GO surface. Afterwards, the solution was stirring for an additional kept 30 min. In order to make the solution acidic (pH = 1-3), few drops of concentrated sulfuric acid were added while the solution was kept stirring for 1 hour. Then, the formed slurry was separated by centrifugation at 6000 rpm for 20 min with washing with DDW several times until the solution pH is neutralized. Finally, the drying of precipitate was conducted at 90 °C for overnight. **Figure1** shows the synthesizing steps of Au-Pd/rGO catalyst.

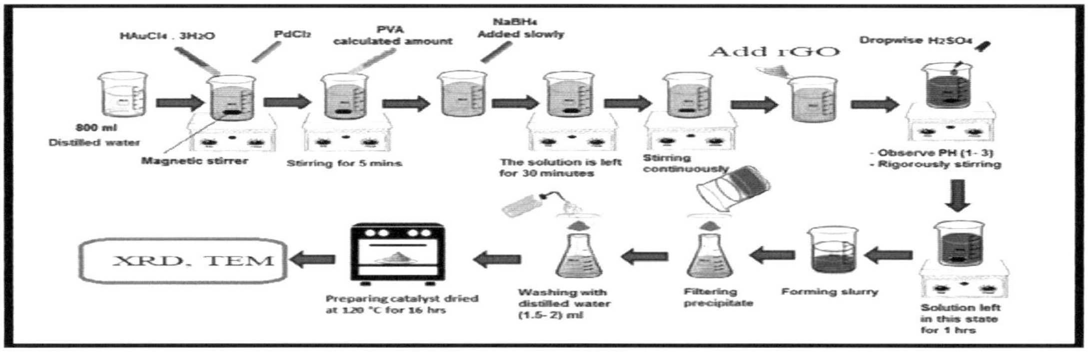

Figure1. Schematic diagram of Au–Pd/rGO synthesis steps.

2.5. Benzyl alcohol oxidation reaction

All benzyl alcohol oxidation processes were carried out in a 50 mL doubled necked round bottom flask as a glass reactor fitted with a reflux condenser which was submerged into an oil bath positioned on a hot plate magnetic stirrer. Typically, 10 mL of benzyl alcohol were poured into flask with a certain amounts of catalyst and hydrogen peroxide. The ratio of benzyl alcohol concentration hydrogen peroxide concentration is 10: 1 (mole ratio). The mixture was heated at the required temperatures (80 or 60 °C) for a period of 240 min with continuous stirring at 600 rpm using a magnetic stirrer. I typical 2 mL samples were taken from the reaction vessel at a particular intervals (30, 120, 240 min). The catalyst particles were removed from samples via centrifugation at 6000 rpm for 20 min. Afterwards, 0.5 mL of mesitylene as a standard substance was added 0.5 mL of clear supernatant solution and kept in the refrigerator until analyzed with gas chromatography (GC) to calculate benzaldehyde (BZ) content [33,34]. The selective efficiency of benzaldehyde was calculated using the following expressions:

Moles of BZ=moles of Mes(area of BnOH)/area of Mes) ……………………………………….. 1

BZ (%)= (Moles of BZ produced/ moles of BnOH reacted)x100…………………………………..2

3. Results and discussion

3.1. Characterization

FT-IR spectroscopy is one of the most necessary techniques for monitoring of functional groups in graphene, graphene oxide and their derivatives. The importance of this analysis highlighted in the follow-up of the groups constituting to graphite structure and the pendant groups which intercalated into the interlayer spacing between the GO layers such as carbonyl, carboxyl, hydroxyl and epoxy groups. As shown in **Figure 2**, a broad band centered at 3422 cm^{-1} can be observed in graphite spectrum, which assigned to the vibration of hydroxyl group that belong to the adsorptive moisture onto graphite surface [35]. The band associated with the C-H stretching vibration of the carbonyl groups can be observed at 2930 cm^{-1} [36].

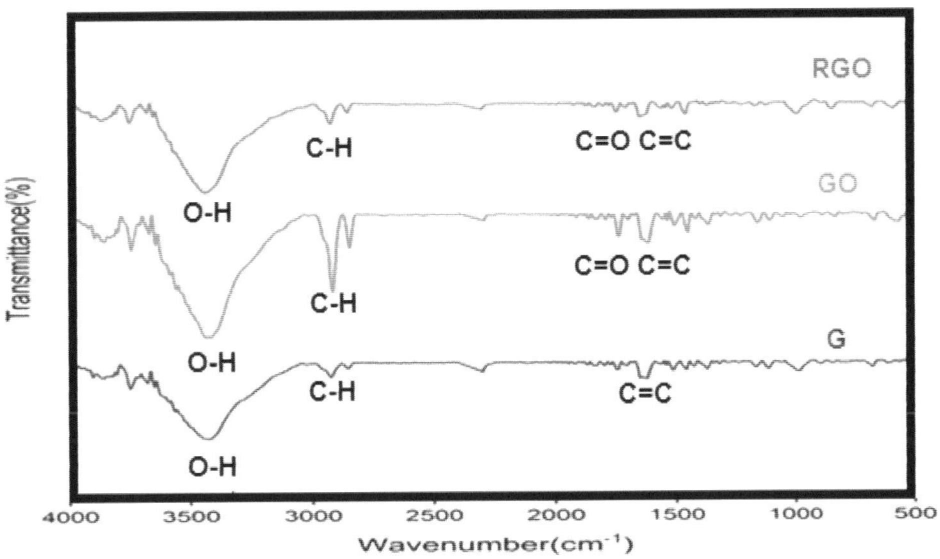

Figure2. FTIR spectra of as synthesized graphite, graphene oxide and reduced graphene oxide.

When G is oxidized to GO, several bands are found in the GO spectrum which is attributed to the formation of oxygen-containing groups after the graphite oxidation process. Accordingly, it can be seen a broad one at 3424 cm^{-1} correspond to the O-H stretching vibration which belongs to carboxyl group as well as to trace moisture adsorbed the GO surface. Additionally, the C-H stretching vibration mode is observed at 2924 cm^{-1}. The medium strength band appeared at 1745 cm^{-1} is assigned to C=O bond which is considered as a clear prove of the success oxidation of graphite. The band at 1624 cm^{-1} refers to stretching vibration mode of C= C bonds which are not completely to C=O groups within the carbon skeleton [37]. In the rGO spectrum, the bands shown at 3439 cm^{-1} and 1747 cm^{-1} are due to existence of hydroxyl and carbonyl groups, respectively. Generally, the weakness of band intensities in rGO spectrum in comparison with their analogues in the GO spectrum clearly indicates the efficiency of the GO reduction process using sodium borohydride as a reductant agent [38].

Raman spectroscopy is an important tool in the analysis of different organic and inorganic compounds, especially carbonaceous materials by tracking the intensities and locations of characteristics bands, G and D. **Figure 3** demonstrates the appearance of G and D graphite bands at 1582 cm^{-1} and 1350 cm^{-1}, respectively. When the graphite is exfoliated to graphene oxide by treating with oxidant agents as described via modified Hummer's method, the oxidation process destroyed the sp^2 structure and red shift occurred in both bands, where G and D bands are shifted to new positions at 1600 cm^{-1} and 1364 cm^{-1}, respectively.

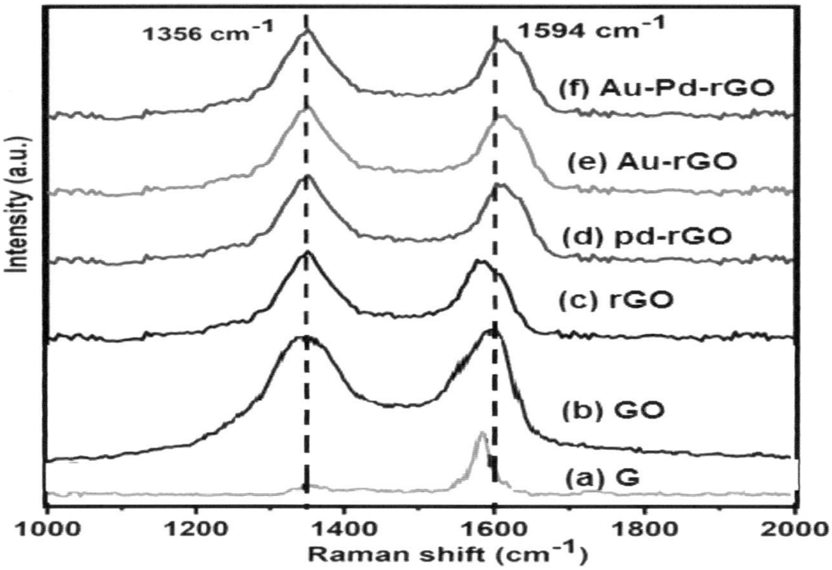

Figure 3. Raman spectra of (a) Graphite, (b) Graphene Oxide, (c) Reduced Graphene Oxide, (d) Pd-rGO,(e)Au-rGO and (f) Au-Pd-rGO.

In the case of Raman spectrum of rGO sample whre GO is to rGO using $NaBH_4$ as reducing agent, the positions of G and D bands have returned to new positions very close to their pristine positions in G spectrum, where the G band appeared at 1585 cm^{-1} and the D band positioned 1352 cm^{-1}. The restoration of the two bands to their original positions confirms the success of the GO reduction process and also confirms the restoration of the graphene sheets their sp^2 conjugative aromatic system [39]. When the Au and Pd ions are reduced on the graphene sheet surfaces, the intensity of D band increased relatively, while the G band is shifted to position with higher wavenumbers. The G band in Au/rGO and Pd/rGO has been shifted to the positions of 1610 cm^{-1} and 1611 cm^{-1}, respectively. The ratio of Raman intensity (ID/IG) of rGO, Au/rGO and Pd/rGO were calculated to be 0.92, 1.14 and 1.11, respectively, which indicates the strong interference between deposited metals and graphene sheets. Regarding the new bimetallic Au-Pd/rGO catalyst, there was no valuable shift in positions nor effective change in the intensities of D and G bands in comparison with Au/rGO and Pd/rGO catalysts.

The surface morphology of as synthesized nanmatrisls was investigted using FE-SEM technique . As shown in **Figure 4**, it can be seen the formation a large number of piled sheets in graphite sample with a smoothness of the surface can also be observed. In the case of GO, it can be seen how the sheets have exfoliated, as well as a high change in surface roughness and morphology. Accordingly, surface roughness increased after enriching the sheets content with high number of oxygen-containing groups.

After GO is trnsformed to rGO, significant number of oxygen –containing groups are remobved where the reduction process causes the layres to be more close to each other when comparing with GO. In addition, as it can be seen from rGO FE-SEM image, the surfsce recovered its roughness via formation of sharp edges [40]. Additionally, As **Figure 5** shows, the Au/rGO, Pd/rGO and Au-Pd/rGO samples have a relatively rough surface with deposition of Au, Pd and Au-Pd NPs in very small sizes and low numbers due to the lower content of matals compared to rGO.

Figure 4. FE-SEM image of graphite , graphene oxide and reduced graphene oxide powders.

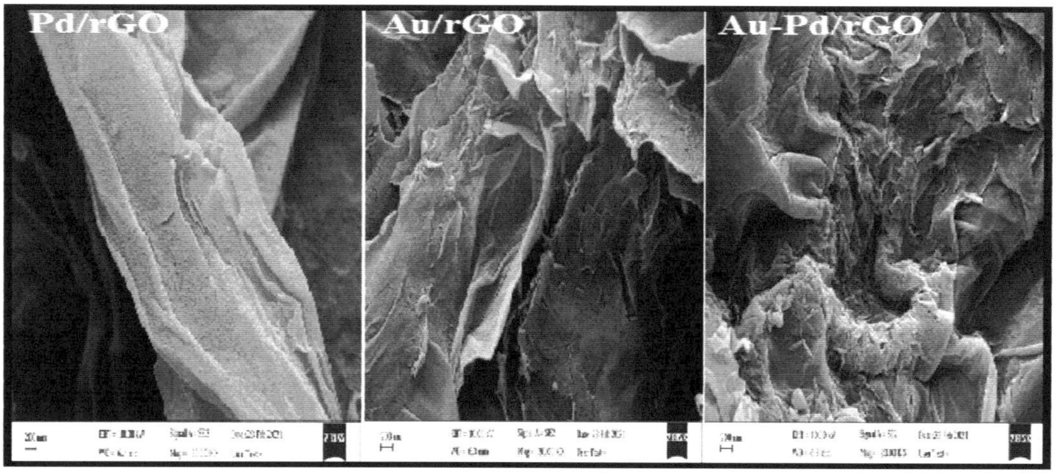

Figure 5. FE-SEM image of Pd/rGO, Au/rGO and Au-Pd/rGO ctalysts.

XRD technique is a critical tool for demonstrating the success of synthesizing different materials by investigation structure phases. **Figure 6** displays the XRD pattern of as synthesized materials in the current work. As noticed, graphite has a sharp and prominent peak at $2\theta = 26.7^o$ and with stacked due to d = 0.33nm, where d is interlayer d-spacing [41]. For GO, a new peak appears at $2\theta = 11.6^o$ (d = 0.79 nm), the high d-spacing value confirms the efficiency of exfoliation process of graphite layers by oxidation reaction. The oxidation process leads to divergence of layers by intercalation of functional groups that contain oxygen between graphite sheets [42].

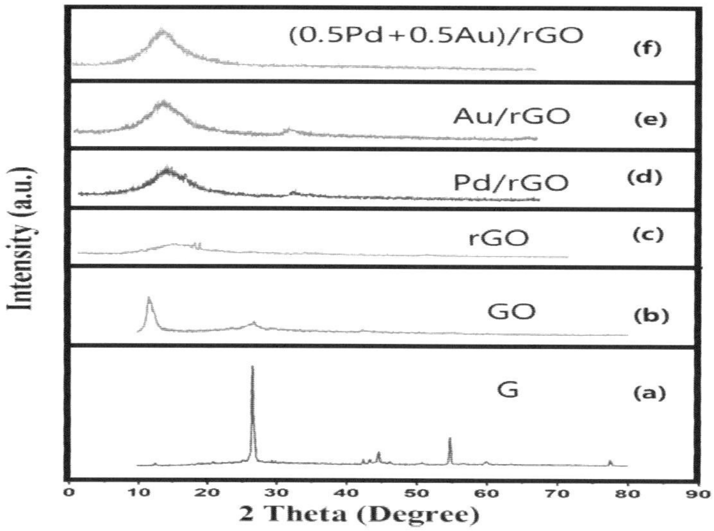

Figure 6. Powder X-Ray diffraction of as synthesized materials

In the diffraction pattern of rGO, as a result of removal most of the pendant groups which intercalated into the interlayer spacing between the GO layers , a wide central peak appears at 2 θ = 23.9° with the value of spacing in between crystal layers of 0.35 nm. The absence of a peak at 10° and the formation of a new peak at 23.9° indicate that GO has been efficiently transformed to rGO. For Au/rGO catalyst, the XRD pattern exhibits a three characteristics diffraction peak positioned at the 2θ = 39.45°, 45.2° and 68.44° which correspond to (111), (200) and (220) lattice planes associated to Pd face-centered cubic. According to XRD patters of Au/rGO, a prominent brag reflections can be observed at 2θ of 42.9°, 44.4° and 77.6°, which corresponds to the e (111), (200) and (311) lattice planes associated with fcc Au crystal [43]. For Au-Pd/rGO catalyst, the appearance of strong peaks was not observed which may be due to the low content and high dispersion of both metals (0.5%) in catalyst [44].

The internal structure and morphology of catalysts was examined with assistance of TEM technique. **Figure 7** presented the TEM images of three catalyst samples. The images show the deposition of Au and Pd metals onto rGO surface and their dispersion almost uniformly with the presence of some agglomerations. It appears that the Au nanoparticles were deposited in very small sizes compared to PD NPs, while the distribution of Pd NPs was more regular. It was found that the average sizes of Au and Pd NPs were 3 and 6 nm respectively, which can be regarded as quantum dots.

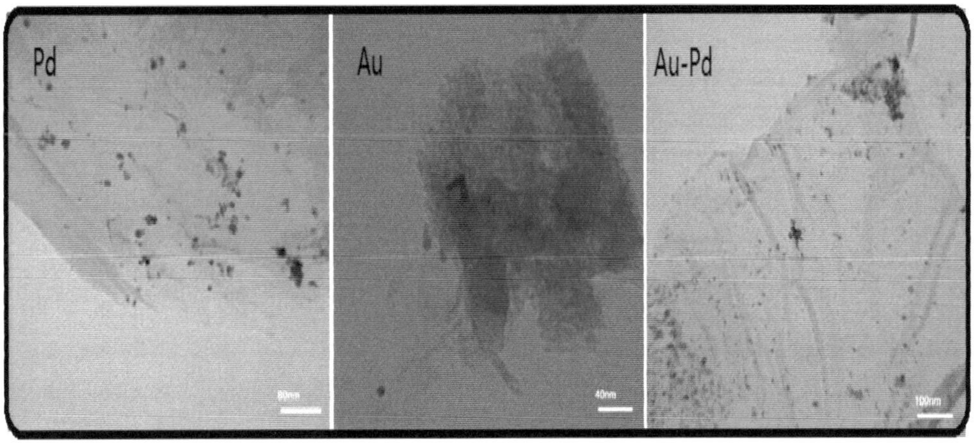

Figure 7. TEM micrographs of Pd/rGO, Au/ rGO and Au-Pd/rGO catalysts.

3.2. Investigation of catalytic activity

To explore the efficiency of as synthesized catalysts in oxidation of benzyl alcohol, the effects of some variables on the efficiency and selectivity were studied. The investigated variables included catalyst mass, reaction temperature and H_2O_2 concentration.

3.2.1. Influence of catalyst mass

The first operator parameter studied was the catalyst masses, where the 5 and 10 mg catalyst masses were investigated, while the H_2O_2 concentration and reaction temperature were kept constant at 30% and 60 ^{0}C. It has been found that concentration of 10 mg gave the highest conversion efficiency of BnOH to BZ, in the same time, slight increase in selectivity efficiency of BZ was found as shown in **Figure 8** (A, B). The reason behind an increase in conversion and selectivity efficiencies is related with rapid increase in H_2O_2 decomposition as the catalyst mass increased which leads to an increase in the oxidizing species [45] [46]. As the results showed, while continuing to increase the reaction time, the direct increase in the efficiency of oxidation process will continue, but in parallel , a continuous decline in selectivity was noticed , which due to the increase in the formation of benzoic acid (BZA).

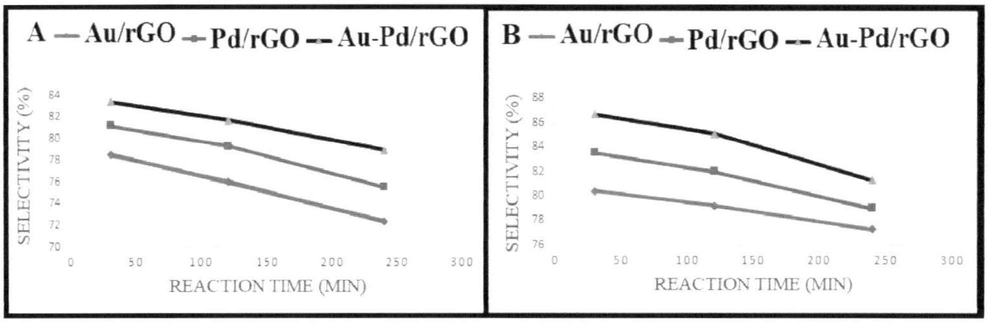

Figure 8. Relationship between reaction time and selectivity of benzaldehyde as a function of catalyst mass (A) 5 mg and (B) 10 mg.

3.2.2. Effect of temperature

After studying the effect of Pd/rGO, Au/rGO and Au-Pd/rGO mass on the conversion efficiency and in order to compare the performance of three catalysts, additional experiments were conducted at 60 °C and 80 °C, where the H_2O_2 concentration and catalyst mass and were held constant at 30% and 10 mg , respectively .The results shown in Figure 9 (A, B) proved an immediate increase in conversion efficiency of BnOH and selectivity of BZ With a raise in temperature, but with continuing to increase the reaction time, the selectivity decreased because the oxidation reaction of BZ to BZA is preferred [47].

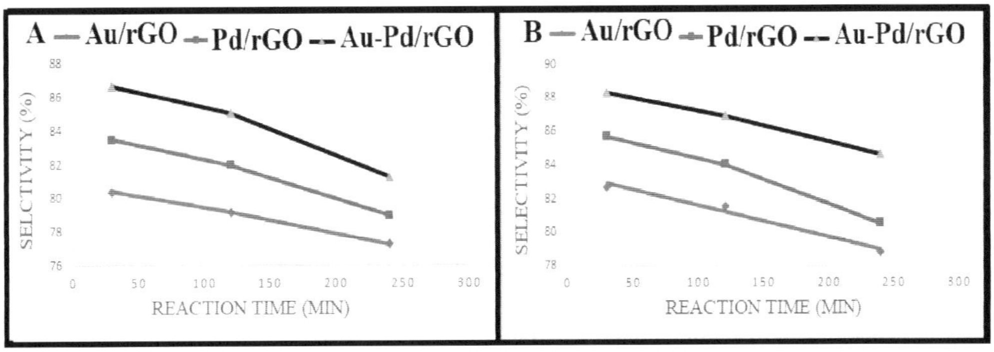

Figure 9. Effect of reaction temperature on selectivity of benzaldehyde (A) 60 °C and (B) 80 °C.

3.2.3. Effect H_2O_2 concentration

Figure 10 (A and B) illustrates the effect of using two different H_2O_2 concentrations (30% and 50%) on the efficiency and selectivity of benzaldehyde. The reaction temperature and catalyst mass were held constant at 80 °C and 10 mg, respectively. The results showed a clear improvement in the selectivity of benzaldehyde when using a high concentration of oxidizing agent (50% H_2O_2). This due to the increase hydroxyl radical concentration via self-decomposition of H_2O_2, where the hydroxyl radicals are driving the transformation of BnOH to BZ. The hydroxyl radicals are known to be one of the most powerful oxidizing agent of organic compounds [48] [49]. As noticed above, by studying the effects of catalyst mass and temperature, the progress of the reaction leads to decline in BZ selectivity due to conversion of most BZ to BZA. **Table 1** summarizes the best for three catalysts for conversion efficiency and selectivity of benzaldehyde. The results clearly indicate that the Au-Pd/rGO catalyst presented the most efficient activity.

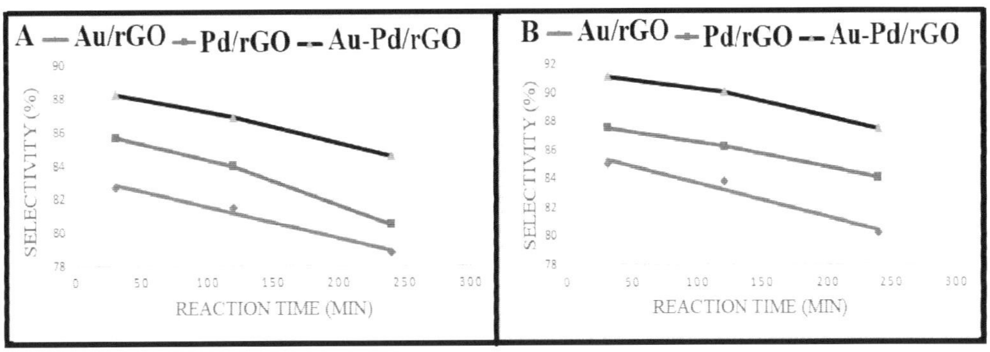

Figure 10. Effect of H_2O_2 Concentration on selectivity of benzaldehyde (A) 30% and (B) 50%.

Table1. Summary of the results obtained using the synthesized mono and bimetallic supported onto rGO catalysts at 10 mg catalyst weight, 50% H_2O_2 concentration temperature of 80 ° C after 240 min treatment with sodium borohydride.

Catalyst	Conversion (%)	Selectivity (%)
Pd/rGO	**3.8**	**83.6**
Au/rGO	**17.6**	**86.0**
Au-Pd/rGO	**24.3**	**87.5**

4. Conclusions

The sol immobilization technique has proven successful in synthesizing the palladium, gold and palladium –gold supported on reduced graphene oxide catalysts. Characterization analyses demonstrated the synthesis of metallic particles in nanoscale and with very small sizes. It was also revealed that the metallic particles have been almost uniformly distributed onto supported rGO surface. Finally, the Au-Pd/rGO proved to be the best among three catalysts adopted in the benzyl alcohol oxidation. At the optimum conditions (10 mg Au-Pd/rGO, 50% H_2O_2 and 80 ^{0}C), the efficiency of conversion benzyl alcohol efficiency and benzaldehyde selectivity were reached 24.3% and 87.5%, respectively.

References

[1] Tareq S, Saiman MI, Hin TYY, Abdullah, Rashid U. The impact of hydrogen peroxide as an oxidant for solvent-free liquid phase oxidation of benzyl alcohol using Au-Pd supported carbon and titanium catalysts. Bulletin of Chemical Reaction Engineering & Catal.
[2] Griffith WP, Jolliffe JM. Ruthenium and osmium carboxylato oxo complexes as organic oxidants. Studies in Surface Science and Catalysis 1991; 66: 395-400..
[3] Muzart J. 2012. Chromium-catalyzed oxidations in organic synthesis. Chem. Rev. 1992: 92:113–140..
[4] Turkevich J, Stevenson PC, Hillier J. A study of the nucleation and growth processes in the synthesis of colloidal gold. Discussions of the Faraday Society1951; 11: 55-75..
[5] Lopez-Sanchez JA, Dimitratos N, Miedziak P, Ntainjua E, Edwards JK, Morgan, D, Carley AF, Tiruvalam R, Kiely CJ, Hutchings GJ. Au–Pd supported nanocrystals prepared by a sol immobilisation technique as catalysts for selective chemical synthesis. Physica.
[6] Bond GC, Thompson DT. Gold catalysis. Catal Rev-Sci Eng 1999: 41: 319-388..
[7] Shan J, Tenhu H. Recent advances in polymer protected gold nanoparticles: synthesis, properties and applications. Chemical Communications 2007:44:4580-4598..
[8] Jeong GH, Choi D, Kang M, Shin J, Kang JG, Kim SW. One-pot synthesis of Au@ Pd/graphene nanostructures: electrocatalytic ethanol oxidation for direct alcohol fuel cells (DAFCs). RSC Advances 2013; 3: 8864-8870..
[9] Morad M, Sankar M, Cao E, Nowicka E, Davies TE, Miedziak PJ, Morgan DJ, Knight DW, Bethell D, Gavriilidis A, Hutchings GJ. Solvent-free aerobic oxidation of alcohols using supported gold palladium nanoalloys prepared by a modified impregnation method. Ca.
[10] Lee EP, Chen J, Yin Y, Campbell CT, Xia Y. Pd-catalyzed growth of Pt nanoparticles or nanowires as dense coatings on polymeric and ceramic particulate supports. Advanced Materials 2006; 18: 3271-3274..

[11] Haruta M. When gold is not noble: catalysis by nanoparticles. The chemical record 2003; 3: 75-87..
[12] Enache DI, Barker D, Edwards JK, Taylor SH, Knight DW, Carley AF, Hutchings GJ. Solvent-free oxidation of benzyl alcohol using titania-supported gold–palladium catalysts: effect of Au–Pd ratio on catalytic performance. Catalysis today 2007; 122: 407-411..
[13] Sheldon RA, Arends IWCE, Brink G-JT, Dijksman A. Green, catalytic oxidations of alcohols. Accounts of Chemical Research 2002; 35: 774-781..
[14] Enache DI, Edwards JK, Landon P, Solsona-Espriu B, Carley AF, Herzing AA, Watanabe M, Kiely CJ, Knight DW, Hutchings GJ. Solvent-free oxidation of primary alcohols to aldehydes using Au-Pd/TiO2 catalysts. Science 2006; 311: 362-365..
[15] Landon P, Collier PJ, Papworth AJ, Kiely CJ, Hutchings GJ, 2002. Direct formation of hydrogen peroxide from H 2/O 2 using a gold catalyst. Chemical Communications 2002; 18: 2058-2059..
[16] Sinha AK, Seelan S, Tsubota S, Haruta M. A three-dimensional mesoporous titanosilicate support for gold nanoparticles: Vapor-phase epoxidation of propene with high conversion. Angewandte Chemie International Edition 2004; 43: 1546-1548..
[17] Hughes MD, Xu YJ, Jenkins P, McMorn P, Landon P, Enache DI, Carley AF, Attard GA, Hutchings GJ, King F, Stitt E. Tunable gold catalysts for selective hydrocarbon oxidation under mild conditions. Nature 2005; 437: 1132-1135.
[18] Abad A, Concepción P, Corma A, García H. A collaborative effect between gold and a support induces the selective oxidation of alcohols. Angewandte Chemie International Edition 2005; 44: 4066-4069..
[19] Dimitratos N, Lopez-Sanchez JA, Morgan D, Carley AF, Tiruvalam R, Kiely CJ, Bethell D, Hutchings GJ. Solvent-free oxidation of benzyl alcohol using Au–Pd catalysts prepared by sol immobilisation. Physical Chemistry Chemical Physics 2009; 11: 5142-5153..
[20] Bond GC, Thompson DT. Gold-catalysed oxidation of carbon monoxide. Gold Bulletin 2000; 33: 41-50..
[21] Stephen A, Hashmi K. 2004. Homogeneous catalysis by gold. Gold Bulletin 2004; 37: 51-65..
[22] Meyer R, Lemire C, Shaikhutdinov SK, Freund HJ. Surface chemistry of catalysis by gold. Gold Bulletin 2004; 37: 72-124.
[23] Zhu X, Guo Q, Sun, Chen S, Wang J-Q, Wu M, Fu W, Tang Y, Duan X, Chen D, Wan Y. Optimising surface d charge of AuPd nanoalloy catalysts for enhanced catalytic activity. Nat Commun 2019; 10: 1428..
[24] Stephen A, Hashmi K, Hutchings GJ. Gold catalysis. Angew. Chem., Int. Ed 2006; 45: 7896-7936..
[25] Choudhary VR, Samanta C. Role of chloride or bromide anions and protons for promoting the selective oxidation of H2 by O2 to H2O2 over supported Pd catalysts in an aqueous medium. Journal of Catalysis 2006; 238: 28-38..
[26] Fang W, Chen J, Zhang Q, Deng W, Wang Y. Hydrotalcite-Supported Gold Catalyst for the Oxidant-Free Dehydrogenation of Benzyl Alcohol: Studies on Support and Gold Size Effects. Chemistry–A European Journal 2011; 17: 1247-1256..
[27] Meenakshisundaram S, Nowicka E, Miedziak PJ, Brett GL, Jenkins RL, Dimitratos N, Taylor SH, Knight DW, Bethell D, Hutchings GJ. Oxidation of alcohols using supported gold and gold–palladium nanoparticles. Faraday Discussions 2010; 145: 341-356..
[28] Bianchi C, Porta F, Prati L, Rossi M. Selective liquid phase oxidation using gold catalysts. Topics in Catalysis 2000; 13: 231-236..
[29] Gil S, Muñoz L, Sánchez-Silva L, Romero A, Valverde JL. Synthesis and characterization of Au supported on carbonaceous material-based catalysts for the selective oxidation of glycerol. Chemical

engineering journal 2011; 172: 418-429..
[30] Dreyer DR, Park S, Bielawski CW, Ruoff RS. The chemistry of graphene oxide. Chemical society reviews 2010; 39: 228-240..
[31] Scheuermann GM, Rumi L, Steurer P, Bannwarth W, Mülhaupt R. Palladium nanoparticles on graphite oxide and its functionalized graphene derivatives as highly active catalysts for the Suzuki–Miyaura coupling reaction. Journal of the American Chemical Socie.
[32] Nie R, Wang J, Wang L, Qin Y, Chen P, Hou Z. Platinum supported on reduced graphene oxide as a catalyst for hydrogenation of nitroarenes. Carbon 2012; 50: 586-596..
[33] Yu X, Huo Y, Yang J, Chang S, Ma Y, Huang W. Reduced graphene oxide supported Au nanoparticles as an efficient catalyst for aerobic oxidation of benzyl alcohol. Applied Surface Science 2013; 280: 450-455..
[34] Altaee H, Alshamsi H A Selective oxidation of benzyl alcohol by reduced graphene oxide supported platinum nanoparticles. Journal of Physics: Conference Series 2020; 1664: 012074..
[35] Altaee H, Alshamsi H A H, Joda BA. Reduced graphene oxide supported palladium nanoparticles as an efficient catalyst for aerobic oxidation of benzyl alcohol. AIP Conference Proceedings 2020; 2290: 030036..
[36] Zhang Y, Ren L, Wang S, Marathe A, Chaudhuri J, Li G. Functionalization of graphene sheets through fullerene attachment. Journal of Materials Chemistry 2011; 21: 5386-5391..
[37] Alshamsi HA, Ali SK, Altaa SHA. Green Synthesis and Characterization of Reduced Graphene Oxide (RGO) using Sabdarriffa L extract and its Solubility Property. J. Phys.: Conf. Ser. 2020; 1664: 012058.
[38] Alshams HAH, Ali SK. Kinetics of Green Reduction of Graphene Oxide via Hibiscus Sabdarriffa L Aqueous Solution. International Journal of ChemTech Research 2016; 9; 321-331..
[39] Al-Nayili A, Albdiry M. AuPd bimetallic nanoparticles supported on reduced graphene oxide nanosheets as catalysts for hydrogen generation from formic acid under ambient temperature. New Journal of Chemistry 2021; 45: 10040-10048..
[40] Kadhem AA, Al-Nayili A. Dehydrogenation of Formic Acid in Liquid Phase over Pd Nanoparticles Supported on Reduced Graphene Oxide Sheets. Catalysis Surveys from Asia 2021; 25: 324–333..
[41] Liu G, Wang L, Wang B, Gao T, Wang D. A reduced graphene oxide modified metallic cobalt composite with superior electrochemical performance for supercapacitors. RSC Advances 2015; 5: 63553-63560..
[42] Blanton TN, Majumdar D. Characterization of X-ray irradiated graphene oxide coatings using X-ray diffraction, X-ray photoelectron spectroscopy, and atomic force microscopy. Powder Diffraction; 2013, 28: 68-71..
[43] Huang Z, Zhou H, Li C, Zeng F, Fu C, Kuang Y. Preparation of well-dispersed PdAu bimetallic nanoparticles on reduced graphene oxide sheets with excellent electrochemical activity for ethanol oxidation in alkaline media. Journal of Materials Chemistry 201.
[44] Wang R, Wu Z, Wang G, Qin Z, Chen C, Dong M, Wang J, Highly active Au–Pd nanoparticles supported on three-dimensional graphene–carbon nanotube hybrid for selective oxidation of methanol to methyl formate. RSC Advances 2015; 5: 44835-44839..
[45] Tareq S, Yap YHT, Saleh TA, Abdullah A, Rashid U, Saiman MI. Synthesis of bimetallic gold-pallidum loaded on carbon as efficient catalysts for the oxidation of benzyl alcohol into benzaldehyde. Journal of Molecular Liquids 2018; 271: 885-891..
[46] Edwards J K, Pritchard J, Piccinini M, Shaw G, He Q, Carley A F, Hutchings GJ. The effect of heat treatment on the performance and structure of carbon-supported Au–Pd catalysts for the direct synthesis of hydrogen peroxide. Journal of catalysis 2012; 292:.

[47] Della Pina C, Falletta E, Prati L, Rossi M. Selective oxidation using gold. Chemical Society Reviews 2008; 37: 2077-2095..
[48] Habeeb HA, Waheeb. Study of chemical oxygen demand (COD) and Removal of azure B dye from aqueous solution by using Advanced Oxidation Processes. Al-Qadisiyah Journal of Pure Science 2013; 18: 162-180..
[49] Habeeb HA, Hussain BS. Optimization of Degradation of Trypan Blue Dye using Advanced Oxidation Processes and Photocatalysis under UV Light. Al-Qadisiyah Journal of Pure Science 2013; 17: 71-87..

AUTHOR INDEX